D1565169

Springer Series in Statistics

Springer
New York
Berlin
Heidelberg
Hong Kong
London
Milan
Paris
Tokyo

Springer Series in Statistics

Andersen/Borgan/Gill/Keiding: Statistical Models Based on Counting Processes.
Atkinson/Riani: Robust Diagnostic Regression Analysis.
Berger: Statistical Decision Theory and Bayesian Analysis, 2nd edition.
Borg/Groenen: Modern Multidimensional Scaling: Theory and Applications
Brockwell/Davis: Time Series: Theory and Methods, 2nd edition.
Chan/Tong: Chaos: A Statistical Perspective.
Chen/Shao/Ibrahim: Monte Carlo Methods in Bayesian Computation.
David/Edwards: Annotated Readings in the History of Statistics.
Devroye/Lugosi: Combinatorial Methods in Density Estimation.
Efromovich: Nonparametric Curve Estimation: Methods, Theory, and Applications.
Eggermont/LaRiccia: Maximum Penalized Likelihood Estimation, Volume I: Density Estimation.
Fahrmeir/Tutz: Multivariate Statistical Modelling Based on Generalized Linear Models, 2nd edition.
Fan/Yao: Nonlinear Time Series: Nonparametric and Parametric Methods.
Farebrother: Fitting Linear Relationships: A History of the Calculus of Observations 1750-1900.
Federer: Statistical Design and Analysis for Intercropping Experiments, Volume I: Two Crops.
Federer: Statistical Design and Analysis for Intercropping Experiments, Volume II: Three or More Crops.
Ghosh/Ramamoorthi: Bayesian Nonparametrics.
Glaz/Naus/Wallenstein: Scan Statistics.
Good: Permutation Tests: A Practical Guide to Resampling Methods for Testing Hypotheses, 2nd edition.
Gouriéroux: ARCH Models and Financial Applications.
Gu: Smoothing Spline ANOVA Models.
Györfi/Kohler/Krzyżak/ Walk: A Distribution-Free Theory of Nonparametric Regression.
Haberman: Advanced Statistics, Volume I: Description of Populations.
Hall: The Bootstrap and Edgeworth Expansion.
Härdle: Smoothing Techniques: With Implementation in S.
Harrell: Regression Modeling Strategies: With Applications to Linear Models, Logistic Regression, and Survival Analysis
Hart: Nonparametric Smoothing and Lack-of-Fit Tests.
Hastie/Tibshirani/Friedman: The Elements of Statistical Learning: Data Mining, Inference, and Prediction
Hedayat/Sloane/Stufken: Orthogonal Arrays: Theory and Applications.
Heyde: Quasi-Likelihood and its Application: A General Approach to Optimal Parameter Estimation.
Huet/Bouvier/Gruet/Jolivet: Statistical Tools for Nonlinear Regression: A Practical Guide with S-PLUS Examples.
Ibrahim/Chen/Sinha: Bayesian Survival Analysis.
Jolliffe: Principal Component Analysis.

(continued after index)

S.N. Lahiri

Resampling Methods for Dependent Data

With 25 Illustrations

S.N. Lahiri
Department of Statistics
Iowa State University
Ames, IA 50011-1212
USA

Library of Congress Cataloging-in-Publication Data
Lahiri, S.N.
Resampling methods for dependent data / S.N. Lahiri.
p. cm. — (Springer series in statistics)
Includes bibliographical references and index.
ISBN 0-387-00928-0 (alk. paper)
1. Resampling (Statistics) I. Title. II. Series.
QA278.8.L344 2003
519.5′2—dc21 2003045455

ISBN 0-387-00928-0 Printed on acid-free paper.

Printed in the United States of America.

9 8 7 6 5 4 3 2 1 SPIN 10922705

Typesetting: Pages created by the author using a Springer TEX macro package.

www.springer-ny.com

Springer-Verlag New York Berlin Heidelberg
A member of BertelsmannSpringer Science+Business Media GmbH

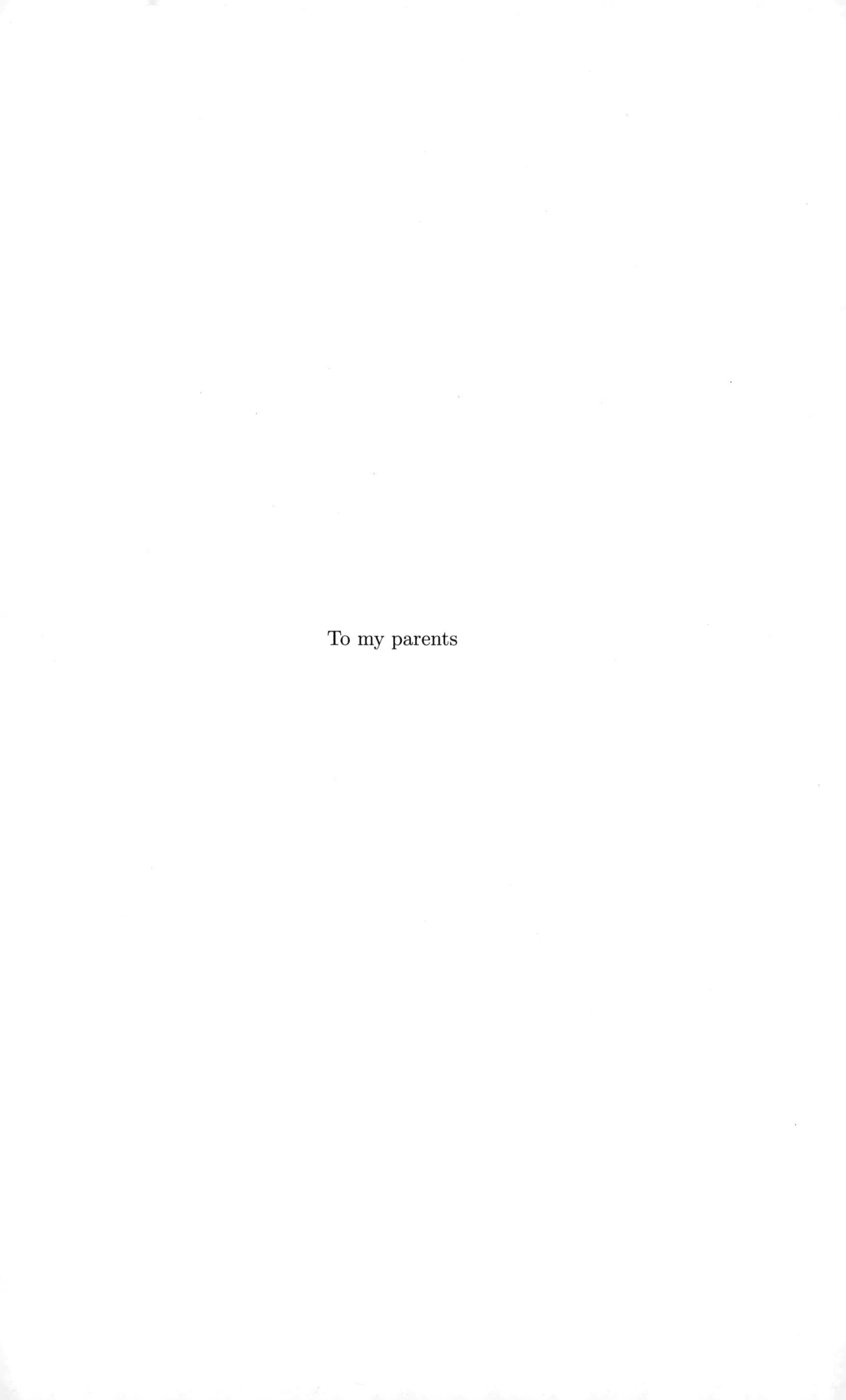

To my parents

Preface

This is a book on bootstrap and related resampling methods for temporal and spatial data exhibiting various forms of dependence. Like the resampling methods for independent data, these methods provide tools for statistical analysis of dependent data without requiring stringent structural assumptions. This is an important aspect of the resampling methods in the dependent case, as the problem of model misspecification is more prevalent under dependence and traditional statistical methods are often very sensitive to deviations from model assumptions. Following the tremendous success of Efron's (1979) bootstrap to provide answers to many complex problems involving independent data and following Singh's (1981) example on the inadequacy of the method under dependence, there have been several attempts in the literature to extend the bootstrap method to the dependent case. A breakthrough was achieved when resampling of single observations was replaced with block resampling, an idea that was put forward by Hall (1985), Carlstein (1986), Künsch (1989), Liu and Singh (1992), and others in various forms and in different inference problems. There has been a vigorous development in the area of resampling methods for dependent data since then and it is still an area of active research. This book describes various aspects of the theory and methodology of resampling methods for dependent data developed over the last two decades.

There are mainly two target audiences for the book, with the level of exposition of the relevant parts tailored to each audience. The first five chapters of the book are written in a pedantic way, giving full details of the proofs of the theoretical results and step-by-step instructions for implementation of the methodology. This part of the book, together with selected material from the later chapters, can be used as a text for a graduate level course. For the first part, familiarity with only basic concepts of theoretical Statistics is assumed. In particular, no prior exposure to Time Series is needed. The second part of the book (Chapters 6–12) is written in the form of a research monograph, with frequent reference to the literature for the proofs and for further ramification of the topics covered. This part is primarily intended for researchers in Statistics and Econometrics, who are interested in learning about the recent advances in this area, or interested in applying the methodology in their own research. A third potential audience is the practitioners, who may go over the descriptions of the resampling methods and the worked out numerical examples, but skip the proofs and other technical discussions. Many of the results presented in the book are from preprints of papers and are yet to appear in a published medium. Furthermore, some (potential) open problems have been pointed out.

Chapter 1 gives a brief description of the "bootstrap principle" and advocates resampling methods, at a heuristic level, as general methods for estimating what are called "level-2" (and "higher-level") parameters in the book. Chapter 2 sketches the historical development of bootstrap methods since Efron's (1979) seminal work and describes various types of bootstrap methods that have been proposed in the context of dependent (temporal) data. Chapter 3 establishes consistency of various block bootstrap methods for estimating the variance and the distribution function of the sample mean. Chapter 4 extends these results to general classes of statistics, including M-estimators and differentiable statistical functionals, and gives a number of numerical examples. Chapter 5 starts with a numerical comparison of different block bootstrap methods and follows it up with some theoretical results. Chapter 6 deals with Edgeworth expansions and second-order properties of block bootstrap methods for normalized and studentized statistics under dependence. Chapter 7 addresses the important problem of selecting the optimal block size empirically. Chapter 8 treats bootstrap based on independent and identically distributed innovations in popular time series models, such as the autoregressive processes. Chapter 9 deals with the frequency domain bootstrap. Chapter 10 describes properties of block bootstrap and subsampling methods for a class of long-range dependent processes. Chapter 11 treats two special topics — viz., extremums of dependent random variables and sums of heavy-tailed dependent random variables. As in the independent case, here the block bootstrap fails if the resample size equals the sample size. A description of the random limit is given in these problems, but the proofs are omitted. Chapter 12 considers resampling methods for spatial data under different spatial sampling designs. It also treats the problem of spatial prediction using resampling methods. A list of important definitions and technical results are given in Appendix A, which a reader may consult to refresh his or her memory.

I am grateful to my colleagues, coauthors, and teachers, A. Bose, K.B. Athreya, G.J. Babu, N. Cressie, A. C. Davison, P. Hall, J. Horowitz, D. Isaacson, B. Y. Jing, H. Koul, D. Politis, and A. Young for their interest, encouragement, and constructive suggestions at various stages of writing the book. Special thanks are due to K. Furukawa for help with the numerical examples and to D. Nordman for carefully going over parts of the manuscript. I also thank J. Fukuchi, Y. D. Lee, S. Sun, and J. Zhu who have enriched my research on the topic as students at various time points. I thank my wife for her moral support and understanding. Many thanks go to Sharon Shepard for converting my scribblings into a typed manuscript with extraordinary accuracy and consistency. I also thank Springer's Editor, John Kimmel, for his patience and good humor over the long time period of this project. I gratefully acknowledge the continuous support of the National Science Foundation for my research work in this area.

Contents

1

Scope of Resampling Methods for Dependent Data

1.1 The Bootstrap Principle

The bootstrap is a computer-intensive method that provides answers to a large class of statistical inference problems without stringent structural assumptions on the underlying random process generating the data. Since its introduction by Efron (1979), the bootstrap has found its application to a number of statistical problems, including many standard ones, where it has outperformed the existing methodology as well as to many complex problems where conventional approaches failed to provide satisfactory answers. However, it is not a panacea for every problem of statistical inference, nor does it apply equally effectively to every type of random process in its simplest form. In this monograph, we shall consider certain classes of dependent processes and point out situations where different types of bootstrap methods can be applied effectively, and also look at situations where these methods run into problems and point out possible remedies, if there is one known.

The bootstrap and other resampling methods typically apply to the statistical inference problems involving what we call *level-2* (and *higher-level*) parameters of the underlying random process. Let $X_1, X_2, \ldots$ be a sequence of random variables with joint distribution P. Suppose that the data at hand can be modeled as a realization of the first n random variables $\{X_1, \ldots, X_n\} \equiv \mathcal{X}_n$. Also suppose that $\theta \equiv \theta(P)$ is a real-valued (say) parameter of interest, which depends on the unknown joint distribution of the sequence $X_1, X_2, \ldots$. A common problem of statistical inference is to define an (point) estimator of θ based on the observations $\mathcal{X}_n$. Many standard and general methods for finding estimators of θ are typically available, such as those based on likelihood theory (maximum likelihood, quasi-likelihood), estimating equations (M-estimators), and nonparametric smoothing (kernel estimators), depending on the form of the parameter θ. Suppose that $\hat{\theta}_n$ is an estimator of θ based on $\mathcal{X}_n$. Having chosen $\hat{\theta}_n$ as an estimator, the statistician needs to answer further questions regarding the accuracy of the estimator $\hat{\theta}_n$ or about the quality of inference based on $\hat{\theta}_n$. Let G_n denote the sampling distribution of the centered estimator $\hat{\theta}_n - \theta$. Because the joint distribution of $X_1, \ldots, X_n$ is unknown, G_n also typically remains unknown. Thus, quantities like the mean squared error of $\hat{\theta}_n$, viz. $\text{MSE}(\hat{\theta}_n) = \int x^2 dG_n(x)$, and the quantiles of $\hat{\theta}_n$, are *unknown* population quantities based on the sampling distribution G_n. We call parameters like θ, *level-1* parameters and parameters like $\text{MSE}(\hat{\theta}_n)$, which relate to the sampling distribution of an estimator of a level-1 parameter, level-2 parameters. Bootstrap and other resampling methods can be regarded as general methods for finding estimators of level-2 parameters. In the same vein, functionals related to the sampling distribution of an estimator of a level-2 parameter are *level-3* parameters, and so on. For estimating such higher-level parameters, one may use a suitable number of iterations of the bootstrap or may successively apply a combination of more than one resampling method, e.g., the Jackknife-After-Bootstrap method of Efron (1992) (see Example 1.3 below).

The basic principle underlying the bootstrap method in various settings and in all its different forms is a simple one; it attempts to recreate the relation between the "population" and the "sample" by considering the sample as an epitome of the underlying population and, by resampling from it (suitably) to generate the "bootstrap sample", which serves as an analog of the given sample. If the resampling mechanism is chosen *appropriately*, then the "resample," together with the sample at hand, is expected to reflect the original relation between the population and the sample. The advantage derived from this exercise is that the statistician can now avoid the problem of having to deal with the unknown "population" directly, and instead, use the "sample" and the "resamples," which are either known or have known distributions, to address questions of statistical inference regarding the unknown population quantities. This (bootstrap) principle is most transparent in the case where $X_1, \ldots, X_n$ are independent and identically distributed (iid) random variables. First we describe the principle for iid random variables, and then describe it for dependent variables.

Suppose, for now, that $X_1, \ldots, X_n$ are iid random variables with common distribution F. Then, the joint distribution of $X_1, \ldots, X_n$ is given by $P_n = F^n$, the n-fold product of F. The level-1 parameter θ is now completely specified by a functional of the underlying marginal distribution F. Hence, suppose that $\theta = \theta(F)$. Let $\hat{\theta}_n = t(X_1, \ldots, X_n)$ be an estimator of θ. Suppose that we are interested in estimating some population characteristic, such as the mean squared error (MSE) of $\hat{\theta}_n$. It is clear that the sampling distribution of the centered estimator $\hat{\theta}_n - \theta$ and, hence, the MSE of $\hat{\theta}_n$, depend on the population distribution function F, which is itself *unknown*. Note that in the present context, we would know F if we could observe all (potential) members of the underlying population from which the sample $X_1, \ldots, X_n$ was drawn. For example, if X_i denoted the hexamine content of the ith pallet produced by a palletizing machine under identical production conditions, then we would know F and the distribution of random variables like $\hat{\theta}_n - \theta$ if all possible pallets were produced using the machine. This may not be possible in a given span of time or may not even be realistically achievable since the long-run performance of the machine is subject to physical laws of deterioration, violating the "identical"-assumption on the resulting observations.

The bootstrap principle addresses this problem without requiring the full knowledge of the population. The first step involves constructing an estimator $\tilde{F}_n$, say, of F from the available observations $X_1, \ldots, X_n$, which presumably provides a representative picture of the population and plays the role of F. The next step involves generating iid random variables $X_1^*, \ldots, X_n^*$ from the estimator $\tilde{F}_n$ (conditional on the observations $\mathcal{X}_n$), which serve the role of the "sample" for the bootstrap version of the original problem. Thus, the "bootstrap version" of the estimator $\hat{\theta}_n$ based on the original sample $X_1, \ldots, X_n$ is given by θ_n^*, obtained by replacing $X_1, \ldots, X_n$ with $X_1^*, \ldots, X_n^*$, and the "bootstrap version" of the level-1 parameter $\theta = \theta(F)$ based on the population distribution function F is given by $\theta(\tilde{F}_n)$. Note that the bootstrap versions of both the population parameter θ and the sample-based estimator $\hat{\theta}_n$ can be defined using the knowledge of the sample $X_1, \ldots, X_n$ only. For a reasonable choice of $\tilde{F}_n$, the bootstrap version accurately mimics those characteristics of the population and the sample that determine the sampling distribution of variables like $\hat{\theta}_n - \theta$. As a result, the bootstrap principle serves as a general method for estimating level-2 parameters related to the unknown distribution of $\hat{\theta}_n - \theta$. Specifically, the bootstrap estimator of the unknown sampling distribution G_n of the random variable $\hat{\theta}_n - \theta$ is given by the conditional distribution $\hat{G}_n$, say, of its bootstrap version $\theta_n^* - \theta(\tilde{F}_n)$. And the bootstrap estimator of the level-2 parameter $\varphi_n \equiv \varphi(G_n)$, derived through a functional $\varphi(\cdot)$ of G_n, is simply given by the "plug-in" estimator $\hat{\varphi}_n \equiv \varphi(\hat{G}_n)$. For example, the bootstrap estimators of the bias and MSE of $\hat{\theta}_n$ (i.e., the first two moments of $\hat{\theta}_n - \theta$), are respectively given by

$$\widehat{BIAS} = \int x \hat{G}_n(dx) = E_*(\theta_n^*) - \theta(\tilde{F}_n)$$

and

$$\widehat{MSE} = \int x^2 \hat{G}_n(dx) = E_*((\theta_n^*) - \theta(\tilde{F}_n))^2 \ ,$$

where E_* denotes the conditional expectation given $X_1, \ldots, X_n$. Note that these formulas are valid for any estimator $\hat{\theta}_n$, and do not presuppose any specific form.

The most common choice of $\tilde{F}_n$ is the empirical distribution function $\hat{F}_n(\cdot) \equiv n^{-1} \sum_{i=1}^{n} \mathbb{1}(X_i \leq \cdot)$, where $\mathbb{1}(A)$ denotes the indicator function of a statement A, and it takes the values 0 or 1 according as the statement A is false or true. In this case, the bootstrap random variables $X_1^*, \ldots, X_n^*$ represent a simple random sample drawn with replacement from the observed sample $X_1, \ldots, X_n$, and as a result, can assume only the n data values observed, thereby justifying the name "resample." There are other situations, e.g., the parametric bootstrap, where $X_1^*, \ldots, X_n^*$ are generated by a different estimated distribution and may take values other than the observed values $X_1, \ldots, X_n$. However, the basic characteristics of the population are captured through the estimated distribution in both cases and are reflected in the respective bootstrap observations $X_1^*, \ldots, X_n^*$. Thus, in spite of such variations due to different choices of $\tilde{F}_n$, all these bootstrap methods fall under the general bootstrap principle described above.

The same can be said about the bootstrap methods that have been proposed in the context of dependent data. Here, the situation is slightly more complicated because the "population" is not characterized entirely by the one-dimensional marginal distribution F alone, but requires the knowledge of the joint distribution of the whole sequence $X_1, X_2, \ldots$. Nonetheless, the basic principle still applies. Here, we consider the block bootstrap methods that are most commonly used in the context of general time series data, and show that these methods fall within the ambit of the bootstrap principle. For simplicity, we restrict the discussion to the case of the nonoverlapping block bootstrap (NBB) method (cf. Carlstein (1986)). The principle behind other block bootstrap methods presented in Chapter 2 can be explained by straightforward modification of our discussion below.

Suppose that $X_1, X_2, \ldots$ is a sequence of stationary and weakly dependent random variables such that the series of the autocovariances of X_i's converges absolutely. To fix ideas, first consider the case where the level-1 parameter θ of interest is the population mean, i.e., $\theta = E(X_1)$. If we use $\hat{\theta}_n \equiv \bar{X}_n$, the sample mean, as an estimator of θ, then the distribution of $\hat{\theta}_n - \theta$ depends on not only on the marginal distribution of X_1, but it is a functional of the joint distribution of $X_1, \ldots, X_n$. For example, $\text{Var}(\hat{\theta}_n) = n^{-1}\big[\text{Var}(X_1) + 2\sum_{i=1}^{n-1}(1 - i/n)\text{Cov}(X_1, X_{1+i})\big]$ depends on the bivariate distribution of X_1 and X_i for all $1 \leq i \leq n$. Note that since the process $\{X_n\}_{n \geq 1}$ is assumed to be weakly dependent, the main contribution to $\text{Var}(\hat{\theta}_n)$ comes from the lower-order lag autocovariances and the total contribution from higher-order lags is negligible. More specifically, if ℓ is a large positive integer (but smaller than n), then the total contribution to $\text{Var}(\hat{\theta}_n)$ from lags of order ℓ or more, viz., $\sum_{i=\ell}^{n-1}(1 - i/n)\text{Cov}(X_1, X_{1+i})$, is bounded in absolute value by $\sum_{i=\ell}^{\infty} |\text{Cov}(X_1, X_{1+i})|$, which tends to zero as ℓ goes to infinity. As a consequence, accurate approximations for the level-2 parameter $\text{Var}(\hat{\theta}_n)$ can be generated from the knowledge of the lag covariances $\text{Cov}(X_1, X_{1+i})$, $0 \leq i < \ell$, which depend on the joint distribution of the *shorter* series $\{X_1, \ldots, X_\ell\}$ of the given sequence of observations $\{X_1, \ldots, X_n\}$.

The block bootstrap methods exploit this fact to recreate the relation between the "population" and the "sample," in a way similar to the iid case. Suppose that we are interested in approximating the sampling distribution of a random variable of the form $T_n = t_n(\mathcal{X}_n; \theta)$, where $\theta \equiv \theta(P)$ is a level-1 parameter based on the joint distribution P of $X_1, X_2, \ldots$ and $t_n(\cdot; \theta)$ is invariant under permutations of $X_1, \ldots, X_n$. For example, we may have $T_n = n^{1/2}(\bar{X}_n - \mu)$, with $\theta = \mu \equiv EX_1$. Next, suppose that ℓ is an integer such that both ℓ and n/ℓ are large, e.g., $\ell = \lfloor n^\delta \rfloor$ for some $0 < \delta < 1$, where for any real number x, $\lfloor x \rfloor$ denotes the largest integer not exceeding x. Also, for simplicity, suppose that $b \equiv n/\ell$ is an integer. Then, for the NBB method, the given sequence of observations $\{X_1, \ldots, X_n\}$ is partitioned into b subseries or "blocks" $\{X_1, \ldots, X_\ell\}, \{X_{\ell+1}, \ldots, X_{2\ell}\}, \ldots, \{X_{(b-1)\ell+1}, \ldots, X_n\}$ of length ℓ, and a set of b blocks are *resampled* from these observed blocks to generate the bootstrap sample $X_1^*, \ldots, X_n^*$. The NBB version of T_n is defined as $T_n^* \equiv t_n(\mathcal{X}_n^*; \tilde{\theta}_n)$, where $\mathcal{X}_n^* \equiv \{X_1^*, \ldots, X_n^*\}$ and $\tilde{\theta}_n$ is an estimator of θ based on the conditional distribution of $X_1^*, \ldots, X_n^*$.

Again, the bootstrap principle underlying the NBB attempts to recreate the relation between the population and the sample, although in a slightly different way. Let P_k denote the joint distribution of $(X_1, \ldots, X_k)$, $k \geq 1$, and let $Y_1, \ldots, Y_b$ denote the b blocks under the NBB, defined by $Y_1 = \{X_1, \ldots, X_\ell\}, \ldots, Y_b = \{X_{(b-1)\ell+1}, \ldots, X_n\}$. Note that because of stationarity, each block has the same (ℓ-dimensional joint) distribution P_ℓ. Furthermore, because of the weak dependence of the original sequence $\{X_n\}_{n\geq1}$, these blocks are approximately independent for large values of ℓ. Thus, $Y_1, \ldots, Y_b$ gives us a collection of "approximately independent" and "identically distributed" random vectors with common distribution P_ℓ. By resampling from the collection $Y_1, \ldots, Y_b$ randomly, the "block" bootstrap method described above actually reproduces the relation between the sample $\{X_1, \ldots, X_n\}$ and the "approximate" population distribution $P_\ell^b \equiv P_\ell \otimes \cdots \otimes P_\ell$, which is "close" to the exact population distribution P_n because of the weak-dependence assumption on $\{X_n\}_{n\geq1}$. Indeed, if $\tilde{P}_\ell$ denotes the empirical distribution of $Y_1, \ldots, Y_b$, then the joint distribution of the bootstrap observations $X_1^*, \ldots, X_n^*$ under the NBB is given by $\tilde{P}_\ell^b$. Thus, the "resampling population" distribution $\tilde{P}_\ell^b$ is close to P_ℓ^b, which is in turn close to the true underlying population P_n. Hence, the relation between $\{X_1, \ldots, X_n\}$ and P_n in the original problem is reproduced (approximately) by the relation between $\{X_1^*, \ldots, X_n^*\}$ and $\tilde{P}_\ell^b$ under the NBB. Let $\mathcal{L}(W; Q)$ denote the probability distribution of a random quantity W under a probability measure Q. For the random quantity $T_n = t_n(\mathcal{X}_n; \theta_n(P_n))$ of interest, the approximations involved in application of the bootstrap principle may be summarized by the following description:

$$\begin{aligned} \mathcal{L}(T_n; P_n) &= P_n(t_n(\mathcal{X}_n; \theta_n(P_n)) \in \cdot) \\ &\approx P_\ell^b(t_n(\mathcal{X}_n; \theta_n(P_l^b)) \in \cdot) \\ &\approx \tilde{P}_\ell^b(t_n(\mathcal{X}_n^*; \theta_n(\tilde{P}_\ell^b)) \in \cdot) \\ &= \mathcal{L}(T_n^*; \tilde{P}_\ell^b) . \end{aligned} \tag{1.1}$$

The justification of the first approximation follows from the weak-dependence assumption on the original process $\{X_n\}_{n\geq 1}$. The second approximation rests on the bootstrap principle, which says the relation $\mathcal{X}_n : P_n$ is reproduced in the relation $\mathcal{X}_n^* : \tilde{P}_\ell^b$, so that for *nice* functions $t_n(\cdot;\cdot)$ and $\theta_n(\cdot)$, the distribution of T_n under P_n is "close" to the distribution of T_n^* under $\tilde{P}_\ell^b$. However, it should be remembered that this is just a heuristic argument, which requires further qualification of the quality and validity of the above approximations in each specific problem. In the subsequent chapters of this book, we look at different situations where this simple principle provides a correct solution, often outperforming alternative traditional inference methods. We also consider certain pathological cases where the simple principle, if applied naively, fails drastically.

Before we look at some illustrative examples, it is worthwhile to point out the important role played by the computer in the *implementation* of the principle. Although the bootstrap principle in (1.1) prescribes an *estimator* (i.e., a function of the data) of the unknown population quantity $\mathcal{L}(T_n; P_n)$ under a very general framework, it does not spell out any general method for evaluating this estimator. An exact computation of the bootstrap estimator is impractical, if not impossible, in most situations. This is due to the fact that for a given set $\mathcal{X}_n$ of n data values, the number of possible values of the NBB observations $\mathcal{X}_n^*$ is $(b)^b$, which grows to infinity very fast. For example, if $\ell = \lfloor n^\delta \rfloor$ for some $0 < \delta < 1$, then this number grows as fast as $n^{(1-\delta)n^{1-\delta}}$ as n goes to infinity. For $\delta = 1/3$, this is comparable to 1.8×10^{19} for $n = 64$ and to $(10)^{200}$ for $n = 1000$. This problem is tackled by using a further Monte-Carlo approximation to the last step in (1.1). It is precisely in this step, where the computer plays an indispensable role. Let $\mathcal{X}_n^{*(j)}$, $j = 1, \ldots, B$ be iid copies of $\mathcal{X}_n^*$ having the common distribution $\tilde{P}_\ell^b$, and let $T_n^{*(j)} = t_n(\mathcal{X}_n^{*(j)}; \tilde{P}_\ell^b)$, $j = 1, \ldots, B$ denote the "bootstrap replicates" of T_n^*. Then, by the strong law of large numbers (or the Glivenko-Cantelli Theorem),

$$\frac{1}{B} \sum_{j=1}^{B} \mathbb{1}(T_n^{*(j)} \in \cdot) \approx \tilde{P}_\ell^b(T_n^* \in \cdot) , \tag{1.2}$$

if B is large. Therefore, the computer can be used to obtain approximations for $\tilde{P}_\ell^b(T_n^* \in \cdot)$ to any degree of accuracy by generating a sufficiently large number of $\mathcal{X}_n^{*(j)}$'s by block-resampling from the given set $\mathcal{X}_n$ of observations and by computing the bootstrap replicates $T_n^{*(j)}$'s.

In the next section, we look at some illustrative examples involving dependent data where the bootstrap principle can be applied effectively.

1.2 Examples

The first example illustrates some of the basic usage of the bootstrap method with a simulated data set.

Example 1.1: Suppose that the observations $X_1, \ldots, X_n$ are generated by a stationary ARMA(1,1) process:

$$X_i = \beta_1 X_{i-1} + \epsilon_i + \alpha_1 \epsilon_{i-1}, \; i \in \mathbb{Z} \;, \tag{1.3}$$

where $|\alpha_1| < 1$, $|\beta_1| < 1$ are constants and $\{\epsilon_i\}_{i \in \mathbb{Z}}$ is a sequence of iid random variables with $E\epsilon_1 = 0$, $E\epsilon_1^2 = 1$. Figure 1.1 shows a simulated data set of size $n = 100$ from the ARMA(1,1) process in (1.3) with $\alpha_1 = 0.2$, $\beta_1 = 0.3$ and with Gaussian variables ϵ_1. Suppose we are interested in estimating the variance and certain other population characteristics of the sample mean $\bar{X}_n = n^{-1}\sum_{i=1}^n X_i$ (which, according to our terminology, are level-2 parameters). For the sake of illustration, suppose that we decided to use the NBB method with block length $\ell = 5$, say, then we first form the blocks $\mathcal{B}_1 = (X_1, \ldots, X_5)$, $\mathcal{B}_2 = (X_6, \ldots, X_{10}), \ldots, \mathcal{B}_{20} = (X_{96}, \ldots, X_{100})$, and then, resample $b = 20$ blocks $\mathcal{B}_1^*, \ldots, \mathcal{B}_{20}^*$ with replacement from $\{\mathcal{B}_1, \ldots, \mathcal{B}_{20}\}$ to generate the block bootstrap observations $X_1^*, \ldots, X_{100}^*$.

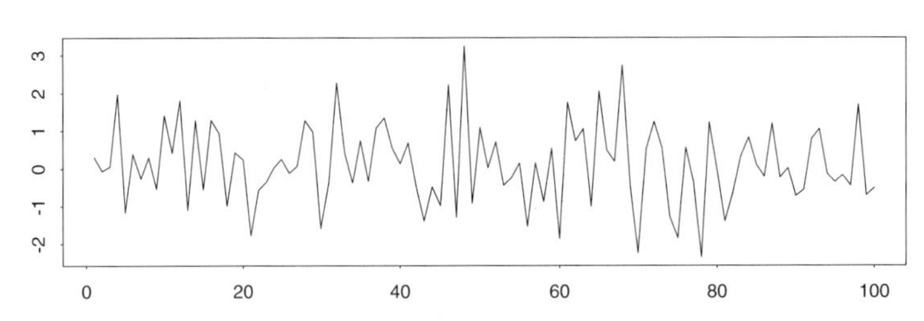

FIGURE 1.1. A simulated data set of size $n = 100$ from model (1.3) with $\beta_1 = 0.2$ and $\alpha_1 = 0.3$.

Example 1.1.1: Variance Estimation
First, we consider estimating the level-2 parameter $\varphi_n = \text{Var}(\bar{X}_n)$. As described before, the bootstrap estimator of $\text{Var}(\bar{X}_n)$ is given by

$$\widehat{\text{Var}}(\bar{X}_n) = E_*(\bar{X}_n^* - E_*\bar{X}_n^*)^2 \;, \tag{1.4}$$

where $\bar{X}_n^* = n^{-1}\sum_{i=1}^n X_i^*$ and $n = 100$. Unlike most problems, this is a special case where the bootstrap estimator $\widehat{\text{Var}}(\bar{X}_n)$ can be evaluated directly, without any recourse to Monte-Carlo simulation. It will be shown in Chapter 3 that we can express the bootstrap estimator as

$$\widehat{\text{Var}}(\bar{X}_n) = b^{-1}\Big[b^{-1}\sum_{j=1}^{b} V_j - \Big(b^{-1}\sum_{j=1}^{b} V_j\Big)^2\Big] , \tag{1.5}$$

where V_j denotes the mean of the jth block. Thus, for our example, $V_1 = (X_1 + \cdots + X_5)/5$, $V_2 = (X_6 + \cdots + X_{10})/5, \ldots$, etc. Applying (1.5) to the data set of Figure 1.1, we obtain $\widehat{\text{Var}}(\bar{X}_n) = 8.77 \times 10^{-3}$ as an *estimate* of $\text{Var}(\bar{X}_n)$. This should be compared to the true value of $\text{Var}(\bar{X}_n)$, which is 11.74×10^{-3} under the assumed specification of model (1.3).

Example 1.1.2: Distribution Function Estimation

Next we consider estimating the sampling distribution of the studentized sample mean $T_n = \sqrt{n}(\bar{X}_n - \mu)/\hat{\sigma}_n$, where $\mu = EX_1$ (which is zero under the model (1.3)) and $\hat{\sigma}_n^2$ is an estimator of the (asymptotic) variance of $\sqrt{n}\bar{X}_n$. For definiteness, suppose that $\hat{\sigma}_n^2 = n\widehat{\text{Var}}(\bar{X}_n)$, where $\widehat{\text{Var}}(\bar{X}_n)$ is given by (1.5). In principle, to obtain the block bootstrap estimator of $G_n(x) \equiv P(T_n \leq x)$, $x \in \mathbb{R}$, we have to form the bootstrap version $T_n^* = \sqrt{n}(\bar{X}_n^* - E_*\bar{X}_n^*)/\sigma_n^*$ of T_n and estimate $G_n(x)$ by the conditional distribution function $\hat{G}_n(x) \equiv P_*(T_n^* \leq x)$, $x \in \mathbb{R}$. Here, the bootstrap variable σ_n^{*2} is defined by replacing $X_1, \ldots, X_n$ in $\hat{\sigma}_n^2$ with $X_1^*, \ldots, X_n^*$. However, unlike the block bootstrap estimator of $\text{Var}(\bar{X}_n)$ in (1.5), an explicit, simple expression for $\hat{G}_n(x)$ is not available. For evaluating the block bootstrap estimator $\hat{G}_n(x)$ for the given data set, we used Monte-Carlo simulation as follows. From the "observed" blocks $\mathcal{B}_1, \ldots, \mathcal{B}_{20}$, formed using the data set in Figure 1.1, we repeated the resampling step (cf. (1.2)) $B = 2000$ times to generate 2000 sets of bootstrap observations $\{X_1^{*(1)}, \ldots, X_{100}^{*(1)}\}, \ldots, \{X_1^{*(2000)}, \ldots, X_{100}^{*(2000)}\}$. Here, the value $B = 2000$ is chosen arbitrarily for the purpose of illustration. A different value may be used, if necessary, depending on the desired level of accuracy. Next, we computed T_n^* for each set of 100 bootstrap observations. The histogram of these 2000 values of T_n^* is shown in Figure 1.2 below. The block bootstrap estimator of $P(T_n \leq x)$ is now given by the proportion of T_n^*-values that are less than or equal to x. For example, for $x = 1.2$, this yields 0.8755 as an estimate of $P(T_n \leq 1.2)$. Similarly, we obtain 0.3170 and 0.4965 as estimates of $P(T_n \leq -0.5)$ and $P(T_n \leq 0)$, respectively, from the histogram of T_n^*-values.

Example 1.1.3: Estimation of Critical Values

If, instead of the distribution function of T_n, it is of interest to find certain quantiles of T_n, bootstrap estimates of quantiles of T_n can also be found

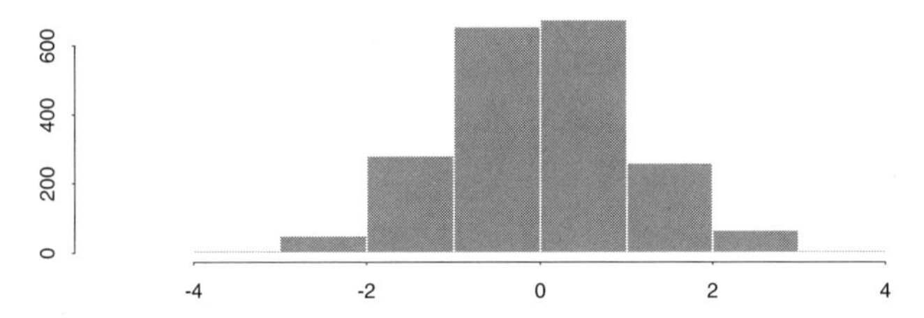

FIGURE 1.2. Histogram of T_n^*-values based on $B = 2000$ block bootstrap replicates.

from the same Monte-Carlo simulation. For $0 < \alpha < 1$, let t_α denote the (smallest) α-quantile of T_n, defined by

$$t_\alpha = \inf\{x : P(T_n \leq x) \geq \alpha\} . \tag{1.6}$$

Then, by the "plug-in principle," the bootstrap estimator of the level-2 parameter $t_\alpha = G_n^{-1}(\alpha)$ is given by $\hat{t}_\alpha = \hat{G}_n^{-1}(\alpha)$, the α-quantile of the conditional distribution $\hat{G}_n(\cdot)$ of T_n^*, given $X_1, \ldots, X_n$. The Monte-Carlo approximation to this is simply the $\lfloor B\alpha \rfloor$th order statistic of the B bootstrap replicates $\{T_n^{*(1)}, \ldots, T_n^{*(B)}\}$ of T_n. Thus, for the data set of Figure 1.1, an *estimate* of the 0.9-quantile of T_n is the $\lfloor (2000)(0.9) \rfloor = 1800$th order statistic of the $B = 2000$ bootstrap replicates of T_n^* summarized in the histogram in Figure 1.2, and is given by 1.3514.

Example 1.1.4: Bootstrap Confidence Intervals
Next we consider the problem of finding a confidence interval (CI) for μ with confidence level, say, 80%. If the quantiles t_α's of T_n were known, an equal-tailed 80% CI for μ would be

$$\left(\bar{X}_n - (t_{0.9}) \cdot \hat{\sigma}_n/\sqrt{n},\ \bar{X}_n - (t_{0.1}) \cdot \hat{\sigma}_n/\sqrt{n}\right) . \tag{1.7}$$

However, since the sampling distribution of T_n is unknown, the quantiles of T_n also remain unknown in practice. Here again, one may use the bootstrap to find an approximate CI for μ. An 80% bootstrap CI for μ is now obtained by replacing the quantiles $t_{0.9}$ and $t_{0.1}$ of T_n by the corresponding quantiles of the conditional distribution of T_n^* as

$$\left(\bar{X}_n - (\hat{t}_{0.9})\hat{\sigma}_n/\sqrt{n},\ \bar{X}_n - (\hat{t}_{0.1})\hat{\sigma}_n/\sqrt{n}\right) . \tag{1.8}$$

For the data set of Figure 1.1, the (Monte-Carlo) values of $\hat{t}_{0.9}$ and $\hat{t}_{0.1}$ are respectively given by 1.3514 and -1.3675, the 1800th and the 200th order statistics of the 2000 bootstrap replicates of T_n^*. The resulting bootstrap CI for μ is thus given by

$$(-0.1258, 0.1287) . \tag{1.9}$$

Note that the true value of the parameter μ in this case is zero. It may be of some interest to compare this CI with the traditional large-sample CI for μ based on asymptotic normality of T_n. Indeed, for the given sample, a 80% approximate normal CI for μ is given by

$$(\bar{x}_n - 1.28\,\hat{\sigma}_n/\sqrt{n},\ \bar{x}_n + 1.28\,\hat{\sigma}_n/\sqrt{n}) = (-0.1191, 0.1206)\ ,$$

which happens to be shorter than the bootstrap confidence interval in (1.9). How do the accuracies of the bootstrap CI in (1.8) and of its one-sided version compare with those of the corresponding large-sample normal CIs? Answers to these questions depend on second- and higher-order properties of the bootstrap approximation $P_*(T_n^* \leq \cdot)$ to the sampling distribution $P(T_n \leq \cdot)$ of T_n. We shall address some of these issues in more details in Chapter 6. □

The next three examples present instances of various inference problems involving level-2 and higher-level parameters, where the bootstrap and other resampling methods can be applied effectively.

Example 1.2: *Smoothing parameter selection.* Consider the model

$$Y_i = m(x_i) + \epsilon_i, \quad 1 \leq i \leq n\ ,$$

where $m : [0,1] \to \mathbb{R}$ is an unknown function, $x_1, \ldots, x_n \in [0,1]$ are nonrandom design points, and $\{\epsilon_n\}_{n \geq 1}$ is a sequence of unobservable zero mean stationary random variables. Here, the function $m(\cdot)$ is a level-1 population parameter that can be estimated from the observations $Y_1, \ldots, Y_n$ by one of the "smoothing" methods. For the discussion here, suppose we use the Nadaraya-Watson kernel estimator (cf. Nadaraya (1964), Watson (1964)):

$$\hat{m}_h(x) = \sum_{i=1}^{n} K_h(x - x_i) Y_i \Big/ \sum_{i=1}^{n} K_h(x - x_i), \ x \in [0,1]$$

to estimate $m(\cdot)$. Here $K(\cdot)$ is a probability density function on $\mathbb{R}$ with a compact support and vanishing first moment, and $h > 0$ is a bandwidth. Performance of $\hat{m}_h(\cdot)$ as an estimator of $m(\cdot)$ critically depends on the bandwidth or smoothing parameter h. A standard measure of global accuracy of $\hat{m}_h(\cdot)$ is the mean integrated squared error (MISE), defined as

$$\text{MISE}(h) = \int_{\mathcal{J}} E(\hat{m}_h(x) - m(x))^2 dx\ ,$$

where $\mathcal{J} \subset [0,1]$. For optimum performance, we need to use the estimator $\hat{m}_h(\cdot)$ with bandwidth $h = h^*$, where h^* minimizes the risk function $\text{MISE}(h)$. Note that the function $\text{MISE}(h)$ and, hence, h^* depend on the sampling distribution of the estimator $\hat{m}_h(\cdot)$ of a level-1 parameter $m(\cdot)$, and thus, are level-2 parameters. In this case, one can apply the bootstrap principle to obtain estimators of the risk function MISE(h) and the optimum bandwidth h^*. For independent error variables $\{\epsilon_n\}_{n\geq 1}$, estimation of these second level parameters by resampling methods has been initiated by Taylor (1989), Faraway and Jhun (1990) (although in the context of density estimation) and has been treated in detail in Hall (1992) and Shao and Tu (1995). For the case of dependent errors, estimation of MISE(h) and h^* becomes somewhat more involved due to the effect of serial correlation of the observations. But the bootstrap principle still works. Indeed, block bootstrap methods can be used to estimate the level-2 parameters MISE(h) and h^* consistently for a wide class of dependent processes including a class of long range dependent processes (cf. Hall, Lahiri and Polzehl (1995)). □

Example 1.3: *Estimation of variogram parameters.* Let $\{Z(s) : s \in \mathbb{R}^d\}$ be a random field. Suppose that $Z(\cdot)$ is intrinsically stationary, i.e.,

$$\begin{aligned} E(Z(s+h) - Z(s)) &= E(Z(h) - Z(0)), \\ \mathrm{Var}(Z(s+h) - Z(s)) &= \mathrm{Var}(Z(h) - Z(0)), \end{aligned} \tag{1.10}$$

for all $s, h \in \mathbb{R}^d$. Intrinsic stationarity of a random field is similar to the concept of second-order stationarity (cf. Cressie (1993), Chapter 2) and is commonly used in geostatistical applications. Like the autocovariance function, the second moment structure of an intrinsic random field may be described in terms of its *variogram* $2\gamma(\cdot)$, defined by

$$2\gamma(h) = \mathrm{Var}(Z(h) - Z(0)), \; h \in \mathbb{R}^d. \tag{1.11}$$

Estimation of the variogram is an important problem in Geostatistics. When the true variogram lies in a parametric model $\{2\gamma(\cdot;\theta) : \theta \in \Theta\}$ of valid variograms, a popular approach is to estimate the variogram parameter θ by minimizing a least-squares criterion of the form

$$Q_n(\theta) \equiv \sum_{i=1}^{K}\sum_{j=1}^{K} v_{ij}(\theta)[2\check{\gamma}_n(h_i) - 2\gamma(h_i;\theta)][2\check{\gamma}_n(h_j) - 2\gamma(h_j;\theta)]$$

where $V(\theta) \equiv \big((v_{ij}(\theta))\big)_{K\times K}$ is a $K \times K$ positive-definite weight matrix and $2\check{\gamma}_n(h_i)$ is a nonparametric estimator of the variogram at lag h_i, $1 \leq i \leq K$. Statistical efficiency of the resulting least-squares estimator, say, $\hat{\theta}_{n,V}$, depends on the choice of the weight matrix V. An optimal choice of V depends on the covariance matrix Σ (say) of the vector of variogram estimators $(2\check{\gamma}_n(h_1), \ldots, 2\check{\gamma}_n(h_K))'$. This again presents an example of a problem where an "optimal" estimator of the level-1 parameter θ requires the knowledge of the level-2 parameter Σ. Thus, one may apply the bootstrap or other resampling methods for spatial data to estimate the level-2 parameter first and minimize the estimated criterion function to obtain an estimator of the level-1 parameter. In Chapter 11, we show that the estimator derived using this approach is asymptotically optimal. □

Example 1.4: *Selection of optimal block size.* Performance of a block bootstrap method critically depends on the particular block length employed in finding the bootstrap estimator. Suppose that $\hat{\theta}_n$ is an estimator of a level-1 parameter of interest, θ, based on observations $X_1, \ldots, X_n$, and that we want to estimate some characteristic φ_n of the sampling distribution of $(\hat{\theta}_n - \theta)$. Thus, φ_n is a level-2 parameter. When the observations are correlated, we may apply a block bootstrap method, based on blocks of length $\ell \in (1, n)$ to estimate φ_n. Let $\hat{\varphi}_n(\ell)$ denote the block bootstrap estimator of φ_n. Typically, $\hat{\varphi}_n(\ell)$ is a consistent estimator of φ_n for a wide range of values of the block length ℓ. If one considers the bias and the variance of $\hat{\varphi}_n(\ell)$ as an estimator of φ_n, in many examples the bias of $\hat{\varphi}_n(\ell)$ decreases as the block length ℓ increases, while the variance of $\hat{\varphi}_n(\ell)$ increases with ℓ. Thus, there is an optimal value, say ℓ^0, of the block length that balances the trade off between the bias and the variance, and thus minimizes the MSE. Therefore, if we define ℓ^0 as $\ell^0 \equiv \arg\min\{\text{MSE}(\hat{\varphi}_n(\ell)) : 1 < \ell < n\}$, it is clear that ℓ^0 is an unknown parameter that depends on the sampling distributions of the estimators $\hat{\varphi}_n(\ell)$, $1 < \ell < n$ of a level-2 parameter φ_n and hence, ℓ^0 is a level-3 parameter. It may certainly be possible to use a plug-in estimator of ℓ^0 for a given φ_n, using analytical calculations. However, a general method that would be applicable more widely and without new analytical calculations for each case may be developed by applying two rounds of resampling methods iteratively. In Chapter 7, we present two such methods that yield nonparametric estimators of the optimal block length ℓ^0. □

1.3 Concluding Remarks

In this chapter, we described the basic principle underlying the bootstrap methods for dependent and independent data. The key point that we try to convey is that the bootstrap and other resampling methods can be viewed as "general" methods for inferring about level-2 and higher-level parameters, just as the Maximum Likelihood and similar methods may be considered as general methods for estimating level-1 parameters. The basic idea underlying different versions of the bootstrap is the *same*, which is to recreate the original "population versus sample" relation for a level-2 inference problem through a suitable resampling mechanism. Section 1.1 attempts to describe this for the (nonoverlapping) block bootstrap method. The arguments apply to the other bootstrap methods described in Chapter 2 with minor modifications. In Section 1.2, we presented examples of some inference problems for dependent data where resampling methods were naturally applicable. In Example 1.1, we illustrated the use of the (nonoverlapping) block bootstrap method with a given data set and showed how it might be used in practice to address various inference issues. Examples 1.2, 1.3, and 1.4 were described in less specific terms. They served as examples of inference problems involving level-2 and higher-level parameters where the bootstrap and other resampling methods might be used effectively, and, thus, illustrated the "scope" of bootstrap methods for dependent data. In subsequent chapters, we describe various forms of bootstrap methods and investigate their properties (success and failure) in similar (level-2+) inference problems under different dependence structures of the data.

1.4 Notation

For reference later on, we collect some common notation to be used in the rest of the book. Let $\mathbb{N}$ and $\mathbb{Z}$, respectively, denote the set of all positive integers and the set of all integers. Also, let $\mathbb{Z}_+ = \{0\} \cup \mathbb{N}$ be the set of all nonnegative integers. Let $\mathbb{R}$ denote the set of all real numbers and $\mathbb{C}$ the set of all complex numbers. The extended real line is denoted by $\bar{\mathbb{R}} = \mathbb{R} \cup \{-\infty, \infty\}$. The Borel σ-field on a metric space $\mathbb{S}$ is denoted by $\mathcal{B}(\mathbb{S})$ and for a set $A \subset \mathbb{S}$, let $\mathrm{cl.}(A)$ and ∂A respectively denote the closure and the boundary of A. For a real number x, let $\lfloor x \rfloor$ denote the largest integer not exceeding x, let $\lceil x \rceil$ denote the smallest integer not less than x, and let $x_+ = \max\{x, 0\}$. For $x, y \in \mathbb{R}$, let $x \wedge y = \min\{x, y\}$ and $x \vee y = \max\{x, y\}$. Let $\mathbb{1}(S)$ denote the indicator function of a statement S, with $\mathbb{1}(S) = 1$ if S is true and $\mathbb{1}(S) = 0$ otherwise. For a subset A of a nonempty set Ω, we also write $\mathbb{1}_A$ for the function $\mathbb{1}_A(w) \equiv \mathbb{1}(w \in A)$, $w \in \Omega$. For a finite set A, we write $|A|$ to denote the size, i.e., the number of elements of A. Also, for a set $A \in \mathcal{B}(\mathbb{R}^k)$, $k \in \mathbb{N}$, we write $\mathrm{vol.}(A)$ to denote the volume or the Lebesgue measure of A. For $k \in \mathbb{N}$, let $\mathbb{I}_k$ denote the identity matrix of order k. Let Γ' denote the transpose of a matrix Γ. We write a $m \times n$ matrix Γ with (i,j)-th element γ_{ij}, $1 \le i \le m$, $1 \le j \le n$ as $\Gamma = ((\gamma_{ij}))$ or as $\Gamma = ((\gamma_{ij}))_{m \times n}$. Let $\det(\Gamma)$ denote the determinant of a square matrix Γ.

As a convention, (random) vectors in $\mathbb{R}^k$, $k \in \mathbb{N}$ are regarded as column vectors in this book. For $x = (x_1, \ldots, x_k)'$, $y = (y_1, \ldots, y_k)' \in \mathbb{R}^k$, and $\alpha = (\alpha_1, \ldots, \alpha_k)' \in \mathbb{Z}_+^k$ ($k \in \mathbb{N}$), write $x^\alpha = \prod_{i=1}^k x_i^{\alpha_i}$, $|x| = |x_1| + \cdots + |x_k|$, $\alpha! = \prod_{i=1}^k \alpha_i!$, and $\|x\| = (x_1^2 + \cdots + x_k^2)^{1/2}$, and write $x \le y$ if $x_i \le y_i$ for all $1 \le i \le k$. For a $m \times n$ matrix A, write $\|A\| = \sup\{\|Ax\| : x \in \mathbb{R}^n, \|x\| = 1\}$ for the spectral norm of A. For a smooth function $h : \mathbb{R}^k \to \mathbb{R}$, let $D_j h$ denote the partial derivative of h with respect to the j-th coordinate, i.e., $D_j h = \frac{\partial h}{\partial x_j}$. For $\alpha \in \mathbb{Z}_+^k$, let D^α denote the differential operator $D^\alpha \equiv D_1^{\alpha_1} \ldots D_k^{\alpha_k}$ $= \frac{\partial^{\alpha_1 + \cdots + \alpha_k}}{\partial x_1^{\alpha_1}, \ldots, \partial x_k^{\alpha_k}}$ on $\mathbb{R}^k$. Write $\iota = \sqrt{-1}$. For a real valued function f defined on a nonempty set A, we write "argmin$\{f(x) : x \in A\}$" to denote a value of the argument $x \in A$ at which f attains its minimum over the set A (assuming that the minimum is attainable on A).

Let C denote a generic constant in $(0, \infty)$ that may assume different values at each appearance and that does not depend on the limit variables like the sample size n. For example, $C \leq C/2$ is a valid inequality under this convention, which should be interpreted as $C_1 \leq C_2/2$ for some constants C_1 and C_2. In some instances, we also use $C(\cdot)$ as a notation for generic constants that may depend on the arguments specified within the parentheses.

For sequences $\{a_n\}_{n\geq 1} \subset \mathbb{R}$, $\{b_n\}_{n\geq 1} \subset (0, \infty)$, we write

$$a_n = o(b_n) \text{ as } n \to \infty \quad \text{if} \quad a_n/b_n \to 0 \text{ as } n \to \infty ,$$

$$a_n = O(b_n) \text{ as } n \to \infty \quad \text{if} \quad \limsup_{n\to\infty} |a_n|/b_n < \infty ,$$

$$a_n \ll b_n \text{ as } n \to \infty \quad \text{if} \quad a_n = o(b_n) \text{ as } n \to \infty$$

and

$$b_n \gg a_n \text{ as } n \to \infty \quad \text{if} \quad a_n = o(b_n) \text{ as } n \to \infty .$$

For $\{a_n\}_{n\geq 1}$, $\{b_n\}_{n\geq 1} \subset (0, \infty)$, we write

$$a_n \sim b_n \text{ as } n \to \infty \quad \text{if} \quad \lim_{n\to\infty} a_n/b_n = 1$$

and

$$a_n \approx b_n \text{ as } n \to \infty \quad \text{if} \quad a_n = O(b_n) \text{ and } b_n = O(a_n), \text{ as } n \to \infty .$$

Similarly, for a sequence of random variables $\{X_n\}_{n\geq 1}$, and a sequence $\{b_n\}_{n\geq 1} \subset (0, \infty)$, we write

$$X_n = o_p(b_n) \text{ as } n \to \infty \quad \text{if} \quad X_n/b_n \to 0 \text{ in probability as } n \to \infty$$

and

$$X_n = O_p(b_n) \text{ as } n \to \infty \quad \text{if} \quad \{X_n/b_n\}_{n\geq 1} \text{ is a tight sequence} ,$$

i.e., for every $\epsilon > 0$, there exists $M \in (0, \infty)$ such that

$$\sup_{n\geq 1} P(|X_n/b_n| > M) < \epsilon .$$

Unless otherwise specified, the limits in order symbols are taken letting the variable "n" tend to infinity. Thus, "$a_n = o(b_n)$" is the same as "$a_n = o(b_n)$ as $n \to \infty$".

Convergence in distribution and convergence in probability of random entities are respectively denoted by $\longrightarrow^d$ and $\longrightarrow_p$. Almost sure convergence with respect to a measure ν is written as a.s. (ν) or simply, a.s., if the relevant measure ν is clear from the context. In the later case, we also use "a.s." as an abbreviation for "almost sure" or "almost surely," as appropriate.

For k-dimensional random vectors X and Y with $E\|X\|^2 + E\|Y\|^2 < \infty$, we define the covariance matrix of X and Y and the variance matrix of X as

$$\text{Cov}(X, Y) = E\{(X - EX)(Y - EY)'\} \quad \text{and} \quad \text{Var}(X) = \text{Cov}(X, X) ,$$

respectively. For a random variable X and for $p \in [1, \infty]$, we define the L^p-norm of X by

$$\|X\|_p = \begin{cases} (E|X|^p)^{1/p} & \text{if} \quad p \in [1, \infty) \\ \text{ess. sup}\{X\} & \text{if} \quad p = \infty . \end{cases}$$

For a collection of σ-fields $\{\mathcal{F}_i : i \in I\}$ on a nonempty set Ω, we write $\vee_{i \in I} \mathcal{F}_i$ to denote the smallest σ-field containing all $\mathcal{F}_i$, $i \in I$. Furthermore, for a collection of random vectors $\{X_i : i \in I\}$ on a probability space $(\Omega, \mathcal{F}, P)$, we write $\sigma\langle\{X_i : i \in I\}\rangle$ to denote the sub σ-field of $\mathcal{F}$ generated by $\{X_i : i \in I\}$. For a random vector X and a σ-field $\mathcal{G}$, we write $\mathcal{L}(X)$ and $\mathcal{L}(X|\mathcal{G})$ to denote the probability distribution of X and the conditional probability distribution of X given $\mathcal{G}$, respectively. For two random vectors X and Y, we write $X =^d Y$ if $\mathcal{L}(X) = \mathcal{L}(Y)$. For a distribution G and for a random vector X, we write $X \sim G$ if $\mathcal{L}(X) = G$. For a nonempty finite set A, we say that a random variable X has the Discrete Uniform distribution on A if

$$P(X = a) = \frac{1}{|A|} \quad \text{for all} \quad a \in A .$$

For a $k \times k$ positive definite matrix Σ, let $\Phi_\Sigma(\cdot)$ and $\Phi(\cdot; \Sigma)$ both denote the Gaussian distribution $N(0, \Sigma)$ on $\mathbb{R}^k$ with mean zero and covariance matrix Σ. Let ϕ_Σ and $\phi(\cdot; \Sigma)$ both denote the density of Φ_Σ with respect to the Lebesgue measure on $\mathbb{R}^k$, given by

$$\phi_\Sigma(x) = \phi(x; \Sigma) = (2\pi)^{-k/2}[\det(\Sigma)]^{-1/2} \exp(-x'\Sigma^{-1}x/2), \ x \in \mathbb{R}^k .$$

Furthermore, we use Φ_Σ and/or $\Phi(\cdot; \Sigma)$ also to denote the distribution function of the $N(0, \Sigma)$ distribution. Thus, Φ_Σ and/or $\Phi(\cdot; \Sigma)$ stands for either of the two functions

$$\Phi(x; \Sigma) \quad = \quad \Phi_\Sigma(x) \equiv \int_{(-\infty, x]} \phi(y; \Sigma) dy, \ x \in \mathbb{R}^k,$$

and

$$\Phi(A; \Sigma) \quad = \quad \Phi_\Sigma(A) \equiv \int_A \phi(y; \Sigma) dy, \ A \in \mathcal{B}(\mathbb{R}^k) .$$

When $\Sigma = \mathbb{I}_k$, we abbreviate Φ_Σ and ϕ_Σ as Φ and ϕ, respectively. The dependence of Φ and ϕ on the dimension k is suppressed in the notation and will be clear from the context.

As a convention, notation for random and nonrandom entities are "local" to a section where they appear, i.e., the same symbol may have different meanings in two different sections. Similarly, the numbering of conditions are "local" to a chapter. Unless otherwise mentioned, the symbols for random and nonrandom entities and the condition labels refer to their local definitions. For referring to a condition introduced in another chapter, we add the chapter number as a prefix. For example, an occurrence of Condition $5.D_r$ in Chapter 6 refers to Condition D_r of Chapter 5, etc. We use the abbreviations cdf (cumulative distribution function), CI (confidence interval), iid (independent and identically distributed), and MSE (mean squared error), as convenient. We also use a box $\square$ to denote the end of a proof or of an example.

2

Bootstrap Methods

2.1 Introduction

In this chapter, we describe various commonly used bootstrap methods that have been proposed in the literature. Section 2.2 begins with a brief description of Efron's (1979) bootstrap method based on simple random sampling of the data, which forms the basis for almost all other bootstrap methods. In Section 2.3, we describe the famous example of Singh (1981), which points out the limitation of this resampling scheme for dependent variables. In Section 2.4, we present bootstrap methods for time-series models driven by iid variables, such as the autoregression model. In Sections 2.5, 2.6, and 2.7, we describe various block bootstrap methods. A description of the subsampling method is given in Section 2.8. Bootstrap methods based on the discrete Fourier transform of the data are described in Section 2.9, while those based on the method of sieves are presented in Section 2.10.

2.2 IID Bootstrap

In this book, we refer to the nonparametric resampling scheme of Efron (1979) , introduced in the context of "iid data," as the IID bootstrap. There are a few alternative terms used in the literature for Efron's (1979) bootstrap, such as "naive" bootstrap, "ordinary" bootstrap, etc. These terms may have a different meaning in this book, since (for example) using the IID bootstrap may not be the "naive" thing to do for data with a *dependence* structure.

We begin with the formulation of the *IID bootstrap method* of Efron (1979). For the discussion in this section, assume that $X_1, X_2, \ldots$ is a sequence of iid random variables with common distribution F. Suppose, $\mathcal{X}_n = \{X_1, \ldots, X_n\}$ generate the data at hand and let $T_n = t_n(\mathcal{X}_n; F)$, $n \geq 1$ be a random variable of interest. Note that T_n depends on the data as well as on the underlying unknown distribution F. Typical examples of T_n include the *normalized* sample mean $T_n \equiv n^{1/2}(\bar{X}_n - \mu)/\sigma$ and the *studentized* sample mean $T_n \equiv n^{1/2}(\bar{X}_n - \mu)/s_n$ where $\bar{X}_n = n^{-1}\sum_{i=1}^n X_i$, $s_n^2 = n^{-1}\sum_{i=1}^n (X_i - \bar{X}_n)^2$, $\mu = E(X_1)$, and $\sigma^2 = \text{Var}(X_1)$. Let G_n denote the sampling distribution of T_n. The goal is to find an accurate approximation to the unknown distribution of T_n or to some population characteristics, e.g., the standard error, of T_n. The bootstrap method of Efron (1979) provides an effective way of addressing these problems without any model assumptions on F.

Given $\mathcal{X}_n$, we draw a simple random sample $\mathcal{X}_m^* = \{X_1^*, \ldots, X_m^*\}$ of size m with replacement from $\mathcal{X}_n$. Thus, conditional on $\mathcal{X}_n$, $\{X_1^*, \ldots, X_m^*\}$ are iid random variables with

$$P_*(X_1^* = X_i) = \frac{1}{n}, \quad 1 \leq i \leq n ,$$

where P_* denotes the conditional probability given $\mathcal{X}_n$. Hence, the common distribution of X_i^*'s is given by the empirical distribution

$$F_n = n^{-1}\sum_{i=1}^n \delta_{X_i} ,$$

where δ_y denotes the probability measure putting unit mass at y. Usually, one chooses the resample size $m = n$. However, there are several known examples where a different choice of m is desirable. See, for example, Athreya (1987), Arcones and Giné (1989, 1991), Bickel, Götze and van Zwet (1997), Fukuchi (1994), and the references therein.

Next define the bootstrap version $T_{m,n}^*$ of T_n by replacing $\mathcal{X}_n$ with $\mathcal{X}_m^*$ and F with F_n as

$$T_{m,n}^* = t_m(\mathcal{X}_n^*; F_n) .$$

Also, let $\hat{G}_{m,n}$ denote the conditional distribution of $T_{m,n}^*$, given $\mathcal{X}_n$. Then the bootstrap principle advocates $\hat{G}_{m,n}$ as an estimator of the unknown sampling distribution G_n of T_n. If, instead of G_n, one is interested in estimating only a certain functional $\varphi(G_n)$ of the sampling distribution of T_n, then the corresponding bootstrap estimator is given by *plugging-in* $\hat{G}_{m,n}$ for G_n, i.e., the bootstrap estimator of $\varphi(G_n)$ is given by $\varphi(\hat{G}_{m,n})$. For example, if $\varphi(G_n) = \text{Var}(T_n) = \int x^2 dG_n(x) - (\int x dG_n(x))^2$, the bootstrap estimator of $\text{Var}(T_n)$ is given by $\varphi(\hat{G}_{m,n}) = \text{Var}(T_{m,n}^* \mid \mathcal{X}_n) =$

$\int x^2 d\hat{G}_{m,n}(x) - (\int x d\hat{G}_{m,n}(x))^2$. Once the variables $\mathcal{X}_n$ have been observed, the common distribution F_n of X_i^*'s becomes known, and, hence, it is possible (at least theoretically) to find the conditional distribution $\hat{G}_{m,n}$ and the bootstrap estimator $\varphi(\hat{G}_{m,n})$ from the *knowledge* of the data. In practice, however, finding $\hat{G}_{m,n}$ *exactly* may be a daunting task, even in moderate samples. This is because the number of possible distinct values of $\mathcal{X}_m^*$ grows very rapidly, at the rate $O(n^m)$ as $n \to \infty$, $m \to \infty$ under the IID bootstrap. Consequently, the conditional distribution of $T_{m,n}^*$ is further approximated by Monte-Carlo simulations as described in Chapter 1.

To illustrate the main ideas, again consider the simplest example where $T_n = \sqrt{n}(\bar{X}_n - \mu)/\sigma$, the centered and scaled sample mean. Here $\mu = EX_1$ is the level-1 parameter we want to infer about. Following the description given above, the bootstrap version $T_{m,n}^*$ of T_n based on a bootstrap sample of size m is given by

$$T_{m,n}^* = \sqrt{m}(\bar{X}_m^* - E_* X_1^*)/(\text{Var}_*(X_1^*))^{1/2}$$

where $\bar{X}_m^* = m^{-1} \sum_{i=1}^m X_i^*$ denotes the bootstrap sample mean based on $X_1^*, \ldots, X_m^*$, and E_* and Var_* respectively denote the conditional expectation and conditional variance, given $\mathcal{X}_n$. It is clear that for any $k \geq 1$,

$$E_*(X_1^*)^k = \int x^k dF_n(x) = n^{-1} \sum_{i=1}^n X_i^k \ . \tag{2.1}$$

In particular, this implies $E_*(X_1^*) = \bar{X}_n$, and $\text{Var}_*(X_1^*) \equiv s_n^2 = n^{-1} \sum_{i=1}^n (X_i - \bar{X}_n)^2$. Hence, we define $T_{m,n}^*$ by replacing $\bar{X}_n$ with $\bar{X}_n^*$ and μ and σ^2 by $E_*(X_1^*)$ and $\text{Var}_*(X_1^*)$, respectively. Thus, the bootstrap version of T_n is given by

$$T_{m,n}^* = \sqrt{m}(\bar{X}_m^* - \bar{X}_n)/s_n \ . \tag{2.2}$$

If, for example, we are interested in estimating $\varphi_\alpha(G_n)$ = the αth quantile of T_n for some $\alpha \in (0,1)$, then the bootstrap estimator of $\varphi_\alpha(G_n)$ is $\varphi_\alpha(\hat{G}_{m,n})$, the αth quantile of the conditional distribution of $T_{m,n}^*$.

As mentioned above, determining $\hat{G}_{m,n}$ exactly is not very easy even in this simple case. However, when $EX_1^2 < \infty$, and $m = n$, we have the following result. Recall that we use the abbreviation a.s. for almost sure or almost surely, as appropriate, and we write $\Phi(\cdot)$ to denote the distribution function of the standard normal distribution on $\mathbb{R}$.

Theorem 2.1 *If $X_1, X_2, \ldots$ are iid with $\sigma^2 = Var(X_1) \in (0, \infty)$, then*

$$\sup_x |P_*(T_{n,n}^* \leq x) - \Phi(x/\sigma)| = o(1) \quad as \quad n \to \infty, \quad \text{a.s.} \tag{2.3}$$

Proof: Since $X_1^*, \ldots, X_n^*$ are iid, by the Berry-Esseen Theorem (see Theorem A.6, Appendix A)

$$\sup_x |P_*(T_{n,n}^* \leq x) - \Phi(x)| \leq (2.75)\hat{\Delta}_n \ , \tag{2.4}$$

where $s_n^2 = E_*(X_1^* - \bar{X}_n)^2$ and $\hat{\Delta}_n = E_*|X_1^* - \bar{X}_n|^3/(s_n^3\sqrt{n})$. Clearly, by the Strong Law of Large Numbers (SLLN) (see Theorem A.3, Appendix A),

$$s_n^2 = n^{-1}\sum_{i=1}^{n} X_i^2 - (\bar{X}_n)^2 \to \sigma^2 \quad \text{a.s.}$$

and by the Marcinkiewicz-Zygmund SLLN (see Theorem A.4, Appendix A),

$$n^{-3/2}\sum_{i=1}^{n} |X_i|^3 \to 0 \quad \text{a.s.}$$

Hence, $\hat{\Delta}_n \to 0$ a.s. as $n \to \infty$, and Theorem 2.1 follows. □

Actually Theorem 2.1 holds for any resample size m_n that goes to infinity at a rate faster than loglog n, but the proof requires a different argument. See Arcones and Giné (1989, 1991) for details.

Note that by the Central Limit Theorem (CLT), T_n also converges in distribution to the $N(0,1)$ distribution. Hence, it follows that

$$\begin{aligned} \tilde{\Delta}_n &\equiv \sup_x |\hat{G}_{n,n}(x) - G_n(x)| \\ &= \sup_x |P_*(T_{n,n}^* \le x) - P(T_n \le x)| = o(1) \quad \text{as} \quad n \to \infty, \quad \text{a.s.}\,, \end{aligned} \tag{2.5}$$

i.e., the conditional distribution $\hat{G}_{n,n}$ of $T_{n,n}^*$ generated by the IID bootstrap method provides a valid approximation for the sampling distribution G_n of T_n. Under some additional conditions, Singh (1981) showed that

$$\tilde{\Delta}_n = O(n^{-1}(\log\log n)^{1/2}) \quad \text{as} \quad n \to \infty, \quad \text{a.s.}$$

Therefore, the bootstrap approximation for $P(T_n \le \cdot)$ is far more accurate than the classical normal approximation, which has an error of order $O(n^{-1/2})$. Similar optimality properties of the bootstrap approximation have been established in many important problems. The literature on bootstrap methods for independent data is quite extensive. By now, there exist some excellent sources that give comprehensive accounts of the theory and applications of the bootstrap methods for independent data. We refer the reader to the monographs by Efron (1982), Hall (1992), Mammen (1992), Efron and Tibshirani (1993), Barbe and Bertail (1995), Shao and Tu (1995), Davison and Hinkley (1997), and Chernick (1999) for the bootstrap methodology for independent data. Here, we have described Efron's (1979) bootstrap for iid data mainly as a prelude to the bootstrap methods for dependent data considered in later sections, as the basic principles in both cases are the same. Furthermore, it provides a historical account of the developments that culminated in formulation of the bootstrap methods for dependent data.

2.3 Inadequacy of IID Bootstrap for Dependent Data

The IID bootstrap method of Efron (1979), being very simple and general, has found application to a hoard of statistical problems. However, the general perception that the bootstrap is an "omnibus" method, giving accurate results in all problems automatically, is misleading. A prime example of this appears in the seminal paper by Singh (1981), which in addition to providing the first theoretical confirmation of the superiority of the IID bootstrap, also pointed out its inadequacy for dependent data.

In this section we consider the aforementioned example of Singh (1981). Suppose $X_1, X_2, ...$ is a sequence of m-dependent random variables with $EX_1 = \mu$ and $EX_1^2 < \infty$. Recall that $\{X_n\}_{n\geq 1}$ is called *m-dependent* for some integer $m \geq 0$ if $\{X_1, \ldots X_k\}$ and $\{X_{k+m+1}, \ldots\}$ are independent for all $k \geq 1$. Thus, an iid sequence of random variables $\{\epsilon_n\}_{n\geq 1}$ is 0-dependent and if we define $X_n = \epsilon_n + 0.5\epsilon_{n+1}$, $n \geq 1$, with this iid sequence $\{\epsilon_n\}_{n\geq 1}$, then $\{X_n\}_{n\geq 1}$ is 1-dependent.

Next, let $\sigma_m^2 = \text{Var}(X_1) + 2\sum_{i=1}^{m-1}\text{Cov}(X_1, X_{1+i})$ and $\bar{X}_n = n^{-1}\sum_{i=1}^{n} X_i$. If $\sigma_m^2 \in (0,\infty)$, then by the CLT for m-dependent variables (cf. Theorem A.7, Appendix A),

$$\sqrt{n}(\bar{X}_n - \mu) \longrightarrow^d N(0, \sigma_m^2) , \tag{2.6}$$

where $\longrightarrow^d$ denotes convergence in distribution. Now, suppose that we want to estimate the sampling distribution of the random variable $T_n = \sqrt{n}(\bar{X}_n - \mu)$ using the IID bootstrap. For simplicity, assume that the resample size equals the sample size, i.e., from $\mathcal{X}_n = (X_1, \ldots, X_n)$, an equal number of bootstrap variables $X_1^*, \ldots, X_n^*$ are generated. Then, the bootstrap version $T_{n,n}^*$ of T_n is given by

$$T_{n,n}^* = \sqrt{n}(\bar{X}_n^* - \bar{X}_n) ,$$

where $\bar{X}_n^* = n^{-1}\sum_{i=1}^{n} X_i^*$. The conditional distribution of $T_{n,n}^*$ under the IID bootstrap method still converges to a normal distribution, but with a "wrong" variance, as shown below.

Theorem 2.2 *Suppose $\{X_n\}_{n\geq 1}$ is a sequence of stationary m-dependent random variables with $EX_1 = \mu$, and $\sigma^2 = Var(X_1) \in (0,\infty)$. Then*

$$\sup_x |P_*(T_{n,n}^* \leq x) - \Phi(x/\sigma)| = o(1) \quad \textit{as} \quad n \to \infty, \quad \text{a.s.} \tag{2.7}$$

Proof: Note that conditional on $\mathcal{X}_n$, $X_1^*, \ldots, X_n^*$ are iid random variables. As in the proof of Theorem 2.1, by the Berry-Esseen Theorem, it is enough to show that

$$s_n^2 \to \sigma^2 \quad \text{as} \quad n \to \infty \quad \text{a.s.}$$

and

$$n^{-3/2}\sum_{i=1}^{n}|X_i|^3 \to 0 \quad \text{as} \quad n\to\infty, \quad \text{a.s.}$$

These follow easily from the following lemma. Hence Theorem 2.2 is proved. □

Lemma 2.1 *Let $\{X_n\}_{n\geq 1}$ be a sequence of stationary m-dependent random variables. Suppose that $f:\mathbb{R}\to\mathbb{R}$ is a Borel measurable function with $E|f(X_1)|^p<\infty$ for some $p\in(0,\infty)$, and that $Ef(X_1)=0$ if $p\geq 1$. Then,*

$$n^{-1/p}\sum_{i=1}^{n}f(X_i)\to 0 \quad \textit{as} \quad n\to\infty, \quad \text{a.s.}$$

Proof: This is most easily proved by splitting the given m-dependent sequence $\{X_n\}_{n\geq 1}$ into $m+1$ iid subsequences $\{Y_{ji}\}_{i\geq 1}$, $j=1,\ldots,m+1$, defined by $Y_{ji}=X_{j+(i-1)(m+1)}$, and then applying the standard results for iid random variables to $\{Y_{ji}\}_{i\geq 1}$'s (cf. Liu and Singh (1992)). For $1\leq j\leq m+1$, let $I_j\equiv I_{jn}=\{1\leq i\leq n: j+(i-1)(m+1)\leq n\}$ and let $N_j\equiv N_{jn}$ denote the size of the set I_j. Note that $N_j/n\to(m+1)^{-1}$ as $n\to\infty$ for all $1\leq j\leq m+1$. Then, by the Marcinkiewicz-Zygmund SLLN (cf. Theorem A.4, Appendix A) applied to each of the sequence of iid random variables $\{Y_{ji}\}_{i\geq 1}$, $j=1,\ldots,m+1$, we get

$$n^{-1/p}\sum_{i=1}^{n}f(X_i)=\sum_{j=1}^{m+1}\Big[N_j^{-1/p}\sum_{i\in I_j}f(Y_{ji})\Big]\cdot(N_j/n)^{1/p}\to 0 \text{ as } n\to\infty, \quad \text{a.s.}$$

This completes the proof of Lemma 2.1. □.

Corollary 2.1 *Under the conditions of Theorem 2.2, if $\sum_{i=1}^{m}Cov(X_1,X_{1+i})\neq 0$ and $\sigma_\infty^2\neq 0$, then for any $x\neq 0$,*

$$\lim_{n\to\infty}[P_*(T_{n,n}^*\leq x)-P(T_n\leq x)]=[\Phi(x/\sigma)-\Phi(x/\sigma_\infty)]\neq 0 \quad \text{a.s.}$$

Proof: Follows from Theorem 2.2 and (2.6). □

Thus, for all $x\neq 0$, the IID bootstrap estimator $P_*(T_{n,n}^*\leq x)$ of the level-2 parameter $P(T_n\leq x)$ has a mean squared error that tends to a *nonzero* number in the limit and the bootstrap estimator of $P(T_n\leq x)$ is not consistent. Therefore, the IID bootstrap method fails drastically for *dependent* data. It follows from the proof of Theorem 2.2 that resampling individual X_i's from the data $\mathcal{X}_n$ ignores the dependence structure of the sequence $\{X_n\}_{n\geq 1}$ completely, and thus, fails to account for the lag-covariance terms (viz., $\text{Cov}(X_1,X_{1+i})$, $1\leq i\leq m$) in the asymptotic variance.

Following this result, there have been several attempts in the literature to extend the IID bootstrap method to the dependent case. In the next section, we first look at extensions of this method to certain dependent models generated by iid random variables. More general resampling schemes (such as the block bootstrap and the frequency domain bootstrap methods), which are applicable without any parametric model assumptions, have been put forward in the literature much later. These are presented in Sections 2.5–2.10.

2.4 Bootstrap Based on IID Innovations

Suppose $\{X_n\}_{n\geq 1}$ is a sequence of random variables satisfying the equation

$$X_n = h(X_{n-1}, \ldots, X_{n-p}; \beta) + \epsilon_n \ , \tag{2.8}$$

$n > p$, where β is a $q \times 1$ vector of parameters, $h : \mathbb{R}^{p+q} \to \mathbb{R}$ is a known Borel measurable function, and $\{\epsilon_n\}_{n>p}$ is a sequence of iid random variables with common distribution F that are independent of the random variables $X_1, \ldots, X_p$. For identifiability of the model (2.8), assume that $E\epsilon_1 = 0$. A commonly used example of model (2.8) is the autoregressive process of order p (cf. (2.9) below). Noting that the process $\{X_n\}_{n\geq 1}$ is driven by the innovations ϵ_i's that are iid, the IID bootstrap method can be easily extended to the dependent model (2.8).

As before, suppose that $\mathcal{X}_n = \{X_1, \ldots, X_n\}$ denotes the sample and that we want to approximate the sampling distribution of a random variable $T_n = t_n(\mathcal{X}_n; F, \beta)$. Let $\hat{\beta}_n$ be an estimator, e.g., the least squares estimator, of β based on $\mathcal{X}_n$. Define the residuals

$$\hat{\epsilon}_i = X_i - h(X_{i-1}, \ldots, X_{i-p}; \hat{\beta}_n), \quad p < i \leq n.$$

Note that, in general,

$$\bar{\epsilon}_n \equiv (n-p)^{-1} \sum_{i=1}^{n-p} \hat{\epsilon}_{i+p} \neq 0 \ .$$

Hence, we center the "raw" residuals $\hat{\epsilon}_i$'s and define the "centered" residuals

$$\tilde{\epsilon}_i = \hat{\epsilon}_i - \bar{\epsilon}_n, \quad p < i \leq n \ .$$

Without such a centering, the resulting bootstrap approximation often has a random bias that does not vanish in the limit and renders the approximation useless. (See, for example, Freedman (1981), Shorack (1982), and Lahiri (1992b) that treat a similar bias phenomenon in regression problems.)

Next draw a simple random sample $\epsilon^*_{p+1}, \ldots, \epsilon^*_m$ of size $(m-p)$ from $\{\tilde{\epsilon}_i : p < i \leq n\}$ with replacement and define the bootstrap pseudo-observations, using the model structure (2.8), as:

$$X^*_i = X_i \quad \text{for} \quad i = 1, \ldots, p, \quad \text{and}$$

$$X_i^* = h(X_{i-1}^*, \ldots, X_{i-p}^*; \hat{\beta}_n) + \epsilon_i^*, \quad p < i \leq m .$$

Note that by construction ϵ_i^*, $p < i \leq m$ are iid and $E_* \epsilon_1^* = 0$. The bootstrap version of the random variable $T_n = t_n(\mathcal{X}_n; F, \beta)$ is defined as

$$T_{m,n}^* = t_m(\mathcal{X}_m^*; F_n, \hat{\beta}_n) ,$$

where $\mathcal{X}_m^* = \{X_1^*, \ldots, X_m^*\}$ and F_n denotes the empirical distribution of the centered residuals $\tilde{\epsilon}_i$, $p < i \leq n$. The sampling distribution of T_n is approximated by the conditional distribution of $T_{m,n}^*$ given $\mathcal{X}_n$. For certain time-series models satisfying (2.8), different versions of this resampling scheme have been proposed by Freedman (1984), Efron and Tibshirani (1986), Swanepoel and van Wyk (1986), and Kreiss and Franke (1992). The IID-innovation-bootstrap method can be applied with some simple modifications to popular parametric models for spatial data as well (e.g., the spatial autoregression model); see Chapter 7, Cressie (1993).

A special case of model (2.8) is the autoregression model of order p (AR(p)), given by

$$X_n = \beta_1 X_{n-1} + \ldots + \beta_p X_{n-p} + \epsilon_n, \; n > p , \tag{2.9}$$

where $\beta = (\beta_1, \ldots, \beta_p)$ is the vector of autoregressive parameters, and $\{\epsilon_n\}_{n>p}$ is an iid sequence satisfying the requirements of model (2.8). For AR(p)-models, validity and the rate of approximation of the IID-Innovation bootstrap have been well-studied in the literature. When the sequence $\{X_n\}_{n\geq 1}$ is *stationary*, Bose (1988) shows that under suitable regularity conditions, a version of the IID-innovation bootstrap approximation to the sampling distribution of the standardized least square estimator is more accurate than the normal approximation. For *nonstationary* cases, performance of this method has been studied by Basawa, Mallik, McCormick and Taylor (1989), Basawa, Mallik, McCormick, Reeves and Taylor (1991), Datta (1995, 1996), Datta and Sriram (1997), and Heimann and Kreiss (1996), among others. It follows from their work that the IID-innovation bootstrap method is very sensitive to the values of the autoregression parameter vector β. Indeed, if the value of β is such that the roots of the characteristic equation $z^p + \beta_1 z^{p-1} + \ldots + \beta_p = 0$ lie on the unit circle, then the IID-innovation bootstrap fails. Because of its dependence on the validity of the model (2.9), and drastic change in the performance with a small change in the parameter value, one needs to be particularly careful when applying the IID-innovation bootstrap method. Properties of the IID-innovation bootstrap and related model based bootstrap methods are described in Chapter 8.

2.5 Moving Block Bootstrap

Bootstrap methods described in the previous sections are applicable either under the hypothesis of independence or under specific model assumptions for dependent data. The main idea in the latter case is to use the approximate independence of the residuals, and then apply the resampling scheme of the IID-bootstrap method to get the right approximation. In a problem where the statistician does not have enough prior knowledge to specify such models, these methods are not very useful. In a significant breakthrough, Künsch (1989) and Liu and Singh (1992) independently formulated a substantially new resampling scheme, called the *moving block bootstrap* (MBB), that is applicable to dependent data without any parametric model assumptions. In contrast to resampling a *single* observation at a time, as has been commonly done under the earlier formulations of the bootstrap, the MBB resamples *blocks* of (consecutive) observations at a time. As a result, the dependence structure of the original observations is preserved within each block. Furthermore, the common length of the blocks increases with the sample size. As a result, when the data are generated by a *weakly* dependent process, the MBB reproduces the underlying dependence structure of the process *asymptotically*. Essentially the same principle was put forward by Hall (1985) in the context of bootstrapping spatial data and by Carlstein (1986) for estimating the variance of a statistic based on time series data. A description of Carlstein's method will be given in the next section. We now turn to a description of the MBB.

Let $X_1, X_2, \ldots$ be a sequence of stationary random variables, and let $\mathcal{X}_n = \{X_1, \ldots, X_n\}$ denote the observations. We shall define the MBB version of estimators of the form $\hat{\theta}_n = T(F_n)$, where F_n denotes the empirical distribution function of $X_1, \ldots, X_n$, and where $T(\cdot)$ is a (real-valued) functional of F_n. Suppose $\ell \equiv \ell_n \in [1, n]$ is an integer. For dependent data, we typically require that

$$\ell \to \infty \quad \text{and} \quad n^{-1}\ell \to 0 \quad \text{as} \quad n \to \infty .$$

However, a description of the MBB can be given without this restriction. Let $\mathcal{B}_i = (X_i, \ldots, X_{i+\ell-1})$ denote the block of length ℓ starting with X_i, $1 \leq i \leq N$ where $N = n - \ell + 1$. (See Figure 2.1 below.) To obtain the MBB samples, we randomly select a suitable number of blocks from the collection $\{\mathcal{B}_1, \ldots, \mathcal{B}_N\}$. Accordingly, let $\mathcal{B}_1^*, \ldots, \mathcal{B}_k^*$ denote a simple random sample drawn with replacement from $\{\mathcal{B}_1, \ldots, \mathcal{B}_N\}$. Note that each of the selected blocks contains ℓ elements. Denote the elements in $\mathcal{B}_i^*$ by $(X_{(i-1)\ell+1}^*, \ldots, X_{i\ell}^*)$, $i = 1, \ldots, k$. Then, $X_1^*, \ldots, X_m^*$ constitute the MBB sample of size $m \equiv k\ell$. The MBB version $\theta_{m,n}^*$ of $\hat{\theta}_n$ is defined as

$$\theta_{m,n}^* = T(F_{m,n}^*) ,$$

where $F_{m,n}^*$ denotes the empirical distribution of $(X_1^*, \ldots, X_m^*)$.

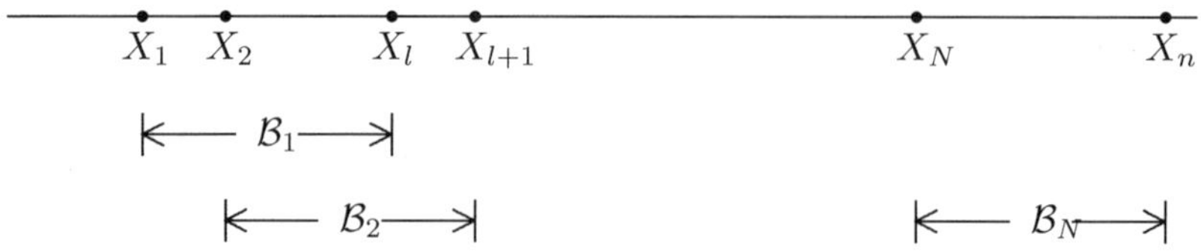

X_1 X_2 X_l X_{l+1} X_N X_n

$\mathcal{B}_1$

$\mathcal{B}_2$ $\mathcal{B}_N$

FIGURE 2.1. The collection $\{\mathcal{B}_1, \ldots, \mathcal{B}_N\}$ of overlapping blocks under the MBB.

An alternative formulation of the MBB can be given as follows. Note that selecting the blocks $\mathcal{B}_i^*$'s randomly from $\{\mathcal{B}_1, \ldots, \mathcal{B}_N\}$ is equivalent to selecting k indices at random from the set $\{1, \ldots, N\}$. Accordingly, let $I_1, \ldots, I_k$ be iid random variables with the discrete uniform distribution on $\{1, \ldots, N\}$. If we set $\mathcal{B}_i^* = \mathcal{B}_{I_i}$ for $i = 1, \ldots, k$, then $\mathcal{B}_1^*, \ldots, \mathcal{B}_k^*$ represent a simple random sample drawn with replacement from $\{\mathcal{B}_1, \ldots, \mathcal{B}_N\}$. The bootstrap sample $X_1^*, \ldots, X_m^*$ can be defined using the resampled blocks $\mathcal{B}_1^*, \ldots, \mathcal{B}_k^*$ as before. Note that conditional on the data $\mathcal{X}_n$, the resampled blocks of observations $(X_1^*, \ldots, X_\ell^*)', (X_{\ell+1)}^*, \ldots, X_{2\ell}^*)', \ldots, (X_{(k-1)\ell+1}^*, \ldots, X_{k\ell}^*)'$ are iid ℓ-dimensional random vectors with

$$
\begin{aligned}
&P_*((X_1^*, \ldots, X_\ell^*)' = (X_j, \ldots, X_{j+\ell-1})') \\
&= P_*(I_1 = j) \\
&= N^{-1}, \quad \text{for} \quad 1 \leq j \leq N,
\end{aligned}
\tag{2.10}
$$

where P_* denotes the conditional probability given $\mathcal{X}_n$. In the special case, when each block consists of a single element (i.e., $\ell = 1$), then by (2.10), $X_1^*, \ldots, X_m^*$ are iid with the common distribution F_n, and hence, the MBB reduces to the IID bootstrap method of Efron (1979) described in Section 2.2. For $\ell > 1$, the ℓ-dimensional joint distribution of the underlying process $\{X_n\}_{n\geq 1}$ is preserved *within* the resampled blocks. Since ℓ tends to infinity with n, any finite-dimensional joint distribution of $\{X_n\}_{n\geq 1}$-process at a given number of finite lag distances can be eventually recovered from the resampled values. As a result, the MBB can effectively capture those characteristics of the underlying process $\{X_n\}_{n\geq 1}$ that are determined by the dependence structure of the observations at short lags.

As in the case of the IID bootstrap, the MBB sample size is typically chosen to be of the same order as the original sample size. If b_1 denotes the smallest integer such that $b_1\ell \geq n$, then one may select $k = b_1$ blocks to generate the MBB samples, and use only the first n values to define the bootstrap version of T_n. However, there are some inference problems where a smaller sample size works better (cf. Chapter 11).

Though estimators of the form $\hat{\theta}_n = T(F_n)$ considered above include many commonly used estimators, e.g., the sample mean, M-estimators of location and scale, von Mises functionals, etc., they are not sufficiently rich for applications in the time series context. This is primarily because $\hat{\theta}_n$ above depends only on the *one-dimensional* marginal empirical distribution F_n, and hence does not cover standard statistics like the sample lag correlations, or the spectral density estimators. We shall now consider a more general version of the MBB that covers such statistics.

Given the observations $\mathcal{X}_n$, let $F_{p,n}$ denote the p-dimensional empirical measure

$$F_{p,n} = (n-p+1)^{-1} \sum_{j=1}^{n-p+1} \delta_{Y_j} \ ,$$

where $Y_j = (X_j, \ldots, X_{j+p-1})$ and where for any $y \in \mathbb{R}^p$, δ_y denotes the probability measure on $\mathbb{R}^p$ putting unit mass on y. The general version of the MBB concerns estimators of the form

$$\hat{\theta}_n = T(F_{p,n}) \ , \tag{2.11}$$

where $T(\cdot)$ is now a functional defined on a (rich) subset of the set of all probability measures on $\mathbb{R}^p$. Here, $p \geq 1$ may be a fixed integer, or it may tend to infinity with n suitably. Some important examples of (2.11) are given below.

Example 2.1: A version of the sample lag covariance of order $k \geq 0$ is given by

$$\hat{\gamma}_n(k) = (n-k)^{-1} \sum_{j=1}^{n-k} (X_{j+k} - \bar{X}_{n,k})(X_j - \bar{X}_{n,k}) \ ,$$

where $\bar{X}_{n,k} = (n-k)^{-1} \sum_{j=1}^{n-k} X_j$. Then, $\hat{\gamma}_n(k)$ is of the form (2.11) with $p = k+1$. □

Example 2.2: Let ψ be a function from $\mathbb{R}^p \times \mathbb{R}^k$ into $\mathbb{R}^k$ such that

$$E\psi(X_1, \ldots, X_p; \theta) = 0.$$

Here, θ is a functional of the p-dimensional joint distribution of $(X_1, \ldots, X_p)$, implicitly defined by the equation above. A *generalized M-estimator* of the parameter $\theta \in \mathbb{R}^k$ is defined (cf. Bustos (1982)) as a solution of the equation

$$\sum_{j=1}^{n-p+1} \psi(X_j, \ldots, X_{j+p-1}; T_n) = 0.$$

The generalized M-estimators can also be expressed in the form (2.11). □

Example 2.3: Let $f(\cdot)$ denote the spectral density of the process $\{X_n\}_{n\geq 1}$. Then, a lag-window estimator of the spectral density (cf., Chapter 6, Priestley (1981)) is given by

$$\hat{f}_n(\lambda) = \sum_{k=-(n-1)}^{(n-1)} w(k/p)\hat{\gamma}_n(k)\cos(k\lambda), \ \lambda \in [0, \pi],$$

where $p \equiv p_n$ tends to infinity at a rate slower than n and where w is a weight function such that $w(0) = (2\pi)^{-1}$ and w vanishes outside the interval $(-1, 1)$. For different choices of w, one gets various commonly used estimators of the spectral density, such as the truncated periodogram estimator, the Bartlett estimator, etc. Since $\hat{f}_n$ is a function of $\hat{\gamma}_n(0), \ldots, \hat{\gamma}_n(p)$, from Example 2.1, it follows that we can express it in the form (2.11). Note that in this example, p tends to infinity with n. □

To define the MBB version of $\hat{\theta}_n$ in (2.11), fix a block size ℓ, $1 < \ell < n - p + 1$, and define the blocks in terms of Y_i's as

$$\tilde{\mathcal{B}}_j = (Y_j, \ldots, Y_{j+\ell-1}), \ 1 \leq j \leq n - p - \ell + 2 \ .$$

For $k \geq 1$, select k blocks randomly from the collection $\{\tilde{\mathcal{B}}_i : 1 \leq i \leq n-p-\ell+2\}$ to generate the MBB observations $Y_1^*, \ldots, Y_\ell^*; Y_{\ell+1}^*, \ldots, Y_{2\ell}^*; \cdots, Y_m^*$, where $m = k\ell$. The MBB version of (2.11) is now defined as

$$\theta^*_{m,n} = T(\tilde{F}^*_{m,n}) \ , \tag{2.12}$$

where $\tilde{F}^*_{m,n} \equiv m^{-1}\sum_{j=1}^{m} \delta_{Y_j^*}$ denotes the empirical distribution of $Y_1^*, \ldots, Y_m^*$. Thus, for estimators of the form (2.11), the MBB version is defined by resampling from blocks of Y-values *instead* of blocks of X-values themselves. This formulation of the MBB was initially given by Künsch (1989) and was further explored by Politis and Romano (1992a). Clearly, the definition (2.12) applies to both the cases where p is fixed and where p tends to infinity with n. In the latter case, Politis and Romano (1992a) called the modified blocking mechanism as the "blocks of blocks" bootstrap, and gave a more general formulation that allows one to control the amount of overlap between the successive blocks of Y-values. We refer the reader to Politis and Romano (1992a) for the other versions of the "blocks of blocks" bootstrap method.

Note that for the more general class of statistics $\hat{\theta}_n$ given by (2.11) for some $p \geq 2$, there is an alternative way of defining the bootstrap version of $\hat{\theta}_n$. Since the estimator $\hat{\theta}_n$ can always be expressed as a function of the given observations $X_1, \ldots, X_n$, one may define the bootstrap version of $\hat{\theta}_n$ by resampling from $X_1, \ldots, X_n$ *directly*. Specifically, suppose that the block bootstrap observations $X_1^*, \ldots, X_m^*$ are generated by resampling from the blocks $\mathcal{B}_i = \{X_i, \ldots, X_{i+\ell-1}\}$, $i = 1, \ldots, N$ of X-values. Then, define bootstrap "analogs" of the p-dimensional variable $Y_i \equiv (X_i, \ldots, X_{i+p-1})'$ in terms of $X_1^*, \ldots, X_m^*$ as $Y_i^{**} \equiv (X_i^*, \ldots, X_{i+p-1}^*)'$, $i = 1, \ldots, m-p+1$. Then, the bootstrap version of $\hat{\theta}_n$ under this alternative approach is defined as

$$\theta_{m,n}^{**} = T(\tilde{F}_{m,n}^{**}) ,$$

where $\tilde{F}_{m,n}^{**} = \sum_{i=1}^{m-p+1} \delta_{Y_i^{**}}$. We call this approach of defining the moving block bootstrap version of $\hat{\theta}_n$ as the "naive" approach, and the other approach leading to $\theta_{m,n}^*$ in (2.12) as the "ordinary" approach of the MBB. We shall also use the terms "naive" and "ordinary" in the context of bootstrapping estimators of the form (2.11) using *other* block bootstrap methods described later in this chapter.

For a comparison of the two approaches, suppose that $\{X_n\}_{n\geq 1}$ is a sequence of stationary random variables. Then, for each i, the random vector $Y_i = (X_i, \ldots, X_{i+p-1})'$ has the same distribution as $(X_1, \ldots, X_p)'$, and hence, the resampled vectors Y_i^* under the "ordinary" approach always retains the dependence structure of $(X_1, \ldots, X_p)'$. However, when the bootstrap blocks are selected by the "naive" approach, the bootstrap observations X_i^*'s, that are at lags less than p and that lie near the boundary of two adjacent resampled blocks $\mathcal{B}_j^*$ and $\mathcal{B}_{j+1}^*$, are *independent*. Thus the components of Y_i^{**} under the "naive" approach do *not* retain the dependence structure of $(X_1, \ldots, X_p)'$. As a result, the naive approach introduces additional bias in the bootstrap version $\theta_{m,n}^{**}$ of $\hat{\theta}_n$. We shall, therefore, always use the "ordinary" form of a block bootstrap method while defining the bootstrap version of estimators $\hat{\theta}_n$ given by (2.11). For a numerical example comparing the naive and the ordinary versions of the MBB and certain other block bootstrap methods, see Section 4.5.

We conclude this section with two remarks. First, it is easy to see that the above description of the MBB and the "blocks of blocks" bootstrap applies almost verbatim if, to begin with, the observations $X_1, \ldots, X_n$ were random *vectors* instead of random variables. Second, performance of a MBB estimator critically depends on the block size ℓ. Since the sampling distribution of a given estimator typically depends on the *joint* distribution of $X_1, \ldots, X_n$, the block size ℓ must *grow to infinity* with the sample size n to capture the dependence structure of the series $\{X_n\}_{n\geq 1}$, eventually. Typical choices of ℓ are of the form $\ell = Cn^\delta$ for some constants $C > 0$, $\delta \in (0, 1/2)$. For more on properties of MBB estimators and effects of block lengths on their performance, see Chapters 3–7.

2.6 Nonoverlapping Block Bootstrap

In this section, we consider the blocking rule due to Carlstein (1986). For simplicity, here we shall consider estimators given by (2.11) with $p = 1$ only. Extension to the case of a general $p \geq 1$ is straightforward. The key feature of Carlstein's blocking rule is to use nonoverlapping segments of the data to define the blocks. The corresponding block bootstrap method will be called the nonoverlapping block bootstrap (NBB). Suppose that $\ell \equiv \ell_n \in [1, n]$ is an integer and $b \geq 1$ is the largest integer satisfying $\ell b \leq n$. Then, define the blocks

$$\mathcal{B}_i^{(2)} = (X_{(i-1)\ell+1}, \ldots, X_{i\ell})', \quad i = 1, \ldots, b .$$

(Here we use the index "2" in the superscipt to denote the blocks for the NBB resampling scheme. We reserve the index 1 for the MBB and we shall use the indices 3, 4, etc. for the other block bootstrap methods described later.) Note that while the blocks in the MBB overlap, the blocks $\mathcal{B}_i^{(2)}$'s under the NBB do not. See Figure 2.2. As a result, the collection of blocks from which the bootstrap blocks are selected is smaller than the collection for the MBB.

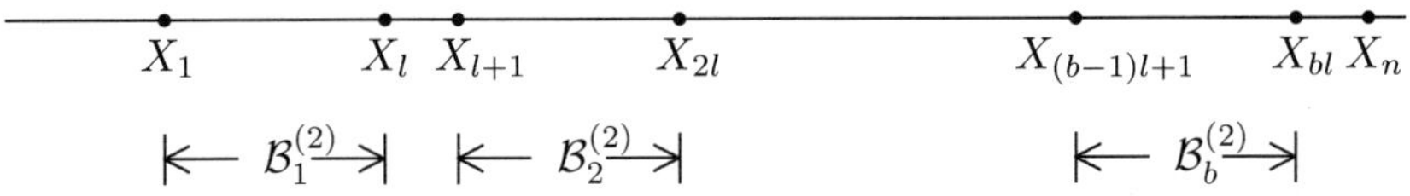

FIGURE 2.2. The collection $\{\mathcal{B}_1^{(2)}, \ldots, \mathcal{B}_b^{(2)}\}$ of nonoverlapping blocks under Carlstein's (1986) rule.

The next step in implementing the NBB is exactly the same as that for the MBB. We select a simple random sample of blocks $\mathcal{B}_1^{*(2)}, \ldots, \mathcal{B}_k^{*(2)}$ with replacement from $\{\mathcal{B}_1^{(2)}, \ldots, \mathcal{B}_b^{(2)}\}$ for some suitable integer $k \geq 1$. With $m = k\ell$, let $F_{m,n}^{*(2)}$ denote the empirical distribution of the bootstrap sample $(X_{2,1}^*, \ldots, X_{2,\ell}^*; \ldots; X_{2,\{(b-1)\ell+1\}}^*, \ldots, X_{2,m}^*)$, obtained by writing the elements of $\mathcal{B}_1^{*(2)}, \ldots, \mathcal{B}_k^{*(2)}$ in a sequence. Then, the bootstrap version of an estimator $\hat{\theta}_n = T(F_n)$ is given by

$$\theta_{m,n}^{*(2)} = T(F_{m,n}^{*(2)}) . \tag{2.13}$$

Even though the definition of the bootstrapped estimators are very similar for the MBB and for the NBB, the resulting bootstrap versions $\theta_{m,n}^*$ and $\theta_{m,n}^{*(2)}$ have different distributional properties. We illustrate the point with the simplest case, where $\hat{\theta}_n = n^{-1}\sum_{j=1}^n X_j$ is the sample mean. The bootstrap version of $\hat{\theta}_n$ under the two methods are respectively given by

$$\theta^*_{m,n} = m^{-1}\sum_{j=1}^{m} X^*_j, \quad \text{and} \quad \theta^{*(2)}_{m,n} = m^{-1}\sum_{j=1}^{m} X^*_{2,j} \, .$$

From (2.10), we get

$$\begin{aligned} E_*(\theta^*_{m,n}) &= E_*(\ell^{-1}\sum_{i=1}^{\ell} X^*_i) \\ &= N^{-1}\sum_{j=1}^{N}\Big(\ell^{-1}\sum_{i=1}^{\ell} X_{j+i-1}\Big) \\ &= N^{-1}\Big\{n\bar{X}_n - \ell^{-1}\sum_{j=1}^{\ell-1}(\ell-j)(X_j + X_{n-j+1})\Big\}. \end{aligned} \tag{2.14}$$

To obtain a similar expression for $E_*(\theta^{*(2)}_{m,n})$, note that under the NBB, the bootstrap variables $(X^*_{2,1},\ldots,X^*_{2,\ell}),\ldots,(X^*_{2,(m-\ell+1)},\ldots,X^*_{2,m})$ are iid, with common distribution

$$P_*\Big((X^*_{2,1},\ldots,X^*_{2,\ell}) = (X_{(j-1)\ell+1},\ldots,X_{j\ell})\Big) = 1/b \tag{2.15}$$

for $j = 1,\ldots,b$. Hence,

$$\begin{aligned} E_*(\theta^{*(2)}_{m,n}) &= E_*(\ell^{-1}\sum_{i=1}^{\ell} X^*_{2,i}) \\ &= b^{-1}\sum_{j=1}^{b}\Big(\ell^{-1}\sum_{i=1}^{\ell} X_{(j-1)\ell+i}\Big) \\ &= (b\ell)^{-1}\Big\{n\bar{X}_n - \sum_{i=b\ell+1}^{n} X_i\Big\}\,, \end{aligned} \tag{2.16}$$

which equals $\bar{X}_n$ if n is a multiple of ℓ. Thus, the bootstrapped estimators have different (conditional) means under the two methods. However, note that if the process $\{X_n\}_{n\geq 1}$ satisfies some standard moment and mixing conditions, then $E\{E_*(\theta^*_{m,n}) - E_*\theta^{*(2)}_{m,n}\}^2 = O(\ell/n^2)$. Hence the difference between the two is negligible for large sample sizes.

2.7 Generalized Block Bootstrap

As follows from its description (cf. Section 2.5), the MBB resampling scheme suffers from an undesirable boundary effect as it assigns lesser weights to the observations toward the beginning and the end of the data set than to the middle part. Indeed, for $\ell \leq j \leq n - \ell$, the jth observation X_j appears in exactly ℓ of the blocks $\{\mathcal{B}_1, \ldots, \mathcal{B}_N\}$, whereas for $1 \leq j \leq \ell - 1$, X_j and X_{n-j+1} appear only in j blocks. Since there is no observation beyond X_n (or prior to X_1), we cannot define new blocks to get rid of this boundary effect. A similar problem also exists under the NBB with the observations near the end of the data sequence when n is not a multiple of ℓ. Politis and Romano (1992b) suggested a simple way out of this boundary problem. Their idea is to wrap the data around a circle and form additional blocks using the "circularly defined" observations. Politis and Romano (1992b, 1994b) put forward two resampling schemes based on circular blocks, called the "circular block bootstrap" (CBB) and the "stationary bootstrap" (SB). Here we describe a generalization of their idea and formulate the generalized block bootstrap method, which provides a unified framework for describing different block bootstrap methods, including the CBB and the SB.

Given the variables $\mathcal{X}_n = \{X_1, \ldots, X_n\}$, first we define a new time series $Y_{n,i}$, $i \geq 1$ by *periodic extension*. Note that for any $i \geq 1$, there are integers $k_i \geq 0$ and $j_i \in [1, n]$ such that $i = k_i n + j_i$. Then, $i = j_i$ (modulo n). We define the variables $Y_{n,i}$, $i \geq 1$ by the relation $Y_{n,i} = X_{j_i}$. Note that this is equivalent to writing the variables $X_1, \ldots, X_n$ repeatedly on a line and labeling them serially as $Y_{n,i}$, $i \geq 1$. See Figure 2.3.

X_1	X_2	$\cdots$	X_n	X_1	$\cdots$	X_n	$\cdots$
$Y_{n,1}$	$Y_{n,2}$	$\cdots$	$Y_{n,n}$	$Y_{n,(n+1)}$	$\cdots$	$Y_{n,(2n)}$	$\cdots$

FIGURE 2.3. The periodically extended time series $Y_{n,i}$, $i \geq 1$.

Next define the blocks

$$\mathcal{B}(i,j) = (Y_{n,i}, \ldots, Y_{n,(i+j-1)})$$

for $i \geq 1$, $j \geq 1$. Let Γ_n be a transition probability function on the set $\mathbb{R}^n \times \bigotimes_{t=1}^{\infty}(\{1, \ldots, n\} \times \mathbb{N})$, i.e., for each $x \in \mathbb{R}^n$, $\Gamma_n(x; \cdot)$ is a probability measure on $\bigotimes_{t=1}^{\infty}(\{1, \ldots, n\} \times \mathbb{N}) \equiv \Big\{\{i_t, \ell_t\}_{t=1}^{\infty} : 1 \leq i_t \leq n, 1 \leq \ell_t < \infty \quad \text{for all} \quad t \geq 1\Big\}$ and for any set $A \subset \bigotimes_{t=1}^{\infty}(\{1, \ldots, n\} \times \mathbb{N})$, $\Gamma_n(\cdot; A)$ is a Borel measurable function from $\mathbb{R}^n$ into [0,1]. Then, the generalized block bootstrap (GBB) resamples blocks from the collection $\{\mathcal{B}(i,j) : i \geq 1, j \geq 1\}$ according to the transition probability function Γ_n as follows. Let $(I_1, J_1), (I_2, J_2), \ldots$ be a sequence of random vectors with conditional joint distribution $\Gamma_n(\mathcal{X}_n; \cdot)$, given $\mathcal{X}_n$. Then, the blocks selected by the GBB are given by $\mathcal{B}(I_1, J_1), \mathcal{B}(I_2, J_2), \ldots$ (which may *not* be independent). Let $X^*_{G,1}, X^*_{G,2}, \ldots$ denote the elements of these resampled blocks. Then, the bootstrap version of an estimator $\hat{\theta}_n = T(F_n)$ under the GBB is defined as $\theta^{*(G)}_{m,n} = T(F^{*(G)}_{m,n})$ for a suitable choice of $m \geq 1$, where $F^{*(G)}_{m,n}$ denotes the empirical distribution of $X^*_{G,1}, \ldots, X^*_{G,m}$.

Almost all block bootstrap methods proposed in the literature can be shown to be special cases of the GBB. For example, for the MBB based on a block length ℓ, $1 \leq \ell \leq n$, the transition probability function Γ_n is given by

$$\Gamma_n(x; \cdot) = \bigotimes_{i=1}^{\infty} \Big((N^{-1} \sum_{j=1}^{N} \delta_j) \times \delta_\ell \Big), \ x \in \mathbb{R}^n$$

where $N = n - \ell + 1$ and δ_y is the probability measure putting mass one at y. In this case, $\Gamma_n(x; \cdot)$ does not depend on $x \in \mathbb{R}^n$, and the random indices $(I_1, J_1), (I_2, J_2), \ldots$ are *conditionally iid* random vectors with conditional distribution

$$P_*(I_1 = j, J_1 = k) = \begin{cases} N^{-1} & \text{if } \ 1 \leq j \leq N \quad \text{and} \quad k = \ell \\ 0 & \text{otherwise} . \end{cases}$$

As a consequence, the resampled blocks $\mathcal{B}(I_1, J_1), \mathcal{B}(I_2, J_2), \ldots$, come from the subcollection $\{\mathcal{B}(i, j) : 1 \leq i \leq N, j = \ell\}$, which is the same as the collection of overlapping blocks $\{\mathcal{B}_1, \ldots, \mathcal{B}_N\}$ defined in Section 2.5. Similarly, the NBB method can also be shown to be a special case of the GBB. Here, we consider a few other examples.

2.7.1 Circular Block Bootstrap

The Circular Block Bootstrap (CBB) method, proposed by Politis and Romano (1992b) resamples overlapping and periodically extended blocks of a given length ℓ (say) satisfying $1 \ll \ell \ll n$ from the subcollection $\{\mathcal{B}(i, \ell), \ldots, \mathcal{B}(n, \ell)\}$. The transition function Γ_n for the CBB is given by

$$\Gamma_n(x; \cdot) = \bigotimes_{i=1}^{\infty} \Big((n^{-1} \sum_{j=1}^{n} \delta_j) \times \delta_\ell \Big), \ x \in \mathbb{R}^n. \tag{2.17}$$

Denote the resampling block indices for the CBB (i.e., the variables I_i's in the collection $(I_1, J_1), (I_2, J_2), \ldots$ whose joint distribution is specified by the $\Gamma_n(\cdot, \cdot)$ of (2.17)) by $I_{3,1}, I_{3,2}, \ldots$. Then, (2.17) implies that the variables $I_{3,1}, I_{3,2}, \ldots$ are *conditionally iid* with $P_*(I_{3,1} = i) = n^{-1}$ and $P_*(J_i = \ell) = 1$ for all $i = 1, \ldots, n$. Since each X_i appears exactly ℓ times in the collection of blocks $\{\mathcal{B}(i, \ell), \ldots, \mathcal{B}(n, \ell)\}$, and since the CBB resamples the blocks from this collection with equal probability, each of the original observations $X_1, \ldots, X_n$ receives equal weight under the CBB. This property distinguishes the CBB from its predecessors, viz., the MBB and the

NBB, which suffer from edge effects. This is also evident from the following observation. Let $X^*_{3,1}, X^*_{3,2}, \ldots$ denote the CBB observations obtained by arranging the elements of the resampled blocks $\{\mathcal{B}(I_{3,i}, \ell) : i \geq 1\}$ and let $\bar{X}^{*(3)}_m$ denote the CBB sample mean based on m bootstrap observations, where $m = k\ell$ for some integer $k \geq 1$. Then, by (2.17),

$$
\begin{aligned}
E_* \bar{X}^{*(3)}_m &= E_* \Big[m^{-1} \sum_{i=1}^{m} X^*_{3,i} \Big] \\
&= \ell^{-1} E_* \Big[\sum_{i=1}^{\ell} X^*_{3,i} \Big] \\
&= \ell^{-1} \Big[n^{-1} \sum_{j=1}^{n} \Big\{ \sum_{i=1}^{\ell} Y_{n,(j+i-1)} \Big\} \Big] \\
&= \ell^{-1} \Big[\ell \bar{X}_n \Big] \\
&= \bar{X}_n . \qquad (2.18)
\end{aligned}
$$

Thus, the conditional expectation of the bootstrap sample mean under the CBB equals the sample mean of the data $\mathcal{X}_n$, a property not shared by the MBB or the NBB. As noted by Politis and Romano (1992b), this makes it easier to define the bootstrap version of a pivotal quantity of the form $T_n = t_n(\bar{X}_n; \mu)$, where $\mu = EX_1$. Under the CBB, $T^*_{m,n} = t_m(\bar{X}^{*(3)}_m; \bar{X}_n)$ gives the appropriate bootstrap version of T_n. However, replacing the population parameter μ simply by $\bar{X}_n$ to define the bootstrap version of T_n under the MBB or the NBB introduces some extra bias and hence, it is no longer the right thing to do (cf. Lahiri (1992a)). We shall look at properties of the CBB method in Chapters 3, 4, and 5.

2.7.2 *Stationary Block Bootstrap*

The stationary bootstrap (SB) of Politis and Romano (1994b) differ from the earlier block bootstrap methods (i.e., from MBB, NBB, and CBB) in that it uses blocks of *random* lengths rather than blocks of a fixed length ℓ. Let $p \equiv p_n \in (0, 1)$ be such that $p \to 0$ and $np \to \infty$ as $n \to \infty$. Then the SB resamples the blocks $\mathcal{B}(I_{4,1}, J_{4,1}), \mathcal{B}(I_{4,2}, J_{4,2}), \ldots$ where the index vectors $(I_{4,1}, J_{4,1}), (I_{4,2}, J_{4,2}), \ldots$ are *conditionally iid* with $I_{4,1}$ having the discrete uniform distribution on $\{1, \ldots, n\}$, and $J_{4,1}$ having the geometric distribution ν_n with parameter p, i.e.,

$$
P_*(J_{4,1} = j) \equiv \nu_n(j) = p(1-p)^{j-1}, \quad j = 1, 2, \ldots . \qquad (2.19)
$$

Furthermore, $I_{4,1}$ and $J_{4,1}$ are independent. Thus, the SB corresponds to the GBB method with the transition function $\Gamma_n(\cdot;\cdot)$ given by

$$\Gamma_n(x;\cdot) = \bigotimes_{i=1}^{\infty} \Big((n^{-1} \sum_{j=1}^{n} \delta_j) \times \nu_n \Big), \; x \in \mathbb{R}^n.$$

Note that here also, $\Gamma_n(x;\cdot)$ does not depend on $x \in \mathbb{R}^n$.

The SB method can be described through an alternative formulation, also given by Politis and Romano (1994b). Suppose, $X_{4,1}^*, X_{4,2}^*, \ldots$ denote the SB observations, obtained by arranging the elements of the resampled blocks $\mathcal{B}(I_{4,1}, J_{4,1}), \mathcal{B}(I_{4,2}, J_{4,2}), \ldots$ in a sequence. The sequence $\{X_{4,i}^*\}_{i\in\mathbb{N}}$ may also be generated by the following resampling mechanism. Let $X_{4,1}^*$ be picked at random from $\{X_1, \ldots, X_n\}$, i.e., let $X_{4,1}^* = Y_{n,I_{4,1}}$ where $I_{4,1}$ is as above. To select the next observation $X_{4,2}^*$, we further randomize and perform a binary experiment with probability of "Success" equal to p. If the binary experiment results in a "Success," then we select $X_{4,2}^*$ again *at random* from $\{X_1, \ldots, X_n\}$. Otherwise, we set $X_{4,2}^* = Y_{n,(I_{4,1}+1)}$, the observation next to $X_{4,1}^* \equiv Y_{n,I_{4,1}}$ in the periodically extended series $\{Y_{n,i}\}_{i\geq 1}$. In general, given that $X_{4,i}^*$ has been chosen and is given by Y_{n,i_0} for some $i_0 \geq 1$, the next SB observation $X_{4,(i+1)}^*$ is chosen as $Y_{n,(i_0+1)}$ with probability $(1-p)$ and is drawn at random from the original data set $\mathcal{X}_n$ with probability p.

To see that these two formulations are equivalent, let W_i denote the variable associated with the binary experiment for selecting $X_{4,i}^*$, $i \geq 2$. Then, conditional on $\mathcal{X}_n$, $W_i, i \geq 2$ are iid random variables with $P_*(W_i = 1) = p = 1 - P_*(W_i = 0)$, and $\{W_i : i \geq 2\}$ is independent of $\{I_{4,i} : i \geq 1\}$. Next define the variables $M_j, j \geq 0$, by

$$\begin{aligned} M_0 &\equiv 1, \\ M_j &= \inf\{i \geq M_{j-1} + 1 : W_i = 1\}, \; j \geq 1. \end{aligned}$$

Thus, M_j, $j \in \mathbb{N}$ denotes the *trial number* in the sequence of trials $\{W_i : i \geq 2\}$ at which the binary experiment resulted in the jth "Success" and has a negative binomial distribution with parameters j and p (up to a translation). Note that the corresponding SB observation, viz., X_{4,M_j}^*, is then selected at random from $\{X_1, \ldots, X_n\}$ as $X_{4,M_j}^* = Y_{n,I_{4,(j+1)}}$, $j \geq 1$. On the other end, for any i between $M_{j-1} + 1$ and $M_j - 1$, the binary experiment resulted in a block of "Failures," and the corresponding SB observations are selected by picking $(M_j - M_{j-1} - 1)$ variables following $Y_{n,I_{4,j}}$ in the sequence $\{Y_{n,i}\}_{i\in\mathbb{N}}$. Thus, the binary trials $\{W_i : i = M_{j-1}, \ldots, M_j - 1\}$ lead to the "SB block" of observations $\{X_{4,M_{j-1}}^*, \ldots, X_{4,(M_j-1)}^*\} = \{Y_{n,I_{4,j}}, \ldots, Y_{n,(I_{4,j}+M_j-M_{j-1}-1)}\}, j \geq 1$. Now, defining $J_{4,j} = M_j - M_{j-1}$, $j \geq 1$ and using the properties of the negative binomial distribution (cf. Section XI.2, Feller (1971a)), we may conclude that $J_{4,1}, J_{4,2}, \ldots$ are (conditionally) iid and follow the geometric distribution with parameter p. Hence, the two formulations of the SB are equivalent.

An important property of the SB method is that conditional on $\mathcal{X}_n$, the bootstrap observations $\{X_{4,i}^*\}_{i\in\mathbb{N}}$ are *stationary* (which is why it is called the "stationary" bootstrap). A simple proof of this fact can be derived using the second formulation of the SB as follows. Let $\{Z_i\}_{i\in\mathbb{N}}$ be a Markov chain on $\{1,\ldots,n\}$ such that conditional on $\mathcal{X}_n$, the initial distribution of the chain is $\boldsymbol{\pi} \equiv (n^{-1},\ldots,n^{-1})'$ and the stationary transition probability matrix of $\{Z_i\}_{i\in\mathbb{N}}$ is $Q \equiv \big((q_{ij})\big)$, where

$$q_{ij} = \begin{cases} p+n^{-1}(1-p) & 1\le i<n, j=i+1 \\ n^{-1}(1-p) & 1\le i<n, j\neq i+1 \\ n^{-1}(1-p) & i=n, 2\le j\le n \\ p+n^{-1}(1-p) & i=n, j=1\ . \end{cases} \tag{2.20}$$

Thus, Z_1 takes the values $1,\ldots,n$ with probability n^{-1} each. Also, for any $k\ge 1$, given that $Z_k = i$, $1\le i\le n$, the next index Z_{k+1} takes the value $i+1$ (modulo n) with probability $p+n^{-1}(1-p)$ and it takes each of the remaining $(n-1)$ values with probability $n^{-1}(1-p)$. Thus, from the second formulation of the SB described earlier, it follows that the SB observations $\{X_{4,i}^*\}_{i\in\mathbb{N}}$ may also be generated by the index variables $\{Z_i\}_{i\in\mathbb{N}}$ as

$$X_{4,i}^* = X_{Z_i},\ i\ge 1\ . \tag{2.21}$$

To see that $\{X_{4,i}^*\}_{i\in\mathbb{N}}$ is stationary, note that by definition, the transition matrix Q is doubly stochastic and that it satisfies the relation $\boldsymbol{\pi}'Q = \boldsymbol{\pi}'$. Therefore, $\boldsymbol{\pi}$ is the *stationary* distribution of $\{Z_i\}_{i\in\mathbb{N}}$ and $\{Z_i\}_{i\in\mathbb{N}}$ is a *stationary* Markov chain. Thus, we have proved the following Theorem.

Theorem 2.3 *Let $\mathcal{F}_{in}$ denote the σ-field generated by Z_i and $\mathcal{X}_n$, $i\ge 1$. Then, conditional on $\mathcal{X}_n$, $\{X_{4,i}^*, \mathcal{F}_{in}\}_{i\in\mathbb{N}}$ is a stationary Markov chain for each $n\ge 1$, i.e.,*

$$\mathcal{L}(X_{4,i}^*|\mathcal{X}_n) = \mathcal{L}(X_{4,1}^*|\mathcal{X}_n) \quad \textit{for all} \quad i\ge 1$$

and

$$\mathcal{L}(X_{4,(i+1)}^*|Z_1,\ldots,Z_i,\mathcal{X}_n) = \mathcal{L}(X_{4,(i+1)}^*|Z_i,\mathcal{X}_n) \quad \textit{for all} \quad i\ge 1\ .$$

In particular, Theorem 2.3 implies that conditional on $\mathcal{X}_n$, $\{X_{4,i}^*\}_{i\ge 1}$ is stationary. Furthermore, by (2.20) and (2.21), for a given resample size m, the conditional expectation of the SB sample mean $\bar{X}_m^{*(4)} \equiv m^{-1}\sum_{i=1}^m X_{4,i}^*$ is given by

$$E_*(\bar{X}_m^{*(4)}) = E_* X_{4,1}^* = \bar{X}_n\ . \tag{2.22}$$

We shall consider other properties of the SB method in Chapters 3–5.

2.8 Subsampling

Use of different subsets of the data to approximate the bias and variance of an estimator is a common practice, particularly in the context of iid observations. For example, the Jackknife bias and variance estimators are computed using subsets of size $n-1$ from the full sample $\mathcal{X}_n = (X_1, \ldots, X_n)$ (cf. Efron (1982)). However, as noted recently (see Carlstein (1986), Politis and Romano (1994a), Hall and Jing (1996), Bickel et al. (1997), and the references therein), subseries of dependent observations can also be used to produce valid estimators of the bias, the variances, and more generally, of the sampling distribution of a statistic under very weak assumptions.

To describe the subsampling method, suppose that $\hat{\theta}_n = t_n(\mathcal{X}_n)$ is an estimator of a parameter θ, such that for some normalizing constant $a_n > 0$, the probability distribution $Q_n(x) = P(a_n(\hat{\theta}_n - \theta) \leq x)$ of the centered and scaled estimator $\hat{\theta}_n$ converges weakly to a limit distribution $Q(x)$, i.e.,

$$Q_n(x) \to Q(x) \quad \text{as} \quad n \to \infty \tag{2.23}$$

for all continuity points x of Q. Furthermore, assume that $a_n \to \infty$ as $n \to \infty$ and that Q is *not degenerate* at zero, i.e., $Q(\{0\}) < 1$. Let $1 \leq \ell \leq n$ be a given integer and let

$$\mathcal{B}_i = (X_i, \ldots, X_{i+\ell-1})' \ ,$$

$1 \leq i \leq N$, denote the overlapping blocks of length ℓ where $N = n - \ell + 1$. Note that the blocks $\mathcal{B}_i$'s are the same as those defined in Section 2.4 for the MBB. Then, the subsampling estimator of Q_n, based on the *overlapping* version of the subsampling method, is given by

$$\hat{Q}_n(x) = N^{-1} \sum_{i=1}^{N} \mathbb{1}(a_\ell(\hat{\theta}_{i,\ell} - \hat{\theta}_n) \leq x), \ x \in \mathbb{R} \ , \tag{2.24}$$

where $\hat{\theta}_{i,\ell}$ is a "copy" of the estimator $\hat{\theta}_n$ on the block $\mathcal{B}_i$, defined by $\hat{\theta}_{i,\ell} = t_\ell(\mathcal{B}_i)$, $i = 1, \ldots, N$. Note that we used $t_\ell(\cdot)$ (in place of $t_n(\cdot)$) to define the subsample copy "$\hat{\theta}_{i,\ell}$," as the ith block $\mathcal{B}_i$ contains only ℓ observations. That is also the reason behind using the scaling constant a_ℓ instead of a_n. From the above description, it follows that the overlapping version of the subsampling method is a special case of the MBB where a *single* block is resampled.

The estimator $\hat{Q}_n$ of the distribution function $Q_n(x)$ can be used to obtain subsampling estimators of the bias and the variance of $\hat{\theta}_n$. Note that the bias of $\hat{\theta}_n$ is given by

$$\text{Bias}(\hat{\theta}_n) = E\hat{\theta}_n - \theta = a_n^{-1} \int x dQ_n(x) \ .$$

The subsampling estimator of $\text{Bias}(\hat{\theta}_n)$ is then obtained by replacing $Q_n(\cdot)$ by $\hat{Q}_n(\cdot)$, viz.,

$$\widehat{\text{Bias}}(\hat{\theta}_n) = a_n^{-1} \int x d\hat{Q}_n(x) = a_\ell a_n^{-1} \Big(N^{-1} \sum_{i=1}^{N} \hat{\theta}_{i,\ell} - \hat{\theta}_n \Big). \tag{2.25}$$

Similarly, the subsampling estimator of the variance of $\hat{\theta}_n$ is given by

$$\widehat{\text{Var}}(\hat{\theta}_n) = a_\ell^2 a_n^{-2} \left[N^{-1} \sum_{i=1}^{N} \hat{\theta}_{i,\ell}^2 - \Big(N^{-1} \sum_{i=1}^{N} \hat{\theta}_{i,\ell} \Big)^2 \right], \tag{2.26}$$

which is the sample variance of $\hat{\theta}_{i,\ell}$'s multiplied by the scaling factor $a_\ell^2 a_n^{-2}$. In (2.25) and (2.26), we need to use the correction factors (a_ℓ / a_n) and $(a_\ell / a_n)^2$ to *scale up* from the level of $\hat{\theta}_{i,\ell}$'s, which are defined using ℓ-observations, to the level of $\hat{\theta}_n$, which is defined using n-observations. In applying a bootstrap method, one typically uses a resample size that is comparable to the original sample size, and therefore, such explicit corrections of the bootstrap bias and variance estimators are usually unnecessary.

In analogy to the bootstrap methods, one may attempt to apply the subsampling method to a centered variable of the form $T_{1n} \equiv (\hat{\theta}_n - \theta)$. However, this may *not* be the right thing to do. Indeed, if instead of using the subsampling method for the scaled random variable $a_n(\hat{\theta}_n - \theta)$, we consider only the centered variable $T_{1n} = (\hat{\theta}_n - \theta)$, then the subsampling estimator of the distribution Q_{1n}, say, of T_{1n} would be given by

$$\hat{Q}_{1n}(x) \equiv N^{-1} \sum_{i=1}^{N} \mathbb{1}((\hat{\theta}_{i,\ell} - \hat{\theta}_n) \leq x), \; x \in \mathbb{R} \, .$$

Since $\text{Bias}(\hat{\theta}_n) = ET_{1n} = \int x d\hat{Q}_{1n}(x)$, using $\hat{Q}_{1n}(x)$, we would get

$$\widehat{\text{Bias}}_{1n}(\hat{\theta}_n) = \int x d\hat{Q}_{1n}(x) = \Big(N^{-1} \sum_{i=1}^{N} \hat{\theta}_{i,\ell} - \hat{\theta}_n \Big) \, ,$$

as an estimator of $\text{Bias}(\hat{\theta}_n)$, and similarly, we would get

$$\widehat{\text{Var}}_{1n}(\hat{\theta}_n) = N^{-1} \sum_{i=1}^{N} \hat{\theta}_{i,\ell}^2 - \Big(N^{-1} \sum_{i=1}^{N} \hat{\theta}_{i,\ell} \Big)^2$$

as an estimator of $\text{Var}(\hat{\theta}_n)$. However, these subsampling estimators of the bias and the variance of $\hat{\theta}_n$, defined using $\hat{Q}_{1n}(x)$, are very "poor" estimators of the corresponding population parameters. To appreciate why, consider the case where $\hat{\theta}_n = \bar{X}_n$ and $\theta = EX_1$ and $n^{1/2}(\hat{\theta}_n - \theta) \longrightarrow^d \text{N}(0, \sigma_\infty^2)$ as $n \to \infty$ with $\sigma_\infty^2 = \sum_{i=-\infty}^{\infty} \text{Cov}\,(X_1, X_{i+1}) \neq 0$. Write

$\bar{X}_{i,\ell}$ for the average of the ℓ observations in $\mathcal{B}_i$, $i = 1, \ldots, N$. Then, $\widehat{\text{Var}}_{1n}(\hat{\theta}_n) = N^{-1}\sum_{i=1}^{N} \bar{X}_{i,\ell}^2 - \hat{\mu}_n^2$, where $\hat{\mu}_n \equiv N^{-1}\sum_{i=1}^{N} \bar{X}_{i,\ell}$ is the average of the N block averages. Then, it is not difficult to show that under some standard moment and weak dependence conditions on the process $\{X_i\}_{i\in\mathbb{Z}}$ and under the assumption that $\ell^{-1} + n^{-1}\ell = o(1)$ as $n \to \infty$,

$$\begin{aligned}
&\widehat{\text{Var}}_{1n}(\hat{\theta}_n) \\
= \quad & \text{Var}(\bar{X}_\ell) + N^{-1}\sum_{i=1}^{N}\Big\{[\bar{X}_{i,\ell} - \theta]^2 - \quad \text{Var}(\bar{X}_\ell)\Big\} - [\hat{\mu}_n - \theta]^2 \\
= \quad & \ell^{-1}\sigma_\infty^2 + O(\ell^{-2}) + O_p([n\ell]^{-1/2}) + O_p(n^{-1}) \;, \qquad (2.27)
\end{aligned}$$

whereas $\text{Var}(\bar{X}_n) = n^{-1}\sigma_\infty^2 + O(n^{-2})$ as $n \to \infty$. Thus, $\widehat{\text{Var}}_{1n}(\hat{\theta}_n)$ indeed overestimates the variance of $\hat{\theta}_n$ by a scaling factor of n/ℓ, which blows up to infinity with n. It is easy to see that the other estimator, viz., $\widehat{\text{Var}}(\hat{\theta}_n)$ is equal to ℓ/n times $\widehat{\text{Var}}_{1n}(\hat{\theta}_n)$ in this case and thus, provides a sensible estimator of $\text{Var}(\bar{X}_n)$. The reason that the subsampling estimator based on T_{1n} does *not* work in this case is that the limit distribution of T_{1n} is *degenerate at zero*, and does not satisfy the nondegeneracy requirement stated above.

Formulas (2.24), (2.25), and (2.26) illustrate a very desirable property of the subsampling method that holds true generally. Computations of $\hat{Q}_n(\cdot)$ and of estimates of other population quantities based on $\hat{Q}_n$ do not involve any resampling and hence, are less demanding. Typically, a simple, closed-form expression can be written down for a subsampling estimator of a level-2 parameter, and it needs computation of the subsampling version $\hat{\theta}_{i,\ell}$ of the estimator $\hat{\theta}_n$ only N times, as compared to a much larger number of times for the resampling methods like the MBB. However, the price paid is the lack of "automatic" second-order correctness of the subsampling method compared to the MBB and other block bootstrap methods.

We conclude this section with an observation. As noted previously, the subsampling method is a special case of the MBB where the number of resampled blocks is identically equal to 1. Exploiting this fact, we may similarly define other versions of the subsampling method based on nonoverlapping blocks or circular blocks. More generally, it is possible to extend the subsampling method in the spirit of the GBB method. We define the "generalized subsampling" method as the GBB method with a single sample (I_1, J_1) of the indices. Thus, the generalized subsampling estimator of $Q_n(x)$ (cf. (2.23)) is given by

$$\hat{Q}_n^{GS}(x) = E_* \mathbb{1}\Big(a_{J_1}[\hat{\theta}_{J_1,n}^* - \hat{\theta}_n] \le x\Big), \; x \in \mathbb{R} \;,$$

where $\hat{\theta}_{J_1,n}^* = t_{J_1}(\mathcal{B}(I_1, J_1))$ is a copy of $\hat{\theta}_n$ based on the GBB samples from a *single* block $\mathcal{B}(I_1, J_1)$ of length J_1.

2.9 Transformation-Based Bootstrap

As described in Chapter 1, the basic idea behind the bootstrap method is to recreate the relation between the population and the sample using the sample itself. For dependent data, the most common approach to this problem is to resample "blocks" of observations instead of single observations, which preserves the dependence structure of the underlying process *within* the resampled blocks and is able to reproduce the effect of dependence at short lags. A quite different approach to the problem was suggested by Hurvich and Zeger (1987). In their seminal work, Hurvich and Zeger (1987) considered the discrete Fourier transform (DFT) of the data and rather than resampling the data values directly, they applied the IID bootstrap method of Efron (1979) to the DFT values. The *transformation based bootstrap* (TBB) described here is a generalization of Hurvich and Zeger's (1987) idea.

To describe it, let $\theta \equiv \theta(P)$ be a parameter of interest, which depends on the underlying probability measure P generating the sequence $\{X_i\}_{i\in\mathbb{Z}}$, and let $T_n \equiv t_n(\mathcal{X}_n)$ be an estimator of θ based on the observations $\mathcal{X}_n = (X_1, \ldots, X_n)$. Our goal is to approximate the sampling distribution of a normalized or studentized statistic $R_n = r_n(\mathcal{X}_n; \theta)$. Let $\mathcal{Y}_n = h_n(\mathcal{X}_n)$ be a (one-to one) transformation of $\mathcal{X}_n$ such that the components of $\mathcal{Y}_n$, say, $\{Y_i : i \in \mathcal{I}_n\}$, are "approximately independent". Also suppose that the variable R_n can be expressed (at least to a close approximation) in terms of $\mathcal{Y}_n$ as $R_n = r_{1n}(\mathcal{Y}_n; \theta)$ for some reasonable function r_{1n}. Then, to approximate the distribution of R_n by the TBB, we resample from a *suitable* subcollection $\{Y_i : i \in \mathcal{J}_n\}$ of $\{Y_i : i \in \mathcal{I}_n\}$ to get the bootstrap observations $\mathcal{Y}_n^* \equiv \{Y_i^* : i \in \mathcal{I}_n\}$ either by selecting a single Y-value at a time as in the IID-bootstrap method of Efron (1979) or by selecting a block of Y-values from $\{Y_i : i \in \mathcal{J}_n\}$ as in the MBB, depending on the dependence structure of $\{Y_i : i \in \mathcal{J}_n\}$. The TBB estimator of the distribution of R_n is then given by the conditional distribution of $R_n^* \equiv r_{1n}(\mathcal{Y}_n^*; \hat{\theta}_n)$ given the data $\mathcal{X}_n$, where $\hat{\theta}_n$ is an estimator of θ based on $\mathcal{X}_n$. Thus, as a principle, the TBB method suggests an *additional* transformation step to reduce the dependence in the data to an iid structure or to a weaker form of dependence.

An important example of the TBB method is the *Frequency Domain Bootstrap* (FDB), which uses the Fourier transform of the data to generate the Y-variables of the TBB. Suppose that $\{X_i\}_{i\in\mathbb{Z}}$ is a sequence of stationary, weakly dependent random variables. The Fourier transform of the observations $\mathcal{X}_n$ is defined as

$$Y_n(w) = n^{-1/2} \sum_{j=1}^{n} X_j \exp(-\iota w j), \;\; w \in (-\pi, \pi] \;,$$

where recall that $\iota = \sqrt{-1}$. Though the X_i's are dependent, a well known result in time-series states (cf. Brockwell and Davis (1991, Chapter 10); Lahiri (2003a)) that for any set of distinct ordinates $-\pi < \lambda_1, \ldots, \lambda_k \leq \pi$, the Fourier transforms $Y_n(\lambda_1), \ldots, Y_n(\lambda_k)$ are *asymptotically independent*. Furthermore, the original observations $\mathcal{X}_n$ admit a representation in terms of the transformed values $\mathcal{Y}_n = \{Y_n(w_j) : j \in \mathcal{I}_n\}$ as (cf. Brockwell and Davis (1991, Chapter 10)),

$$X_t = n^{-1/2} \sum_{j \in \mathcal{I}_n} Y_n(w_j) \exp(\iota t w_j), \quad t = 1, \ldots, n \tag{2.28}$$

where $w_j = 2\pi j/n$ and $\mathcal{I}_n = \{-\lfloor (n-1)/2 \rfloor, \ldots, \lfloor n/2 \rfloor\}$. Thus, using the *inversion formula* (2.28), we can express a given variable $R_n = r_n(\mathcal{X}_n; \theta)$ also in terms of the transformed values $\mathcal{Y}_n$. Since the variables in $\mathcal{Y}_n$ are approximately independent, we may (suitably) resample these Y-values to define the FDB version of R_n. Here, however, some care must be taken since the (asymptotic) variance of the Y-variables are not necessarily identical. A more complete description of the FDB method and its properties are given in Chapter 9.

2.10 Sieve Bootstrap

Let $\{X_i\}_{i \in \mathbb{Z}}$ be a stationary time series and let $T_n = t_n(X_1, \ldots, X_n)$ be an estimator of a level-1 parameter of interest $\theta = \theta(P)$, where P denotes the (unknown) joint distribution of $\{X_i\}_{i \in \mathbb{Z}}$. Then, the sampling distribution of T_n is given by

$$G_n(B) = P(T_n \in B) = P \circ t_n^{-1}(B) \tag{2.29}$$

for Borel sets B in $\mathbb{R}$, where $P \circ t_n^{-1}$ denotes the probability distribution on $\mathbb{R}$ induced by the transformation $t_n(\cdot)$ under P. As described in Chapter 1, the bootstrap and other resampling methods are general estimation methods for estimating the level-2 parameters like $G_n(B)$, $\text{Var}(T_n)$, etc. When the X_i's are iid with a common distribution F, we may write $P = F^\infty$ and an estimator of $G_n(B)$ in (2.29) may be generated by replacing P with $\hat{P}_n = \hat{F}_n^\infty$ in (2.28), where $\hat{F}_n$ is an estimator of F. However, when the X_i's are dependent, such a factorization of P does not hold. In this case, estimation of the level-2 parameter $G_n(B)$ can be thought of as a two-step procedure where, in the first step, P is approximated by a "simpler" probability distribution $\tilde{P}_n$ and in the next step, $\tilde{P}_n$ is estimated using the data $\{X_1, \ldots, X_n\}$. The idea of the sieve bootstrap is to choose $\{\tilde{P}_n\}_{n \geq 1}$ to be a sieve approximation to P, i.e., $\{\tilde{P}_n\}_{n \geq 1}$ is a sequence of probability measures on $\big(\mathbb{R}^\infty, \mathcal{B}(\mathbb{R}^\infty)\big)$ such that for each n, $\tilde{P}_{n+1}$ is a finer approximation to P than $\tilde{P}_n$ and $\tilde{P}_n$ converges to P (in some suitable sense) as $n \to \infty$.

For the block bootstrap methods like the NBB or the MBB, the first step approximation $\tilde{P}_n$ is taken to be $P_\ell \otimes P_\ell \otimes \ldots$, where P_ℓ denotes the joint distribution of the block $\{X_1, \ldots, X_\ell\}$ of length ℓ. In the second step, P_ℓ is estimated by the empirical distribution of all overlapping (under MBB) or nonoverlapping (under NBB) blocks of length ℓ contained in the data. For a large class of stationary processes, Bühlmann (1997) presents a sieve bootstrap method based on a sieve of autoregressive processes of increasing order, which we shall briefly describe here. However, other choices of $\{\tilde{P}_n\}_{n \geq 1}$ is possible. See Bühlmann (2002) for another interesting proposal based on variable length Markov chains for finite state space categorical time series. In general, there is a trade-off between the accuracy and the range of validity of a given sieve bootstrap method. Typically, one may choose a sieve to obtain a more accurate bootstrap estimator, but only at the expense of restricting the applicability to a smaller class of processes (cf. Lahiri (2002b)).

Let $\{X_i\}_{i \in \mathbb{Z}}$ be a stationary process with $EX_1 = \mu$ such that it admits the one-sided moving average representation

$$X_i - \mu = \sum_{j=0}^{\infty} \alpha_j \epsilon_{i-j} \ , \quad i \in \mathbb{Z} \tag{2.30}$$

where $\{\epsilon_i\}_{i \in \mathbb{Z}}$ is a sequence of zero mean uncorrelated random variables and where $\alpha_0 = 1$, $\sum_{i=1}^{\infty} \alpha_i^2 < \infty$. Suppose that $\{X_i\}_{i \in \mathbb{Z}}$ satisfies the standard invertibility conditions for a linear process (cf. Theorem 7.6.9, Anderson (1971)). Then, we can represent $\{X_i - \mu\}_{i \in \mathbb{Z}}$ as a one-sided infinite order autoregressive process

$$(X_i - \mu) = \sum_{j=1}^{\infty} \beta_j (X_{i-j} - \mu) + \epsilon_i \ , \quad i \in \mathbb{Z} \tag{2.31}$$

with $\sum_{j=1}^{\infty} \beta_j^2 < \infty$. The representation (2.31) suggests that autoregressive processes of finite orders p_n, $n \geq 1$, may be used to define a sieve approximation for the joint distribution P of $\{X_i\}_{i \in \mathbb{Z}}$. To describe the sieve bootstrap based on autoregression, let $\mathcal{X}_n = \{X_1, \ldots, X_n\}$ denote the observations from the process $\{X_i\}_{i \in \mathbb{Z}}$. Let $\{p_n\}_{n \geq 1}$ be a sequence of positive integers such that $p_n \uparrow \infty$ as $n \to \infty$, but $n^{-1} p_n \to 0$ as $n \to \infty$. The sieve approximation $\tilde{P}_n$ to P is determined by the autoregressive process

$$X_i - \mu = \sum_{j=1}^{p_n} \beta_j (X_{i-j} - \mu) + \epsilon_i \ , \quad i \in \mathbb{Z} \ . \tag{2.32}$$

Next, we fit the AR(p_n) model (2.32) to the data $\mathcal{X}_n$ to obtain estimators of the autoregression parameters $\hat{\beta}_{1n}, \ldots, \hat{\beta}_{p_n n}$ (for example, by the least squares method). This yields the residuals

$$\hat{\epsilon}_{in} = (X_i - \bar{X}) - \sum_{j=1}^{p_n} \hat{\beta}_{jn}(X_{i-j} - \bar{X}_n) \ , \quad p_n + 1 \leq i \leq n$$

where $\bar{X}_n = n^{-1} \sum_{i=1}^{n} X_i$. As in Section 2.4, we center the residuals at $\bar{\epsilon}_n = (n - p_n)^{-1} \sum_{i=p_n+1}^{n} \hat{\epsilon}_{in}$ and resample from the centered residuals $\{\hat{\epsilon}_{in} - \bar{\epsilon}_n : p_n + 1 \leq i \leq n\}$ to generate the sieve bootstrap error variables ϵ_i^*, $i \geq p_n + 1$. Then, the sieve bootstrap observations are generated by the recursion relation

$$(X_i^* - \bar{X}_n) = \sum_{j=1}^{p_n} \hat{\beta}_{jn}(X_{i-j}^* - \bar{X}_n) + \epsilon_i^* \ , \quad i \geq p_n + 1$$

by setting the initial p_n-variables $X_1^*, \ldots, X_{p_n}^*$ equal to $\bar{X}_n$. The autoregressive sieve bootstrap version of the estimator $T_n = t_n(X_1, \ldots, X_n)$ is now given by

$$T_{m,n}^* = t_m(X_1^*, \ldots, X_m^*) \ , \quad m > p_n \ .$$

Under some regularity conditions on the variables $\{\epsilon_i\}_{i \in \mathbb{Z}}$ of (2.30) and the sieve parameter p_n, Bühlmann (1997) establishes consistency of the autoregressive sieve bootstrap. It follows from his results that the autoregressive sieve bootstrap provides a more accurate variance estimator for the class of estimators given by (2.11) than the MBB and the NBB. However, consistency of the autoregressive sieve bootstrap variance estimators holds for a more restricted class of processes than the block bootstrap methods. See Bühlmann (1997), Choi and Hall (2000), and the references therein for more about the properties of the autoregressive sieve bootstrap.

3

Properties of Block Bootstrap Methods for the Sample Mean

3.1 Introduction

In this chapter, we study the first-order properties of the MBB, the NBB, the CBB, and the SB for the sample mean. Note that for the first three block bootstrap methods, the block length is nonrandom. In Section 3.2, we establish consistency of these block bootstrap methods for variance and distribution function estimations for the sample mean. The SB method uses a random block length and hence, requires a somewhat different treatment. We study consistency properties of the SB method for the sample mean in Section 3.3.

For later reference, we introduce some standard measures of weak dependence for time series. Let $(\Omega, \mathcal{F}, P)$ be a probability space and let $\mathcal{A}$ and $\mathcal{B}$ be two sub σ-fields of $\mathcal{F}$. When $\mathcal{A}$ and $\mathcal{B}$ are independent, for any $A \in \mathcal{A}$ and any $B \in \mathcal{B}$, we have the relations $\Delta_1 \equiv [P(A \cap B) - P(A) \cdot P(B)] = 0$ and $\Delta_2 \equiv [P(B|A) - P(B)] = 0$, provided $P(A) \neq 0$. When $\mathcal{A}$ and $\mathcal{B}$ are not independent, we may quantify the degree of dependence of $\mathcal{A}$ and $\mathcal{B}$ by looking at the maximal values of Δ_1 or Δ_2 or of some other similar quantities. This leads to the following coefficients of dependence:

Strong mixing or α-mixing:

$$\alpha(\mathcal{A}, \mathcal{B}) = \{|P(A \cap B) - P(A) \cdot P(B)| : A \in \mathcal{A}, B \in \mathcal{B}\} . \tag{3.1}$$

ϕ-mixing:

$$\phi(\mathcal{A},\mathcal{B}) = \left\{\left|\frac{P(A\cap B)}{P(A)} - P(B)\right| : A\in\mathcal{A}, P(A)\neq 0, B\in\mathcal{B}\right\}. \tag{3.2}$$

Ψ-mixing:

$$\Psi(\mathcal{A},\mathcal{B}) = \sup\left\{\left|\frac{P(A\cap B)}{P(A)P(B)} - 1\right| : A\in\mathcal{A}, B\in\mathcal{B}, P(A)\neq 0, P(B)\neq 0\right\}. \tag{3.3}$$

ρ-mixing:

$$\rho(\mathcal{A},\mathcal{B}) = \sup\left\{\frac{|\mathrm{Cov}(X,Y)|}{\sqrt{\mathrm{Var}(X)}\sqrt{\mathrm{Var}(Y)}} : X\in L^2(\mathcal{A}), Y\in L^2(\mathcal{B})\right\}, \tag{3.4}$$

where $\mathrm{Cov}(X,Y) = EXY - EXEY$, $\mathrm{Var}(X) = \mathrm{Cov}(X,X)$ and $L^2(\mathcal{A}) = \{X : X$ is a random variable on $(\Omega,\mathcal{A},P)$ with $EX^2<\infty\}$. In general,

$$\begin{gathered} 4\,\alpha(\mathcal{A},\mathcal{B}) \leq \rho(\mathcal{A},\mathcal{B}) \leq 2\,\phi^{1/2}(\mathcal{A},\mathcal{B})\cdot\phi^{1/2}(\mathcal{B},\mathcal{A}) \\ \text{and} \\ \rho(\mathcal{A},\mathcal{B}) \leq \Psi(\mathcal{A},\mathcal{B}). \end{gathered} \tag{3.5}$$

See Chapter 1 of Doukhan (1994) for the properties of these mixing coefficients. For an index set $I\subset\mathbb{Z}$, $I\neq\emptyset$, the mixing coefficients of a time series $\{X_i\}_{i\in I}$ at lag $m\geq 1$ are defined by considering the maximal values of these coefficients over the σ-fields $\mathcal{A} = \sigma\langle\{X_i : i\leq k,\ i\in I\}\rangle$ and $\mathcal{B} = \sigma\langle\{X_i : i\geq k+m,\ i\in I\}\rangle$ for all $k\in I$. Specifically, we have the following definition for the α-mixing and the ρ-mixing cases.

Definition 3.1 *Let $\{X_n\}_{n\in\mathbb{N}}$ be a sequence of random variables on $(\Omega,\mathcal{A},P)$. Let $\mathcal{F}_a^b = \sigma\langle\{X_i : a\leq i<b\}\rangle$, $1\leq a\leq b\leq\infty$.*

(1) The strong mixing (or α-mixing) coefficient of $\{X_i\}_{i=1}^\infty$ is defined by

$$\alpha(m) = \sup\left\{\alpha\big(\mathcal{F}_1^{k+1},\mathcal{F}_{k+m+1}^\infty\big) : k\in\mathbb{N}\right\},\ m\geq 1 \tag{3.6}$$

where $\alpha(\cdot,\cdot)$ is as defined in (3.1). The process $\{X_i\}_{i\geq 1}$ is called strongly mixing if $\alpha(m)\to 0$ as $m\to\infty$.

(2) The ρ-mixing coefficient of $\{X_i\}_{i=1}^\infty$ is defined by

$$\rho(m) = \sup\left\{\rho\big(\mathcal{F}_1^{k+1},\mathcal{F}_{k+m+1}^\infty\big) : k\in\mathbb{N}\right\},\ m\geq 1 \tag{3.7}$$

where $\rho(\cdot,\cdot)$ is as defined in (3.4). The process $\{X_i\}_{i\geq 1}$ is called ρ-mixing if $\rho(m)\to 0$ as $m\to\infty$.

For a doubly-infinite time series $\{X_i\}_{i\in\mathbb{Z}}$, the coefficients $\alpha(m)$ and $\rho(m)$, $m \geq 1$ are defined by replacing the σ-field $\mathcal{F}_1^{k+1}$ by $\mathcal{F}_{-\infty}^{k+1} \equiv \sigma\langle\{X_i : i \leq k\}\rangle$ and the set $\mathbb{N}$ by $\mathbb{Z}$ in (3.6) and (3.7), respectively. The ϕ-mixing and the Ψ-mixing coefficients of $\{X_i\}_{i\in\mathbb{Z}}$ are defined similarly. Each of the four mixing conditions says that the dependence of the process $\{X_i\}_{i\in\mathbb{Z}}$ decreases as the distance m between the two segments $\{X_i : i \leq k\}$ and $\{X_i : i \geq k+m+1\}$ increases. Note that if a process is m_0-dependent for some $m_0 \geq 0$, then all these mixing coefficients are zero for all $m > m_0$. Thus, the notion of m-dependence is the most stringent than the four measures of weak dependence described above. And by (3.5), the notion of strong mixing is the least stringent.

The following is a useful inequality for the covariance of mixing random variables. Recall that for a random variable W and $p \in [1,\infty]$, we define the p-norm of W by

$$\|W\|_p = \begin{cases} (E|W|^p)^{1/p}\ , & p \in [1,\infty) \\ \inf\{x : P(|W| > x) = 0\}\ , & p = \infty\ . \end{cases} \tag{3.8}$$

Proposition 3.1 *Let X and Y be two random variables on a probability space $(\Omega, \mathcal{F}, P)$.*

(a) If $P(|X| \leq a_1) = 1,\ P(|Y| \leq a_2) = 1$ for some $a_1, a_2 \in (0,\infty)$, then

$$|Cov(X,Y)| \leq 4a_1a_2\ \alpha(\sigma\langle\{X\}\rangle, \sigma\langle\{Y\}\rangle)\ .$$

(b) Let $p,q,r \in (1,\infty)$ be any real numbers satisfying $p^{-1}+q^{-1}+r^{-1} = 1$. Then,

$$|Cov(X,Y)| \leq 8[\alpha(\sigma\langle\{X\}\rangle, \sigma\langle\{Y\}\rangle)]^{1/r}\ \|X\|_p\|Y\|_q\ .$$

(c)

$$|Cov(X,Y)| \leq \rho(\sigma\langle\{X\}\rangle, \sigma\langle\{Y\}\rangle)\ \|X\|_2\|Y\|_2\ .$$

Proof: See Section 1.2.2 of Doukhan (1994). □

In the next section, we establish consistency of the MBB, the NBB and the CBB method for the sample mean.

3.2 Consistency of MBB, NBB, CBB: Sample Mean

Let $\{X_i\}_{i\in\mathbb{Z}}$ be a sequence of stationary random vectors taking values in $\mathbb{R}^d$. Let T_n denote the centered and scaled sample mean

$$T_n = \sqrt{n}(\bar{X}_n - \mu)\ ,$$

where $\mu = E(X_1)$ and $\bar{X}_n = n^{-1}\sum_{i=1}^n X_i$. In this section, we establish consistency of the MBB, the NBB, and the CBB estimators of the (asymptotic) covariance matrix

$$\mathrm{Var}(T_n) \equiv ET_nT_n'$$

of T_n, and also, of the sampling distribution

$$G_n(x) \equiv P(T_n \leq x), \ x \in \mathbb{R}^d$$

of T_n. For simplicity, we suppose that for each of the block bootstrap methods, $b \equiv \lfloor n/\ell \rfloor$ blocks are resampled and thus, the resample size is $n_1 = b\ell$. Write $\bar{X}_n^{*(1)}$, $\bar{X}_n^{*(2)}$, and $\bar{X}_n^{*(3)}$ for the sample means of the n_1 bootstrap observations based on the MBB, the NBB, and the CBB, respectively. The bootstrap versions of T_n are then given by

$$T_n^{*(j)} \equiv \sqrt{n_1}\left(\bar{X}_n^{*(j)} - E_*\bar{X}_n^{*(j)}\right), \ j = 1,2,3 \ .$$

The bootstrap estimators of $\mathrm{Var}(T_n)$ are given by $\mathrm{Var}_*(T_n^{*(j)})$, $j = 1,2,3$, where, as before, E_* and Var_* respectively denote the conditional expectation and the conditional variance, given $\mathcal{X}_n$. Similarly, the bootstrap estimators of $G_n(\cdot)$ are given by the conditional distribution of $T_n^{*(j)}$, given $\mathcal{X}_n$, $j = 1,2,3$. First, we consider properties of the variance estimators.

3.2.1 Consistency of Bootstrap Variance Estimators

The bootstrap variance estimators $\mathrm{Var}_*(T_n^{*(j)})$, $j = 1,2,3$ have a very desirable property, namely, that they can be expressed by simple, closed-form formulas involving the observations $\mathcal{X}_n$, and thus, may be computed directly without any Monte-Carlo simulations. This is possible because of the linearity of the bootstrap sample mean in the resampled observations. Let $U_i = (X_i + \cdots + X_{i+\ell-1})/\ell$ denote the average of the block $(X_i, \ldots, X_{i+\ell-1})$, $i \geq 1$, let $U_i^{(2)} \equiv U_{(i-1)\ell+1} = (X_{(i-1)\ell+1} + \cdots + X_{i\ell})/\ell$, $i \geq 1$ be the average of the nonoverlapping blocks, and similarly, let $U_i^{(3)} = (Y_{n,i} + \cdots + Y_{n,(i+\ell-1)})/\ell$, $i \geq 1$ be the block averages for the periodically extended series $\{Y_{n,i}\}_{i\geq 1}$. Then, using the independence of the resampled blocks, we get

$$\begin{aligned}
\mathrm{Var}_*(T_n^{*(1)}) &= \ell\Big[\frac{1}{N}\sum_{i=1}^N U_iU_i' - \hat{\mu}_n\hat{\mu}_n'\Big] , \\
\mathrm{Var}_*(T_n^{*(2)}) &= \ell\Big[\frac{1}{b}\sum_{i=1}^b U_i^{(2)}U_i^{(2)'} - \hat{\mu}_{n,2}\hat{\mu}_{n,2}'\Big] , \\
&\text{and} \\
\mathrm{Var}_*(T_n^{*(3)}) &= \ell\Big[n^{-1}\sum_{i=1}^n U_i^{(3)}U_i^{(3)'} - \bar{X}_n\bar{X}_n'\Big] , \qquad (3.9)
\end{aligned}$$

where $N = n - \ell + 1$, $\hat{\mu}_n = N^{-1} \sum_{i=1}^{n} U_i$, and $\hat{\mu}_{n,2} = b^{-1} \sum_{i=1}^{b} U_i^{(2)}$.

When the process $\{X_i\}_{i \in \mathbb{Z}}$ satisfies certain standard moment and strong mixing conditions (such as those of Theorem 3.1 below), the asymptotic covariance matrix of T_n is given by the infinite (matrix) series

$$\Sigma_\infty \equiv \lim_{n \to \infty} \mathrm{Var}(T_n) = \sum_{i=-\infty}^{\infty} E Z_1 Z'_{1+i} ,$$

where $Z_i = X_i - \mu$, $i \in \mathbb{Z}$. Thus, the bootstrap estimators $\mathrm{Var}_*(T_n^{*(j)})$, $j = 1, 2, 3$ may be viewed also as estimators of the population parameter Σ_∞. The following result proves consistency of the bootstrap estimators for the level-2 parameter $\mathrm{Var}(T_n)$ or, equivalently, for Σ_∞.

Theorem 3.1 *Suppose that there exists a $\delta > 0$ such that $E\|X_1\|^{2+\delta} < \infty$ and that $\sum_{n=1}^{\infty} \alpha(n)^{\delta/2+\delta} < \infty$. If, in addition, $\ell^{-1} + n^{-1}\ell = o(1)$ as $n \to \infty$, then for $j = 1, 2, 3$,*

$$Var_*(T_n^{*(j)}) \longrightarrow_p \Sigma_\infty \quad as \quad n \to \infty . \tag{3.10}$$

Theorem 3.1 shows that under mild moment and strong mixing conditions on the process $\{X_i\}_{i \in \mathbb{Z}}$, the bootstrap variance estimators $\mathrm{Var}_*(T_n^{*(j)})$, $j = 1, 2, 3$ are consistent for a wide range of bootstrap block sizes ℓ, so long as ℓ tends to infinity with n but at a rate slower than n. Thus, block sizes given by $\ell = \log\log n$ or $\ell = n^{1-\epsilon}$, $0 < \epsilon < 1$, are all admissible block lengths for the consistency of $\mathrm{Var}_*(T_n^{*(j)})$, $j = 1, 2, 3$. We shall show later (cf. Chapters 5 and 7), an *optimal* choice of ℓ, that asymptotically minimizes the mean squared error of the block bootstrap variance estimator $\mathrm{Var}_*(T_n^{*(j)})$, is of the form $\ell = C_j n^{1/3}(1 + o(1))$ as $n \to \infty$ where $C_j > 0$ is a suitable constant that depends on certain population parameters.

For proving the theorem, we need the following lemma. Let $U_{1i} = \sqrt{\ell}\, U_i = (X_i + \cdots + X_{i+\ell-1})/\sqrt{\ell}$, and $U_{1i}^{(2)} = \sqrt{\ell}\, U_i^{(2)}$, $i \geq 1$.

Lemma 3.1 *Let $f : \mathbb{R}^d \to \mathbb{R}$ be a Borel measurable function and let $\{X_i\}_{i \in \mathbb{Z}}$ be a (possibly nonstationary) sequence of random vectors with strong mixing coefficient $\alpha(\cdot)$. Define $\|f\|_\infty = \sup\{|f(x)| : x \in \mathbb{R}^d\}$ and $\zeta_{2+\delta,n} = \max\big\{\big(E|f(U_{1i})|^{2+\delta}\big)^{1/(2+\delta)} : 1 \leq i \leq N\big\}$, $\delta > 0$. Let $\{w_{in} : i \geq 1, n \geq 1\} \subset [-1, 1]$ be a collection of real numbers. Then, there exist numerical constants C_1, C_2, and constants $C_3(\delta)$ and $C_4(\delta)$ (none depending on $f(\cdot)$, ℓ, n, and w_{in}'s), such that for any $1 < \ell < n/2$ and any $n > 2$,*

(a)

$$Var\Big(\sum_{i=1}^{N} w_{in} f(U_{1i})\Big)$$

$$\leq \min\Big\{C_1\|f\|_\infty^2 n\ell\Big[1+\sum_{1\leq k\leq n/\ell}\alpha(k\ell)\Big],$$

$$C_3(\delta)\zeta_{2+\delta,n}^2 n\ell\Big[1+\sum_{1\leq k\leq n/\ell}\alpha(k\ell)^{\delta/(2+\delta)}\Big]\Big\};$$

(b)

$$Var\Big(\sum_{i=1}^{b} w_{in}f(U_{1i}^{(2)})\Big)$$

$$\leq \min\Big\{C_2\|f\|_\infty^2 b\Big[1+\sum_{1\leq k\leq n/\ell}\alpha(k\ell)\Big],$$

$$C_4(\delta)\zeta_{2+\delta,n}^2 b\Big[1+\sum_{1\leq k\leq n/\ell}\alpha(k\ell)^{\delta/(2+\delta)}\Big]\Big\}.$$

Proof: First we consider part (a). We group the summands into "blocks of blocks" such that alternate "blocks of blocks" sums are approximately independent. Define the variables

$$R(j) = \sum_{i=2(j-1)\ell+1}^{2j\ell} w_{in}f(U_{1i}),\quad 1\leq j\leq J, \tag{3.11}$$

where $J = \lfloor N/2\ell \rfloor$ and set $R(J+1) = \sum_{i=1}^{n} w_{in}f(U_{1i}) - \sum_{j=1}^{J} R(j)$. Also, let $\sum^{(1)}$ and $\sum^{(2)}$ respectively denote summation over even and odd $j \in \{1,\ldots,J+1\}$. Note that for any $1 \leq j, j+k \leq J+1$, $k \geq 2$, the random variables $R(j)$ and $R(j+k)$ depend on disjoint sets of X_i's that are separated by $(k-1)2\ell - \ell$ observations in between. Hence, noting that $|R(j)| \leq 2\ell\|f\|_\infty$ for all $1 \leq j \leq J+1$, by Proposition 3.1, we get $|\mathrm{Cov}(R(j), R(j+k))| \leq 4\alpha((k-1)2\ell-\ell)(4\ell\|f\|_\infty)^2$ for all $k \geq 2$, $j \geq 1$. Therefore, using the inequalities $(a+b)^2 \leq 2(a^2+b^2)$ and $\mathrm{Var}(Z_1+\cdots+Z_m) \leq \sum_{k=1}^{m} EZ_j^2 + m\sum_{k=1}^{m-1}\max\{|\mathrm{Cov}(Z_i,Z_{i+k})| : 1 \leq i, i+k \leq m\}$ for any set $\{Z_1,\ldots,Z_m\}$ of bounded random variable, we have

$$\begin{aligned}
&\mathrm{Var}\Big(\sum_{i=1}^{N} w_{in}f(U_{1i})\Big)\\
&= \mathrm{Var}\big(\textstyle\sum^{(1)} R(j) + \sum^{(2)} R(j)\big)\\
&\leq 2\Big[\mathrm{Var}(\textstyle\sum^{(1)} R(j)) + \mathrm{Var}(\sum^{(2)} R(j))\Big]
\end{aligned}$$

$$\begin{aligned}
&\leq 2\Big[\sum_{i=1}^{J+1} ER(j)^2 + (J+1)\cdot \sum_{1\leq k\leq J/2} \alpha((2k-1)2\ell-\ell)\cdot 4(4\ell\|f\|_\infty)^2\Big] \\
&\leq 2\Big[(\frac{n}{2\ell}+1)4\ell^2\|f\|_\infty^2 + 64(n\ell)\|f\|_\infty^2 \sum_{k=1}^{J}\alpha(k\ell)\Big] \\
&\leq C_1\|f\|_\infty^2 \cdot \Big[n\ell + (n\ell)\sum_{k=1}^{J}\alpha(k\ell)\Big] .
\end{aligned}$$

This yields the first term in the upper bound in part (a). The second term in the bound is obtained similarly by using the inequalities $|\text{Cov}(R(j), R(j+k))| \leq C(\delta)\big(ER(j)^{2+\delta}\big)^{1/(2+\delta)}\big(ER(j+k)^{2+\delta}\big)^{1/(2+\delta)}\alpha((k-1)2\ell-\ell)^{\delta/(2+\delta)}$ and $\big(ER(j)^{2+\delta}\big)^{1/(2+\delta)} \leq 2\ell\zeta_{2+\delta,n}$, and retracing the steps above.

For proving part (b), splitting the sum over odd and even indices, we get

$$\begin{aligned}
&\text{Var}\Big(\sum_{i=1}^{b} w_{in} f(U_{1i}^{(2)})\Big) \\
&\quad\leq 2\Big[\text{Var}\big(\textstyle\sum_{\text{even}\, i} w_{in} f(U_{1i}^{(2)})\big) + \text{Var}\big(\textstyle\sum_{\text{odd}\, i} w_{in} f(U_{1i}^{(2)})\big)\Big] \\
&\quad\leq 2\Big[\textstyle\sum_i Ef(U_{1i}^{(2)})^2 + b\cdot \sum_{1\leq k\leq b/2}\alpha((2k-1)2\ell-\ell)(2\|f\|_\infty)^2\Big] \\
&\quad\leq 2\Big[b(\|f\|_\infty)^2 + 4b\|f\|_\infty^2 \textstyle\sum_{1\leq k\leq b}\alpha(k\ell)\Big] .
\end{aligned}$$

The other term in the bound may be obtained as in part (a). Hence, the proof of the lemma is complete. □

Next we prove the theorem.

Proof of Theorem 3.1: Without loss of generality, let $\mu = 0$. We prove the theorem for $j = 1$, i.e., for the MBB estimator first. Let $U_i^* = (X_{(\ell-1)i+1}^* + \cdots + X_{\ell i}^*)/\ell$ denote the average of the ith resampled block under the MBB, $1 \leq i \leq b$. Then, from (2.10), it follows that conditional on $\mathcal{X}_n$, $U_1^*, \ldots, U_b^*$ are iid and

$$P_*(U_1^* = U_i) = 1/N\ , \quad 1 \leq i \leq N,$$

where, recall that, $N = n-\ell+1$ and $U_i = (X_i + \cdots + X_{i+\ell-1})/\ell$, $i \geq 1$. Also, note that $\bar{X}_n^{*(1)} = b^{-1}\sum_{i=1}^{b} U_i^*$. Hence, by the conditional independence of $U_1^*, \ldots, U_b^*$,

$$\begin{aligned}
\text{Var}_*(T_n^*) &= n_1 \text{Var}_*\Big(\frac{1}{b}\sum_{i=1}^{b} U_i^*\Big) \\
&= n_1 b^{-1}\text{Var}_*(U_1^*)
\end{aligned}$$

$$= \ell \Big[N^{-1} \sum_{i=1}^{N} U_i U_i' - \hat{\mu}_n \hat{\mu}_n' \Big], \tag{3.12}$$

where $\hat{\mu}_n \equiv E_* \bar{X}_n^{*(1)} = E_* U_1^* = N^{-1} \sum_{i=1}^{N} U_i$. Next note that by Proposition 3.1, in the $d = 1$ case,

$$\begin{aligned} E\Big(\sum_{i=1}^{\ell} w_{in} X_i \Big)^2 &\leq \sum_{i=1-\ell}^{\ell-1} (\ell - |i|) |E X_1 X_{1+i}| \\ &\leq \ell \Big[E X_1^2 + 16 \sum_{i=1}^{\ell-1} \alpha(i)^{\delta/(2+\delta)} \Big(E|X_1|^{2+\delta} \Big)^{2/(2+\delta)} \Big] \\ &= O(\ell) \quad \text{as} \quad n \to \infty \end{aligned}$$

for any constants $w_{1n}, \ldots, w_{\ell n} \in [-1, 1]$. For $d > 1$, using this bound component-wise and using the stationarity of the X_i's, from (2.14), we get

$$\begin{aligned} nE\|\hat{\mu}_n - \bar{X}_n\|^2 &\leq nE \Big\{ \Big| 1 - \frac{n}{N} \Big| \, \|\bar{X}_n\| \\ &\quad + (N\ell)^{-1} \Big\| \sum_{i=1}^{\ell} (\ell - i)(X_i + X_{n-i+1}) \Big\| \Big\}^2 \\ &\leq 2 \Big\{ (\ell/N)^2 nE\|\bar{X}_n\|^2 + 2nN^{-2} \Big[E \Big\| \sum_{i=1}^{\ell} (i/\ell) X_i \Big\|^2 \\ &\quad + E \Big\| \sum_{i=1}^{\ell} (i/\ell) X_{\ell-i} \Big\|^2 \Big] \Big\} \\ &= O([\ell/n]^2) + O([\ell/n]) \\ &= O(\ell/n) \quad \text{as} \quad n \to \infty , \end{aligned}$$

as $E\|\sqrt{n}\bar{X}_n\|^2 = O(1)$. This implies

$$\begin{aligned} E\{\ell \|\hat{\mu}_n \hat{\mu}_n'\|\} &\leq C(d) \cdot \ell E \|\hat{\mu}_n\|^2 \\ &\leq C(d) \cdot \ell \cdot \{2E\|\bar{X}\|^2 + 2E\|\bar{X}_n - \hat{\mu}_n\|^2\} \\ &= O(\ell/n) . \end{aligned} \tag{3.13}$$

Hence, by (3.12) and (3.13), it remains to show that

$$\ell N^{-1} \sum_{i=1}^{N} U_i U_i' \longrightarrow_p \Sigma_\infty .$$

Let $V_{in} = U_{1i} U_{1i}' \, \mathbb{1}(\|U_{1i}\| < (n/\ell)^{1/8})$ and $W_{in} = U_{1i} U_{1i}' - V_{in}$, $1 \leq i \leq N$. Then, applying Lemma 3.1 component-wise, for large n we get,

$$E \Big\| N^{-1} \sum_{i=1}^{N} \Big(V_{in} - EV_{in} \Big) \Big\|^2$$

$$
\begin{aligned}
&\leq C(d)(n/\ell)^{1/2}\Big[n\ell + n\ell \textstyle\sum_{1\leq k<n/\ell} \alpha(k\ell)\Big]/N^2 \\
&\leq C(d)(n/\ell)^{-1/2}\Big[1 + \textstyle\sum_{k\geq 1} \alpha(k)\Big] \\
&= o(1) .
\end{aligned}
$$

Next, note that by definition, $U_{11} = \sqrt{\ell}\bar{X}_\ell$, and that under the conditions of Theorem 3.1 (cf. Appendix A), $\sqrt{n}\bar{X}_n \longrightarrow^d N(0, \Sigma_\infty)$. Hence, by the (extended) dominated convergence theorem,

$$
\lim_{n\to\infty} E\|U_{11}\|^2 \mathbb{1}\Big(\|U_{11}\| > (n/\ell)^{1/8}\Big) = 0 . \tag{3.14}
$$

Therefore, $\|EV_{in} - \Sigma_\infty\| \leq C(d)E\|U_{11}\|^2\mathbb{1}(\|U_{11}\|^8 > n/\ell) + \|EU_{11}U_{11}' - \Sigma_\infty\| = o(1)$. Hence, for any $\epsilon > 0$, by Markov's inequality,

$$
\begin{aligned}
&\lim_{n\to\infty} P\Big(\Big\|\ell N^{-1}\sum_{i=1}^{N} U_i U_i' - \Sigma_\infty\Big\| > 3\epsilon\Big) \\
\leq\; &\lim_{n\to\infty} P\Big(\Big\|N^{-1}\sum_{i=1}^{N}(V_{in} - EV_{in})\Big\| + \|EV_{in} - \Sigma_\infty\| \\
&\quad + \Big\|N^{-1}\sum_{i=1}^{N} W_{in}\Big\| > 3\epsilon\Big) \\
\leq\; &\lim_{n\to\infty} P\Big(\Big\|N^{-1}\sum_{i=1}^{N}(V_{in} - EV_{in})\Big\| > \epsilon\Big) \\
&\quad + \lim_{n\to\infty} P\Big(\Big\|N^{-1}\sum_{i=1}^{N} W_{in}\Big\| > \epsilon\Big) \\
\leq\; &\lim_{n\to\infty} \epsilon^{-2} E\Big\|N^{-1}\sum_{i=1}^{N}(V_{in} - EV_{in})\Big\|^2 + \lim_{n\to\infty} \epsilon^{-1}E\|W_{11}\| \\
\leq\; &0 + \lim_{n\to\infty} \epsilon^{-1}C(d)E\|U_{11}\|^2\mathbb{1}\Big(\|U_{11}\| > (n/\ell)^{1/8}\Big) \\
=\; &0 .
\end{aligned} \tag{3.15}
$$

This proves Theorem 3.1 for the MBB. Next, consider the NBB. Write $U_i^{*(2)}$ for the ith resampled block average under the NBB. Then, by (2.15),

$$
\begin{aligned}
\mathrm{Var}_*(T_n^{*(2)}) &= n_1 b^{-1} \mathrm{Var}_*(U_1^{*(2)}) \\
&= \ell\Big[b^{-1}\sum_{i=1}^{b} U_{(i-1)\ell+1}U'_{(i-1)\ell+1} - \bar{X}_{n_1}\bar{X}'_{n_1}\Big] .
\end{aligned} \tag{3.16}
$$

Since $\sqrt{n}\bar{X}_n \longrightarrow^d N(0, \Sigma_\infty)$, it follows that $\ell\|\bar{X}_{n_1}\bar{X}'_{n_1}\| = O_p(\ell/n)$. Now, using Lemma 3.1(b) and (3.14), and retracing the steps in (3.15), we get

$\text{Var}_*(T_n^{*(2)}) \longrightarrow_p \Sigma_\infty$ as $n \to \infty$. Finally, for the CBB, note that $E_*\bar{X}_n^{*(3)} = \bar{X}_n$ for any block length $\ell \in [1, n]$. Hence,

$$\begin{aligned}\text{Var}_*(T_n^{*(3)}) &= n_1 b^{-1}\Big[\text{Var}_*(U_1^{*(3)})\Big] \\ &= \ell\Big[n^{-1}\sum_{i=1}^{n} U_i^{(3)} U_i^{(3)'} - \bar{X}_n\bar{X}_n'\Big] \qquad (3.17)\end{aligned}$$

where, recall that, $U_i^{(3)} \equiv (Y_{n,i} + \cdots + Y_{n,(i+\ell-1)})/\ell$. Noting that for $1 \le i \le N$, $U_i^{(3)} = U_i$ and that under the conditions of Theorem 3.1, $E\|X_1 + \cdots + X_m\|^2 \le C(d)m$ for all $m \ge 1$, we get

$$\begin{aligned}&E\Big\|\Big(\ell N^{-1}\sum_{i=1}^{N} U_i U_i'\Big) - \Big(\ell n^{-1}\sum_{i=1}^{n} U_i^{(3)} U_i^{(3)'}\Big)\Big\| \\ &\le E\Big[(N^{-1} - n^{-1})\Big\|\sum_{i=1}^{N} U_{1i}U_{1i}'\Big\| + n^{-1}\ell \sum_{i=N+1}^{n} \|U_i^{(3)}\|^2\Big] \\ &\le \ell(Nn)^{-1}C(d)\sum_{i=1}^{N} E\|U_{1i}\|^2 + 2n^{-1}\ell^{-1}\sum_{i=N+1}^{n}\Big\{E\|X_i + \cdots + X_n\|^2 \\ &\quad + E\|X_1 + \cdots + X_{\ell-(n-i+1)}\|^2\Big\} \\ &\le C(d)(n^{-1}\ell)E\|\sqrt{\ell}\bar{X}_\ell\|^2 \\ &\quad + 2n^{-1}\ell^{-1}\Big[\ell\Big\{2\max\{E\|X_1 + \cdots + X_i\|^2 : 1 \le i \le \ell\Big\}\Big] \\ &= O(n^{-1}\ell)\,. \qquad (3.18)\end{aligned}$$

Hence, by (3.12), (3.13), (3.15), (3.17) and (3.18), it follows that $\text{Var}_*(T_n^{*(3)}) - \text{Var}_*(T_n^*) = o_p(1)$, proving the theorem for the CBB.

3.2.2 Consistency of Distribution Function Estimators

In this section, we establish consistency of the bootstrap methods for distribution function estimation. Recall that G_n denotes the sampling distribution of T_n. Let $\hat{G}_n^{(j)}$, $j = 1, 2, 3$ denote the conditional distribution of $T_n^{*(j)}$ under the MBB, the NBB, and the CBB methods, respectively. We may think of $\{G_n\}_{n\ge 1}$ as a sequence of points in the space of all probability measures on $\mathbb{R}^d$. When G_n weakly converges to a limit distribution G_∞, say, the classical approach is to approximate G_n using the limit distribution G_∞. In contrast, the bootstrap methods attempt to approximate G_n by generating *random* probability measures $\hat{G}_n^{(j)}$'s that change with n. The next theorem establishes consistency of these random measures $\hat{G}_n^{(j)}$'s when T_n is asymptotically normal. For stating the result, recall the convention that for $x = (x_1, \ldots, x_d)'$, $y = (y_1, \ldots, y_d)' \in \mathbb{R}^d$, $x \leq y$ if $x_i \leq y_i$ for all $i = 1, \ldots, d$.

Theorem 3.2 *Suppose that there exists a $\delta > 0$ such that $E\|X_1\|^{2+\delta} < \infty$ and $\sum_{n=1}^{\infty} \alpha(n)^{\delta/(2+\delta)} < \infty$. Also, suppose that $\Sigma_\infty = \sum_{i \in \mathbb{Z}} \text{Cov}(X_1, X_{1+i})$ is nonsingular and that $\ell^{-1} + n^{-1}\ell = o(1)$ as $n \to \infty$. Then, for $j = 1, 2, 3$,*

$$\sup_{x \in \mathbb{R}^d} \left| P_*(T_n^{*(j)} \leq x) - P(T_n \leq x) \right| \longrightarrow_p 0 \quad as \quad n \to \infty \ .$$

Theorem 3.2 shows that like the bootstrap variance estimators, the distribution function estimators $\hat{G}_n^{(j)}$'s are consistent estimators of G_n for a wide range of values of the block length parameter ℓ. Indeed, the conditions on ℓ presented in both Theorem 3.1 and Theorem 3.2 are also *necessary* for consistency of these bootstrap estimators. If ℓ remains bounded, then the block bootstrap methods fail to capture the dependence structure of the original data sequence and converge to a *wrong* normal limit as in the example of Singh (1981) (cf. Section 2.3). On the other hand, if ℓ goes to infinity at a rate comparable to the sample size n (violating the condition $n^{-1}\ell = o(1)$ as $n \to \infty$), then there are not enough number of *distinct* blocks to recreate a representative image of the "infinite population". It can be shown (cf. Lahiri (2001)) that, in this case, the estimators $\hat{G}_n^{(j)}$ converge to certain *random* probability measures.

Consistency of the bootstrap estimators $\hat{G}_n^{(j)}$'s remains valid over a much larger class of sets than asserted in the statement of Theorem 3.2. Let ϱ be a metric on the set of all probability measures on $\mathbb{R}^d$, metricizing the topology of weak convergence of probability measures (see (A.2) of Appendix A or Parthasarathi (1967)). The proof actually shows that under the conditions of Theorem 3.2,

$$\varrho\Big(\mathcal{L}(T_n^{*(j)} | \mathcal{X}_n), N(0, \Sigma_\infty)\Big) \longrightarrow_p 0 \quad \text{as} \quad n \to \infty$$

for $j = 1, 2, 3$, where, recall that $\mathcal{L}(T_n^{*(j)} | \mathcal{X}_n)$ denotes the conditional distribution of $T_n^{*(j)}$ given $\mathcal{X}_n$. Since $N(0, \Sigma_\infty)$ is an absolutely continuous distribution on $\mathbb{R}^d$, a result of Rao (1962) on uniformity classes (see Section 1.2 of Bhattacharya and Rao (1986)) implies that the convergence of $P_*(T_n^{*(j)} \in \cdot)$ and of $P(T_n \in \cdot)$ to $\Phi(\cdot; \Sigma_\infty)$ is uniform over the collection $\mathcal{C}$ of all Borel-measurable convex subsets of $\mathbb{R}^d$. Hence, it follows that under the conditions of Theorem 3.2,

$$\sup_{B \in \mathcal{C}} \left| P_*(T_n^{*(j)} \in B) - P(T_n \in B) \right| \longrightarrow_p 0 \quad \text{as} \quad n \to \infty$$

for $j = 1, 2, 3$.

Consistency of the block bootstrap estimators of $\mathcal{L}(T_n)$ continues to hold when a resample size of an order different from the sample size is chosen (cf.

Lahiri (2001)). Furthermore, under some additional conditions, the convergence of bootstrap estimators of the variance and the sampling distribution of T_n may be strengthened to almost sure convergence. For such extensions, see Künsch (1989), Peligrad and Shao (1995), Radulović (1996), and the references therein.

The rest of this section is devoted to the proof of Theorem 3.2. Recall the $\Phi(\cdot;\Sigma_\infty)$ denotes the $N(0,\Sigma_\infty)$ probability measure on the Borel σ-field $\mathcal{B}(\mathbb{R}^d)$ on $\mathbb{R}^d$. For simplicity of notation, we also write $\Phi(x;\Sigma_\infty)$, $x \in \mathbb{R}^d$ for the distribution function of $N(0,\Sigma_\infty)$ on $\mathbb{R}^d$.

Proof: We begin with the case $j = 1$. Since T_n converges in distribution to $N(0,\Sigma_\infty)$ and $N(0,\Sigma_\infty)$ is a continuous distribution, by a multivariate version of Polyā's Theorem (cf. Section 1.2, Bhattacharya and Rao (1986)),

$$\sup_{x\in\mathbb{R}^d}\Big|P(T_n \le x) - \Phi(x;\Sigma_\infty)\Big| \to 0 \quad \text{as} \quad n \to \infty\,.$$

Hence, it is enough to show that

$$\sup_{x\in\mathbb{R}^d}\Big|P_*(T_n^{*(1)} \le x) - \Phi(x;\Sigma_\infty)\Big| \longrightarrow_p 0 \quad \text{as} \quad n \to \infty\,. \tag{3.19}$$

Let $\hat{\Delta}_n(a) = \ell b^{-1}\sum_{i=1}^{b} E_*\|U_i^* - \hat{\mu}_n\|^2 \mathbb{1}\Big(\sqrt{\ell}\|U_i^* - \hat{\mu}_n\| > 2a\Big)$, $a > 0$. Note that conditional on $\mathcal{X}_n$, $U_1^*,\ldots,U_b^*$ are iid random vectors, and that for any two random variables X and Y and any $\eta > 0$,

$$\begin{aligned}
&E|X+Y|^2\mathbb{1}(|X+Y|>\eta)\\
&\le\ 4E(|X|^2 \vee |Y|^2)\mathbb{1}(2|X|\vee|Y| > \eta)\\
&\le\ 4\Big[E|X|^2\mathbb{1}(|X|>\eta/2) + E|Y|^2\mathbb{1}(|Y|>\eta/2)\Big]\,,
\end{aligned}$$

where recall that $x \vee y = \max\{x,y\}$, $x,y \in \mathbb{R}$. Hence, by (3.13) and (3.14) and the inequality above, for any $\epsilon > 0$,

$$\begin{aligned}
&P\Big(\hat{\Delta}_n\big((n/\ell)^{1/4}\big) > \epsilon\Big) \le \epsilon^{-1}E\hat{\Delta}_n\Big((n/\ell)^{1/4}\Big)\\
&=\ \epsilon^{-1}E\Big\{\ell E_*\|U_1^* - \hat{\mu}\|^2\mathbb{1}\Big(\sqrt{\ell}\|U_1^* - \hat{\mu}_n\| > 2(n/\ell)^{1/4}\Big)\Big\}\\
&=\ \epsilon^{-1}E\|U_{11} - \sqrt{\ell}\hat{\mu}_n\|^2\mathbb{1}\Big(\|U_{11} - \sqrt{\ell}\hat{\mu}_n\| > 2(n/\ell)^{1/4}\Big)\\
&\le\ 4\epsilon^{-1}\Big[E\|U_{11}\|^2\mathbb{1}\Big(\|U_{11}\| > (n/\ell)^{1/4}\Big) + \ell E\|\hat{\mu}_n\|^2\Big]\\
&\qquad \to 0 \quad \text{as} \quad n \to \infty\,.
\end{aligned} \tag{3.20}$$

Thus,

$$\hat{\Delta}_n\Big((n/\ell)^{1/4}\Big) \longrightarrow_p 0 \quad \text{as} \quad n \to \infty\,. \tag{3.21}$$

Next, note that (3.19) would follow if for any subsequence $\{n_i\}$, there is a further subsequence $\{n_k\} \subset \{n_i\}$ such that

$$\lim_{k\to\infty} \sup_{x\in\mathbb{R}^d} \Big| P_*(T_{n_k}^{*(1)} \le x) - \Phi(x;\Sigma_\infty) \Big| = 0 \quad \text{a.s.} \tag{3.22}$$

Fix a subsequence $\{n_i\}$. Then, by (3.21) and Theorem 3.1, there exists a subsequence $\{n_k\}$ of $\{n_i\}$ such that as $k \to \infty$

$$\mathrm{Var}_*(T_{n_k}^{*(1)}) \to \Sigma_\infty \quad \text{a.s. and} \quad \hat{\Delta}_{n_k}\Big((n_k/\ell_{n_k})^{1/4} \Big) \to 0 \quad \text{a.s.} \tag{3.23}$$

Note that $T_n^{*(1)} = \sum_{i=1}^b (U_i^* - \hat{\mu}_n)\sqrt{\ell/b}$ is a sum of conditionally iid random vectors $(U_1^* - \hat{\mu}_n)\sqrt{\ell/b}, \ldots, (U_b^* - \hat{\mu}_n)\sqrt{\ell/b}$, which, by (3.23), satisfy Lindeberg's condition along the subsequence n_k, almost surely. Hence, by the CLT for independent random vectors (cf. Theorem A.5, Appendix A), the conditional distribution $\mathcal{L}(T_{n_k}^{*(1)} | \mathcal{X}_{n_k})$ of $T_{n_k}^{*(1)}$ converges to $N(0, \Sigma_\infty)$ as $k \to \infty$, almost surely. Hence, by a multivariate version of Polyā's Theorem, (3.22) follows. This proves Theorem 3.2 for the case $j = 1$. The proof is similar for $j = 2, 3$. The reader is invited to supply the details. □

3.3 Consistency of the SB: Sample Mean

In this section, we consider consistency of the SB method for estimating the variance and the distribution function of the sample mean. As before, let $\{X_i\}_{i\in\mathbb{Z}}$ be a sequence of stationary $\mathbb{R}^d$-valued random vectors with mean μ. Also, for $n \ge 1$, let $\{Y_{n,i}\}_{i\ge1}$ denote the periodically extended time series, defined by $Y_{n,i} = X_j$ if $i = j$ (modulo) n (cf. Section 2.7). First we consider the SB estimator of the asymptotic covariance matrix Σ_∞ of the sample mean. From Section 2.7, note that the SB resamples blocks of random lengths to generate the bootstrap sample. Let $\bar{X}_n^{*(4)}$ denote the mean of the first n bootstrap values under the SB. As noted in Section 2.7, $E_*\bar{X}_n^{*(4)} = \bar{X}_n$ and, hence, the SB version of $T_n = \sqrt{n}(\bar{X}_n - \mu)$ is given by $T_n^{*(4)} = \sqrt{n}(\bar{X}_n^{*(4)} - \bar{X}_n)$.

3.3.1 Consistency of SB Variance Estimators

For the centered and scaled sample mean T_n, the SB variance estimator admits a closed form expression and hence, it can be calculated without recourse to any resampling of the data. We note this in the following proposition.

Proposition 3.2 *Let* $\hat{\Gamma}_n(k) = n^{-1}\sum_{i=1}^{n-k} X_i X_{i+k}' - \bar{X}_n \bar{X}_n'$, $0 \le k < n$, $q = 1 - p$, $q_{n0} = 1/2$, *and* $q_{nk} = (1 - n^{-1}k)q^k + (n^{-1}k)q^{n-k}$, $1 \le k < n$. *If*

$0 < p < 1$, *then*

$$Var_*(T_n^{*(4)}) = \sum_{k=0}^{n-1} q_{nk}\left(\hat{\Gamma}_n(k) + \hat{\Gamma}_n(k)'\right) .$$

Proof: Note that conditional on $\mathcal{X}_n$, L_1 has the Geometric distribution with parameter p. Also, under the SB resampling scheme, X_1^* and X_{1+k}^*, $k \geq 1$, lie in the same resampled block if and only if $1 + k \leq L_1$. Hence, writing $\mathcal{T}_n = \sigma\langle \mathcal{X}_n,\ L_1, \ldots, L_n\rangle$, the σ-field generated by $\mathcal{X}_n$ and $L_1, \ldots, L_n$, for any $1 \leq k \leq n-1$, we get

$$\begin{aligned}
E_* X_1^* X_{1+k}^{*'} &= E\Big(X_1^* X_{1+k}^{*'} \mid \mathcal{X}_n\Big) = E\Big\{E\Big(X_1^* X_{1+k}^{*'} \mid \mathcal{T}_n\Big) \mid \mathcal{X}_n\Big\} \\
&= E\bigg\{\bigg(n^{-1}\sum_{i=1}^{n} Y_{n,i} Y_{n,i+k}'\bigg) \cdot \mathbb{1}(L_1 \geq 1+k) \mid \mathcal{X}_n\bigg\} \\
&\quad + E\bigg\{\bigg(n^{-1}\sum_{i=1}^{n} Y_{n,i}\bigg)\bigg(n^{-1}\sum_{i=1}^{n} Y_{n,i}'\bigg)\mathbb{1}(L_1 \leq k) \mid \mathcal{X}_n\bigg\} \\
&= \bigg(n^{-1}\sum_{i=1}^{n} Y_{n,i} Y_{n,i+k}'\bigg) P(L_1 > k \mid \mathcal{X}_n) \\
&\quad + \bar{X}_n \bar{X}_n' \, P(L_1 \leq k \mid \mathcal{X}_n) \\
&= n^{-1}\bigg\{\sum_{i=1}^{n-k} X_i X_{i+k}' + \sum_{i=n-k+1}^{n} X_i X_{i+k-n}'\bigg\} q^k \\
&\quad + \bar{X}_n \bar{X}_n' (1 - q^k) \\
&= \Big\{\hat{\Gamma}_n(k) + \hat{\Gamma}_n(n-k)'\Big\} q^k + \bar{X}_n \bar{X}_n' . \qquad (3.24)
\end{aligned}$$

Next, noting that the bootstrap samples under the SB form a stationary sequence, we have

$$\begin{aligned}
\mathrm{Var}_*(T_n^{*(4)}) &= \bigg[\Big\{E_* X_1^* X_1^{*'} - \bar{X}_n \bar{X}_n'\Big\} + \\
&\quad \sum_{k=1}^{n-1}(1 - n^{-1}k)\Big\{E_* X_1^* X_{1+k}^{*'} + E_* X_1^{*'} X_{1+k}^{*} - 2\bar{X}_n \bar{X}_n'\Big\}\bigg]. \qquad (3.25)
\end{aligned}$$

Hence, the proposition follows from (3.24) and (3.25). □

Next we prove consistency of the SB variance estimator. For this, we assume a stronger set of conditions than those in Theorem 3.1.

Theorem 3.3 *Assume that* $E\|X_1\|^{4+\delta} < \infty$ *and* $\sum_{n=1}^{\infty} n^3 \alpha(n)^{\delta/(4+\delta)} < \infty$ *for some* $\delta > 0$. *Also, assume that* $p + (n^{1/2}p)^{-1} \to 0$ *as* $n \to \infty$. *Then,*

$$Var_*(T_n^{*(4)}) \longrightarrow_p \Sigma_\infty \quad as \quad n \to \infty .$$

For proving the theorem, we need two auxiliary results. The first one is a standard bound on the cumulants of strongly mixing random vectors. For later reference, we state it in a slightly more general form than the set up of Theorem 3.3, allowing nonstationarity of the random vectors $\{X_i\}_{i\in\mathbb{Z}}$. For any random variables $Z_1, \ldots, Z_r$, $(r \geq 1)$, we define the rth-order cumulant $\mathcal{K}_r(Z_1, \ldots, Z_r)$ by

$$\begin{aligned}
&\mathcal{K}_r(Z_1, \ldots, Z_r) \qquad (3.26)\\
&= (\iota)^r \frac{\partial^r}{\partial t_1 \cdots \partial t_r} (\log E \exp(\iota[t_1 Z_1 + \cdots + t_r Z_r]))\Big|_{t_1=\cdots=t_r=0},
\end{aligned}$$

where $\iota = \sqrt{-1}$. Also, for a random vector $W = (W_1, \ldots, W_d)'$ and an integer-vector $\nu = (\nu_1, \ldots, \nu_d)' \in \mathbb{Z}_+^d$, we set $\mathcal{K}_\nu(W) = \mathcal{K}_{|\nu|}(W_1, \ldots, W_1; \cdots; W_d, \ldots, W_d)$, where the jth component W_j of W is repeated ν_j times, $1 \leq j \leq d$. We may express the cumulant $\mathcal{K}_r(Z_1, \ldots, Z_r)$ in terms of the moments of the Z_i's by the formula

$$\mathcal{K}_r(Z_1, \ldots, Z_r) = \sum_{j=1}^{r} \sum^{(*j)} c(I_1, \ldots, I_j) \prod_{k=1}^{j} E \prod_{i \in I_k} Z_i , \qquad (3.27)$$

where $\sum^{(*j)}$ extends over all partitions $\{I_1, \ldots, I_j\}$ of $\{1, \ldots, r\}$ and where $c(I_1, \ldots, I_j)$'s are combinatorial coefficients (cf. Zhurbenko (1972)). It is easy to check that cumulants are multilinear forms, i.e.,

$$\begin{aligned}
&\mathcal{K}_r(Z_1, \ldots, Z_{1i} + Z_{2i}, \ldots, Z_r)\\
&\quad = \mathcal{K}_r(Z_1, \ldots, Z_{1i}, \ldots, Z_r) + \mathcal{K}_r(Z_1, \ldots, Z_{2i}, \ldots, Z_r)
\end{aligned}$$

for every $1 \leq i \leq r$.

Note that if $\{Z_1, \ldots, Z_r\}$ and $\{W_1, \ldots, W_s\}$ are independent, then

$$\begin{aligned}
&\mathcal{K}_{r+s}(Z_1, \ldots, Z_r;\ W_1, \ldots, W_s)\\
&= (\iota)^{r+s} \frac{\partial^{r+s}}{\partial t_1 \ldots \partial t_{r+s}} \Big[\log E \exp(\iota[t_1 Z_1 + \cdots + t_r Z_r])\\
&\qquad + \log E \exp(\iota[t_{r+1} W_1 + \cdots + t_{r+s} W_s])\Big]\Big|_{t_1=\cdots=t_{r+s}=0}\\
&= 0 \qquad (3.28)
\end{aligned}$$

for any $r \geq 1$, $s \geq 1$. This identity plays an important role in the proof of the lemma below.

Lemma 3.2 *Let $\{X_i\}_{i\in\mathbb{Z}}$ be a sequence of (possibly nonstationary) $\mathbb{R}^d$-valued random vectors with* $\sup\Big\{(E\|X_i\|^{2r+\delta})^{\frac{1}{2r+\delta}} : i \in \mathbb{Z}\Big\} \equiv \zeta_{2r+\delta} < \infty$ *and* $\Delta(r, \delta) \equiv 1 + \sum_{i=1}^{\infty} i^{2r-1}[\alpha(i)]^{\delta/(2r+\delta)} < \infty$ *for some integer* $r \geq 1$ *and*

$\delta > 0$. Also, let $a_1, \ldots, a_m$ be any m unit vectors in $\mathbb{R}^d$ (i.e., $\|a_i\| = 1$, $1 \leq i \leq m$) for some $2 \leq m \leq 2r$. Then,

$$\left|\mathcal{K}_m(a_1' S_n, \ldots, a_m' S_n)\right| \leq C(d, r)\Delta(r; \delta)\zeta_{2r+\delta}^m \cdot n$$

for all $n \geq 1$, where $S_n = X_1 + \cdots + X_n$ and where $C(d, r)$ is a constant that depends only on d and r but not on n. Furthermore, for any $\nu \in \mathbb{Z}_+^d$ with $|\nu| \leq 2r$,

$$E|S_n^\nu| \leq C(d, r)\Delta(r; d)\zeta_{2r+\delta}^{|\nu|} n^{|\nu|/2}$$

for all $n \geq 1$.

Proof: Using the multilinearity property of cumulants, we get

$$\begin{aligned}
&\left|\mathcal{K}_m(a_1' S_n, \ldots, a_m' S_n)\right| \\
&\leq \sum_{1 \leq j_1, \ldots, j_m \leq n} \left|\mathcal{K}_m(a_1' X_{j_1}, \ldots, a_m' X_{j_m})\right| .
\end{aligned} \tag{3.29}$$

Next, for any set of indices $1 \leq j_1 \leq \cdots \leq j_m \leq n$, consider the maximal gap in the sequence $j_1, \ldots, j_m$. Suppose that $j_{k+1} - j_k = \max\{j_{i+1} - j_i : 1 \leq i < m\}$. Let $J_1 = \{j_1, \ldots, j_k\}$ and $J_2 = \{j_{k+1}, \ldots, j_m\}$. Also, let $\{\tilde{X}_i : i \in J_2\}$ be an independent copy of $\{X_i : i \in J_2\}$. Then, by (3.28), $\mathcal{K}_m(a_1' X_{j_1}, \ldots, a_k' X_{j_k}, a_{k+1}' \tilde{X}_{j_{k+1}}, \ldots, a_m' \tilde{X}_{j_m}) = 0$. Hence, by (3.27), the strong mixing condition and Proposition 3.1,

$$\begin{aligned}
&\left|\mathcal{K}_m(a_1' X_{j_1}, \ldots, a_m' X_{j_m})\right| \\
&= \Big|\mathcal{K}_m(a_1' X_{j_1}, \ldots, a_m' X_{j_m}) \\
&\quad - \mathcal{K}_m(a_1' X_{j_1}, \ldots, a_k' X_{j_k}, a_{k+1}' \tilde{X}_{j_{k+1}}, \ldots, a_m' \tilde{X}_{j_m})\Big| \\
&\leq \sum_{j=1}^{m} \sum^{(*j)} c(I_1, \ldots, I_j) \Big| \prod_{i=1}^{j} E \prod_{s \in I_i} a_s' X_{j_s} \\
&\quad - \prod_{i=1}^{j} \Big\{ \Big(E \prod_{s \in I_i \cap J_1} a_s' X_{j_s}\Big) \Big(E \prod_{s \in I_i \cap J_2} a_s' X_{j_s}\Big) \Big\} \Big| \\
&\leq C(m)\zeta_{m+\delta,n}^m \big[\alpha(j_{k+1} - j_k)\big]^{\delta/m+\delta}
\end{aligned}$$

where $\zeta_{s,n} \equiv \sup\Big\{(E\|X_j\|^s)^{1/s} : 1 \leq j \leq n\Big\}$ for all $s > 0$, $n \geq 1$, and $C(m)$ is a constant that depends only on m, not on n. Next, note that for any $0 \leq t \leq n-1$, there are at most $n \cdot t^{m-1}$ sets of indices $\{j_1, \ldots, j_m\} \subset \{1, \ldots, n\}$ that has maximal gap t. Hence,

$$\sum_{1 \leq j_1, \ldots, j_m \leq n} \left|\mathcal{K}_m(a_1' X_{j_1}, \ldots, a_m' X_{j_m})\right|$$

$$\begin{aligned} &\leq \sum_{t=0}^{n-1}(nt^{m-1})\cdot C(m)\cdot\zeta_{m+\delta,n}^{m}\big[\alpha(t)\big]^{\delta/m+\delta} \\ &\leq C(m)\Delta(r,\delta)\zeta_{2r+\delta}^{m}\cdot n\,. \end{aligned} \tag{3.30}$$

This, together with (3.29), completes the proof of the first inequality.

The bound on $E|S_n^\nu|$ readily follows by using cumulant expansions for moments (cf. Section 6, Bhattacharya and Rao (1986)) and the first inequality. Hence, the lemma is proved. □

The next lemma gives an expression for the covariance of the "uncentered" sample cross-covariance estimators

$$\hat{\sigma}_n(j;\alpha,\beta)\equiv n^{-1}\sum_{i=1}^{n-j}X_i^\alpha X_{i+j}^\beta\,,$$

where $0\leq j\leq n-1$ and $\alpha,\beta\in\mathbb{Z}_+^d,\ |\alpha|=|\beta|=1$.

Lemma 3.3 *Suppose that the conditions of Theorem 3.3 hold and that $EX_1=0$. Then, for any $\alpha,\beta,\gamma,\nu\in\mathbb{Z}_+^d$ with $|\alpha|=|\beta|=|\gamma|=|\nu|=1$, and any $0\leq j,\ k\leq n-1$,*

$$\begin{aligned} &Cov\Big(\hat{\sigma}(j;\alpha,\beta),\hat{\sigma}(k;\gamma,\nu)\Big) \\ &= n^{-1}\sum_{m=-(n-j)+1}^{(n-j-v-1)}\Big\{1-n^{-1}(\eta_{jv}(m)+k)\Big\} \\ &\quad\cdot\Big\{(EX_1^\alpha X_{1+m}^\gamma)(EX_1^\beta X_{1+m+v}^\nu)+(EX_1^\alpha X_{m+k+1}^\nu)(EX_1^\beta X_{1+m-j}^\gamma)\Big\} \\ &\quad+R_n(j,k;\alpha,\beta,\gamma,\nu) \end{aligned}$$

where $v=k-j$, $\eta_{jv}(m)=m\cdot\mathbb{1}(m>0)-(m+v)\mathbb{1}(-(n-j)+1\leq m<-v)$, and the remainder terms $R_n(j,k;\alpha,\beta,\gamma,\nu)$'s satisfy the inequality

$$\begin{aligned} &\sum_{j=0}^{n-1}\sum_{k=0}^{n-1}q_{nj}q_{nk}|R_n(j,k;\alpha,\beta,\gamma,\nu)| \\ &\quad\leq Cn^{-2}\sum_{j=0}^{n-1}\sum_{k=0}^{n-1}\sum_{s=1}^{n-j}\sum_{t=1}^{n-k} \\ &\qquad\cdot\sup\Big\{\big|\mathcal{K}_4(z_1'X_0,z_2'X_{t-s},z_3'X_j,z_4'X_{k-j})\big|:\|z_i\|\leq 1,\ i=1,2,3,4\Big\}\,. \end{aligned} \tag{3.31}$$

Proof: This is a variant of Bartlett's (1946) formula for the covariance of sample autocovariance estimators. See, for example, Section 5.3 of Priestley (1981) for a derivation in the one-dimensional case. Inequality

(3.31) is obtained from Bartlett's (1946) bound on $R_n(j,k;\alpha,\beta,\gamma,\nu)$'s upon using the fact that $q_{nj} \le 1$ for all $0 \le j < n$. □

Proof of Theorem 3.3: Let $\tilde{\Gamma}_n(k) = n^{-1}\sum_{i=1}^{n-k} X_i X'_{i+k}$, $0 \le k < n$ and $\tilde{\Sigma}_n = \tilde{\Gamma}_n(0) + \sum_{k=1}^{n-1} q_{nk}\big(\tilde{\Gamma}_n(k) + \tilde{\Gamma}_n(k)'\big)$. Then, by Proposition 3.2,

$$\mathrm{Var}_*(T_n^{*(4)}) = \tilde{\Sigma}_n - \bar{X}_n \bar{X}'_n \left(1 + 2\sum_{k=1}^{n-1} q_{nk}\right).$$

Since $\sum_{n=1}^{n-1} q_{nk} \le 2\sum_{k=0}^{\infty} q^k = 2p^{-1}$ and $p^{-1}\|\bar{X}_n\|^2 = O_p\big((np)^{-1}\big) = o_p(1)$, it is enough to show that

$$\tilde{\Sigma}_n \longrightarrow_p \Sigma_\infty \quad \text{as} \quad n \to \infty. \tag{3.32}$$

We prove this by showing that the bias and the variance of each element of the matrix $\tilde{\Sigma}_n$ go to zero as $n \to \infty$. For this, we label the elements of a $d \times d$ matrix A by the d-dimensional *unit* vectors $\alpha, \beta \in \mathbb{Z}_+^d$. For example, if $\alpha = (1,0,\ldots,0)'$ and $\beta = (0,1,0,\ldots,0)'$, then $A(\alpha,\beta)$ would denote the (1,2)-*th* element of the matrix A. With this notation, for any $\alpha, \beta \in \mathbb{Z}_+^d$ with $|\alpha| = |\beta| = 1$,

$$\begin{aligned}
&E\tilde{\Sigma}_n(\alpha,\beta) - \Sigma_\infty(\alpha,\beta) \\
&= E\left\{\hat{\sigma}_n(0;\alpha,\beta) + \sum_{k=1}^{n-1} q_{nk}\Big(\hat{\sigma}_n(k;\alpha,\beta) + \hat{\sigma}_n(k;\beta,\alpha)\Big)\right\} \\
&\quad - \left\{EX_1^{\alpha+\beta} + \sum_{k=1}^{\infty}\big(EX_1^\alpha X_{1+k}^\beta + EX_1^\beta X_{1+k}^\alpha\big)\right\} \\
&= \sum_{k=1}^{n-1}\Big\{q_{nk}(1-n^{-1}k) - 1\Big\}\Big(EX_1^\alpha X_{1+k}^\beta + EX_1^\beta X_{1+k}^\alpha\Big) \\
&\quad - \sum_{k=n}^{\infty}\Big(EX_1^\alpha X_{1+k}^\beta + EX_1^\beta X_{1+k}^\alpha\Big).
\end{aligned} \tag{3.33}$$

Note that $|q_{nk}(1-n^{-1}k)-1| = |\{q^k - n^{-1}k(q^k - q^{n-k} + q_{nk}) - 1| \le |1-q^k| + 3n^{-1}k \le kp + 3n^{-1}k$ for all $1 \le k^2 \le p^{-1}$. Since $\sum_{k=1}^{\infty}|k|\,\|EX_1X'_{1+k}\| < \infty$, from (3.33), we get

$$\begin{aligned}
&\Big|E\tilde{\Sigma}_n(\alpha,\beta) - \Sigma_\infty(\alpha,\beta)\Big| \\
&\le 2\sum_{1\le k^2 \le p^{-1}} (kp + 3n^{-1}k)\|EX_1X'_{1+k}\| + 8\sum_{k^2p>1}\|EX_1X'_{1+k}\| \\
&\to 0 \quad \text{as} \quad n \to \infty.
\end{aligned} \tag{3.34}$$

Hence, the bias part goes to zero. Next, we consider the variance part. Note that by Lemma 3.3,

$$\begin{aligned}&\Big|\mathrm{Cov}\Big(\hat{\sigma}_n(j;\alpha,\beta),\hat{\sigma}_n(k;\gamma,\nu)\Big)\Big|\\&\le\ 6n^{-1}\Big(\sum_{m=-\infty}^{\infty}\|EX_1X'_{1+m}\|\Big)E\|X_1\|^2\\&\quad+|R_n(j,k;\alpha,\beta,\gamma,\nu)|\end{aligned}\tag{3.35}$$

for all $|\alpha|=|\beta|=|\gamma|=|\nu|=1$ and for all $0\le j,k\le n-1$. Hence, using (3.31), (3.35) and the arguments in the proof of Lemma 3.2 (cf. (3.30)), for any $\alpha,\beta\in\mathbb{Z}^d$, $|\alpha|=|\beta|=1$, we get

$$\begin{aligned}&\mathrm{Var}\Big(\tilde{\Sigma}_n(\alpha,\beta)\Big)\\&=\ \mathrm{Var}\Big(\sum_{k=0}^{n-1}q_{nk}\Big(\hat{\sigma}_n(k;\alpha,\beta)+\hat{\sigma}_n(k;\beta,\alpha)\Big)\Big)\\&\le\ 6n^{-1}\sum_{j=0}^{n-1}\sum_{k=0}^{n-1}q_{nk}q_{nj}\Big(\sum_{m=-\infty}^{\infty}\|EX_1X'_{1+m}\|\Big)\cdot E\|X_1\|^2\\&\quad+\sum_{j=0}^{n-1}\sum_{k=0}^{n-1}q_{nj}q_{nk}\Big(\max_{\alpha_i\in\{\alpha,\beta\}}|R_n(j,k;\alpha_1,\alpha_2,\alpha_3,\alpha_4)|\Big)\\&\le\ Cn^{-1}\Big(\sum_{k=0}^{n-1}q_{nk}\Big)^2\\&\quad+C(d)n^{-2}\sum_{j=0}^{n-1}\sum_{k=0}^{n-1}\sum_{s=1}^{n-j}\sum_{t=1}^{n-k}\sup\Big\{\big|\mathcal{K}_4(z_1'X_0,z_2'X_{t-s},z_3'X_j,z_4'X_{k-j})\big|:\\&\qquad\qquad\|z_i\|\le1,\ i=1,2,3,4\Big\}\\&\le\ Cn^{-1}p^{-2}+C(d)\cdot n^{-2}\cdot n\Big(\sum_{i=1}^{\infty}i^3\alpha(i)^{\delta/(4+\delta)}\Big)\Big(E\|X_1\|^{4+\delta}\Big)^{4/(4+\delta)}\\&\quad\to0\quad\text{as}\quad n\to\infty\,.\end{aligned}\tag{3.36}$$

This completes the proof of Theorem 3.3. □

3.3.2 *Consistency of SB Distribution Function Estimators*

Next we prove consistency of the SB estimator for estimating the distribution function of the sample mean. For simplicity of exposition, we consider a variant of the bootstrapped statistic $T_n^{*(4)}$ where all of the SB samples from the last resampled block are retained. Let $K\equiv\inf\{i\ge1:L_1+\cdots+L_i\ge n\}$ denote the minimum number of blocks necessary to generate n SB samples, and let $N_1 = L_1 + \cdots + L_K$ denote the total number of SB observations in the first K resampled blocks. Define the SB version of the centered and scaled sample mean T_n based on a resample of size N_1 by

$$\tilde{T}_n^{*(4)} = N_1^{-1/2} \sum_{i=1}^{N_1} (X_i^{*(4)} - \bar{X}_n).$$

Thus, $\tilde{T}_n^{*(4)}$ differs from $T_n^{*(4)}$ only by the inclusion of the additional bootstrap observations, if any, in the last block that lie beyond the first n resampled values. It is easy to see that $N_1 - L_K \leq n \leq N_1$, so that the difference between n and N_1 is at most L_K. Since we assume that the expected value of the block lengths is negligible compared to n, the difference between these two versions can be shown to be negligible under the conditions of Theorem 3.4. Theorem 3.4 establishes consistency of the SB method for estimating the distribution function of T_n.

Theorem 3.4 *Suppose that* $E\|X_1\|^{6+\delta} < \infty$, Σ_∞ *is nonsingular, and* $\sum_{n=1}^{\infty} n^5 \alpha(n)^{\delta/(6+\delta)} < \infty$ *for some* $\delta > 0$. *Also, assume that* $p + (n^{1/2}p)^{-1} \to 0$ *as* $n \to \infty$. *Then,*

$$\sup_{x \in \mathbb{R}^d} \Big| P_*(\tilde{T}_n^{*(4)} \leq x) - P(T_n \leq x) \Big| \longrightarrow_p 0 \quad as \quad n \to \infty\,. \tag{3.37}$$

From the discussion following Theorem 3.2, it follows that (3.37) is equivalent to

$$\varrho\Big(\mathcal{L}(\tilde{T}_n^{*(4)} \mid \mathcal{X}_n),\ N(0, \Sigma_\infty)\Big) \longrightarrow_p 0 \quad as \quad n \to \infty\,,$$

where ϱ *is a metric metricizing weak convergence of probability measures on* $\mathbb{R}^d$.

Proof: We prove the result only for the one-dimensional case to keep the proof simple. Write σ_∞ for Σ_∞ under $d = 1$ and set $S(i;\ell) \equiv \sum_{j=i}^{i+\ell-1} Y_{n,j}$, $i \geq 1, \ell \geq 1$. Note that conditional on $\mathcal{T}_n \equiv \sigma\langle \mathcal{X}_n, L_1, \ldots, L_n \rangle$, $S(I_{4,1}; L_1), \ldots, S(I_{4,K}; L_K)$ are independent, but not necessarily identically distributed random variables with

$$P\Big(S(I_{4,j}; L_j) = S(i; L_j) \mid \mathcal{T}_n\Big) = \frac{1}{n} \quad \text{for} \quad i = 1, \ldots, n,\ 1 \leq j \leq n\,.$$

Hence, $E\big(S(I_{4,j}; L_j) \mid \mathcal{T}_n\big) = n^{-1} \sum_{i=1}^{n} S(i; L_j) = L_j \bar{X}_n$ for all $1 \leq j \leq n$. By the Berry-Esseen Theorem for independent random variables (cf. Theorem A.6, Appendix A),

$$\sup_{x \in \mathbb{R}} \Big| P_*(\tilde{T}_n^{*(4)} \leq x) - \Phi(x/\sigma_\infty) \Big|$$

$$
\begin{aligned}
&= \sup_{x\in\mathbb{R}} \Big| E\Big\{ P\Big(\frac{1}{\sqrt{N_1}} \sum_{j=1}^{K} (S(I_{4,j}; L_j) - L_j \bar{X}_n) \le x \mid \mathcal{T}_n \Big) \mid \mathcal{X}_n \Big\} \\
&\qquad\qquad - \Phi(x/\sigma_\infty) \Big| \\
&\le E\Big[\Big\{ \sup_{x\in\mathbb{R}} \Big| P\Big(\frac{1}{\sqrt{N_1}} \sum_{j=1}^{K} (S(I_{4,j}; L_j) - L_j \bar{X}_n) \le x \mid \mathcal{T}_n \Big) \\
&\qquad\qquad - \Phi(x/\hat{\sigma}_{n,p}) \Big| + \Big| \Phi(x/\hat{\sigma}_{n,p}) - \Phi(x/\sigma_\infty) \Big| \Big\} \mathbb{1}(A_n) \mid \mathcal{X}_n \Big] \\
&\quad + 2P(A_n^c \mid \mathcal{X}_n) \\
&\le C \cdot E\Big[\Big\{ \sum_{j=1}^{K} E\Big(\Big| S(I_{4,j}; L_j) - L_j \bar{X}_n \Big|^3 N_1^{-3/2} \mid \mathcal{T}_n \Big) \Big/ \hat{\sigma}_{n,p}^3 \Big\} \\
&\qquad\qquad\qquad\qquad \times \mathbb{1}(A_n) \mid \mathcal{X}_n \Big] \\
&\quad + E\Big\{ \sup_{x\in\mathbb{R}} \Big| \Phi(x/\hat{\sigma}_{n,p}) - \Phi(x/\sigma_\infty) \Big| \mathbb{1}(A_n) \mid \mathcal{X}_n \Big\} \\
&\quad + 2P(A_n^c \mid \mathcal{X}_n) ,
\end{aligned} \tag{3.38}
$$

where $\hat{\sigma}_{n,p}^2 = \sum_{j=1}^{K} \mathrm{Var}\big((S(I_{4j}; L_j) \mid \mathcal{T}_n\big)/N_1$, $A_n = \{|\hat{\sigma}_{n,p} - \sigma_\infty| \le a_n \sigma_\infty\}$, and $a_n = \max\{p^{1/2}, (n^{1/2}p)^{-1/2}\}$. Next, note that $N_1 \ge n$ and that K is a stopping time with respect to $\sigma\langle L_1, \ldots, L_j\rangle$, $1 \le j \le n$ (cf. Definition A.3, Appendix A). Hence, with $\hat{\rho}_3(\ell) = E\Big(|S(I_{4,1}; \ell)|^3 \mid \mathcal{T}_n \Big)$, $\ell \ge 1$, by (3.38) and Wald's Lemmas (cf. Theorem A.2, Appendix A), we get

$$
\begin{aligned}
&\sup_{x\in\mathbb{R}} \Big| P_*(\tilde{T}_n^{*(4)} \le x) - \Phi(x/\sigma_\infty) \Big| \\
&\le C \cdot n^{-3/2} \sigma_\infty^{-3} E\Big\{ \sum_{j=1}^{K} \hat{\rho}_3(L_j) \mid \mathcal{X}_n \Big\} \\
&\quad + \Big\{ \sup_{x\in\mathbb{R}} |x\phi(x)| \Big\} \cdot E\Big\{ [|\hat{\sigma}_{n,p} - \sigma_\infty| (\hat{\sigma}_{n,p}\sigma_\infty)^{-1} \mathbb{1}(A_n)] \mid \mathcal{X}_n \Big\} \\
&\quad + 2P(A_n^c \mid \mathcal{X}_n) \\
&\le C(\sigma_\infty) n^{-3/2} E(K \mid \mathcal{X}_n) E(\hat{\rho}_3(L_1) \mid \mathcal{X}_n) + C(\sigma_\infty) \cdot a_n + 2P(A_n^c \mid \mathcal{X}_n) \\
&= O_p\big(n^{-3/2} \cdot (np) \cdot (p^{-3/2})\big) + O(a_n) \\
&\quad + O_p\big(p^{1/2} + n^{-1/4} (\log n)^3\big) \\
&\qquad\qquad \to 0 \quad \text{as} \quad n \to \infty ,
\end{aligned}
$$

by the following Lemma. Hence, Theorem 3.4 is proved. □

To complete the proof of Theorem 3.4 and for later reference, here we establish some basic properties of the stopping time K and of the (random length) block sums for the SB method in the following result.

Lemma 3.4 *Assume that $p + (np)^{-1} \to 0$ as $n \to \infty$. Let $\{t_n\}_{n\geq 1}$ be a sequence of positive number such that $t_n \to \infty$ as $n \to \infty$. Also, let $r \geq 1$ be a given integer. Then,*

(a) *(i)* $P(L_1 > t_n p^{-1}) = O\big(\exp(-t_n)\big)$;

(ii) $P(|K - np| > (np)^{1/2}(\log n)) = O\big(\exp(-C(\log n)^2)\big)$;

(iii) $E(L_K)^r = O\big(p^{-(r+1)}\big)$;

(iv) $E(K)^r = O\big((np)^r\big)$;

(v) $E\Big(K^{r-1}\sum_{i=1}^{K} L_i^r\Big) = O(n^r)$.

(b) Suppose that the conditions of Theorem 3.4 hold. Let $\hat{\tau}^2(\ell) \equiv Var\big(S(I_{4,1};\ell) \mid \mathcal{T}_n\big)$, $\ell \geq 1$. Then,

(i) $E\{E_*|S(I_{4,1};L_1)|^4\} = O(p^{-2})$;

(ii) $E\{pE_*\hat{\tau}^2(L_1) - \sigma_\infty^2\}^2 = O\Big(p^2 + (np)^{-1}(\log n)^6\Big)$;

(iii) $P(|\hat{\sigma}_{n,p} - \sigma_\infty| > u_n\sigma_\infty \mid \mathcal{X}_n) = O_p\Big(u_n^{-1}[p + (np)^{-1/2}(\log n)^3]\Big)$

for any $\{u_n\}_{n\geq 1}$ satisfying $u_n + u_n^{-1}(np)^{-1/2}(\log n)^3 = o(1)$ as $n \to \infty$.

Proof of Lemma 3.4: Let $q = 1 - p$. Since L_1 has the Geometric distribution with parameter p, we have

$$\begin{aligned} P(L_1 > t_n p^{-1}) &= \sum_{i > p^{-1}t_n} pq^{i-1} \\ &\leq q^{(t_n/p)-1} \\ &= \exp([p^{-1}t_n - 1]\log q) \\ &= O\big(\exp(-t_n)\big) , \end{aligned}$$

proving part a(i). Next consider a(ii). Let $k_0 \equiv k_{0n} = \lfloor np - (np)^{1/2}\log n \rfloor$. Then, by the definition of K,

$$\begin{aligned} P(K \leq k_0) &= P(L_1 + \cdots + L_{k_0} \geq n) \\ &= P(\exp(t(L_1 + \cdots + L_{k_0})) > \exp(tn)) \\ &\leq e^{-tn}(pe^t/[1 - qe^t])^{k_0} \end{aligned}$$

for all $0 < t < -\log q$.

Next, let $f(t) = \log\{e^{-tn}(pe^t(1-qe^t)^{-1})^{k_0}\}, 0 < t < -\log q$. It is easy to see that $f(t)$ attains its minimum at $t_0 \equiv \log[(n-k_0)/n] - \log q \in (0, -\log q)$. Now using Taylor's expansion, after some algebra,we get

$$\begin{aligned} P(K \le k_0) &\le \exp(f(t_0)) \\ &= \exp\Big(-\frac{3}{2}np\cdot\eta^2 + O(n(p\eta)^2 + np\eta^3)\Big) , \end{aligned}$$

where $\eta \equiv (np)^{-1/2}\log n$. By similar arguments,

$$P(K > np + (np)^{1/2}\log n) = O\Big(\exp\big(-C(\log n)^2\big)\Big).$$

This completes the proof of a(ii).

Next we consider a(iii). Using the definition of K, we have for $m \ge 1$,

$$\begin{aligned} P(L_K = m) &= \sum_{k=1}^{n} P(L_K = m, K = k) \\ &= \sum_{k=1}^{n} P(L_k = m, \sum_{i=1}^{k-1} L_i < n \le \sum_{i=1}^{k} L_i) \\ &= P(L_1 = m, L_1 \ge n) + P(L_1 = m)\sum_{k=2}^{n} P(n - m \le \sum_{i=1}^{k-1} L_i < n) \\ &\le P(L_1 = m)\Big[1 + \sum_{k=2}^{n} P(k \le K < k + m)\Big] \\ &= P(L_1 = m)\Big[1 + \sum_{k=2}^{n}\sum_{j=k}^{k+m-1} P(K = j)\Big] \\ &\le (m+1)P(L_1 = m) . \end{aligned}$$

Hence, $E(L_K)^r \le \sum_{m=1}^{\infty} m^r(m+1)P(L_1 = m) \le C(r)p^{-(r+1)}$. For a(iv), noting that $1 \le K \le n$, by part a(ii), we have,

$$\begin{aligned} E(K)^r &= E(K)^r \mathbb{1}(|K - np| \le \sqrt{np}\ \log n) + O\Big(n^r \exp(-C(\log n)^2)\Big) \\ &= O\big((np)^r\big) . \end{aligned}$$

To prove part a(v), note that K is a stopping time with respect to the σ-fields $\{\sigma\langle L_1, \ldots, L_k\rangle : 1 \le k \le n\}$. Hence, using part a(iv), Wald's Lemmas (cf. Theorem A.2, Appendix A), and Hölder's inequality, we get,

$$\begin{aligned} &E\Big(K^{r-1}\sum_{i=1}^{K} L_i^r\Big) \\ &\le \{E(K^{2r-2})\}^{1/2}\Big\{E\Big(\sum_{i=1}^{K} L_i^r\Big)^2\Big\}^{1/2} \end{aligned}$$

$$
\begin{aligned}
&\le C(r)(np)^{r-1}\Big[E\Big\{\sum_{i=1}^{K}(L_i^r - EL_1^r)\Big\}^2 + (EK^2)(EL_1^r)^2\Big]^{1/2} \\
&\le C(r)(np)^{r-1}[E(K)\cdot EL_1^{2r} + (EK^2)(EL_1^{2r})]^{1/2} \\
&\le C(r)(np)^r p^{-r} .
\end{aligned}
$$

Next consider part (b). Part b(i) follows from a more general result (cf. Lemma 5.3) proved in Chapter 5. To prove b(ii), first note that by Lemmas 3.1 and 3.2, for any $1 \le \ell < n/2$,

$$
\begin{aligned}
&\mathrm{Var}\Big(\sum_{i=1}^{n} S(i;\ell)^2\Big) \\
&\le 2\Big[\ell^2 \mathrm{Var}\Big(\sum_{i=1}^{n-\ell+1} S(i;\ell)^2/\ell\Big) + E\Big(\sum_{i=n-\ell+2}^{n} S(i;\ell)^2\Big)^2\Big] \\
&\le C\ell^2(n\ell)\Big(E|S(1;\ell)/\sqrt{\ell}|^6\Big)^{2/3}\cdot\Big(1+\sum_{1\le k\le n/\ell}\alpha(k\ell)^{1/3}\Big) \\
&\quad + 2\ell^2 \max\Big\{ES(i;\ell)^4 : n-\ell+2 \le i \le n\Big\} \\
&\le Cn\ell\cdot\Big[E\Big(S(1;\ell)\Big)^6\Big]^{2/3} \\
&\quad + 2^4\cdot\ell^2 \max\Big\{E(X_1+\cdots+X_k)^4 \\
&\quad + E(X_1+\cdots+X_{\ell-k})^4 : 1\le k<\ell\Big\} \\
&\le C(\delta)\Big[\zeta_{6+\delta}^6\Delta(3;\delta)\Big]^{2/3} n\ell^3 . \qquad (3.39)
\end{aligned}
$$

Next write $\tau^2(k) = E(X_1+\cdots+X_k)^2$ and $\omega_{kn} = pq^{k-1} = P(L_1 = k)$, $k \ge 1$. Then, by (3.39) and Cauchy-Schwarz inequality, we have,

$$
\begin{aligned}
&E\Big\{E_*\Big(n^{-1}\sum_{i=1}^{n}\Big[S(i;L_1)^2 - \tau^2(L_1)\Big]\Big)\Big\}^2 \\
&= E\Big\{\sum_{k=1}^{\infty}\Big(n^{-1}\sum_{i=1}^{n}\Big[S(i;k)^2-\tau^2(k)\Big]\Big)\omega_{kn}\Big\}^2 \\
&\le \sum_{k=1}^{\infty}\omega_{kn}\cdot E\Big(n^{-1}\sum_{i=1}^{n}\Big[S(i;k)^2-\tau^2(k)\Big]\Big)^2 \\
&\le \sum_{k\le p^{-1}(\log n)^2}\omega_{kn}\,\mathrm{Var}\Big(n^{-1}\sum_{i=1}^{n}S(i;k)^2\Big) \\
&\quad + 2\sum_{k>p^{-1}(\log n)^2}\omega_{kn}\Big\{E\Big(n^{-1}\sum_{i=1}^{n}S(i;k)^4\Big) + \tau(k)^4\Big\}
\end{aligned}
$$

$$
\begin{aligned}
&\le \Big(\sum_{k\ge1}\omega_{kn}\Big)\cdot n^{-2}\cdot C(\delta)\Big[\zeta_{6+\delta}^{6}\Delta(3;\delta)\Big]^{2/3}\cdot n\Big(p^{-1}(\log n)^2\Big)^3 \\
&\quad + C \sum_{k>p^{-1}(\log n)^2} \omega_{kn}\cdot k^4 EX_1^4 \\
&= O\Big(n^{-1}p^{-3}(\log n)^6\Big) + O\Big(\Big(\sum_{k\ge1}k^8\omega_{kn}\Big)^{1/2}\Big(\sum_{k>p^{-1}(\log n)^2}\omega_{kn}\Big)^{1/2}\Big) \\
&= O\Big(n^{-1}p^{-3}(\log n)^6\Big) + O\Big(p^{-4}\exp\big(-\frac{1}{2}(\log n)^2\big)\Big) . \qquad (3.40)
\end{aligned}
$$

Also, with $\gamma(i) \equiv EX_1X_{1+i},\ i \ge 1$, we have

$$
\begin{aligned}
&\big|p\cdot E_*\tau^2(L_1) - \sigma_\infty^2\big| \\
&= \Big|pE_*\Big(\tau^2(L_1) - L_1\sigma_\infty^2\Big)\Big| \\
&= \Big|pE_*\Big\{\sum_{i=1-L_1}^{L_1-1}(L_1-|i|)\gamma(i) - \sum_{i=-\infty}^{\infty}L_1\gamma(i)\Big|\Big\} \\
&\le 2p\sum_{i=1}^{\infty}|i|\,|\gamma(i)| + 2pE_*\Big\{L_1\sum_{i\ge L_1}|\gamma(i)|\Big\} \\
&\le C\cdot p + 2p\cdot\Big[P(L_1\le p^{-1/2})\cdot p^{-1/2}\cdot\Big(\sum_{i\ge1}|\gamma(i)|\Big) \\
&\qquad + \{E(L_1)\}\cdot\sum_{i\ge p^{-1/2}}|\gamma(i)|\Big] \\
&\le Cp + Cp\cdot(p^{1/2})p^{-1/2} + Cp\cdot(p^{-1})p\cdot\Big(\sum_{i\ge p^{-1/2}}i^2|\gamma(i)|\Big) \\
&\le Cp , \qquad (3.41)
\end{aligned}
$$

since $P(L_1 \le t) = 1 - q^t \le Ctp$ for all $1 \le t \le p^{-1}/2$. Hence, by (3.40) and (3.41), it follows that

$$
\begin{aligned}
&E\Big\{pE_*\hat\tau^2(L_1) - \sigma_\infty^2\Big\}^2 \\
&= E\Big\{p\,E_*\Big(n^{-1}\sum_{i=1}^{n}S(i;L_1)^2 - L_1^2\bar X_n^2\Big) - \sigma_\infty^2\Big\}^2 \\
&\le 4\Big[p^2E\Big\{E_*(L_1^2\bar X_n^2)\Big\}^2 + E\Big\{pE_*\tau^2(L_1) - \sigma_\infty^2\Big\}^2 \\
&\qquad + p^2E\Big\{E_*\Big(n^{-1}\sum_{i=1}^{n}\Big[S(i;L_1)^2 - \tau^2(L_1)\Big]\Big)\Big\}^2\Big] \\
&= O(p^2p^{-4}n^{-2}) + O(p^2) + O\Big((np)^{-1}(\log n)^6\Big) .
\end{aligned}
$$

This proves b(ii).

Next we consider b(iii). Using a(i) and arguments similar to (3.39) and (3.40), we get

$$\begin{aligned}
& E\Big[E_*\Big\{n^{-1}\sum_{i=1}^{n} S(i;L_1)^4\Big\}\Big] \\
\leq\ & \max\Big\{n^{-1}\sum_{i=1}^{n} ES(i;k)^4 : 1 \leq k \leq p^{-1}(\log n)^2\Big\} \\
& + (E\|X_1\|^4)\cdot \sum_{k>p^{-1}(\log n)^2} k^4\omega_{kn} \\
=\ & O\Big(\Big[p^{-1}(\log n)^2\Big]^2\Big) \qquad (3.42)
\end{aligned}$$

and by a(iv), Wald's lemmas and the fact that $N_1 \geq n$,

$$\begin{aligned}
& E_*|N_1^{-1}K - p| = E|N_1^{-1}K - p| \\
\leq\ & \Big\{E|Kp^{-1} - N_1|\Big\}(p/n) \\
=\ & n^{-1}p\cdot E\Big|\sum_{j=1}^{K}(L_j - EL_j)\Big| \\
\leq\ & (n^{-1}p)\cdot \Big[E\Big\{\sum_{j=1}^{K}(L_j - EL_j)\Big\}^2\Big]^{1/2} \\
=\ & (n^{-1}p)[(EK)\mathrm{Var}(L_1)]^{1/2} \\
\leq\ & C\cdot(n^{-1}p)\cdot[np\cdot p^{-1}]^{1/2} \\
=\ & O(n^{-1/2}p)\ . \qquad (3.43)
\end{aligned}$$

Next, note that $\hat{\tau}^2(L_j) = n^{-1}\sum_{i=1}^{n} S(i,L_j)^2 - L_j^2\bar{X}_n^2$, $1 \leq j \leq n$ are conditionally iid given $\mathcal{X}_n$, and $\hat{\sigma}_{n,p}^2 = N_1^{-1}\sum_{i=1}^{k}\hat{\tau}^2(L_j)$. Now, using b(ii), (3.42), and (3.43), we have

$$\begin{aligned}
& P\Big(|\hat{\sigma}_{n,p} - \sigma_\infty| > u_n\sigma_\infty \mid \mathcal{X}_n\Big) \\
\leq\ & P\Big(|\hat{\sigma}_{n,p}^2 - \sigma_\infty^2| > u_n\sigma_\infty^2 \mid \mathcal{X}_n\Big) \\
\leq\ & P\Big(\Big|\sum_{j=1}^{K}\Big\{\hat{\tau}^2(L_j) - E\Big(\hat{\tau}^2(L_j)\mid\mathcal{X}_n\Big)\Big\}\Big| > \frac{1}{2}\cdot nu_n\sigma_\infty^2 \mid \mathcal{X}_n\Big) \\
& + P\Big(\Big|N_1^{-1}KE\Big(\hat{\tau}^2(L_1)\mid\mathcal{X}_n\Big) - \sigma_\infty^2\Big| > \frac{1}{2}\cdot u_n\sigma_\infty^2 \mid \mathcal{X}_n\Big) \\
\leq\ & C\cdot(nu_n\sigma_\infty^2)^{-2}\Big\{E_*(K)\Big\}\Big\{\mathrm{Var}_*\Big(\hat{\tau}^2(L_1)\Big)\Big\} \\
& + 2(u_n\sigma_\infty^2)^{-1}E_*\Big|N_1^{-1}KE_*\Big(\hat{\tau}^2(L_1)\Big) - \sigma_\infty^2\Big|
\end{aligned}$$

$$
\begin{aligned}
&\leq C(\sigma_\infty)\cdot(nu_n)^{-2}(np)E_*\left[n^{-1}\sum_{i=1}^{n}S(i;L_1)^4\right] \\
&\quad + C(\sigma_\infty)\cdot(u_n)^{-1}\Big\{\Big(E_*|N_1^{-1}K-p|\Big)\Big(E_*\hat{\tau}^2(L_1)\Big) \\
&\qquad\qquad + \big|pE_*\hat{\tau}^2(L_1)-\sigma_\infty^2\big|\Big\} \\
&= O_p\left(n^{-1}pu_n^{-2}\left[p^{-1}(\log n)^2\right]^2\right) \\
&\quad + O_p\left(u_n^{-1}\left\{(n^{-1/2}p)p^{-1}+\left[p+(np)^{-1/2}(\log n)^3\right]\right\}\right).
\end{aligned}
$$

This completes the proof of part b(iii), and hence of the lemma. □

4

Extensions and Examples

4.1 Introduction

In this chapter, we establish consistency of different block bootstrap methods for some general classes of estimators and consider some specific examples illustrating the theoretical results. Section 4.2 establishes consistency of estimators that may be represented as smooth functions of sample means. Section 4.3 deals with (generalized) M-estimators, including the maximum likelihood estimators of parameters, which are defined through estimating equations. Some special considerations are required while defining the bootstrap versions of such estimators. We describe the relevant issues in detail in Section 4.3. Section 4.4 gives results on the bootstrapped empirical process, and establishes consistency of bootstrap estimators for certain differentiable statistical functionals. Section 4.5 contains three numerical examples, illustrating the theoretical results of Sections 4.2–4.4.

4.2 Smooth Functions of Means

Results of Sections 3.2 and 3.3 allow us to establish consistency of the MBB, the NBB, the CBB, and the SB methods for some general classes of estimators. In this section, we consider the class of estimators that fall under the purview of the *Smooth Function Model* (cf. Bhattacharya and Ghosh (1978); Hall (1992)). Suppose that $\{X_{0i}\}_{i \in \mathbb{Z}}$ is a $\mathbb{R}^{d_0}$-valued stationary process. Let $f : \mathbb{R}^{d_0} \rightarrow \mathbb{R}^d$ be a Borel measurable function, and let $H : \mathbb{R}^d \to \mathbb{R}$ be a smooth function. Suppose that the level-1 parameter of interest is given by $\theta = H(Ef(X_{01}))$. A natural estimator of θ is given by $\hat{\theta}_n = H(n^{-1}\sum_{i=1}^n f(X_{0i}))$. Thus, the parameter θ and its estimator $\hat{\theta}_n$ are both smooth functions, respectively, of the population and the sample means of the transformed sequence $\{f(X_{0i})\}_{i\in\mathbb{Z}}$. Many level-1 parameters and their estimators may be expressed as smooth functions of means as above. Some common examples of estimators satisfying this 'Smooth Function Model' formulation are given below.

Example 4.1: Let $\{X_{0i}\}_{i\in\mathbb{Z}}$ be a stationary real-valued time series with autocovariance function $\gamma(k) = \mathrm{Cov}(X_{0i}, X_{0(i+k)})$, $i, k \in \mathbb{Z}$. An estimator of $\gamma(k)$ based on a sample $X_{01}, \ldots, X_{0n}$ of size n is given by

$$\hat{\gamma}_n(k) = (n-k)^{-1}\sum_{i=1}^{n-k} X_{0i}X_{0(i+k)} - \bar{X}_{0(n-k)}^2 \ , \tag{4.1}$$

where $\bar{X}_{0(n-k)} = (n-k)^{-1}\sum_{i=1}^{n-k} X_{0i}$. Note that $\hat{\gamma}_n(k)$ is a version of the sample autocovariance at lag k. We now show that the estimator $\hat{\gamma}_n(k)$ and the parameter $\gamma(k)$ admit the representation specified by the Smooth Function Model. Define a new sequence of bivariate random vectors

$$X_i = (X_{0i}, X_{0i}X_{0(i+k)})', \ i \in \mathbb{Z} \ .$$

Then, the level-1 parameter of interest $\theta \equiv \gamma(k)$ is given by

$$\theta = EX_{01}X_{0(1+k)} - (EX_{01})^2 = H(EX_1)$$

where $H : \mathbb{R}^2 \to \mathbb{R}$ is given by $H((x,y)') = y - x^2$. And similarly, its estimator $\hat{\theta}_n \equiv \hat{\gamma}_n(k)$ is given by

$$\hat{\theta}_n = \hat{\gamma}_n(k) = H(\bar{X}_{n-k})$$

where $\bar{X}_{n-k} = (n-k)^{-1}\sum_{i=1}^{n-k} X_i$. Thus, this is an example that falls under the purview of the Smooth Function Model. □

Example 4.2: Let $\{X_{0i}\}_{i\in\mathbb{Z}}$ be a stationary time series with $\mathrm{Var}(X_{01}) \in (0,\infty)$, and the level-1 parameter of interest is the lag-k autocorrelation coefficient

$$r(k) = \mathrm{Cov}(X_{01}, X_{0(1+k)})/\mathrm{Var}(X_{01}) \ ,$$

for some fixed integer $k \geq 0$. As an estimator of $r(k)$, we consider the following version of the sample autocorrelation coefficient

$$\hat{r}_n(k) = \frac{\left\{(n-k)^{-1}\sum_{i=1}^{n-k} X_{0i}X_{0(i+k)} - \bar{X}_{0(n-k)}^2\right\}}{\left\{(n-k)^{-1}\sum_{i=1}^{n-k} X_{0i}^2 - \bar{X}_{0(n-k)}^2\right\}} \ ,$$

where $\bar{X}_{0(n-k)} = (n-k)^{-1} \sum_{i=1}^{n-k} X_{0i}$. Then, as in Example 4.1, $r(k)$ and $\hat{r}_n(k)$ may be expressed as smooth functions of sample means of certain *lag-product* variables. Define the function $H : \mathbb{R}^3 \to \mathbb{R}$ by

$$H(x, y, z) = \{(z - x^2)/(y - x^2)\}\mathbb{1}(y > x^2)$$

and set $Y_i = (X_{0i}, X_{0i}^2, X_{0i}X_{0(i+k)})'$, $i \in \mathbb{Z}$. Then, it is easy to see that the function $H(\cdot)$ is smooth in a neighborhood of EY_1 and that $r(k)$ and $\hat{r}_n(k)$ can be expressed as

$$r(k) = H(EY_1)$$

and

$$\hat{r}_n(k) = H(\bar{Y}_{(n-k)}) ,$$

where $\bar{Y}_m = m^{-1} \sum_{i=1}^{m} Y_i$, $m \geq 1$. □

Example 4.3: Let $\{X_{0i}\}_{i\in\mathbb{Z}}$ be a zero-mean stationary autoregressive process of order $p \in \mathbb{N}$ (AR(p), in short), satisfying

$$X_{0i} = \sum_{j=1}^{p} \beta_j X_{0(i-j)} + \epsilon_i \ , \quad i \in \mathbb{Z} \ , \tag{4.2}$$

where $\{\epsilon_i\}_{i\in\mathbb{Z}}$ is an uncorrelated stationary process with $E\epsilon_1 = 0$ and $E\epsilon_1^2 = \sigma^2 \in (0, \infty)$, and $\beta_1, \ldots, \beta_p \in \mathbb{R}$ are the autoregressive parameters. Suppose that $(\beta_1, \ldots, \beta_p)$ are such that the polynomial

$$\beta(z) \equiv 1 - \beta_1 z - \cdots - \beta_p z^p, \ z \in \mathbb{C} \tag{4.3}$$

has no zero on the closed unit disc $\{|z| \leq 1\}$. Then, the AR(p)-process $\{X_{0i}\}_{i\in\mathbb{Z}}$ admits a representation of the form

$$X_{0i} = \sum_{j=0}^{\infty} a_j \epsilon_{i-j}, \ i \in \mathbb{Z} \tag{4.4}$$

where the sequence of constants $\{a_i\}_{i\geq 0}$ is determined by $\sum_{j=0}^{\infty} a_j z^j = 1/\beta(z)$, $|z| \leq 1$ (see Theorem 3.1.1, Brockwell and Davis (1991)). This property of $\{X_{0i}\}_{i\in\mathbb{Z}}$ is referred to as "causality" and it yields the Yule-Walker estimators of the parameters $\beta_1, \ldots, \beta_p$ and σ^2.

Let $\gamma(k)$ and $\hat{\gamma}_n(k)$, $1 \leq k \leq p$ be as in Example 4.1 with $(n-k)$ replaced by $(n-p)$. Also, let $\hat{\gamma}_{p,n} \equiv \big(\hat{\gamma}_n(1), \ldots, \hat{\gamma}_n(p)\big)'$ and let $\hat{\Gamma}_{p,n}$ be the $p \times p$ matrix with (i,j)-th element $\hat{\gamma}_n(i-j)$, $1 \leq i,j \leq p$. When $\hat{\Gamma}_{p,n}$ is nonsingular, a version of the Yule-Walker estimators $\hat{\beta}_{1n}, \ldots, \hat{\beta}_{pn}$, $\hat{\sigma}_n^2$ of $\beta_1, \ldots, \beta_p$, σ^2, is given by

$$(\hat{\beta}_{1n}, \ldots, \hat{\beta}_{pn})' = \hat{\Gamma}_{p,n}^{-1} \hat{\gamma}_{p,n} \ , \tag{4.5}$$

$$\hat{\sigma}_n^2 = \hat{\gamma}_n(0) - (\hat{\beta}_{1n}, \ldots, \hat{\beta}_{pn})' \hat{\gamma}_{p,n} \ . \tag{4.6}$$

We claim that the parameter vector $\theta = (\beta_1, \ldots, \beta_p; \sigma^2)'$ and the estimator $\hat{\theta}_n = (\hat{\beta}_{1n}, \ldots, \hat{\beta}_{pn}; \hat{\sigma}_n^2)'$ satisfy the requirements of the Smooth Function Model. To see this, define a new $\mathbb{R}^{p+2}$-valued process $\{X_i\}_{i\in\mathbb{Z}}$ by

$$X_i = \Big(X_{0i}, X_{0i}^2, X_{0i}X_{0(i+1)}, \ldots, X_{0i}X_{0(i+p)}\Big)', \ i \in \mathbb{Z} .$$

Let $h_1(x) = (x_2 - x_1^2)$ and $h_2(x) = (x_3 - x_1^2, \ldots, x_{p+2} - x_1^2)'$, $x = (x_1, \ldots, x_{p+2})' \in \mathbb{R}^{p+2}$. Then, writing $\bar{X}_m = m^{-1}\sum_{i=1}^m X_i$, $m \geq 1$, we have $\hat{\gamma}_n(0) = h_1(\bar{X}_{n-p})$ and $\hat{\gamma}_{p,n} = h_2(\bar{X}_{n-p})$.

Next, let $\mathcal{S}_p^*$ (and $\mathcal{S}_p^{*+}$, respectively) denote the collection of all symmetric (and symmetric nonsingular) matrices of order p, and let $g_1 : \mathcal{S}_p^* \to \mathcal{S}_p^*$ be defined by

$$g_1(A) = \begin{cases} A^{-1} & \text{if} \quad A \in \mathcal{S}_p^{*+} \\ 0 & \text{otherwise} . \end{cases}$$

Since, for $A \in \mathcal{S}_p^{*+}$, the elements of A^{-1} are given by the ratios of the cofactors of A and the determinant of A, and the determinant of a matrix is a polynomial in its elements, the components of the function $g_1(\cdot)$ are rational functions (and, hence, infinitely differentiable functions) of the elements of its argument at any $A \in \mathcal{S}_p^{*+}$. Also, let $g_2 : \mathbb{R}^p \to \mathcal{S}_p^*$ be defined by

$$g_2\big((y_1, \ldots, y_p)'\big) = \big((y_{|i-j|})\big)_{p\times p} .$$

Then the estimator $\hat{\theta}_n = (\hat{\beta}_{1n}, \ldots, \hat{\beta}_{pn}; \hat{\sigma}_n^2)'$ can be expressed as

$$\hat{\theta}_n = H(\bar{X}_{n-p}) \equiv (H^{(1)}(\bar{X}_{n-p})';\ H^{(2)}(\bar{X}_{n-p}))' ,$$

where

$$\begin{aligned} (\hat{\beta}_{1n}, \ldots, \hat{\beta}_{pn})' &= g_1\big(g_2(h_2(\bar{X}_{n-p}))\big)h_2(\bar{X}_{n-p}) \equiv H^{(1)}(\bar{X}_{n-p}) , \\ &\text{and} \\ \hat{\sigma}_n^2 &= h_1(\bar{X}_{n-p}) - H^{(1)}(\bar{X}_{n-p})'h_2(\bar{X}_{n-p}) \equiv H^{(2)}(\bar{X}_{n-p}) . \end{aligned}$$

The corresponding representation for the parameter θ as $\theta = H(EX_1)$ holds since by (4.4) (cf. p. 239, Brockwell and Davis (1991)),

$$(\beta_1, \ldots, \beta_p)' = \Gamma_p^{-1}\gamma_p , \quad \sigma^2 = \gamma(0) - (\beta_1, \ldots, \beta_p)\gamma_p$$

where $\gamma_p = \big(\gamma(1), \ldots, \gamma(p)\big)'$ and Γ_p is the $p \times p$ matrix with (i,j)-th element $\gamma(i-j)$, $1 \leq i, j \leq p$. Thus, the Yule-Walker estimators also fall under the purview of the Smooth Function Model. □

Now we consider the general framework of the Smooth Function Model described in the first paragraph of this section and establish consistency of different block bootstrap estimators of the sampling distribution of the centered and scaled estimator $\hat{\theta}_n$, given by

$$T_{1n} = \sqrt{n}(\hat{\theta}_n - \theta) ,$$

with $\theta = H\big(Ef(X_{01})\big)$ and $\hat{\theta}_n = H\big(n^{-1}\sum_{i=1}^n f(X_{0i})\big)$. Block bootstrap versions of T_{1n} can be defined following the descriptions given in Chapter 2 and Sections 3.2 and 3.3. Let $X_i = f(X_{0i}),\ i \in \mathbb{Z}$. Note that in terms of the transformed variables X_i's, we may rewrite θ and $\hat{\theta}_n$, respectively, as $\theta = H(EX_1)$ and $\hat{\theta}_n = H(\bar{X}_n)$. Next, let $\mathcal{X}_n^{*(j)},\ j = 1, 2, 3$ denote the set of n_1 bootstrap samples based on b blocks of length ℓ from the transformed variables $\mathcal{X}_n = \{X_1, \ldots, X_n\}$ under the MBB, the NBB, and the CBB, respectively, and let $\mathcal{X}_n^{*(4)}$ denote the set of N_1 bootstrap observations under the SB, with expected block length $p^{-1} \in (1, n)$, where $b = \lfloor n/\ell \rfloor$, $n_1 = b\ell$ and $N_1 = L_1 + \cdots + L_K$ are as in Chapter 3. Also, with a slight abuse of notation, let $\bar{X}_n^{*(j)}$, $j = 1, 2, 3, 4$ denote the means of the corresponding bootstrap samples. Then, the block bootstrap versions of T_{1n} are given by

$$T_{1n}^{*(j)} = \sqrt{n_1}(\theta_n^{*(j)} - \tilde{\theta}_{n,j}),\ j = 1, 2, 3 ,$$

and

$$T_{1n}^{*(j)} = \sqrt{N_1}(\theta_n^{*(j)} - \tilde{\theta}_{n,j}),\ j = 4 ,$$

where $\theta_n^{*(j)} = H(\bar{X}_n^{*(j)})$ and $\tilde{\theta}_{n,j} = H(E_* \bar{X}_n^{*(j)})$, $1 \le j \le 4$.

Then, we have the following result.

Theorem 4.1 *Suppose that the function H is differentiable in a neighborhood $N_H \equiv \{x \in \mathbb{R}^d : \|x - EX_1\| < 2\eta\}$ of EX_1 for some $\eta > 0$, $\sum_{|\alpha|=1} |D^\alpha H(EX_1)| \ne 0$, and that the first-order partial derivatives of H satisfy a Lipschitz condition of order $\kappa > 0$ on N_H. Assume that the conditions of Theorem 3.2 hold for $j = 1, 2, 3$ and that the conditions of Theorem 3.4 hold for $j = 4$ (with the transformed sequence $\{X_i\}_{i \in \mathbb{Z}}$). Then,*

$$\sup_{x \in \mathbb{R}} \Big| P_*(T_{1n}^{*(j)} \le x) - P(T_{1n} \le x) \Big| \longrightarrow_p 0 \quad as \quad n \to \infty$$

for $j = 1, 2, 3, 4$.

For proving Theorem 4.1, we shall use a suitable version of the well known Slutsky's theorem for conditional distributions.

Lemma 4.1 *(*A CONDITIONAL SLUTSKY'S THEOREM*). For $n \in \mathbb{N}$, let b_n^* and T_n^* be r-dimensional ($r \in \mathbb{N}$) and s-dimensional ($s \in \mathbb{N}$) random vectors, and let A_n^* be a $r \times s$ random matrix, all defined on a common probability space $(\Omega, \mathcal{F}, P)$. Suppose that $\mathcal{X}_\infty$ is a sub-σ-field of $\mathcal{F}$ and that there exist $\mathcal{X}_\infty$-measurable variables $\hat{A}$ and $\hat{b}$, such that*

$$P(\|A_n^* - \hat{A}\| > \epsilon \mid \mathcal{X}_\infty) + P(\|b_n^* - \hat{b}\| > \epsilon \mid \mathcal{X}_\infty) \longrightarrow_p 0 \quad as \quad n \to \infty , \tag{4.7}$$

for every $\epsilon > 0$. *Also, suppose that there is a probability distribution* ν *on* $\mathbb{R}^s$ *such that* $\mathcal{L}(T_n^* \mid \mathcal{X}_\infty) \longrightarrow^d \nu$ *in probability, as* $n \to \infty$, *i.e.,* $\varrho_s(\mathcal{L}(T_n^* \mid \mathcal{X}_\infty),\ \nu) \longrightarrow_p 0$ *as* $n \to \infty$, *where* $\varrho_k(\cdot)$ *metricizes convergence in distribution of random vectors on* $\mathbb{R}^k$, $k \in \mathbb{N}$. *Then,*

$$\mathcal{L}(A_n^* T_n^* + b_n^* \mid \mathcal{X}_\infty) \longrightarrow^d \nu o \hat{g}^{-1} \quad \textit{in probability,}$$

i.e.,

$$\varrho_r(\mathcal{L}(A_n^* T_n^* + b_n^* \mid \mathcal{X}_\infty), \nu o \hat{g}^{-1}) \longrightarrow_p 0 \quad \textit{as} \quad n \to \infty\ ,$$

where $\hat{g} : \mathbb{R}^s \to \mathbb{R}^s$ *is the mapping* $\hat{g}(t) = \hat{A}t + \hat{b}$, $t \in \mathbb{R}^s$, *and where* $\nu o \hat{g}^{-1}$ *denotes the probability distribution induced by the mapping* $\hat{g}$ *on* $\mathbb{R}^r$ *under* ν, *i.e.,* $\nu o \hat{g}^{-1}(B) = \nu(\hat{g}^{-1}(B))$, $B \in \mathcal{B}(\mathbb{R}^r)$.

Note that if T is a random vector with distribution ν, then $\nu o \hat{g}^{-1}$ is the distribution of the transformed random vector $\hat{g}(T)$. Hence, the lemma can be restated in a less formal way as follows:

Lemma 4.1′ *If* $\mathcal{L}(T_n^* \mid \mathcal{X}_\infty) \longrightarrow^d \mathcal{L}(T)$ *in probability and (4.7) holds, then*

$$\mathcal{L}(A_n^* T_n^* + b_n^* \mid \mathcal{X}_\infty) \longrightarrow^d \mathcal{L}(\hat{A}T + \hat{b}) \quad \textit{in probability}\ .$$

For proving the lemma, we shall use the following equivalent form of condition (4.7):

There exists a sequence $\{\epsilon_n\}_{n \geq 1}$ of positive real numbers such that $\epsilon_n \downarrow 0$ as $n \to \infty$ and

$$\begin{aligned} & P(\|A_n^* - \hat{A}\| > \epsilon_n \mid \mathcal{X}_\infty) + P(\|b_n^* - \hat{b}\| > \epsilon_n \mid \mathcal{X}_\infty) \\ & \longrightarrow_p 0 \quad \text{as} \quad n \to \infty\ . \end{aligned} \tag{4.8}$$

It is clear that (4.8) implies (4.7). To prove the converse, note that by (4.7), for each $k \geq 2$, there exists a positive integer $m_k > m_{k-1}$ such that

$$P(\hat{a}_n(k^{-1}) > k^{-1}) \leq k^{-1} \quad \text{for all} \quad n \geq m_k$$

where

$$\hat{a}_n(\epsilon) \equiv P(\|A_n^* - \hat{A}\| > \epsilon \mid \mathcal{X}_\infty) + P(\|b_n^* - \hat{b}\| > \epsilon \mid \mathcal{X}_\infty),\ 0 < \epsilon \leq 1$$

and $m_1 = 1$. Define $\epsilon_n = k^{-1}$ for $m_k \leq n < m_{k+1}$, $k \geq 1$. Then, $\epsilon_n \downarrow 0$ as $n \to \infty$ and $P(\hat{a}_n(\epsilon_n) > \epsilon_n) \leq \epsilon_n$ for all $n \geq 1$, which implies (4.8).

Proof of Lemma 4.1: It is enough to show that given any subsequence $\{n_i\}$, there is a further subsequence $\{n_k\} \subset \{n_i\}$ such that

$$\varrho_r(\mathcal{L}(A_{n_k}^* T_{n_k}^* + b_{n_k}^* \mid \mathcal{X}_\infty), \nu o \hat{g}^{-1}) \to 0 \quad \text{as} \quad k \to \infty \quad \text{a.s.} \tag{4.9}$$

Fix a subsequence $\{n_i\}$. Let $\hat{a}_n(\epsilon)$ be as defined above. Then, by (4.8) and the conditions of the lemma, there exists a subsequence $\{n_k\} \subset \{n_i\}$ such that

$$\hat{a}_{n_k}(\epsilon_{n_k}) \to 0 \quad \text{as} \quad k \to \infty, \quad \text{a.s. } (P) \tag{4.10}$$

and

$$\varrho_s\big(\mathcal{L}(T^*_{n_k} \mid \mathcal{X}_\infty), \nu\big) \to 0 \quad \text{as} \quad k \to \infty, \quad \text{a.s. } (P)\ . \tag{4.11}$$

We shall show that (4.9) holds for this choice of the subsequence $\{n_k\}$. For any vector $x \in \mathbb{R}^r$, write

$$x'[A^*_n T^*_n + b^*_n] = (x'\hat{A}T^*_n + x'\hat{b}) + R^*_n(x)\ , \tag{4.12}$$

where $R^*_n(x) = x'(A^*_n - \hat{A})T^*_n + x'(b^*_n - \hat{b})$.

Note that by (4.11) and the continuous mapping theorem (applied pointwise on a set of P-probability one that does not depend on $x \in \mathbb{R}^r$),

$$\mathcal{L}\big(x'[\hat{A}T^*_{n_k} + \hat{b}] \mid \mathcal{X}_\infty\big) \longrightarrow^d \hat{\nu}_x \quad \text{as} \quad k \to \infty, \quad \text{a.s. } (P) \tag{4.13}$$

where $\hat{\nu}_x = \nu o \hat{g}_x^{-1}$ and $\hat{g}_x : \mathbb{R}^s \to \mathbb{R}$ is defined by $\hat{g}_x(t) = x'(\hat{A}t + \hat{b}),\ t \in \mathbb{R}^s$. Also, by (4.10) and (4.11),

$$\begin{aligned} &P\Big(\,\big|R^*_{n_k}(x)\big| > 2\epsilon_{n_k}^{1/2}\|x\| \mid \mathcal{X}_\infty\Big) \\ &\leq \quad P(\|T^*_{n_k}\| > \epsilon_{n_k}^{-1/2} \mid \mathcal{X}_\infty) + a_{n_k}(\epsilon_{n_k}) \\ &\quad \to 0 \quad \text{as} \quad k \to \infty, \quad \text{a.s. } (P)\ . \end{aligned} \tag{4.14}$$

Hence, from (4.12), (4.13), and (4.14), it follows that

$$\varrho\big(\mathcal{L}(x'[A^*_{n_k} T^*_{n_k} + b^*_{n_k}] \mid \mathcal{X}_\infty), \hat{\nu}_x\big) \to 0 \quad \text{as} \quad n_k \to \infty, \quad \text{a.s. } (P)$$

for all $x \in \mathbb{R}^r$. Thus, by the Cramer-Wold device, (4.9) holds and the lemma is proved. □

Proof of Theorem 4.1: Write $T_n^{*(j)} = a_n(j)\left(\bar{X}_n^{*(j)} - E_*\bar{X}_n^{*(j)}\right)$, $j = 1,2,3,4$ and $T_n = \sqrt{n}\big(\bar{X}_n - EX_1\big)$, where $a_n(j)^2 = n_1$ for $j = 1,2,3$ and $a_n(4)^2 = N_1$. Note that for $j = 4$, $T_n^{*(j)}$ was denoted by $\tilde{T}_n^{*(4)}$ in Chapter 3. For notational simplicity, we shall use the updated definitions of $T_n^{*(4)}$ and $\bar{X}_n^{*(j)}$'s in this chapter. Also, set $\Sigma_\infty = \lim_{n\to\infty} n \cdot \text{Var}(\bar{X}_n)$ and $\mathcal{X}_\infty = \sigma\langle\{X_i : i \geq 1\}\rangle$. Then, by Theorems 3.2 and 3.4, it follows that for $j = 1,2,3,4$,

$$\varrho_d\left(\mathcal{L}(T_n^{*(j)} \mid \mathcal{X}_\infty), N(0, \Sigma_\infty)\right) \longrightarrow_p 0 \quad \text{as} \quad n \to \infty\ . \tag{4.15}$$

Also, note that

$$T_{1n}^{*(j)} = AT_n^{*(j)} + R_n^{*(j)}, \quad j = 1,2,3,4,$$

where A is the $1 \times d$ dimensional matrix (row vector) with elements $D_1H(EX_1), \ldots, D_dH(EX_1)$, where $D_iH(x) = \frac{\partial}{\partial x_i}H(x)$, and where the remainder term $R_n^{*(j)}$ is defined by subtraction. We now show that

$$P\left(|R_n^{*(j)}| > 2\epsilon \mid \mathcal{X}_\infty\right) = o_p(1) \quad \text{as} \quad n \to \infty, \tag{4.16}$$

for any $\epsilon > 0$ and for all $j = 1, 2, 3, 4$.

Let $\hat{t}_n(j) = \|E_*\bar{X}_n^{*(j)} - EX_1\|$, $j = 1, 2, 3, 4$. Then, on the set $\{\hat{t}_n(j) \le \eta\} \cap \{\|\bar{X}_n^{*(j)} - E_*\bar{X}_n^{*(j)}\| \le \eta\}$, using a one-term Taylor's expansion of the function H around EX_1, we get

$$\begin{aligned} |R_n^{*(j)}| &= \left|a_n(j)\left(H(\bar{X}_n^{*(j)}) - H(E_*\bar{X}_n^{*(j)})\right) - AT_n^{*(j)}\right| \\ &\le C \cdot \left(\left\|\bar{X}_n^{*(j)} - E_*\bar{X}_n^{*(j)}\right\|^\kappa + \hat{t}_n(j)^\kappa\right) \cdot \|T_n^{*(j)}\| \,. \end{aligned}$$

Hence, it follows that

$$\begin{aligned} & P\left(|R_n^{*(j)}| > 2\epsilon \mid \mathcal{X}_\infty\right) \\ \le\; & P_*\left(|R_n^{*(j)}| > 2\epsilon,\ \|\bar{X}_n^{*(j)} - E_*\bar{X}_n^{*(j)}\| \le \eta\right) \cdot \mathbb{1}(\hat{t}_n(j) \le \eta) \\ & + \mathbb{1}(\hat{t}_n(j) > \eta) + P_*\left(\|\bar{X}_n^{*(j)} - E_*\bar{X}_n^{*(j)}\| > \eta\right) \\ \le\; & P_*\left(C\|T_n^{*(j)}\|^{1+\kappa} a_n(j)^{-\kappa} > \epsilon\right) + P_*\left(\hat{t}_n(j)^\kappa \|T_n^{*(j)}\| > \epsilon\right) \\ & + P_*\left(\|\bar{X}_n^{*(j)} - E_*\bar{X}_n^{*(j)}\| > \eta\right) + \mathbb{1}\left(\hat{t}_n(j) > \eta\right) \\ \le\; & \left[3P_*\left(\|T_n^{*(j)}\| > C(\epsilon, \kappa, \eta) \cdot \log n\right) + 2\mathbb{1}\left(\hat{t}_n(j) > \eta(\log n)^{-1}\right)\right] \\ =\; & o_p(1) \end{aligned}$$

by (4.15) and the fact that $E\mathbb{1}\left(\hat{t}_n(j) > \eta(\log n)^{-1}\right) = P\big(\hat{t}_n(j) > \eta(\log n)^{-1}\big) \le (\eta^{-1}\log n)^2 E\hat{t}_n(j)^2 = O\big(n^{-1}(\log n)^2\big)$ as $n \to \infty$. Hence, by Lemma 4.1, the theorem now follows from (4.15) and (4.16). □

Remark 4.1 In many applications, we may be interested in an estimator of the parameter $\theta = H(EX_1)$ of the form

$$\bar{\theta}_n = H\left((n-k_1)^{-1}\sum_{i=1}^{n-k_1} f_1(X_{0i}), \ldots, (n-k_d)^{-1}\sum_{i=1}^{n-k_d} f_d(X_{0i})\right)$$

for some *fixed* integers $k_1, \ldots, k_d$, not depending on n, where $f_1, \ldots, f_d$ denote the components of the function $f : \mathbb{R}^{d_0} \to \mathbb{R}^d$. It can be easily shown that the conclusions of Theorem 4.1 continue to hold for $\bar{\theta}_n$ if we replace $\hat{\theta}_n$ by $\bar{\theta}_n$ and $\theta_n^{*(j)}$ by the corresponding block bootstrap versions. In the same vein, it is easy to check that consistency of the block bootstrap distribution function estimators continue to hold if the function is vector valued and each component of H satisfies the conditions of Theorem 4.1.

Remark 4.2 As in Theorems 3.1 and 3.3, the block bootstrap methods also provide consistent estimators of the asymptotic variance of the statistic $\hat{\theta}_n$ considered in Theorem 4.1. However, we need to assume some stronger moment and mixing conditions than those of Theorem 4.1 to establish the consistency of bootstrap variance estimators. A set of sufficient conditions will be given in Chapter 5, which guarantee (mean squared error or L^2-) consistency of these bootstrap estimators of the asymptotic variance of $\hat{\theta}_n$.

4.3 M-Estimators

Suppose that $\{X_i\}_{i \in \mathbb{Z}}$ is a stationary process taking values in $\mathbb{R}^d$. Also, suppose that the parameter of interest θ is defined implicitly as a solution to the equation

$$E\Psi(X_1, \ldots, X_m; \theta) = 0 \tag{4.17}$$

for some function $\Psi : \mathbb{R}^{dm+s} \to \mathbb{R}^s$, $m, s \in \mathbb{N}$. An M-estimator $\hat{\theta}_n$ of θ is defined as a solution of the 'estimating equation'

$$(n - m + 1)^{-1} \sum_{i=1}^{(n-m+1)} \Psi(X_i, \ldots, X_{i+m-1}; \hat{\theta}_n) = 0 \; . \tag{4.18}$$

Estimators defined by an estimating equation of the form (4.18) are called *generalized M-estimators* (cf. Bustos (1982)). This class of estimators contains the maximum likelihood estimators and certain robust estimators of parameters in many popular time series models, including the autoregressive moving average models of order p, q (ARMA (p, q)), $p, q \in \mathbb{Z}_+$. See Bustos (1982), Martin and Yohai (1986) and the references therein.

To define the bootstrap version of $\hat{\theta}_n$, let $Y_i = (X_i', \ldots, X_{i+m-1}')'$, $1 \leq i \leq n_0$ denote the $(m-1)$-th order lag vectors, where $n_0 = n - m + 1$. Next suppose that $Y_1^{*(j)}, \ldots, Y_{n_0}^{*(j)}$ denote the "ordinary" block bootstrap sample of size n_0 drawn from the observed Y-vectors under the jth method, $j = 1, 2, 3, 4$. Because of the structural restriction (4.17), there appears to be more than one way of defining the bootstrap version of the generalized M-estimator $\hat{\theta}_n$ and its centered and scaled version

$$T_{2n} \equiv \sqrt{n}(\hat{\theta}_n - \theta) \; .$$

Following the description of the block bootstrap methods in Chapter 2, we may define the bootstrap version $\theta_n^{*(j)}$ of $\hat{\theta}_n$ based on the jth method, $j = 1, 2, 3, 4$, as a solution of the equation

$$n_0^{-1} \sum_{i=1}^{n_0} \Psi\big(Y_i^{*(j)}; \theta_n^{*(j)}\big) = 0 \; . \tag{4.19}$$

The bootstrap version of T_{2n} is then given by

$$T_{2n}^{*(j)} = \sqrt{n_0}\big(\theta_n^{*(j)} - \tilde{\theta}_n^{(j)}\big) , \tag{4.20}$$

where the centering value $\tilde{\theta}_n^{(j)}$ is defined as a solution of

$$n_0^{-1} \sum_{i=1}^{n_0} E_* \Psi\big(Y_i^{*(j)}; \tilde{\theta}_n^{(j)}\big) = 0 \tag{4.21}$$

to ensure the bootstrap analog of (4.17) at the centering value $\tilde{\theta}_n^{(j)}$ in the definition of $T_{2n}^{*(j)}$. Note that for the CBB or the SB applied to the series $\{Y_1, \ldots, Y_{n_0}\}$, equation (4.21) reduces to (4.18) and, hence, $\tilde{\theta}_n^{(j)} = \hat{\theta}_n$ for $j = 3, 4$. Thus, the original estimator $\hat{\theta}_n$ itself may be employed for centering its bootstrap version $\theta_n^{*(j)}$ for the CBB and the SB. However, for the MBB and the NBB, $\tilde{\theta}_n^{(j)}$ need not be equal to $\hat{\theta}_n$ and, hence, computation of the bootstrap version $T_{2n}^{*(j)}$ in (4.20) requires solving an *additional* set of equations for the "right" centering constant $\tilde{\theta}_n^{(j)}$. It may be tempting to replace $\tilde{\theta}_n^{(j)}$ with $\hat{\theta}_n$ and define

$$\check{T}_{2n}^{*(j)} = \sqrt{n_0}\big(\theta_n^{*(j)} - \hat{\theta}_n\big) \tag{4.22}$$

as a bootstrap version of T_{2n} for $j = 1, 2$. However, for the MBB and the NBB, centering $\theta_n^{*(j)}$ at $\hat{\theta}_n$ introduces some *extra* bias, which typically leads to a worse rate of approximation of $\mathcal{L}(T_{2n})$ by $\mathcal{L}(\check{T}_{2n}^{*(j)} \mid \mathcal{X}_n)$ compared to the classical normal approximation (cf. Lahiri (1992a)). Indeed, this "naive centering" can render the bootstrap approximation totally invalid for M-estimators in *linear regression* models as noted by several authors in the independent case (cf. Freedman (1981), Shorack (1982), and Lahiri (1992b)).

An altogether different approach to defining the bootstrap version of T_{2n} is to reproduce the structural relation between equations (4.17) and (4.18) in the definition of the bootstrap version of the M-estimator itself. Note that if we replaced $\hat{\theta}_n$ in (4.18) by θ, then the expected value of the left side of (4.18) would be zero. As a result, the estimating function defining $\hat{\theta}_n$ is *unbiased* at the centering value θ. However, in the definition of the bootstrapped M-estimator in (4.19), this unbiasedness property of the estimating function does not always hold. A simple solution to this problem has been suggested by Shorack (1982) in the context of bootstrapping M-estimators in a linear regression model with *iid* errors. Following his approach, here we define an alternative bootstrap version $\theta_n^{**(j)}$ of $\hat{\theta}_n$ as a solution to the modified equation

$$n_0^{-1} \sum_{i=1}^{n_0} \Big[\Psi\big(Y_i^{*(j)}; \theta_n^{**(j)}\big) - \hat{\psi}_j\Big] = 0 , \tag{4.23}$$

where $\hat{\psi}_j = n_0^{-1} E_* \{ \sum_{i=1}^{n_0} \Psi(Y_i^{*(j)}; \hat{\theta}_n) \}$. Note that for all $j = 1, 2, 3, 4$, the (conditional) expectation of the estimating function $\sum_{i=1}^{n_0} [\Psi(Y_i^{*(j)}; t) - \hat{\psi}_j]$ is zero at $t = \hat{\theta}_n$. Thus, $\hat{\psi}_j$ is the appropriate constant that makes the estimating function in (4.23) *unbiased* if we are to center the bootstrapped M-estimator at $\hat{\theta}_n$. The bootstrap version of T_{2n} under this alternative approach is given by

$$T_{2n}^{**(j)} = \sqrt{n_0}(\theta_n^{**(j)} - \hat{\theta}_n) . \tag{4.24}$$

An advantage of using (4.24) over (4.20) is that for finding the bootstrap approximation under the MBB or the NBB, we need to solve only one set of equations (viz. (4.23)), as compared to solving two sets of equations (viz., (4.19) and (4.21)) under the first approach. Since $\hat{\psi}_j = n_0^{-1} \sum_{i=1}^{n_0} \Psi(Y_i; \hat{\theta}_n) = 0$ for $j = 3, 4$, the centering is *automatic* for the CBB and the SB. As a consequence, both approaches lead to the *same* bootstrap version of T_{2n} under the CBB and the SB.

The following result establishes validity of the block bootstrap approximations for the two versions $T_{2n}^{*(j)}$ and $T_{2n}^{**(j)}$, $j = 1, 2, 3, 4$. Write $\Sigma_\Psi = \lim_{n\to\infty} \text{Var}(n_0^{-1/2} \sum_{i=1}^{n_0} \Psi(Y_i; \theta))$ and let D_Ψ be the $s \times s$ matrix with (i, j)-th element $E[\frac{\partial}{\partial \theta_j} \Psi_i(Y_1; \theta)]$, $1 \le i, j \le s$. Also, assume that the solutions to the estimating equations are measurable and unique.

Theorem 4.2 *Assume that*

(i) *$\Psi(y; t)$ is differentiable with respect to t for almost all y (under F_Y) and the first-order partial derivatives of Ψ (in t) satisfy a Lipschitz condition of order $\kappa \in (0, 1]$, a.s. (F_Y), where F_Y denotes the probability distribution of Y_1.*

(ii) *$E\Psi(Y_1; \theta) = 0$, and Σ_Ψ and D_Ψ are nonsingular.*

(iii) *There exists a $\delta > 0$ such that $E\|D^\alpha \Psi(Y_1; \theta)\|^{2r_j + \delta} < \infty$ for all $\alpha \in \mathbb{Z}_+^s$ with $|\alpha| = 0, 1$, and $\Delta(r_j; \delta) < \infty$, where $r_j = 1$ for $j = 1, 2, 3$ and $r_j = 3$ for $j = 4$.*

(iv) *$\ell^{-1} + n^{-1/2}\ell = o(1)$ and $p + (n^2 p)^{-1} = o(1)$ as $n \to \infty$.*

Then,

(a) *$\{\hat{\theta}_n\}_{n \ge 1}$ is consistent for θ and*

$$\sqrt{n}(\hat{\theta}_n - \theta) \longrightarrow^d N(0, D_\Psi \Sigma_\Psi D_\Psi') .$$

(b) *For $j = 1, 2, 3, 4$,*

$$\sup_{x \in \mathbb{R}^s} \left| P_*(T_{2n}^{**(j)} \le x) - P(T_{2n} \le x) \right| \longrightarrow_p 0 \quad \text{as} \quad n \to \infty .$$

(c) Part (b) remains valid if we replace $T_{2n}^{**(j)}$ by $T_{2n}^{*(j)}$ of (4.20).

To prove Theorem 4.2, we shall follow an approach used in Bhattacharya and Ghosh (1978). The following result is useful for establishing the consistency of $\{\hat{\theta}_n\}_{n\geq 1}$.

Proposition 4.1 *(Brouwer's Fixed Point Theorem). Let $\breve{\mathcal{U}} = \{x \in \mathbb{R}^s : \|x\| \leq 1\}$ denote the unit ball in $\mathbb{R}^s$ and let $f : \breve{\mathcal{U}} \to \breve{\mathcal{U}}$ be a continuous function. Then, f has a fixed point in $\breve{\mathcal{U}}$, i.e., there exists $x_0 \in \breve{\mathcal{U}}$ such that*

$$f(x_0) = x_0 \ .$$

Proof: See page 14, Milnor (1965). □

Lemma 4.2 *Suppose that A and B are two $d \times d$ matrices for some $d \in \mathbb{N}$ and A is nonsingular.*

(a) If $\|A - B\| < \delta/\|A^{-1}\|$ for some $\delta \in (0,1)$, then B is nonsingular and

$$\|B^{-1}\| < \|A^{-1}\|/(1-\delta) \ .$$

(b) If B is also nonsingular, then

$$\|B^{-1} - A^{-1}\| \leq \|A - B\| \cdot \|A^{-1}\| \, \|B^{-1}\| \ .$$

Proof:

(a) Let $(\Gamma)^0 = \mathbb{I}_d$ for any $d \times d$ matrix Γ. Since $\sum_{k=0}^{\infty} \|\mathbb{I}_d - A^{-1}B\|^k \leq \sum_{k=0}^{\infty} (\|A^{-1}\| \, \|A - B\|)^k < \infty$, (each of the d^2 components of) the matrix-valued series $\sum_{k=0}^{\infty} (\mathbb{I}_d - A^{-1}B)^k$ is absolutely convergent. Write $Q = \sum_{k=0}^{\infty} (\mathbb{I}_d - A^{-1}B)^k$. Then,

$$\begin{aligned} Q(A^{-1}B) &= Q[\mathbb{I}_d - (\mathbb{I}_d - A^{-1}B)] \\ &= Q - \sum_{k=1}^{\infty} (\mathbb{I}_d - A^{-1}B)^k = \mathbb{I}_d \ , \end{aligned}$$

and similarly, $(A^{-1}B)Q = \mathbb{I}_d$, so that $A^{-1}B$ is nonsingular and $Q = (A^{-1}B)^{-1}$. Now, premultiplication by A and postmultiplication by A^{-1} of the identity $(A^{-1}B)Q = \mathbb{I}_d$ implies that $BQA^{-1} = \mathbb{I}_d$. Hence $(QA^{-1})B = Q(A^{-1}B) = \mathbb{I}_d = B(QA^{-1})$. This proves that B is nonsingular, with $B^{-1} = QA^{-1}$, and

$$\begin{aligned} \|B^{-1}\| &\leq \|A^{-1}\| \, \|Q\| \\ &\leq \|A^{-1}\| \sum_{k=0}^{\infty} \|\mathbb{I}_d - A^{-1}B\|^k \\ &\leq \|A^{-1}\| \sum_{k=0}^{\infty} \delta^k = \|A^{-1}\|(1-\delta)^{-1} \ . \end{aligned}$$

(b) Follows from the identity

$$B^{-1} - A^{-1} = B^{-1}(A - B)A^{-1} . \tag{4.25}$$

Hence, the proof of Lemma 4.2 is complete. □

Proof of Theorem 4.2: For notational simplicity, without loss of generality, we suppose that $m = 1$ and $EX_1 = 0$. Then, $n_0 = n$, $Y_i = X_i$ and $Y_i^{*(j)} = X_i^{*(j)}$ for all i, and for all $j = 1, 2, 3, 4$.
(a) By (4.18) and Taylor's expansion,

$$0 = n^{-1} \sum_{i=1}^{n} \Big[\Psi(X_i; \theta) + \sum_{|\alpha|=1} D^\alpha \Psi(X_i; \theta)(\hat{\theta}_n - \theta) \Big] + R_{1n} , \tag{4.26}$$

where, by condition (i),

$$\begin{aligned} \|R_{1n}\| \quad &\leq \quad n^{-1} \sum_{i=1}^{n} \sum_{|\alpha|=1} \Big\{ \|\hat{\theta}_n - \theta\| \\ &\quad \times \Big\| \int_0^1 D^\alpha \Psi\big(X_i; \theta + u(\hat{\theta}_n - \theta)\big) du - D^\alpha \Psi(X_i; \theta) \Big\| \Big\} \\ &\leq \quad C \|\hat{\theta}_n - \theta\|^{1+\kappa} . \end{aligned}$$

Note that, by Markov's inequality,

$$\begin{aligned} & P\Big(\Big\| n^{-1} \sum_{i=1}^{n} \Psi(X_i; \theta) \Big\| > n^{-1/2} \log n \Big) \\ & \leq E \Big\| n^{-1} \sum_{i=1}^{n} \Psi(X_i; \theta) \Big\|^2 \big(n (\log n)^{-2} \big) \\ & = O\big((\log n)^{-2} \big) \end{aligned} \tag{4.27}$$

and, similarly, for all $|\alpha| = 1$,

$$\begin{aligned} & P\Big(\Big\| n^{-1} \sum_{i=1}^{n} \Big\{ D^\alpha \Psi(X_i; \theta) - E D^\alpha \Psi(X_1; \theta) \Big\} \Big\| > (n^{-1/2} \log n) \Big) \\ & = O\big((\log n)^{-2} \big) . \end{aligned} \tag{4.28}$$

Next, define $A_{1n} \equiv \Big\{ \|n^{-1} \sum_{i=1}^{n} \Psi(X_i; \theta)\| \leq n^{-1/2} \log n$ and $\|n^{-1} \sum_{i=1}^{n} \big(D^\alpha \Psi(X_i; \theta) - E D^\alpha \Psi(X_i; \theta) \big)\| \leq n^{-1/2} \log n$ for all $|\alpha| = 1 \Big\}$. Then, by (4.27) and (4.28), $P(A_{1n}) \to 1$ as $n \to \infty$. Let $\hat{D}_{\Psi,n}$ be the $s \times s$ matrix with (i,j)-th element $n^{-1} \sum_{t=1}^{n} \big(\partial \Psi_i(X_t; \theta) / \partial \theta_j \big)$, $1 \leq i, j \leq s$. By

Lemma 4.2 and condition (ii), $\hat{D}_{\Psi,n}$ is nonsingular on the set A_{1n}, for n large. Hence, for n large, on the set A_{1n}, we can write (4.26) as

$$(\hat{\theta}_n - \theta) = \hat{D}_{\Psi,n}^{-1}\left[n^{-1}\sum_{i=1}^{n}\Psi(X_i;\theta) + R_{1n}\right] . \tag{4.29}$$

Note that the right side of (4.29) is a continuous function of $(\hat{\theta}_n - \theta)$; call it $g(\hat{\theta}_n - \theta)$. Now, using (4.27), (4.29), and the bound on R_{1n}, we see that there exists a $C_1 \in (1,\infty)$ such that $\|g(\hat{\theta}_n - \theta)\| \leq C_1 n^{-1/2}(\log n)$ for all $\|\hat{\theta}_n - \theta\| \leq C_1 n^{-1/2}(\log n)$. Thus, setting $f(x) = [C_1 n^{-1/2}\log n]^{-1} g([C_1 n^{-1/2}\log n]x)$, $x \in \breve{\mathcal{U}}$, we have a continuous function $f : \breve{\mathcal{U}} \to \breve{\mathcal{U}}$. Hence, by Proposition 4.1, there exists a $x_0 \in \breve{\mathcal{U}}$ such that $f(x_0) = x_0$, or equivalently, $g([C_1 n^{-1/2}\log n]x_0) = [C_1 n^{-1/2}\log n]x_0$. Since, by assumption, $(\hat{\theta}_n - \theta)$ is the *unique* solution to (4.29), we must have $\hat{\theta}_n - \theta = [C_1 n^{-1/2}\log n]x_0$. Therefore, $\|\hat{\theta}_n - \theta\| \leq C_1 n^{-1/2}(\log n)$ on the set A_{1n}, for n large. Since $P(A_{1n}) \to 1$ as $n \to \infty$, this implies that

$$\|\hat{\theta}_n - \theta\| = O_p(n^{-1/2}\log n) \quad \text{as} \quad n \to \infty . \tag{4.30}$$

In particular, $\{\hat{\theta}_n\}_{n\geq 1}$ is consistent for θ.

Next, multiplying both sides of (4.29) by $\sqrt{n}$ and using (4.28), the bound on R_{1n}, and Slutsky's Theorem, we get

$$\begin{aligned}\sqrt{n}(\hat{\theta}_n - \theta) &= \big(D_\Psi + o_p(1)\big)^{-1}\Bigg[\frac{1}{\sqrt{n}}\sum_{i=1}^{n}\Psi(X_i;\theta) \\ &\qquad + O_p\big(n^{-\kappa/2}(\log n)^{1+\kappa}\big)\Bigg] \\ &\longrightarrow^d N(0, D_\Psi\Sigma_\Psi D_\Psi') \quad \text{as} \quad n \to \infty .\end{aligned}$$

This proves part (a).

Next we prove the results on the bootstrapped M-estimators. Note that for $m = 1$, $Y_i^{*(j)} = X_i^{*(j)}$ and $n_0 = n$. To simplify the proofs, we shall consider the case where resamples are based on "complete" blocks; the proof for the resample size n is similar, but is somewhat more involved due to the effect of the "incomplete" segment from the last resampled block. Accordingly, suppose that the resample size is $n_1 \equiv b\ell$ for $j = 1,2,3$ with $b = \lfloor n\ell \rfloor$ for some integer $\ell \in (1,n)$ and it is $N_1 \equiv L_1 + \cdots + L_K$, with $K \equiv \inf\{k \geq 1 : L_1 + \cdots + L_k \geq n\}$ for $j = 4$. The bootstrap equations (4.19)–(4.21) and (4.23)–(4.24) are now redefined by replacing n_0 with n_1 for $j = 1,2,3$ and by replacing n_0 with N_1 for $j = 4$.

For proving the results, we now introduce some notation. Let $Z_{\alpha i}^{*(j)} = D^\alpha\Psi(X_i^{*(j)};\theta)$, $Z_{\alpha i} = D^\alpha\Psi(X_i;\theta)$, $\alpha \in \mathbb{Z}_+^s$, $i \geq 1$, $j = 1,2,3,4$. Also, define $A_n^{(j)} = \Big\{\|\hat{\theta}_n - \theta\| \leq Cn^{-1/2}\log n,\ \sum_{|\alpha|=1}\Big\|E_*\sum_{i=1}^{n_1}(Z_{\alpha i}^{*(j)} - EZ_{\alpha 1})\Big\| \leq$

$n^{1-\kappa/4}$, and $\sum_{|\alpha|=0}^{1} E_* \Big\| \sum_{i=1}^{n_1} (Z_{\alpha i}^{*(j)} - E_* Z_{\alpha i}^{*(j)}) \Big\|^2 \leq C(s)n \Big\}$, $1 \leq j \leq 3$. And for $j = 4$, let $A_n^{(4)}$ be defined as $A_n^{(1)}$ above, but with n_1 replaced by N_1 and $j = 1$ replaced by $j = 4$..

First we consider the case $j \in \{1,2,3\}$. Expanding both $\Psi(\cdot; \theta_n^{**(j)})$ and $\Psi(\cdot; \hat{\theta}_n)$ on the left side of the equation (4.23) in Taylor's series around θ and using condition (i) on the set $A_n^{(j)}$, we get

$$
\begin{aligned}
0 \;=\; & n_1^{-1} \sum_{i=1}^{n_1} \Bigg[\Big\{ Z_{0i}^{*(j)} + \sum_{|\alpha|=1} (\theta_n^{**(j)} - \theta)^\alpha Z_{\alpha i}^{*(j)} \Big\} \\
& - E_* \Big\{ Z_{0i}^{*(j)} + \sum_{|\alpha|=1} (\hat{\theta}_n - \theta)^\alpha Z_{\alpha i}^{*(j)} \Big\} \Bigg] \\
& + R_{1n}^{**(j)} \,,
\end{aligned}
$$

where $\|R_{1n}^{**(j)}\| < C\big[\|\theta_n^{**(j)} - \theta\|^{1+\kappa} + \|\hat{\theta}_n - \theta\|^{1+\kappa}\big]$. We can rewrite this as

$$
(\theta_n^{**(j)} - \hat{\theta}_n) = \Big[D_{\Psi,n}^{**(j)}\Big]^{-1} \Big[n_1^{-1} \sum_{i=1}^{n} \Big(Z_{0i}^{*(j)} - E_* Z_{0i}^{*(j)} \Big) + R_{2n}^{**(j)} \Big] \,, \tag{4.31}
$$

provided $D_{\Psi,n}^{**(j)}$ is nonsingular, where $D_{\Psi,n}^{**(j)}$ is the $s \times s$ matrix with (k,r)-th element $n_1^{-1} \sum_{i=1}^{n_1} \partial \Psi_k(X_i^{*(j)}; \theta)/\partial \theta_r$, $1 \leq k, r \leq s$ and where the remainder term $R_{2n}^{**(j)}$ admits the bound

$$
\begin{aligned}
\Big\| R_{2n}^{**(j)} \Big\| \;\leq\; & \|\hat{\theta}_n - \theta\| \sum_{|\alpha|=1} \Big\| n_1^{-1} \sum_{i=1}^{n_1} \Big(Z_{\alpha i}^{*(j)} - E_* Z_{\alpha i}^{*(j)} \Big) \Big\| \\
& + \Big\| R_{1n}^{**(j)} \Big\| \,.
\end{aligned} \tag{4.32}
$$

Next define the set $A_n^{**(j)}$, $j \in \{1,2,3\}$ by

$$
A_n^{**(j)} = \Big\{ \sum_{|\alpha|=0}^{1} \Big\| n_1^{-1} \sum_{i=1}^{n_1} Z_{\alpha i}^{*(j)} - E_* Z_{\alpha i}^{*(j)} \Big\| \leq n^{-1/2} \log n \Big\} \,.
$$

Note that by condition (ii) and Lemma 4.2, $D_{\Psi,n}^{**(j)}$ is nonsingular on $A_n^{**(j)} \cap A_n^{(j)}$ for large n. Hence, as in the proof of part (a), by (4.31), (4.32), and Proposition 4.1, on the set $A_n^{(j)}$, there exists a constant $C > 0$ such that

$$
\begin{aligned}
& P_*\big(\|\theta_n^{**(j)} - \hat{\theta}_n\| \leq C n^{-1/2} \log n\big) \\
& \geq\; P_*(A_n^{**(j)}) \\
& \geq\; 1 - Cn(\log n)^{-2} \sum_{|\alpha|=0}^{1} E_* \Big\| n_1^{-1} \sum_{i=1}^{n_1} Z_{\alpha i}^{*(j)} - E_* Z_{\alpha i}^{*(j)} \Big\|^2 \\
& \geq\; 1 - C \cdot (\log n)^{-2} \,,
\end{aligned} \tag{4.33}
$$

for n large.

Next write $\bar{Z}^{*(j)}_{\alpha n} = n_1^{-1}\sum_{i=1}^{n_1} Z^{*(j)}_{\alpha i}$ and $\bar{Z}_{\alpha k} = k^{-1}\sum_{i=1}^{k} Z_{\alpha i}$, $k \in \mathbb{N}$, $|\alpha| = 0, 1$. Note that $E_* \bar{Z}^{*(2)}_{\alpha n} = \bar{Z}_{\alpha n_1}$, $E_* \bar{Z}^{*(3)}_{\alpha n} = \bar{Z}_{\alpha n}$, and as in (3.13),

$$E\big\| E_* \bar{Z}^{*(1)}_{\alpha n} - EZ_{\alpha 1}\big\|^2 = O(n^{-1}) .$$

Hence, noting that $\|x\|^2 \le C(d)\|xx'\|$ for any $x \in \mathbb{R}^d$ and using (4.30), Lemma 3.2, and Theorem 3.1, we get

$$\begin{aligned}
&P\Big(\big(A_n^{(j)}\big)^c\Big) \\
&\le P\big(\|\hat{\theta}_n - \theta\| > Cn^{-1/2}\log n\big) \\
&\quad + Cn^{\kappa/2} \sum_{|\alpha|=1} E\Big\| E_*\big(\bar{Z}^{*(j)}_{\alpha n}\big) - EZ_{\alpha 1}\Big\|^2 \\
&\quad + \sum_{|\alpha|=0}^{1} P\Big(\Big\|\mathrm{Var}_*\Big(\sqrt{n_1}\bar{Z}^{*(j)}_{\alpha n}\Big) - \mathrm{Var}\Big(\sqrt{n}\bar{Z}_{\alpha n}\Big)\Big\| > C\Big) \\
&= o(1) + O(n^{\kappa/2} n^{-1}) + o(1) \to 0 \quad \text{as} \quad n \to \infty . \qquad (4.34)
\end{aligned}$$

Hence, using the definitions of the sets $A_n^{**(j)}$ and $A_n^{(j)}$, it follows from (4.32)–(4.34) that

$$P_*\Big(\Big\|D^{**(j)}_{\Psi,n} - D_\Psi\Big\| > Cn^{-\kappa/4}\Big) \longrightarrow_p 0 \quad \text{as} \quad n \to \infty \qquad (4.35)$$

and

$$P_*\Big(\Big\|\sqrt{n}R^{**(j)}_{2n}\Big\| > Cn^{-\kappa/2}(\log n)^2\Big) \longrightarrow_p 0 \quad \text{as} \quad n \to \infty . \qquad (4.36)$$

Hence, for $j = 1, 2, 3$, part (b) of the theorem now follows from Theorem 3.2, (4.31), (4.35), (4.36), and Lemma 4.1.

Next consider $j = 4$. In this case, by (4.18) and Wald's lemmas (cf. Appendix A),

$$\begin{aligned}
\hat{\psi}_j &= N_1^{-1}E_*\Big(\sum_{i=1}^{N_1} \Psi(X_i^{*(j)}; \hat{\theta}_n)\Big) \\
&= N_1^{-1}(E_*K)E_*\Big(\sum_{i=1}^{L_1} \Psi(X_i^{*(j)}; \hat{\theta}_n)\Big) \\
&= N_1^{-1}(E_*K)(E_*L_1)\Big(n^{-1}\sum_{i=1}^{n} \Psi(X_i; \hat{\theta}_n)\Big) \\
&= 0 .
\end{aligned}$$

Hence, the bootstrapped M-estimator $\theta_n^{**(j)}$ is a solution of

$$\sum_{i=1}^{N_1} \Psi\left(X_i^{*(j)}; \theta_n^{**(j)}\right) = 0 . \qquad (4.37)$$

Now, as in (4.31), using Taylor's expansion of the function $\Psi(x;\cdot)$ in (4.37) around θ, we get

$$\left(\theta_n^{**(j)} - \hat{\theta}_n\right) = \left[D_{\Psi,n}^{**(j)}\right]^{-1}\left[N_1^{-1}\sum_{i=1}^{N_1}\left(Z_{0i}^{*(j)} - E_* Z_{0i}^{*(j)}\right) + R_{2n}^{**(j)}\right] \tag{4.38}$$

where, with $j = 4$, $\|R_{2n}^{**(j)}\| < C[\|\theta_n^{**(j)} - \theta\|^{1+\kappa} + \|\hat{\theta}_n - \theta\|^{1+\kappa}] + \|\hat{\theta}_n - \theta\|\sum_{|\alpha|=1}\|N_1^{-1}\sum_{i=1}^{N_1}(Z_{\alpha i}^{*(j)} - E_* Z_{\alpha i}^{*(j)})\|$ and where $D_{\Psi,n}^{**(4)}$ has (k,r)-th element $N_1^{-1}\sum_{i=1}^{N_1}\partial\Psi_k(X_i^{**(4)};\theta)/\partial\theta_r$, $1 \le k,r \le s$. Next, define the set $A_n^{**(4)}$ by replacing n_1 with N_1 in the definition of $A_n^{**(1)}$. Since $N_1 \ge n$, it follows that on the set $A_n^{(4)}$,

$$\begin{aligned}
&1 - P_*\left(A_n^{**(4)}\right) \\
&\le\; Cn(\log n)^{-2}\sum_{|\alpha|=0}^{1} E_*\left\{N_1^{-2}\left\|\sum_{i=1}^{N_1}\left(Z_{\alpha i}^{*(4)} - E_* Z_{\alpha i}^{*(4)}\right)\right\|^2\right\} \\
&\le\; Cn(\log n)^{-2}n^{-2}\sum_{|\alpha|=0}^{1} E_*\left\|\sum_{i=1}^{N_1}\left(Z_{\alpha i}^{*(4)} - E_* Z_{\alpha i}^{*(4)}\right)\right\|^2 \\
&\le\; C(s)(\log n)^{-2}\,.
\end{aligned} \tag{4.39}$$

Let $S_{\alpha k}^{*(4)}$ is the kth block sum of the $Z_{\alpha i}^{*(4)}$'s, $\alpha \in \mathbb{Z}_+^s$, $|\alpha| = 0, 1$. Then, using an iterated conditioning argument similar to the one employed in the proof of Theorem 3.4 (cf. (3.38)), and using Wald's lemmas (cf. Appendix A), we have

$$\begin{aligned}
&E_*\sum_{i=1}^{N_1}\left(Z_{\alpha i}^{*(4)} - EZ_{\alpha 1}\right) \\
&= E\left[\sum_{k=1}^{K} E\left\{\left(S_{\alpha k}^{*(4)} - L_k EZ_{\alpha 1}\right) \mid \mathcal{T}_n\right\} \mid \mathcal{X}_n\right] \\
&= E\left\{\sum_{k=1}^{K} L_k(\bar{Z}_{\alpha n} - EZ_{\alpha 1}) \mid \mathcal{X}_n\right\} \\
&= \left\{(E_* K)p^{-1}\right\}\left(\bar{Z}_{\alpha n} - EZ_{\alpha 1}\right)
\end{aligned}$$

and

$$\begin{aligned}
&E_*\left\|\sum_{i=1}^{N_1}\left(Z_{\alpha i}^{*(4)} - E_* Z_{\alpha i}^{*(4)}\right)\right\|^2 \\
&= E\left\{\sum_{k=1}^{K} E\left(\left\|S_{\alpha k}^{*(4)} - L_k\bar{Z}_{\alpha n}\right\|^2 \mid \mathcal{T}_n\right) \mid \mathcal{X}_n\right\} \\
&= (E_* K)E_*\left\|S_{\alpha 1}^{*(4)} - L_1\bar{Z}_{\alpha n}\right\|^2
\end{aligned}$$

for all $\alpha \in \mathbb{Z}_+^s$, $|\alpha| = 0, 1$. Let $\hat{\tau}^2(\ell; r, \alpha)$ denote the conditional variance of the rth component of $S_{\alpha 1}^{*(4)}$ given $\mathcal{T}_n$, $|\alpha| = 0, 1$, $r = 1, \ldots, s$. Then, noting that $E(S_{\alpha 1}^{*(4)} \mid \mathcal{T}_n) = L_1 \bar{Z}_{\alpha n}$, we have

$$
\begin{aligned}
& E_* \left\| S_{\alpha 1}^{*(4)} - L_1 \bar{Z}_{\alpha n} \right\|^2 \\
= \; & E\Big(\Big\{ E \left\| S_{\alpha 1}^{*(4)} - L_1 \bar{Z}_{\alpha n} \right\|^2 \mid \mathcal{T}_n \Big\} \mid \mathcal{X}_n \Big) \\
= \; & \sum_{r=1}^{s} E_* \hat{\tau}^2(L_1; r, \alpha) \; .
\end{aligned}
$$

Hence, using Lemma 3.4 (parts a(iv) and b(ii)) and the above identities, we get

$$
\begin{aligned}
& 1 - P\Big(A_n^{(n)}\Big) \\
\leq \; & P\left(\|\hat{\theta}_n - \theta\| > n^{-1/2} \log n \right) \\
& + P\Big(\sum_{|\alpha|=1} \|\bar{Z}_{\alpha n} - EZ_{\alpha 1}\| > Cn^{-\kappa/4} \Big) \\
& + P\Big(\sum_{|\alpha|=0}^{1} E_* \left\| S_{\alpha 1}^{*(4)} - L_1 \bar{Z}_{\alpha n} \right\|^2 > C(s) C p^{-1} \Big) \\
\to \; & 0 \quad \text{as} \quad n \to \infty \; .
\end{aligned} \tag{4.40}
$$

Now part (b), $j = 4$ follows from (4.39), (4.40), Theorem 3.4 and Lemma 4.1.

Part (c) of the theorem can be proved using similar arguments. □

4.4 Differentiable Functionals

In this section, we consider consistency of the MBB approximation for estimators that are smooth functions of the empirical process. Let $\{X_i\}_{i \in \mathbb{Z}}$ be a sequence of stationary $\mathbb{R}^d$-valued random vectors. Define the empirical distribution of m-dimensional subseries of the data $X_1, \ldots, X_n$ as

$$
F_n^{(m)} = (n - m + 1)^{-1} \sum_{i=1}^{n-m+1} \delta_{Y_i} \; ,
$$

where $Y_i \equiv (X_i', \ldots, X_{i+m-1}')'$, $i \geq 1$, and where δ_y denotes the probability measure putting mass one at y. The probability distribution $F_n^{(m)}$ serves as an estimator of the marginal distribution $F^{(m)}$ of the m-dimensional subseries $(X_1, \ldots, X_m)$.

Suppose that the level-1 parameter of interest θ is a s-dimensional functional of $F^{(m)}$, given by

$$\theta = T(F^{(m)}) . \tag{4.41}$$

Then, a natural estimator of θ is given by

$$\hat{\theta}_n = T(F_n^{(m)}) . \tag{4.42}$$

Many commonly used estimators, including the generalized M-estimators of Section 4.3, may be expressed as (4.42). For the generalized M-estimator, the relevant functional $T(\cdot)$ is defined *implicitly* by the relation

$$\int \Psi\big(x_1, \ldots, x_m; T(G^{(m)})\big) dG^{(m)} = 0, \quad G^{(m)} \subset \mathcal{G}^{(m)}$$

for a suitable family $\mathcal{G}^{(m)}$ of probability measures on $\mathbb{R}^{dm}$, depending on the function Ψ. Below we describe another important class of robust estimators that can be expressed in this form.

Example 4.4: Suppose that the process $\{X_i\}_{i\in\mathbb{Z}}$ is *real-valued.* Then, for any $1/2 < \alpha < 1$, the α-trimmed mean is given by

$$\hat{\theta}_n = \frac{1}{n(1-2\alpha)} \sum_{i=\lfloor n\alpha \rfloor+1}^{\lfloor n(1-\alpha)\rfloor} X_{i:n} , \tag{4.43}$$

where $X_{1:n} \le \cdots \le X_{n:n}$ denote the order-statistics corresponding to $X_1, \ldots, X_n$. Write $F_n^{(1)} = F_n$ and $F^{(1)} = F$ for notational simplicity. Then, the estimator $\hat{\theta}_n$ in (4.43) is asymptotically equivalent to a functional $T(\cdot)$ of the one-dimensional empirical distribution F_n, given by

$$T(F_n) \equiv \int_\alpha^{1-\alpha} F_n^{-1}(u)du \Big/ (1-2\alpha) ,$$

where, for any probability distribution G on $\mathbb{R}$, $G^{-1}(u) = \inf\{x \in \mathbb{R} : G((-\infty, x]) \ge u\}$, $0 < u < 1$, denotes the quantile transform of G. The estimator $\hat{\theta}_n$ or $T(F_n)$ is used for estimating the parameter

$$\theta = T(F) \equiv \int_\alpha^{1-\alpha} F^{-1}(u)du \Big/ (1-2\alpha) .$$

The α-trimmed mean $\hat{\theta}_n$ is a robust estimator of θ that guards against the influence of extreme values and outliers. Note that as $\alpha \to \frac{1}{2}-$, the limiting form of the parameter θ is the population median $F^{-1}(1/2)$ (provided $F^{-1}(u)$ is continuous at $u = 1/2$) while for $\alpha = 0$, we get $\theta = EX_1 =$ the population mean. Thus, for different values of α, θ provides some of the most commonly used measures of central tendency.

Somewhat more generally, the class of L-estimators of location parameters can be expressed in the form (4.42). Let $J : (0,1) \to \mathbb{R}$ be a Borel measurable function. Then, define the L-estimator $\hat{\theta}_n$ with weight function $J(\cdot)$ as

$$\hat{\theta}_n = \int_0^1 J(u) F_n^{-1}(u) du.$$

In this case, the functional $T(\cdot)$ is given by $T(G) = \int_0^1 J(u) G^{-1}(u) du$. □

Now we return to the discussion of a general estimator (4.42) that is a functional of $F_n^{(m)}$. If the functional $T(\cdot)$ is smooth, then asymptotic distribution of $\hat{\theta}_n$ can be derived from the asymptotic distribution of the empirical process $\left\{\sqrt{n}\left(F_n^{(m)}(\cdot) - F^{(m)}(\cdot)\right)\right\}$ on a suitable metric space, using a version of the Delta method. To prove consistency of the block bootstrap methods for the distribution of such estimators, we need to know the asymptotic behavior of the bootstrapped empirical process. In Section 4.4.1, we describe some results on the bootstrapped empirical process, and establish consistency of bootstrap for estimators of the form (4.42) in Section 4.4.2.

4.4.1 Bootstrapping the Empirical Process

Results on asymptotic distribution of empirical processes for dependent random variables have been obtained by several authors; see, for example, Billingsley (1968), Deo (1973), Sen (1974), Yoshihara (1975), Arcones and Yu (1994), and the references therein. Similar results for the bootstrapped empirical process are known only for the MBB and the CBB. For the ease of exposition, in this section, we shall suppose that $m = 1$. Indeed, if we set $Y_i = (X_i', \ldots, X_{i+m-1}')'$, $i \in \mathbb{Z}$ and $n_0 = n - m + 1$, then, in terms of the Y-variates,

$$F_n^{(m)} = n_0^{-1} \sum_{i=1}^{n_0} \delta_{Y_i}$$

and this essentially reduce the general problem to the case $m = 1$ for the dm-dimensional random vectors $Y_1, \ldots, Y_{n_0}$. Hence, without loss of generality, we set $m = 1$. Also, for notational simplicity, write $F_n^{(1)} = F_n$ and $F^{(1)} = F$.

Let $\mathbb{D}_d$ denote the space of all real-valued functions on $[-\infty, \infty]^d$ that are continuous from above and have limits from below. We equip $\mathbb{D}_d$ with the (extended) Skorohod metric (cf. Bickel and Wichura (1971)). Write $F_n(x)$ and $F(x)$ for the distribution functions corresponding to the probability measures F_n and F, respectively. Thus,

$$F_n(x) = n^{-1} \sum_{i=1}^{n} \mathbb{1}(X_i \le x) \tag{4.44}$$

and

$$F(x) = P(X_1 \leq x) , \tag{4.45}$$

$x \in [-\infty, \infty]^d$. Recall that for any two vectors $x \equiv (x_1, \ldots, x_d)'$ and $y \equiv (y_1, \ldots, y_d)'$, $x \leq y$ means $x_i \leq y_i$ for all $1 \leq i \leq d$. Define the empirical process

$$W_n(x) = \sqrt{n}\big(F_n(x) - F(x)\big), \quad x \in [-\infty, \infty]^d .$$

Then, under some regularity conditions (cf. Yoshihara (1975)), the empirical process W_n converges weakly (as $\mathbb{D}_d$-valued random elements) to a Gaussian process W on $[-\infty, \infty]^d$ satisfying

$$\begin{aligned} &EW(x) = 0 \\ &EW(x)W(y) = \sum_{k=-\infty}^{\infty} \mathrm{Cov}\big(\mathbb{1}(X_1 \leq x), \mathbb{1}(X_{1+k} \leq y)\big) , \\ &\qquad\text{and} \\ &P(W \in \mathbb{C}_d^0) = 1 , \end{aligned} \tag{4.46}$$

for all $x, y \in [-\infty, \infty]^d$, where $\mathbb{C}_d^0$ is the collection of continuous functions on $[-\infty, \infty]^d$ that vanish at $(-\infty, \ldots, -\infty)'$ and $(\infty, \ldots, \infty)'$. The next theorem shows that a similar result holds for the bootstrapped empirical process. Let $F_n^*(x)$ denote the empirical process of n_1 MBB samples based on blocks of length ℓ, and let $W_n^*(x) = \sqrt{n_1}\big(F_n^*(x) - E_* F_n^*(x)\big),\ x \in \mathbb{R}^d$.

Theorem 4.3 *Suppose that* $\{X_n\}_{n \geq 1}$ *is a sequence of stationary strong-mixing* $\mathbb{R}^d$*-valued random vectors with* $\sum_{i=1}^{\infty} i^{8d+7}[\alpha(i)]^{1/2} < \infty$, *and that* X_1 *has a continuous distribution. Let* $\ell \to \infty$ *and* $\ell = O(n^{1/2-\epsilon})$ *as* $n \to \infty$ *for some* $0 < \epsilon < 1/2$. *Then,*

$$W_n^* \longrightarrow^d W \quad \textit{as} \quad n \to \infty \quad \textit{almost surely} .$$

Proof: See Bühlmann (1994).

Note that $\mathbb{D}_d$ is a complete separable metric space and, hence, there is a metric, say, ϱ, that metricizes the topology of weak convergence on $\mathbb{D}_d$ (cf. Parthasarathi (1967), Billingsley (1968)). Thus, Theorem 4.3 implies that

$$\varrho\big(\mathcal{L}(W_n^* \mid \mathcal{X}_n), \mathcal{L}(W)\big) \longrightarrow 0 \quad \text{as} \quad n \to \infty \quad \text{a.s.} \tag{4.47}$$

Results in Yoshihara (1975) yield weak convergence of W_n to W on $\mathbb{D}_d$ under the conditions of Theorem 4.3. Hence, combining the two, we have the following corollary.

Corollary 4.1 *Under the conditions of Theorem 4.3,*

$$\varrho\big(\mathcal{L}(W_n^* \mid \mathcal{X}_n), \mathcal{L}(W_n)\big) \longrightarrow 0 \quad \textit{as} \quad n \to \infty \quad \textit{a.s.}$$

A variant of Theorem 4.3 in the one-dimensional case (i.e., $d = 1$) has been proved by Naik-Nimbalkar and Rajarshi (1994), also for the MBB. In the $d = 1$ case, Peligrad (1998) proves a version of Theorem 4.3 for the CBB under a significantly weaker condition on the strong mixing coefficient, but assuming a restrictive condition on the block length. A rigorous proof of weak convergence of empirical processes for the SB does not seem to exist in the literature at this point. Note that by Prohorov's Theorem (cf. Billingsley (1968)), proving this would involve showing (i) weak convergence of finite dimensional distributions and (ii) tightness of the bootstrapped empirical processes. Theorem 3.4 shows that the finite-dimensional distributions under the SB method has the appropriate limits. Thus, the main technical problem that needs to be resolved here is the tightness of the bootstrapped empirical process.

4.4.2 Consistency of the MBB for Differentiable Statistical Functionals

Let $\hat{\theta}_n$ be an estimator that can be represented as a *smooth* functional of the m-dimensional empirical distribution $F_n^{(m)}$ as in (4.42) for some $m \geq 1$. A general approach to deriving asymptotic distributions of such statistical functionals using differentiability was initiated by von Mises (1947), and further developed by Filippova and Brunswick (1962) and Reeds (1976), among others. For a systematic treatment of the topic and some applications, see Boos (1979), Serfling (1980), Huber (1981), and Fernholz (1983). Here we consider statistical functionals that are *Fréchet differentiable*. Let $\mathbb{P}_k$ denote the set of all probability measures on $\mathbb{R}^k$ and let $\mathbb{S}_k$ denote the set of all finite signed measures on $\mathbb{R}^k$, $k \geq 1$. Let $\|\cdot\|_{(k)}$ be a norm on $\mathbb{S}_k$ and let $\|\cdot\|$ denote the Euclidean norm on $\mathbb{R}^k$. Then, we say that a functional $T : \mathbb{P}_k \to \mathbb{R}^s$ is *Fréchet differentiable at* $F \in \mathbb{P}_k$ *under* $\|\cdot\|_{(k)}$ if there exists a function $T^{(1)}(F;\cdot) : \mathbb{S}_k \to \mathbb{R}^s$ such that

(i) $T^{(1)}(F;\cdot)$ is linear, i.e.,

$$T^{(1)}(F; a\nu_1 + b\nu_2) = aT^{(1)}(F;\nu_1) + bT^{(1)}(F;\nu_2)$$

for all $a, b \in \mathbb{R}$, and $\nu_1, \nu_2 \in \mathbb{S}_k$;

and

(ii) for $G \in \mathbb{P}_k$,

$$\frac{\|T(G) - T(F) - T^{(1)}(F; G - F)\|}{\|G - F\|_{(k)}} \to 0$$

as $\|G - F\|_{(k)} \to 0$.

The linear functional $T^{(1)}(F;\cdot)$ is called the *Fréchet derivative* of T at F. This definition of Fréchet differentiability is slightly different from the standard Functional Analysis definition (cf. Diéudonne (1960)), where the functional T is assumed to be defined on a *normed vector space*, like $\mathbb{S}_k$, rather than just on $\mathbb{P}_k$. Thus, for the standard definition, one has to extend the definition of a functional T from $\mathbb{P}_k$ to $\mathbb{S}_k$. However, the given definition is adequate for our purpose, since we are only interested in the values of the functional at probability measures, not at signed measures. A similar definition of Fréchet differentiability has been used by Huber (1981).

Now suppose that the parameter θ and its estimator $\hat{\theta}_n$ are given by (4.41) and (4.42), i.e., $\theta = T(F^{(m)})$ and $\hat{\theta}_n = T(F_n^{(m)})$ for some functional $T : \mathbb{P}_{dm} \to \mathbb{R}^s$, for some $m, s \in \mathbb{N}$. If T is Fréchet differentiable at $F^{(m)}$ with Fréchet derivative $T^{(1)}(F^{(m)};\cdot)$, then by the linearity property of $T^{(1)}(F^{(m)};\cdot)$,

$$\begin{aligned}
\hat{\theta}_n - \theta &= T(F_n^{(m)}) - T(F^{(m)}) \\
&= T^{(1)}(F^{(m)}; F_n^{(m)} - F^{(m)}) + R_n \\
&= n_0^{-1} \sum_{i=1}^{n_0} T^{(1)}\big(F^{(m)}; \delta_{Y_i} - F^{(m)}\big) + R_n \\
&\equiv n_0^{-1} \sum_{i=1}^{n_0} h(Y_i) + R_n, \quad \text{say}\ , \qquad (4.48)
\end{aligned}$$

where $n_0 = (n - m + 1)$, $h(y) = T^{(1)}\big(F^{(m)}; \delta_y - F^{(m)}\big)$, $y \in \mathbb{R}^{dm}$, and R_n is the remainder term satisfying $\|R_n\| = o\big(\|F_n^{(m)} - F^{(m)}\|_{(dm)}\big)$ as $\|F_n^{(m)} - F^{(m)}\|_{(dm)} \to 0$. Therefore, we may obtain asymptotic distribution of $\sqrt{n}(\hat{\theta}_n - \theta)$ from (4.48), provided that $R_n = o_p(n^{-1/2})$. The latter condition holds if $\sqrt{n}\|F_n^{(m)} - F^{(m)}\|_{(dm)}$ is stochastically bounded under the norm $\|\cdot\|_{(dm)}$. Here, we shall take $\|\cdot\|_{(dm)}$ to be Kolmogorov's half-space norm defined by

$$\|\nu\|_\infty = \sup\{|\nu((-\infty, x])| : x \in \mathbb{R}^{dm}\},\ \nu \in \mathbb{S}_{dm}\ . \qquad (4.49)$$

Then, we have the following result.

Theorem 4.4 *Assume that $T(\cdot)$ is Fréchet differentiable at $F^{(m)}$ in the $\|\cdot\|_\infty$-norm and that $E\|h(Y_1)\|^3 < \infty$, $Eh(Y_1) = 0$ and $\Sigma_\infty^{(m)} \equiv \lim_{n\to\infty} Var\big(n^{-1/2}\sum_{i=1}^n h(Y_i)\big)$ is nonsingular. Also, assume that the conditions of Theorem 4.3 hold. Then,*

(a) $\sqrt{n}(\hat{\theta}_n - \theta) \longrightarrow^d N(0, \Sigma_\infty^{(m)})$ *as* $n \to \infty$.

(b) Let $\theta_n^* = T(F_n^{(m)*})$ and $\tilde{\theta}_n = T(E_* F_n^{(m)*})$, where $F_n^{(m)*}$ is the empirical distribution corresponding to a MBB sample based on $b_0 \equiv \lfloor n_0/\ell \rfloor$

resampled blocks of length ℓ. Then,

$$\sup_{x\in\mathbb{R}^s}\Big|P_*\Big(\sqrt{n_0}(\theta_n^* - \tilde{\theta}_n) \le x\Big) - P\Big(\sqrt{n}(\hat{\theta}_n - \theta) \le x\Big)\Big| \\ \longrightarrow_p 0 \quad as \quad n \to \infty \; .$$

Thus, the MBB approximation to the sampling distribution of $\sqrt{n}(\hat{\theta}_n - \theta)$ is consistent under the conditions of Theorem 4.4. Here we remark that Theorem 4.4 remains valid under a weaker moment condition on $h(Y_1)$ than what we have assumed above. Indeed, conclusions of Theorem 4.4 hold under the moment condition '$E\|h(Y_1)\|^{2+\delta} < \infty$ for some $\delta > 0$', provided $\sum_{i=1}^{\infty}\alpha(i)^{\delta/(2+\delta)} < \infty$ and the remaining conditions of Theorem 4.4 hold. We have used the stronger moment condition on $h(Y_1)$ only to simplify the proof.

Proof: For a function $f : [-\infty,\infty]^k \to \mathbb{R}$, let $\|f\|_\infty = \sup\{|f(x)| : x \in [-\infty,\infty]^k\}$, $k \ge 1$. Note that this definition is consistent with (4.49), in the sense that $\|\nu\|_\infty$ can be interpreted either as the Kolmogorov norm for the signed measure ν or as the sup norm for the corresponding distribution function.

Under the conditions of Theorem 4.3, $W_n^{(m)}(x) \equiv \sqrt{n}\big(F_n^{(m)}((-\infty,x]) - F^{(m)}((-\infty,x])\big)$, $x \in [-\infty,\infty]^{dm}$ converges weakly (as random elements of $\mathbb{D}_{dm}$ equipped with the extended Skorohod topology) to a Gaussian process $W^{(m)}(x)$, where $W^{(m)}(x)$ has continuous sample paths with probability one. Note that for a *continuous* function f on $[-\infty,\infty]^{dm}$, if $f_n \in \mathbb{D}_{dm}$ converges to f in the Skorohod metric, then $\sup\{|f_n(x) - f(x)| : x \in [-\infty,\infty]^{dm}\} \to 0$ as $n \to \infty$. Thus, the mapping $g : \mathbb{D}_{dm} \to \mathbb{R}$, defined by $g(f) = \|f\|_\infty$ is continuous under the Skorohod metric at $f = f_0$ if f_0 is continuous on $[-\infty,\infty]^d$. Hence, by the continuous mapping Theorem (cf. Theorem 5.1, Billingsley (1968)),

$$\|\sqrt{n}\big(F_n^{(m)} - F^{(m)}\big)\|_\infty = O_p(1) \; .$$

Now part (a) of the theorem follows from (4.48) and the Central Limit Theorem for dependent random vectors (cf. Theorem A.8, Appendix A).

To prove the second part, for notational simplicity, we assume that $n_0 = b_0\ell$. Also, let Z be a random vector having the $N(0,\Sigma_\infty^{(m)})$ distribution on $\mathbb{R}^s$. Then, we need to show that

$$\sup_{x\in\mathbb{R}^s}\Big|P_*\Big(\sqrt{n_0}(\theta_n^* - \tilde{\theta}_n) \le x\Big) - P(Z \le x)\Big| \longrightarrow_p 0 \quad \text{as} \quad n \to \infty \; .$$

Note that by the Fréchet differentiability of $T(\cdot)$ at $F^{(m)}$ under $\|\cdot\|_\infty$, we can write

$$T(G^{(m)}) - T(F^{(m)}) = T^{(1)}(F^{(m)}; G^{(m)} - F^{(m)}) + R(G^{(m)}) \; , \qquad (4.50)$$

where, for any $\epsilon > 0$, there exists a $\delta > 0$ such that

$$\|R(G^{(m)})\| \leq \epsilon \cdot \|G^{(m)} - F^{(m)}\|_\infty \tag{4.51}$$

whenever $\|G^{(m)} - F^{(m)}\|_\infty < \delta$. Also, by the linearity of $T^{(1)}(F^{(m)}; \cdot)$,

$$\begin{aligned}
& T^{(1)}\Big(F^{(m)}; E_* F_n^{(m)*} - F^{(m)}\Big) \\
&= T^{(1)}\bigg(F^{(m)}; \big[(n_0 - \ell + 1)\ell\big]^{-1} \sum_{j=1}^{n_0-\ell+1} \sum_{i=j}^{j+\ell-1} \big(\delta_{Y_i} - F^{(m)}\big)\bigg) \\
&= \big[(n_0 - \ell + 1)\ell\big]^{-1} \sum_{j=1}^{n_0-\ell+1} \sum_{i=j}^{j+\ell-1} h(Y_i) \\
&= E_*\bigg\{l^{-1} \sum_{i=1}^{\ell} h(Y_i^*)\bigg\} = E_*\bigg\{{n_0}^{-1} \sum_{i=1}^{n_0} h(Y_i^*)\bigg\} .
\end{aligned}$$

Hence, by (4.48) and (4.50), we get

$$\begin{aligned}
\sqrt{n_0}(\hat{\theta}_n^* - \tilde{\theta}_n) &= \sqrt{n_0}\Big[\Big(T(F_n^{(m)*}) - T(F^{(m)})\Big) \\
&\quad - \Big(T(E_* F_n^{(m)*}) - T(F^{(m)})\Big)\Big] \\
&= \frac{1}{\sqrt{n_0}} \sum_{i=1}^{n_0} \Big[h(Y_i^*) - E_* h(Y_i^*)\Big] + R_n^* \\
&\equiv T_n^* + R_n^*, \quad \text{say} ,
\end{aligned} \tag{4.52}$$

where $n_0^{-1/2} R_n^* = R\big(F_n^{(m)*}\big) + R\big([E_* F_n^{(m)*}]\big)$.

Let $A(x, \epsilon)$ be the ϵ-neighborhood of the boundary of $(-\infty, x]$, defined by $A(x, \epsilon) = (-\infty, x + \epsilon\mathbf{1}] \backslash (-\infty, x - \epsilon\mathbf{1}]$, $\epsilon > 0$, $x \in \mathbb{R}^s$, where $\mathbf{1} = (1, \ldots, 1)' \in \mathbb{R}^s$. Then, for any $\epsilon > 0$,

$$\begin{aligned}
\Delta_n &\equiv \Big\|P_*\Big(\sqrt{n}(\theta_n^* - \tilde{\theta}_n) \leq x\Big) - P(Z \leq x)\Big\|_\infty \\
&\leq \|P_*(T_n^* \leq x) - P(Z \leq x)\|_\infty + P_*(\|R_n^*\| > \epsilon) \\
&\quad + \sup_x P(Z \in A(x, \epsilon)) \\
&\equiv \Delta_{1n} + \Delta_{2n}(\epsilon) + \Delta_3(\epsilon), \quad \text{say.}
\end{aligned} \tag{4.53}$$

Since Z has a normal distribution, there exist $C_0 > 1$ and $\epsilon_0 > 0$ such that for all $0 < \epsilon < \epsilon_0$,

$$\Delta_3(\epsilon) \leq C_0 \epsilon . \tag{4.54}$$

Next, note that $\|W_n^{(m)}\|_\infty = O_p(1)$ and that by (an extension of) Theorem 4.3 and the continuous mapping theorem (cf. Theorem 5.1, Billingsley

(1968)), $\varrho\big(\mathcal{L}(\|W_n^{*(m)}\|_\infty \mid \mathcal{X}_n), \mathcal{L}(\|W^{(m)}\|_\infty)\big) \longrightarrow_p 0$ as $n \to \infty$. Hence, given $\eta \in (0,1)$, there exists $M > 1$ such that for all $n \geq M$,

$$\begin{aligned} &P(\|W^{(m)}\|_\infty > M) < \eta/12 \\ &P(\|W_n^{(m)}\|_\infty > M) < \eta/6 \\ &P\big(P_*(\|W_n^{*(m)}\|_\infty > M) > \eta/3\big) < \eta/6 \; . \end{aligned} \tag{4.55}$$

Now, fix $\eta \in (0, \epsilon_0)$. Let $\epsilon_1 = \eta/(3C_0)$. Then, by (4.51) (with $\epsilon = \epsilon_1/6M$), there exists $M_1 \geq M$ such that for all $n \geq M_1$,

$$\|R(F_n^{(m)*})\| \leq \epsilon_1/(2\sqrt{n_0})$$

on $\big\{\|F_n^{(m)*} - F^{(m)}\|_\infty \leq 3M/\sqrt{n_0}\big\}$ and

$$\|R(E_* F_n^{(m)*})\| \leq \epsilon_1/(2\sqrt{n_0})$$

on $A_n \equiv \big\{\|E_* F_n^{(m)*} - F^{(m)}\|_\infty \leq 2M/\sqrt{n}\big\}$. Hence, using (4.52) and (4.55), and the arguments leading to (3.13), for $n \geq M_1$ from (4.53), we get

$$\begin{aligned} &P(\Delta_{2n}(\epsilon_1) > \eta/3) \\ &\leq P\Big(\big\{P_*(\|R_n^*\| > \epsilon_1) > \eta/3\big\} \cap A_n \cap \big\{\|R(E_* F_n^{(m)*})\| \leq \epsilon_1/(2\sqrt{n_0})\big\}\Big) \\ &\quad + P\big(\|E_* F_n^{(m)*} - F^{(m)}\|_\infty > 2M/\sqrt{n}\big) \\ &\leq P\Big(\big\{P_*(\|R(F_n^{(m)*})\| > \epsilon_1/(2\sqrt{n_0})) > \eta/3\big\} \cap A_n\Big) \\ &\quad + P\big(\|E_* F_n^{(m)*} - F_n^{(m)}\|_\infty > M/\sqrt{n}\big) + P(\|W_n^{(m)}\|_\infty > M) \\ &\leq P\Big(\big\{P_*(\|(F_n^{(m)*} - F^{(m)})\|_\infty > 3M/\sqrt{n_0}) > \eta/3\big\} \cap A_n\Big) \\ &\quad + M^{-1} n^{1/2} E\big(\|E_* F_n^{(m)*} - F_n^{(m)}\|_\infty\big) + \eta/6 \\ &\leq P\big(P_*(\|W_n^{*(m)}\|_\infty > M) > \eta/3\big) + M^{-1} n^{1/2}(2\ell/n) + \eta/6 \\ &\leq \frac{\eta}{3} + 2M^{-1} n^{-1/2} \ell \; . \end{aligned} \tag{4.56}$$

Also, note that by Theorem 3.2, (4.52), and (4.53), $\Delta_{1n} \longrightarrow_p 0$. Hence, for any $0 < \eta < \epsilon_0$, by (4.54) and (4.56), for sufficiently large n,

$$\begin{aligned} P(\Delta_n > \eta) &\leq P(\Delta_{1n} > \eta/3) + P(\Delta_{2n}(\epsilon_1) > \eta/3) \\ &\leq \eta/3 + \big(\eta/3 + 2\ell/(M n^{1/2})\big) \\ &< \eta \; . \end{aligned}$$

This completes the proof of the theorem. □

The proof of Theorem 4.4 can be simplified significantly, if instead of Fréchet differentiability, we assume a stronger version of it, known as the *strong Fréchet differentiability* (cf. Liu and Singh (1992)). A functional T is called *strongly Fréchet differentiable* at $F \in \mathbb{P}_k$ under $\|\cdot\|_{(k)}$ if there exists a linear function $T^{(1)}(F;\cdot) : \mathbb{S}_k \to \mathbb{R}^s$ such that

$$\|T(G) - T(H) - T^{(1)}(F; G - H)\| / \|G - H\|_{(k)} \to 0$$

as $\|G - F\|_{(k)} \to 0$ and $\|H - F\|_{(k)} \to 0$. While Fréchet differentiability of many robust estimators is known, the notion of strong Fréchet differentiability for statistical functionals is not very well-studied in the literature. Hence, we have established validity of the bootstrap approximation assuming regular Fréchet differentiability only, so that Theorem 4.4 can be readily applied in such known cases. For results under a further weaker notion of differentiability, viz., Hadamard differentiability, see Chapter 12.

4.5 Examples

Example 4.5: Let $\{X_{0i}\}_{i\in\mathbb{Z}}$ be a stationary real-valued time series with autocovariance function $\gamma(k) = \text{Cov}(X_{0i}, X_{0(i+k)})$, $i, k \in \mathbb{Z}$. For $0 \le u < n$, let $\hat{\gamma}_n(k) = (n-k)^{-1}\sum_{i=1}^{n-k} X_{0i}X_{0(i+k)} - \bar{X}^2_{0(n-k)}$ be the estimator of $\gamma(k)$ introduced in Example 4.1. Then $\hat{\theta}_n = \hat{\gamma}_n(k)$ and $\theta = \gamma(k)$ admit a representation satisfying the requirements of the "Smooth Function Model". Since the function $H(\cdot)$ is infinitely many times differentiable, conclusions of Theorem 4.1 hold for $\hat{\gamma}_n(k)$ and $\gamma(k)$, provided the time series $\{X_{0i}\}_{i\in\mathbb{Z}}$ satisfies the relevant moment and strong mixing conditions.

For the purpose of illustration, now suppose that $\{X_{0i}\}_{i\in\mathbb{Z}}$ is an ARMA (3,4) process specified by

$$\begin{aligned} X_{0i} &- 0.4X_{0(i-1)} - 0.2X_{0(i-2)} - 0.1X_{0(i-3)} \\ &= \epsilon_i + 0.2\epsilon_{i-1} + 0.3\epsilon_{i-2} + 0.2\epsilon_{i-3} + 0.1\epsilon_{i-4}\ , \end{aligned} \tag{4.57}$$

where $\{\epsilon_i\}_{i\in\mathbb{Z}}$ is a sequence of iid $N(0,1)$ random variables. Then, $\{X_{0i}\}_{i\in\mathbb{Z}}$ is strongly mixing, with an exponentially decaying mixing coefficient (cf. Doukhan (1994)) and it has finite moments of all orders. Thus, the conditions of Theorem 4.1 hold for this ARMA (3,4) process. By Theorem 4.1 and Remark 4.2, all four block bootstrap methods provide consistent estimators of the sampling distribution and the second moment of

$$T_{1n} \equiv \sqrt{n-k}(\hat{\gamma}_n(k) - \gamma(k)) = \sqrt{n-k}(\hat{\theta}_n - \theta)\ .$$

Here, we look at the finite-sample performance of different block bootstrap methods for estimating the *mean squared error* (MSE) of $\hat{\gamma}_n(k)$ when $k = 2$ and the sample size $n = 102$. Thus, the level-2 parameter of interest here is given by

$$\varphi_n \equiv ET^2_{1n} = (n-2)\cdot \text{MSE}\big(\hat{\gamma}_n(2)\big)\ .$$

For the process (4.57), the value of φ_n, found by 10,000 simulation runs, is given by 1.058, and the value of the level-1 parameter θ is given by -0.0131. Figure 4.1 below presents a realization of a sample of size $n = 102$ from the ARMA process (4.57). We now apply the MBB, the NBB, the CBB, and the SB to this data set.

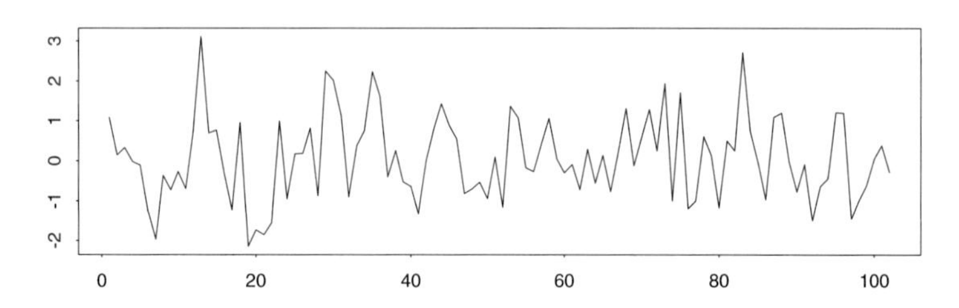

FIGURE 4.1. A simulated data set of size $n = 102$ from the ARMA process (4.57).

First consider the MBB. From the $n = 102$ original observations $\{X_{0i} : i = 1, \ldots, 102\}$, we define the vectors X_i's by the relation $X_i = (X_{0i}, X_{0i}X_{0(i+2)})'$ for $i = 1, \ldots, 100$. For the purpose of illustration, we suppose that the block length ℓ is equal to 8. Then, we define the overlapping blocks in terms of the X_i's as $\mathcal{B}_i = \{X_i, \ldots, X_{i+7}\}$, $i = 1, \ldots, 93$ and draw a simple random sample of size k_0 blocks from $\{\mathcal{B}_1, \ldots, \mathcal{B}_{93}\}$, where k_0 is the smallest integer $\geq 100/\ell$. Thus, for $\ell = 8$, $k_0 = 13$. Let $\mathcal{B}_1^*, \ldots, \mathcal{B}_{k_0}^*$ denote the resampled blocks. Then, writing down the $k_0\ell$ elements of $\mathcal{B}_1^*, \ldots, \mathcal{B}_{k_0}^*$ in a series, we get a MBB sample of size $k_0\ell = 13 \times 8 = 104$. We use the first 100 of these values to define the (ordinary) bootstrap version of T_n under the MBB as

$$T_{1n}^{*(1)} = \sqrt{100}\Big(H(\bar{X}_{100}^{*(1)}) - H(\hat{\mu}_{n,1})\Big) , \tag{4.58}$$

where $\bar{X}_{100}^{*(1)}$ is sample mean of the first 100 MBB samples and where $\hat{\mu}_{n,1} \equiv \hat{\mu}_{n,1}(\ell) = E_* \bar{X}_{100}^{*(1)}$. The centering variable $\hat{\mu}_{n,1}$ may be evaluated without any resampling by using the formula

$$\hat{\mu}_{n,1} = 100^{-1}\left[(k_0 - 1)(93)^{-1}\sum_{i=1}^{93} V_i(\ell) + (93)^{-1}\sum_{i=1}^{93} V_i(a)\right] , \tag{4.59}$$

where $a \equiv 100-(k_0-1)\ell = 4$, and $V_i(0) = 0$, and $V_i(m) = X_i + \ldots + X_{i+m-1}$ is the sum of the first m elements of the block $\mathcal{B}_i$, $1 \leq m \leq \ell$. This easily follows by noting that $X_1^*, \ldots, X_{100}^*$ consist of $(k_0 - 1) = 12$ *complete* blocks of length 8 and the *first* $a = 4$ elements from the k_0th resampled block. The MBB estimator of the level-2 parameter φ_n based on blocks of size ℓ is given by

$$\hat{\varphi}_n(1; \ell) = E_* \left[T_{1n}^{*(1)}\right]^2 .$$

A closed-form expression for $\hat{\varphi}_n(1;\ell)$ is intractable because of the nonlinearity of the estimator $\hat{\theta}_n = H(\bar{X}_{100})$. Therefore, we evaluate $\hat{\varphi}_n(1,\ell)$ by Monte-Carlo simulation as follows. Let B be a large positive integer, denoting the number of bootstrap replicates. For each $r = 1,\ldots,B$, generate a set of $k_0 = 13$ iid random variables $\{{}_rI_1,\ldots,{}_rI_{k_0}\}$ with the Discrete Uniform distribution on $\{1,\ldots,93\}$, the index set of all overlapping blocks of size $\ell = 8$. Then, for each r, $\{\mathcal{B}_i : i = {}_rI_1,\ldots,{}_rI_{k_0}\}$ represents a random sample of size k_0 from $\{\mathcal{B}_1,\ldots,\mathcal{B}_{100-\ell+1}\}$, also called a *replicate* of $\mathcal{B}_1^*,\ldots,\mathcal{B}_{k_0}^*$. Let ${}_r\bar{X}_{100}^{*(1)}$ denote the sample mean of first 100 values in the resampled blocks $\{\mathcal{B}_i : i = {}_rI_1,\ldots,{}_rI_{k_0}\}$. Then, for $r = 1,\ldots,B$, the rth *replicate* of $T_{1n}^{*(1)}$ based on the resampled MBB blocks $\{\mathcal{B}_i : i = {}_rI_1,\ldots,{}_rI_{k_0}\}$ is given by

$$ {}_rT_{1n}^{*(1)} = \sqrt{100}\Big(H\big({}_r\bar{X}_{100}^{*(1)}\big) - H(\hat{\mu}_{n,1})\Big) , \qquad (4.60) $$

where $\hat{\mu}_{n,1}$ is computed (only once) using formula (4.59). The Monte-Carlo approximation to the MBB estimator $\hat{\varphi}_n(1,\ell)$ is now given by

$$ \hat{\varphi}_n(1;\ell)^{\mathrm{MC}} = B^{-1}\sum_{r=1}^{B}\Big[{}_rT_{1n}^{*(1)}\Big]^2 . \qquad (4.61) $$

Note that as $B \to \infty$, the average of the $\big[{}_rT_{1n}^{*(1)}\big]^2$-values tends to the corresponding expected value $E_*\big[{}_rT_{1n}^{*(1)}\big]^2 \equiv \hat{\varphi}_n(1;\ell)$. Thus, by choosing B appropriately large, one can get an approximation to the MBB estimator $\hat{\varphi}_n(1;\ell)$ to any given degree of accuracy. In Table 4.1 below, we report the MBB estimators (along with the other block bootstrap estimators) of φ_n for the data set of Figure 4.1 for different block sizes, including $\ell = 8$. As mentioned earlier, the "true" value of the target parameter is given by $\varphi_n = 1.058$. The number of bootstrap replicates used here is $B = 800$. (This value of B is chosen only for the purpose of illustration. In practice, a much larger value of B may be desirable depending on the parameter φ_n.)

TABLE 4.1. Block bootstrap estimates of the level-2 parameter $\varphi_n = ET_{1n}^2$ based on different (expected) block sizes, for the data set of Figure 4.1. The true value of φ_n is given by 1.058.

Block Size	4	6	8	10	15	20
MBB	1.159	1.085	0.881	0.820	1.078	0.884
NBB	1.299	0.904	1.093	0.763	0.879	1.030
CBB	1.020	1.106	0.951	0.812	0.968	0.808
SB	0.935	0.941	0.898	0.810	0.746	0.642

The steps involved the implementation of the other block bootstrap methods are similar. For the NBB, we consider the nonoverlapping blocks

$\{\mathcal{B}_i^{(2)} : i = 1, \ldots, b\}$ where $\mathcal{B}_i^{(2)} \equiv \mathcal{B}_{(i-1)\ell+1}$ and $b = \lfloor 100/\ell \rfloor$. Next, we generate the NBB samples by resampling k_0 blocks from this collection. For $\ell = 8$, this amounts to resampling $k_0 = 13$ blocks from the collection of $b = 12$ disjoint blocks $\{\mathcal{B}_1^{(2)}, \ldots, \mathcal{B}_{12}^{(2)}\} = \{\{X_1, \ldots, X_8\}, \ldots, \{X_{89}, \ldots, X_{96}\}\}$. The NBB estimator of the parameter φ_n is given by

$$\hat{\varphi}_n(2; \ell) = E_* \left[T_{1n}^{*(2)} \right]^2 .$$

where $T_{1n}^{*(2)} = \sqrt{100}\big(H(\bar{X}_{100}^{*(2)}) - H(\hat{\mu}_{n,2})\big)$, $\bar{X}_{100}^{*(2)}$ is NBB sample mean of the first 100 resampled data values, and $\hat{\mu}_{n,2} \equiv E_* \bar{X}_{100}^{*(2)} = \frac{1}{96} \sum_{i=1}^{96} X_i$. Note that for this choice of ℓ, the last 4 X_i-values *never* appear in the definition of the NBB estimator $\hat{\varphi}_n(2; \ell)$. For Monte-Carlo approximation to the NBB estimator $\hat{\varphi}_n(2; \ell)$, generate B sets of iid random variables $\{{}_rI_{2,1}, \ldots, {}_rI_{2,k_0}\}$ with the Discrete Uniform distribution on $\{1, \ldots, b\}$ and define the replicates of $T_n^{*(2)}$ as

$${}_rT_{1n}^{*(2)} = \sqrt{100}\Big(H\big({}_r\bar{X}_{100}^{*(2)}\big) - H(\hat{\mu}_{n,2}) \Big)$$

for $r = 1, \ldots, B$. The Monte-Carlo approximation to the NBB estimator $\hat{\varphi}_n(2; \ell)$ is now given by

$$\hat{\varphi}_n(2; \ell)^{\mathrm{MC}} = B^{-1} \sum_{r=1}^{B} \left[{}_rT_{1n}^{*(2)} \right]^2 . \tag{4.62}$$

The NBB estimates of φ_n for different block lengths ℓ for the data set of Figure 4.1 are given in the second row of Table 4.1 above.

For the CBB and the SB, we need to consider blocks defined in terms of the *periodically extended* time series $\{Y_{n,i}\}_{i \in \mathbb{Z}}$, where we define $Y_{n,(kn+i)} = X_i$ for all $k \in \mathbb{Z}$ and all $1 \le i \le 100$. As described in Chapter 2, for a *given* block length ℓ, the CBB resamples k_0 blocks from the collection $\{\mathcal{B}(i; \ell) : i = 1, \ldots, 100\}$, where, as before, k_0 is the smallest integer not less than $100/\ell$ and where

$$\mathcal{B}(i; k) = \{Y_{n,i}, \ldots, Y_{n,(i+k-1)}\}, \quad i \ge 1, k \ge 1 .$$

The CBB version of T_n based on these circular blocks of length ℓ is given by

$$T_{1n}^{*(3)} = \sqrt{100}\Big(H\big(\bar{X}_{100}^{*(3)}\big) - H(\bar{X}_{100}) \Big) ,$$

where $\bar{X}_{100}^{*(3)}$ is the sample mean of the first 100 elements of k_0 resampled blocks. Note that $E_* \bar{X}_{100}^{*(3)} = \bar{X}_{100}$, the sample mean of $\{X_1, \ldots, X_{100}\}$, for any choice of ℓ and hence, we do *not* need an additional formula like (4.59) to find $E_* \bar{X}_{100}^{*(3)}$. The CBB estimator of φ_n is now given by

$$\hat{\varphi}_n(3; \ell) \equiv E_* \big[T_{1n}^{*(3)} \big]^2 .$$

For Monte-Carlo evaluation of $\hat{\varphi}_n(3;\ell)$, with $\ell = 8$, we generate $B = 800$ sets of iid random variables $\{{}_rI_{3,1},\ldots,{}_rI_{3,k_0}\}$ with the Discrete Uniform distribution on $\{1,\ldots,100\}$, setting k_0 equal to 13 as before. Let ${}_rT_{1n}^{*(3)}$ be the replicate of $T_{1n}^{*(3)}$ based on the resampled blocks$\{\mathcal{B}({}_rI_{3,1};8),\ldots,\mathcal{B}({}_rI_{3,k_0};8)\}$ with $k_0 = 13$. Then, the Monte-Carlo value of the CBB estimator is given by $\hat{\varphi}_n(3;\ell)^{\mathrm{MC}} = B^{-1}\sum_{r=1}^{B}\left[{}_rT_{1n}^{*(3)}\right]^2$. The CBB estimates of φ_n, computed using the data set of Figure 4.1 are given in the third row of Table 4.1 for $\ell = 8$, and also for other block lengths.

Finally, we consider the SB. For notational consistency, we write ℓ for the *expected* block length under the SB. Thus, following the description of the SB given in Chapter 2, we resample $K \equiv \inf\{1 \le k \le 100 : L_1 + \ldots + L_k \ge 100\}$ circular blocks $\{\mathcal{B}(I_{4,i}, L_i) : i = 1,\ldots,K\}$, where L_i's are iid random variables having the Geometric (p) distribution with $p = \ell^{-1}$ and where $I_{4,i}$'s are iid random variables having the Discrete Uniform distribution on $\{1,\ldots,100\}$. Furthermore, $I_{4,i}$'s and L_i's are independent. The SB version of T_{1n} is given by

$$T_{1n}^{*(4)} = \sqrt{100}\Big(H\big(\bar{X}_{100}^{*(4)}\big) - H(\bar{X}_{100})\Big) .$$

Note that like the CBB, $E_*\bar{X}_{100}^{*(4)} = \bar{X}_{100}$ for any choice of ℓ and hence, centering the SB version $H(\bar{X}_{100}^{*(4)})$ of the estimator $\hat{\theta}_n$ is simpler compared to centering the MBB and the NBB versions. The SB estimator of φ_n is given by

$$\hat{\varphi}_n(4;\ell) \equiv E_*\left[T_{1n}^{*(4)}\right]^2 .$$

For Monte-Carlo evaluation of $\hat{\varphi}_n(4,\ell)$, say, again with $\ell = 8$, for each $r = 1,\ldots,B$, first we generate iid Geometric$(\frac{1}{8})$ random variables $\{{}_rL_1,\ldots,{}_rL_{{}_rK}\}$, where ${}_rK = \inf\{1 \le k \le 100 : {}_rL_1 + \ldots + {}_rL_k \ge 100\}$. Note that ${}_rK$, the number of resampled blocks under the SB method at the rth replication, is *random* and unlike the first three block bootstrap methods, ${}_rK$'s take on different values for different r's. Next, having generated the ${}_rL_i$'s, we *independently* generate iid Discrete Uniform $\{1,\ldots,100\}$ random variables ${}_rI_{4,i}, i = 1,\ldots,{}_rK$ for each $r = 1,\ldots,B$. This yields B sets of SB resampled blocks $\{\mathcal{B}({}_rI_{4,i};{}_rL_i) : i = 1,\ldots,{}_rK\}$, for $r = 1,\ldots,B$, where each set of resampled blocks contains *at least* 100 SB observations. We compute a replicate ${}_rT_{1n}^{*(4)}$ of $T_{1n}^{*(4)}$ using the first 100 SB observations from the rth set, and combine these to get the Monte-Carlo value of the SB estimate as $\hat{\varphi}_n(4;\ell)^{\mathrm{MC}} = B^{-1}\sum_{r=1}^{B}\left[{}_rT_{1n}^{*(4)}\right]^2$. The estimates for the SB method for the data set of Figure 4.1 are given in row 4 of Table 4.1. □

Example 4.6: (*Comparison of "naive" and "ordinary" versions of block bootstraps*). We continue with the set up of the last example. Suppose that $\{X_{0i}\}_{i\in\mathbb{Z}}$ is a stationary process and we are interested in estimating population characteristics of $T_{1n} = \sqrt{n-k}(\hat{\gamma}_n(k) - \gamma(k))$, where $\gamma(k) = \text{Cov}(X_{01}, X_{0(k+1)})$ and $\hat{\gamma}_n(k)$ is as defined in Example 4.5. For estimators like $\hat{\gamma}_n(k)$, that are defined in terms of kth order lag-vectors of the original observations $\{X_{0i}\}_{i\in\mathbb{Z}}$, an alternative approach, called the "*naive approach*," of defining block bootstrap estimators was described in Section 2.5. Here we compare the performance of the "naive" approach and the "blocks of blocks" (i.e., the "ordinary") approach in the context of Example 4.5.

Suppose that $\{X_{0i}\}_{i\in\mathbb{Z}}$ is the ARMA(3,4) process given by (4.57) and the level-2 parameter of interest is given by $\varphi_n = ET_n^2$ with $n = 102$ and $k = 2$. Under the "naive approach," one applies the block bootstrap methods to the observations $X_{01}, \ldots, X_{0(102)}$ *directly*. For the purpose of illustration, here we describe the "naive" MBB estimator of φ_n based on blocks of length $\ell = 8$. Define the MBB blocks $\mathcal{B}_{0i} = \{X_{0i}, \ldots, X_{0(i+7)}\}$, $i = 1, \ldots, 95$ based on X_{0i}'s (but *not* defined in terms of the vectors X_i's). Let b_1 be the smallest integer $\geq 102/\ell$. Thus, for our example, $b_1 = 13$. Then, we resample b_1 blocks at random from the collection $\{\mathcal{B}_{0i} : i = 1, \ldots, 95\}$ to generate $b_1\ell = 13 \times 8 = 104$ MBB samples $\{X_{0i}^* : i = 1, \ldots, 104\}$, and use the first $n = 102$ of these variables to define a "naive" version of T_n as

$$T_{1n}^{*(1)\text{nv}} = \sqrt{n-2}\Big(\gamma_n^{*(1)}(2) - \hat{\gamma}_n(2)\Big) , \tag{4.63}$$

where $\gamma_n^{*(1)}(2) = (100)^{-1}\sum_{i=1}^{100} X_{0i}^* X_{0(i+2)}^* - [\bar{X}_{n-2}^{*(1)}]^2$ and $\bar{X}_{n-2}^{*(1)} = (100)^{-1}\sum_{i=1}^{100} X_{0i}^*$. The "naive" MBB estimator of φ_n is given by

$$\hat{\varphi}_n(1;\ell)^{\text{nv}} \equiv E_*\left[T_{1n}^{*(1)\text{nv}}\right]^2 .$$

Similarly, we can define the naive versions of the other three block bootstrap estimators. In Table 4.2, we report the "naive" block bootstrap estimators of $\varphi_n = 1.058$ based on (expected) block sizes $\ell = 4, 6, 8, 10, 15, 20$.

TABLE 4.2. Block bootstrap estimates of $\varphi_n = ET_{1n}^2$ for the data set of Figure 4.1 based on the "naive" approach. Here, the true value of φ_n is 1.058.

Block Size	4	6	8	10	15	20
MBB	1.768	1.242	1.230	1.094	1.297	0.986
NBB	1.275	1.113	0.672	1.578	0.905	0.852
CBB	1.402	1.349	1.077	1.187	0.888	0.937
SB	1.571	1.183	1.126	1.151	0.902	0.760

As explained in Section 2.5, the "naive" bootstrap estimators tend to have *larger* biases compared to the "ordinary" or the "blocks of blocks" versions. For this example, contributions to the biases of the "naive" bootstrap estimators result from the "within-block-independence" of the components of $(X_{0i}^*, X_{0i}^* X_{0(i+2)}^*)$ near the boundary of the resampled blocks. Further, this bias effect is more pronounced when the (expected) block size ℓ is small. The result of Table 4.2 appears to lend some support to this observation for the data set of Figure 4.1. From the table, we find that the "naive" estimators tend to *overestimate* the level-2 parameter $\varphi_n = 1.058$ for values of $\ell \leq 10$ for the MBB, the CBB, and the SB methods, while the estimates based on the NBB method fluctuates around the true value $\varphi_n = 1.058$. The discrepancy for each of the four methods is indeed larger for smaller values of ℓ. □

Example 4.7: *(Estimation of Distribution Function).* Next we look at block bootstrap estimators of the distribution function. Let $\{X_{0i}\}_{i\in\mathbb{Z}}$ be a stationary time series with $\text{Var}(X_{01}) \in (0, \infty)$, and the level-1 parameter of interest is the lag-k autocorrelation coefficient

$$r(k) = \text{Cov}(X_{01}, X_{0(1+k)})/\text{Var}(X_{01}) ,$$

for some given integer $k \geq 0$. As an estimator of $r(k)$, we consider the following version of the sample autocorrelation coefficient

$$\hat{r}_n(k) = \left\{(n-k)^{-1}\sum_{i=1}^{n-k} X_i X_{i+k} - \bar{X}_n^2\right\}\left\{n^{-1}\sum_{i=1}^{n} X_i^2 - \bar{X}_n^2\right\}^{-1} ,$$

which is slightly different from the estimator considered in Example 4.2. With

$$H(x, y, z) = \{(z - x^2)/(y - x^2)\}\mathbb{1}(y > x^2), \quad (x, y, z)' \in \mathbb{R}^3$$

and $Y_i \equiv (Y_{1i}, Y_{2i}, Y_{3i})' = (X_{0i}, X_{0i}^2, X_{0i}X_{0(i+k)})'$, $i \in \mathbb{Z}$, it is easy to see that $r(k)$ and $\hat{r}_n(k)$ can be expressed as

$$r(k) = H(EY_1)$$

and

$$\hat{r}_n(k) = H\big(\bar{Y}_{1n}, \bar{Y}_{2n}, \bar{Y}_{3(n-k)}\big) ,$$

where $\bar{Y}_{jm} = m^{-1}\sum_{i=1}^{m} Y_{ji}$, $m \geq 1$, $1 \leq j \leq 3$. Note that in this case, the estimator $\hat{r}_n(k)$ does not directly fall in the framework of the Smooth Function Model treated in Section 4.2, since it is a function of averages of *different* numbers of X-variables in the first, the second, and the third co-ordinates. However, by Remark 4.1 of Section 4.2, the block bootstrap approximations for the sampling distribution of

$$\check{T}_{1n} \equiv \sqrt{n-k}(\hat{r}_n(k) - r(k))$$

remain valid, whenever the regularity conditions of Theorem 4.1 on $H(\cdot)$, $\alpha(\cdot)$ and $\{X_{0i}\}_{i\in\mathbb{Z}}$ are satisfied. Since $\sigma^2 = \text{Var}(X_{01}) \in (0,\infty)$, the function H, being a rational function with a nonvanishing denominator at EY_1, is infinitely differentiable in a neighborhood of EY_1. Hence, if the sequence $\{X_{0i}\}_{i\in\mathbb{Z}}$ satisfies the moment and mixing conditions of Theorem 4.1, then for $j = 1, 2, 3, 4$,

$$\sup_x \left| P_*\left(\sqrt{n-k}\left(r_n^{*(j)}(k) - \tilde{r}_n^{(j)}(k)\right) \leq x\right) - P(\check{T}_{1n} \leq x)\right|$$
$$\longrightarrow_p 0 \quad \text{as} \quad n \to \infty ,$$

where, as usual, we write $j = 1$ for the MBB, $j = 2$ for the NBB, $j = 3$ for the CBB, and $j = 4$ for the SB, and define the variables $r_n^*(j)$'s and $\tilde{r}_n(j)$'s as $r_n^{*(j)} = H\big(\bar{Y}_{1n_2}^{*(j)}; \bar{Y}_{2n_2}^{*(j)}; \bar{Y}_{3n_2}^{*(j)}\big)$ and $\tilde{r}_n^{(j)}(k) = H\big(E_*\bar{Y}_{1n_2}^{*(j)}, E_*\bar{Y}_{2n_2}^{*(j)}, E_*\bar{Y}_{3n_2}^{*(j)}\big)$ for $j = 1, 2, 3$ and $r_n^{*(4)}$ and $\tilde{r}_n^{(4)}$ similarly, with n_2 replaced by $N_2 = L_1 + \cdots + L_{K_2}$. Here n_2 and K_2 are defined by replacing n with $n - k$, e.g., $n_2 = \lfloor (n-k)/\ell \rfloor \ell$ for a given block size $\ell \in (1, n-k)$ for the Y-variables, and $K_2 = \inf\{j \in \mathbb{N} : L_1 + \cdots + L_j \geq n - k\}$.

Again for the purpose of illustration, suppose that $\{X_{0i}\}_{i\in\mathbb{Z}}$ is the ARMA (3,4) process given by (4.57). We now consider the block bootstrap estimators of the distribution function of

$$\check{T}_{1n} \equiv \sqrt{n-k}\big(\hat{r}_n(k) - r(k)\big)$$

for the data set of Figure 4.1 with $k = 2$ and $n = 102$. As in Example 4.5, we define the blocks in terms of the *transformed* vectors $Y_1, \ldots, Y_{100}$ for each of the block bootstrap methods. Suppose, for example, that the (expected) block size ℓ is chosen to be 6. Thus, for the MBB, the blocks are $\{\mathcal{B}_i \equiv (Y_i, \ldots, Y_{i+5}) : i = 1, \ldots, 95\}$, for the NBB the blocks are $\{(Y_1, \ldots, Y_6), (Y_7, \ldots, Y_{12}), \ldots, (Y_{91}, \ldots, Y_{96})\}$, and for the CBB and the SB, the blocks are defined using the periodic extension of $Y_1, \ldots, Y_{100}$. To generate the bootstrap samples, we resample k_0 blocks for the first three methods, where k_0 is the smallest integer not less than $100/\ell$. For $\ell = 6$, $k_0 = 17$. Similarly, for the SB, we resample a random number of blocks of lengths $L_1, \ldots, L_K$ where $L_1, L_2, \ldots$ are iid Geometric($\frac{1}{6}$) variables and $K = \inf\{k \geq 1 : L_1 + \cdots + L_k \geq 100\}$. Let $Y_1^{*(j)}, \ldots, Y_{100}^{*(j)}$ denote the first 100 bootstrap samples under the jth method, $j = 1, 2, 3, 4$. Although in Theorem 4.1 we have proved validity of the four block bootstrap methods for resample sizes n_1 for $j = 1, 2, 3$ and N_1 for $j = 4$ mainly to simplify proofs, consistency of bootstrap approximations continue to hold if the resample size is set equal to n. Hence, in practice, a resample size equal to the size of the observed Y-vectors may be employed for the "ordinary" block bootstrap estimators. Accordingly, *in practice*, we define the block bootstrap versions of $\check{T}_{1n}$ as

$$\check{T}_{1n}^{*(j)} = \sqrt{100}\left(r_n^{*(j)}(2) - \tilde{r}_n^{(j)}(2)\right),\ j = 1, 2, 3, 4 ,$$

where $r_n^{*(j)}(2) = H(\bar{Y}_n^{*(j)})$ and $\tilde{r}_n^{(j)}(2) = H\Big(E_*(\bar{Y}_n^{*(j)})\Big)$ with $\bar{Y}_n^{*(j)} \equiv (100)^{-1}\sum_{i=1}^{100} Y_i^{*(j)}$. Strictly speaking, this definition of the bootstrapped statistic does not reflect the difference in the number of variables averaged in different components of Y_j's in the definition of $r_n^{*(j)}(2)$, but the effect is negligible. The bootstrap estimators of the distribution function $G_{1n}(x) \equiv P(\check{T}_{1n} \leq x)$, $x \in \mathbb{R}$ is given by

$$\hat{G}_{1n}^{(j)}(x) = P_*(\check{T}_{1n}^{*(j)} \leq x), \ x \in \mathbb{R}, \ j = 1, 2, 3, 4 \ . \tag{4.64}$$

For Monte-Carlo evaluation of $\hat{G}_n^{(j)}$'s, as in Example 4.5, we generate B sets of block bootstrap samples and compute the replicates ${}_r\check{T}_{1n}^{*(j)}$ for $r = 1, \ldots, B$. Then, the Monte-Carlo approximation to the jth block bootstrap estimator is given by

$$\hat{G}_{1n}^{(j)\mathrm{MC}}(x) = B^{-1}\sum_{r=1}^{B} \mathbb{1}({}_r\check{T}_{1n}^{*(j)} \leq x), \ x \in \mathbb{R}, \ j = 1, 2, 3, 4 \ . \tag{4.65}$$

As an example, we computed the block bootstrap distribution function estimates of (4.65) for the data set of Figure 4.1 with $B = 800$. (In practice, a higher value of B should be used for distribution function estimation.) Figure 4.2 below gives the corresponding histograms. As follows from the discussion of Section 4.2, the variable $\check{T}_{1n}$ is asymptotically normal. The block bootstrap estimates also show a similar shape with slightly higher masses to the left of the origin.

The bootstrap estimator $\hat{G}_{1n}^{(j)}(x)$ of the sampling distribution of T_{1n} can also be used to obtain estimators of the quantiles of T_{1n}. For $\alpha \in (0, 1)$, define

$$q_{1n}(\alpha) = G_{1n}^{-1}(\alpha)$$

where, recall that, for any distribution function G on $\mathbb{R}$, $G^{-1}(\alpha) = \inf\{x \in \mathbb{R} : G(x) \geq \alpha\}$. Thus, $q_{1n}(\alpha)$ is the αth quantile of G_{1n}. The block-bootstrap estimators of the level-2 parameter $q_{1n}(\alpha)$ are given by "plugging-in" $\hat{G}_{1n}^{(j)}$ in place of G_{1n} as

$$\hat{q}_{1n}^{(j)}(\alpha) = \big[\hat{G}_{1n}^{(j)}\big]^{-1}(\alpha), \ j = 1, 2, 3, 4 \ . \tag{4.66}$$

Monte-Carlo evaluation of $\hat{q}_{1n}^{(j)}(\alpha)$ is rather simple, once the bootstrap replicates ${}_r\check{T}_{1n}^{*(j)}$, $r = 1, \ldots, B$ have been computed. Arrange the replicates ${}_r\check{T}_{1n}^{*(j)}$, $r = 1, \ldots, B$ in an increasing order to get the corresponding *order statistics* ${}_{(1)}\check{T}_{1n}^{*(j)} \leq \cdots \leq {}_{(B)}\check{T}_{1n}^{*(j)}$. Then, the Monte-Carlo approximation to $\hat{q}_{1n}^{(j)}(\alpha)$ is given by the $\lfloor B\alpha \rfloor$th order statistic, i.e., by

$$\hat{q}_{1n}^{(j)\mathrm{MC}}(\alpha) = {}_{(\lfloor B\alpha \rfloor)}\check{T}_{1n}^{*(j)}, \ j = 1, 2, 3, 4 \ .$$

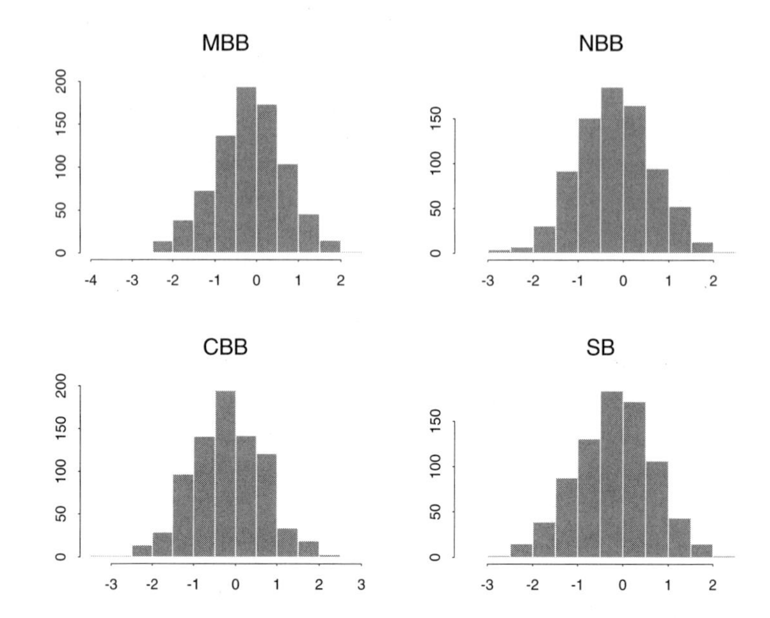

FIGURE 4.2. Histograms of block bootstrap distribution function estimates $\hat{G}_{1n}^{(j)}, j = 1, 2, 3, 4$ of (4.64), based on (expected) block length 6 and $B = 800$ bootstrap replicates.

As an example, consider the computation of $\hat{q}_{1n}^{(j)\text{MC}}(\alpha)$ for $\alpha = 0.05$, based on $B = 800$ block bootstrap replicates. Then, the Monte-Carlo values of the block bootstrap estimates of $q_{1n}(\alpha)$ are given by the $\lfloor(800)(0.05)\rfloor = 40$th order statistic of the bootstrap replicates ${}_1\check{T}_{1n}^{*(j)}, \ldots, {}_{800}\check{T}_{1n}^{*(j)}$. For the data set of Figure 4.1, the block bootstrap estimates are, respectively, given by -1.635 (MBB), -1.535 (NBB), -1.583 (CBB), and -1.624 (SB).

Confidence intervals (CIs) for the level-1 parameter $r(2)$ may also be constructed using the bootstrap methods. Note that if the quantiles of $\check{T}_{1n}$ were known, an equal-tailed two-sided $(1-\alpha)$ CI for $r(2)$ would be given by

$$I_\alpha \equiv \Big(\hat{r}_n(2) - \frac{1}{\sqrt{100}}\, q_{1n}(1 - \frac{\alpha}{2}),\ \hat{r}_n(2) - \frac{1}{\sqrt{100}}\, q_{1n}(\frac{\alpha}{2})\Big)\,.$$

A *percentile* block bootstrap CI for $r(2)$ is obtained by replacing the "true" quantiles $q_{1n}(\alpha/2)$ and $q_{1n}(1-\frac{\alpha}{2})$ by their bootstrap estimators, viz.,

$$I_{\alpha,\text{percentile}}^{(j)} = \Big(\hat{r}_n(2) - \frac{1}{\sqrt{100}}\, \hat{q}_{1n}^{(j)}(1 - \frac{\alpha}{2}), \hat{r}_n(2) - \frac{1}{\sqrt{100}}\, \hat{q}_{1n}^{(j)}(\frac{\alpha}{2})\Big)\,. \tag{4.67}$$

For computing the interval estimates of $r(2)$ using formula (4.67), the bootstrap quantiles $\hat{q}_{1n}^{(j)}(\cdot)$'s in (4.67) are further replaced with their Monte-

Carlo approximations. As an example, we construct 90% *equal-tailed percentile bootstrap CIs* for $r(2) = \text{Cov}(X_{01}, X_{03})$ under model (4.57). By (4.67), these are given by $\Big(\hat{r}_n(2) - \hat{q}_{1n}^{(j)}(0.95)/10,\ \hat{r}_n(2) - \hat{q}_{1n}^{(j)}(0.05)/10\Big)$, $j = 1, 2, 3, 4$. The block bootstrap quantiles $\hat{q}_{1n}^{(j)}(0.05)$ and $\hat{q}_{1n}^{(j)}(0.95)$, $j = 1, 2, 3, 4$ may be found as described in the previous paragraph. They are, respectively, given by the 40th and the $\lfloor (800)(0.95) \rfloor = 760$th order statistics of the replicates ${}_1\check{T}_{1n}^{*(j)}, \ldots, {}_{800}\check{T}_{1n}^{*(j)}$, $j = 1, 2, 3, 4$. For the data-set of Figure 4.1, 90% percentile block bootstrap CIs for $r(2)$ based on (expected) block size $\ell = 6$ are given by

$(-1.059,\ 1.740)$	(MBB)
$(-1.079,\ 1.633)$	(NBB)
$(-1.085,\ 1.687)$	(CBB)
$(-1.093,\ 1.728)$	(SB)

Note that for constructing these bootstrap interval estimates, we do not have to estimate the standard error of $\hat{r}_n(2)$ explicitly. All four block bootstrap methods provide consistent estimators of the standard error implicitly through the bootstrap quantiles $\hat{q}_{1n}^{(j)}(\cdot)$'s. In comparison, the user must separately derive and compute an estimate of the standard error of $\hat{r}_n(2)$ for constructing a large sample interval estimate for $r(2)$ using the traditional normal approximation. □

Example 4.8: (*Differentiable Statistical Functionals*). Suppose that $\{X_i\}_{i \in \mathbb{Z}}$ is a sequence of stationary random variables with (one-dimensional) marginal distribution function F. Suppose that we are interested in estimating the parameter $\theta = \int J(u) F^{-1}(u) du$ using the L-estimator

$$\hat{\theta}_n = \int J(u) F_n^{-1}(u) du$$

for a given function $J : (0, 1) \to \mathbb{R}$, where F_n is the empirical distribution function of $X_1, \ldots, X_n$, and F^{-1} and F_n^{-1} are as defined in Section 4.4. As discussed there, $\hat{\theta}_n$ may be represented as a statistical functional, say $\hat{\theta}_n = T(F_n)$. Fréchet differentiability of T at F depends on the joint behavior of the functions $J(\cdot)$ and $F(\cdot)$ and may be guaranteed through different sets of conditions on $J(\cdot)$ and $F(\cdot)$. Here, we state one set of sufficient conditions. For variations of these conditions, see Serfling (1980) and Fernholz (1983), and the references therein.

Note that the function F^{-1} is nondecreasing and left continuous. Hence, F^{-1} corresponds to a measure on $\mathbb{R}$. We assume that

(i) J is bounded and continuous almost everywhere with respect to F^{-1} and with respect to the Lebesgue measure, and

(ii) there exists $0 < a < b < 1$ such that $J(u) = 0$ for all $u \notin [a, b]$.

By Boos (1979), under assumptions (i) and (ii), the functional $T(\cdot)$ is Fréchet differentiable at F under the sup-norm $\|\cdot\|_\infty$. Hence, if we write

$$\bar{T}_{2n} = \sqrt{n}(\hat{\theta}_n - \theta)$$

and $\bar{T}_{2n}^*$ for the MBB version of $\bar{T}_{2n}$ based on blocks of size ℓ, then under the conditions of Theorem 4.4 and under assumptions (i) and (ii) above,

$$\sup_{x \in \mathbb{R}} \left| P_*(\bar{T}_{2n}^* \leq x) - P(\bar{T}_{2n} \leq x) \right| \longrightarrow_p 0 \quad \text{as} \quad n \to \infty .$$

Note that for the α-trimmed mean $(0 < \alpha < 1/2)$, the level-1 parameter of interest is given by (cf. Example 4.4)

$$\theta = \int_\alpha^{1-\alpha} F^{-1}(u)du/(1-2\alpha) , \tag{4.68}$$

which corresponds to the function $J(u) = (1-2\alpha)^{-1} \cdot \mathbb{1}_{[\alpha,1-\alpha]}(u)$, $u \in (0,1)$. Clearly, $J(\cdot)$ satisfies assumptions (i) and (ii), if the two-point set $\{\alpha, 1-\alpha\}$ has F^{-1}-measure zero. It is easy to check that this holds, provided the function F is strictly increasing in some neighborhoods of the quantiles $F^{-1}(\alpha)$ and $F^{-1}(1-\alpha)$.

As an example, we now consider the stationary process $\{X_i\}_{i \in \mathbb{Z}}$ given by

$$X_i = X_{1,i}^2 \cdot \text{sgn}(X_{1,i}), \ i \in \mathbb{Z} , \tag{4.69}$$

where the sgn$(\cdot)$ function is defined as $\text{sgn}(x) = \mathbb{1}(x \geq 0) - \mathbb{1}(x \leq 0)$, $x \in \mathbb{R}$ and $\{X_{1,i}\}_{i \in \mathbb{Z}}$ is an ARMA(3,3) process satisfying

$$\begin{aligned} & X_{1,i} - 0.4X_{1,i-1} - 0.2X_{1,i-2} - 0.1X_{1,i-3} \\ = \ & \epsilon_i + 0.2\epsilon_{i-1} + 0.3\epsilon_{i-2} + 0.2\epsilon_{i-3}, \ i \in \mathbb{Z} , \end{aligned}$$

where ϵ_i's are iid $N(0,1)$ variables. Note that the marginal distribution F of X_i is symmetric (about the origin), continuous and is strictly increasing over $\mathbb{R}$. Furthermore, X_i has finite moments of all order and $\{X_i\}_{i \in \mathbb{Z}}$ is strongly mixing with an exponentially decaying mixing coefficient. Thus, the conditions and conclusions of Theorem 4.4 hold for the centered and scaled α-trimmed mean $\bar{T}_{2n} = \sqrt{n}(\hat{\theta}_n - \theta)$, where θ is given by (4.68), $0 < \alpha < \frac{1}{2}$.

Now we consider the performance of the MBB in a finite sample situation. Figure 4.3 below gives a realization of X_i, $i = 1, \ldots, 250$ from the process (4.69). We apply the MBB with block size $\ell = 10$ to estimate the distribution of $\bar{T}_{2n}$ for different values of α. As in the previous examples, we resample $b_1 \equiv \lceil n/\ell \rceil = 25$ blocks from the collection of overlapping blocks $\mathcal{B}_i = (X_i, \ldots, X_{i+9})$, $i = 1, \ldots, 241$ to generate the MBB observations $X_1^*, \ldots, X_{10}^*; \ldots; X_{241}^*, \ldots, X_{250}^*$. Let $X_{(1)}^* \leq \cdots \leq X_{(250)}^*$ denote the order statistics corresponding to $X_1^{*(1)}, \ldots, X_{250}^{*(1)}$. Then, define the MBB version of $\check{T}_{2n}$ as

$$\bar{T}_{2n}^* = \sqrt{250}(\theta_n^* - \tilde{\theta}_n)$$

where $\theta_n^* = \sum_{n\alpha \le i \le n(1-\alpha)} X_{(i)}^*/[n(1-2\alpha)]$ and where $\tilde{\theta}_n = (1-2\alpha)^{-1}\int_\alpha^{1-\alpha} \tilde{F}_n^{-1}(u)du$ and $\tilde{F}_n(x) = E_*[n^{-1}\sum_{i=1}^n \mathbb{1}(X_i^* \le x)]$, $x \in \mathbb{R}$. Using arguments similar to (4.59), we can express $\tilde{F}_n(\cdot)$ as

$$\tilde{F}_n(x) = \sum_{i=1}^n \omega_{in} \mathbb{1}(X_i \le x), \ x \in \mathbb{R}$$

where, with $N = n - \ell + 1$,

$$\omega_{in} = \begin{cases} N^{-1} & \text{if } \ \ell \le i \le N \\ i/(N\ell) & \text{if } \ 1 \le i \le \ell - 1 \\ (n-i+1)/(N\ell) & \text{if } \ N+1 \le i \le n \ . \end{cases}$$

With the help of this formula, we may further simplify the definition of $\tilde{\theta}_n$, and write down an explicit expression for $\tilde{\theta}_n$ that may be evaluated without any resampling. Let $X_{(1)} \le \cdots \le X_{(n)}$ denote the order statistics of X_i, $i = 1, \ldots, n$. Also, let $w_{(i)}$ denote the weight associated with the order statistic $X_{(i)}$. For example, if $X_{(1)} = X_{10}$ and $X_{(2)} = X_3$, then $\omega_{(1)} = \omega_{10n}$ and $\omega_{(2)} = \omega_{3n}$. Then, the centering variable $\tilde{\theta}_n$ may be taken as

$$\tilde{\theta}_n = \sum_{i=L_\alpha+1}^{U_\alpha} \omega_{(i)} X_{(i)} \Big/ \sum_{i=L_\alpha+1}^{U_\alpha} \omega_{(i)}$$

where $L_\alpha = \max\{k : 1 \le k \le n, \ \sum_{i=1}^k \omega_{(i)} < \alpha\}$ and $U_\alpha = \min\{k : 1 \le k \le n, \ \sum_{i=1}^k \omega_{(i)} \ge 1 - \alpha\}$.

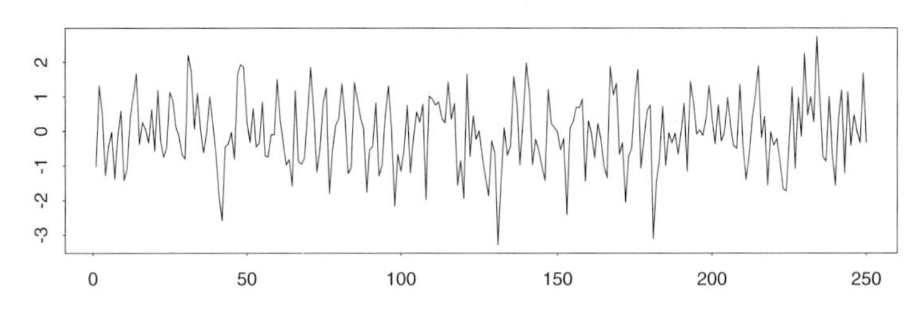

FIGURE 4.3. A simulated data set of $n = 250$ X_i-values from model (4.69).

Figure 4.4 below gives the histograms of the MBB estimates of the distribution function $G_{2n}(x) \equiv P(\bar{T}_{2n} \le x)$, $x \in \mathbb{R}$ based on $B = 800$ bootstrap replicates for $\alpha = 0, 0.08, 0.2, 0.4$, and 0.5. Note that $\alpha = 0$ represents the case where $\hat{\theta}_n$ is the sample mean and $\alpha = .5$ represents the case where $\hat{\theta}_n$ is the sample median. Although we have verified the conditions of Theorem

4.4 only for $0 < \alpha < \frac{1}{2}$, here we include these limiting α-values to obtain a more complete picture of how the MBB performs under varying degrees of trimming. It follows from Figure 4.4 that the bootstrap estimates of the sampling distribution are more skewed for larger values of α. Although $\bar{T}_{2n}$ is asymptotically normal for all these α-values, the "exact" distribution of $\bar{T}_{2n}$ is *not* symmetric for finite sample sizes. The limiting normal distribution fails to reveal this feature of the true sampling distribution. But the bootstrap estimates of the sampling distribution functions of $\bar{T}_{2n}$ for different α-values provide useful information on the skewness of the true distributions of $\bar{T}_{2n}$.

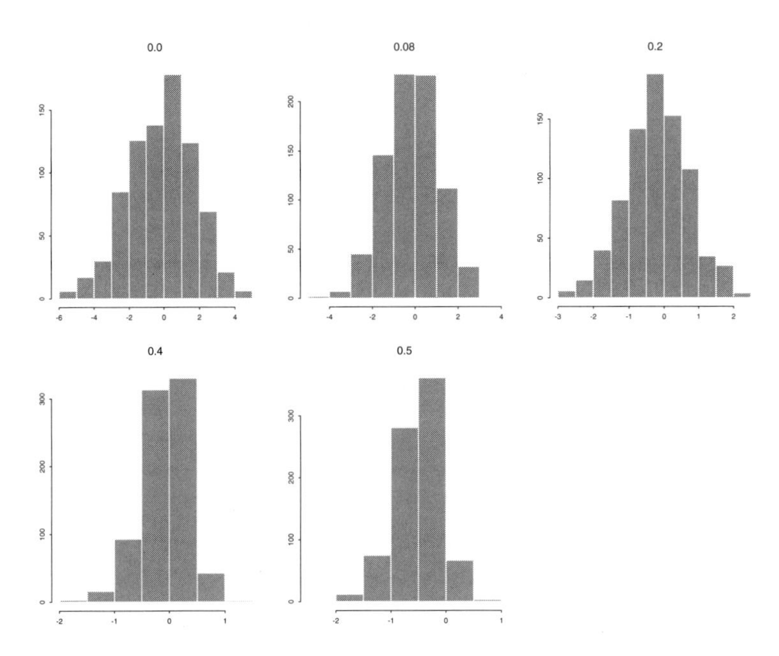

FIGURE 4.4. Histograms of the MBB distribution function estimates of the centered and scaled α-trimmed mean $\bar{T}_{2n}$ for $\alpha = 0.0, 0.08, 0.2, 0.4, 0.5$.

We may also use the bootstrap replicates of $\bar{T}^*_{2n}$ to construct *percentile* MBB CIs for the level-1 parameter $\theta \equiv (1-2\alpha)^{-1} \int_\alpha^{1-\alpha} F^{-1}(u)du$ as in the last example. Since the marginal distribution F of X_i is symmetric about the origin, the true value of θ is equal to 0 for all α. An equal-tailed two-sided 80% MBB percentile CI for θ is given by

$$\left(\hat{\theta}_n - \hat{q}_{2n}(.90)/\sqrt{n},\ \hat{\theta}_n - \hat{q}_{2n}(.10)/\sqrt{n}\right),$$

where $\hat{q}_{2n}(\beta)$, $0 < \beta < 1$, is the βth quantile of the conditional distribution of $\bar{T}^*_{2n}$. For the data set of Figure 4.3, 80% (equal-tailed) CIs based on the MBB with block size $\ell = 10$ are given by

$$
\begin{array}{lll}
(-2.215,\ 2.615) & \text{for} & \alpha = 0.00 \\
(-1.468,\ 1.715) & \text{for} & \alpha = 0.08 \\
(-0.936,\ 1.310) & \text{for} & \alpha = 0.2 \\
(-0.423,\ 0.575) & \text{for} & \alpha = 0.4 \\
(+0.022,\ 1.021) & \text{for} & \alpha = 0.5
\end{array}
$$

Note that all interval estimates except the one for $\alpha = 0.5$ (the median) contains the true value $\theta = 0$.

5

Comparison of Block Bootstrap Methods

5.1 Introduction

In this chapter, we compare the performance of the MBB, the NBB, the CBB, and the SB methods considered in Chapters 3 and 4. In Section 5.2, we present a simulated data example and illustrate the behavior of the block bootstrap methods under some simple time series models. Although the example treats the simple case of the sample mean, it provides a representative picture of the properties of the four methods in more general problems. In the subsequent sections, the empirical findings of Section 5.2 are substantiated through theoretical results that provide a comparison of the methods in terms of the (asymptotic) MSEs of the bootstrap estimators. In Section 5.3, we describe the framework for the theoretical comparison. In Section 5.4, we obtain expansions for the MSEs of the relevant bootstrap estimators as a function of the block size (expected block size, for the SB). These expansions provide the basis for the theoretical comparison of the sampling properties of the bootstrap methods. In Section 5.5, the main theoretical findings are presented. Here, we compare the bootstrap methods using the leading terms in the expansions of the MSEs derived in the previous section. In Section 5.5, we also derive theoretical optimal (expected) block lengths for each of the block bootstrap estimators and compare the methods at the corresponding optimal block lengths. Some conclusions and implications of the theoretical and finite sample simulation results are discussed in Section 5.6. Proofs of two key results from Section 5.4 are separated out into Section 5.7.

5.2 Empirical Comparisons

First we consider the behavior of the block bootstrap methods across different block lengths for a fixed set of observations. In Section 4.5, we considered two numerical examples where the four block bootstrap methods have been applied to the data set of Figure 4.1 for variance estimation and also for distribution function estimation. As Table 4.1 shows, the block bootstrap estimates of the level-2 parameter $\varphi_n = ET_n^2$ (cf. Example 4.5) produced by the various methods exhibit different patterns of variations across the (expected) block lengths considered. The SB method produced variance estimates that (nearly monotonically) decreased in value as the expected block length increased and resulted in the biggest "underestimation" of the target parameter $\varphi_n = 1.058$ among all four methods. The MBB and the CBB tended to have a similar pattern across the block lengths and were of comparable value. The NBB estimates fluctuated around the true value $\varphi_n = 1.058$, having both over- and underestimates at different block sizes. Similar comments apply to the bootstrap distribution function estimates as well.

The observations, noted above, on the behavior of the block bootstrap methods are based on a *single* data set only, and as such do not say much about the properties of these methods across different realizations of the variables $X_1, \ldots, X_n$, i.e., about their sampling properties. To get some idea about the sampling properties of these methods, we need to compare suitable population measures of accuracy (e.g., the MSE) of the resulting estimators. More precisely, let φ_n be a level-2 parameter of interest, which is to be estimated by the various block bootstrap methods. For $j = 1, 2, 3, 4$ and $\ell \in (1, n)$, write $\hat{\varphi}_n(j; \ell)$ for the bootstrap estimator of φ_n obtained by using the jth block bootstrap method with (expected) block length ℓ. Then, from the statistical decision-theoretic point of view, one effective way of comparing the performance of the block bootstrap methods is to compare the values of $\text{MSE}(\hat{\varphi}_n(\cdot; \ell))$'s. For the sake of illustration, we now suppose that $\varphi_n = n\text{Var}(\bar{X}_n)$, where $\bar{X}_n$ denote the sample mean of the first n observations from a stationary time series $\{X_i\}_{i \in \mathbb{Z}}$. We compare the performance of the four block bootstrap methods under the following models for $\{X_i\}_{i \in \mathbb{Z}}$:

$$\text{ARMA (1,1) model:} \quad X_i - 0.3X_{i-1} = \epsilon_i + 0.4\epsilon_{i-1}, \; i \in \mathbb{Z}, \tag{5.1}$$

$$\text{AR(1) model:} \quad X_i = 0.3X_{i-1} + \epsilon_i, \; i \in \mathbb{Z}, \tag{5.2}$$

$$\text{MA(1) model:} \quad X_i = \epsilon_i + 0.4\epsilon_{i-1}, \; i \in \mathbb{Z}, \tag{5.3}$$

where, in each of the three models, the innovations ϵ_i's are iid $N(0, 1)$ random variables.

Figure 5.1 below shows a plot of the MSEs of the block bootstrap estimators of φ_n, produced by the MBB, the NBB, and the SB under each of the models (5.1)–(5.3) and for a sample of size $n = 100$. The MSEs are computed using $K = 2000$ simulation runs. In each simulation run, the block bootstrap estimators at a given (expected) block length are computed using $B = 500$ bootstrap replicates. From the plots, it appears that the SB estimators have larger MSEs than the MSEs of the MBB and the NBB estimators under all three models and at all levels of the block length parameter ℓ considered, starting from $\ell = 2$. In the plots, the MSE curves for the CBB estimators have been left out because of the almost identical performance of the CBB estimators compared to the MBB estimators over the range of values of ℓ considered. Note that when considered as a function of the block length, the MSEs of the NBB estimators lie between the corresponding MSE curves of the MBB and the SB. Indeed, a similar pattern continues to hold for larger sample sizes. Figure 5.2 gives the MSE curves for the three methods for a sample size $n = 400$ for models (5.1)–(5.3), using the same values of the simulation parameters K and B as above.

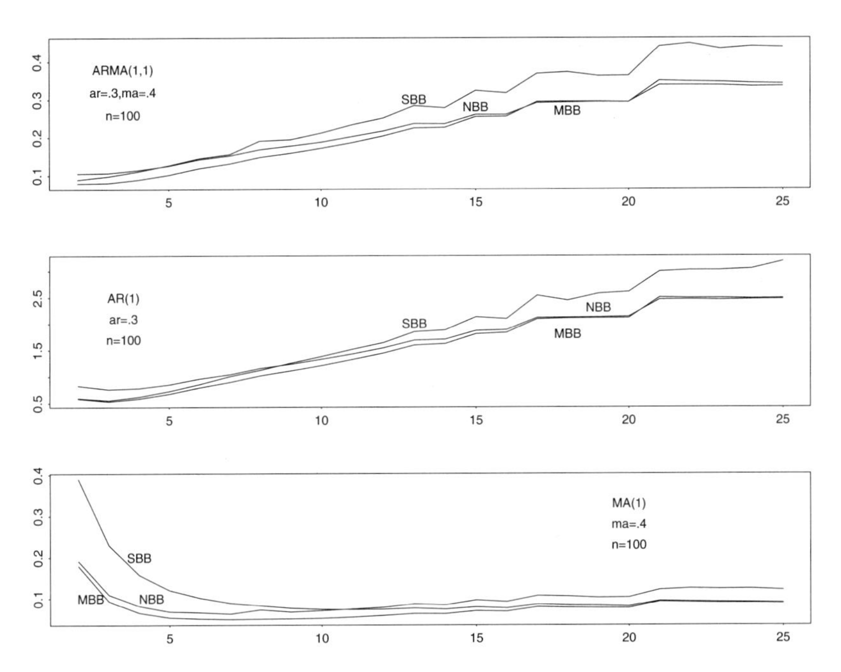

FIGURE 5.1. Mean square errors of the block bootstrap estimators of the level-2 parameter $\varphi_n = n\text{Var}(\bar{X}_n)$ at $n = 100$ under models (5.1)–(5.3). MSEs are computed using $K = 2000$ simulation runs, with $B = 500$ bootstrap iterations at each (expected) block length under each simulation run.

The numerical example considered above naturally leads us to the following questions. Is it true that the MBB outperforms the NBB and the SB in terms of the MSE criterion for more general processes and for more general level-2 parameters? If so, can we quantify the relative efficiency of the MBB with respect to the other block bootstrap methods, at least for

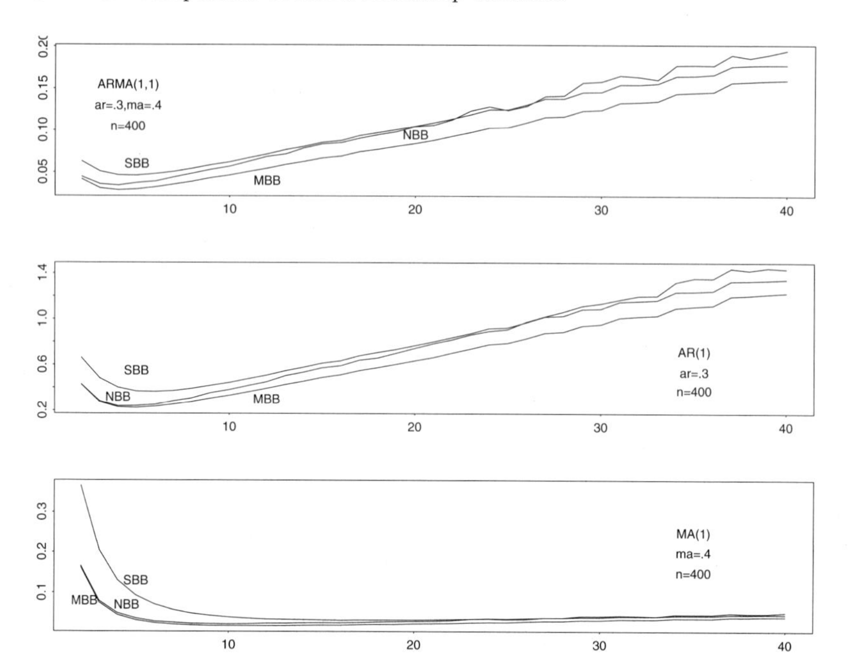

FIGURE 5.2. Mean square errors of the block bootstrap estimators of $\varphi_n = n\text{Var}(\bar{X}_n)$ at $n = 400$ under models (5.1)–(5.3). The same values of K and B as in Figure 5.1 are used.

large sample sizes? In the same vein, if the SB has the largest MSE, how large can the relative magnitude of the MSE of a SB estimator compared to a MBB estimator be? Also, as the performance of each block bootstrap method depends on the blocking parameter and there is a different "optimal" block length in each case, how do the MSEs of the "best" estimators from each of the four methods compare against one another? In the next few sections, we describe some theoretical results that provide answers to some of these questions. As we will see, the empirical results of the numerical examples above are in close agreement with our theoretical findings described below.

5.3 The Theoretical Framework

Let $\{X_i\}_{i\in\mathbb{Z}}$ be a $\mathbb{R}^d$-valued stationary process with $EX_1 = \mu$. For the theoretical development, we shall work under the "Smooth Function Model" introduced in Section 4.2. Without loss of generality, we may then suppose that the level-1 parameter of interest θ and its estimator $\hat{\theta}_n$ can be expressed as a smooth function of the population and the sample means, μ and $\bar{X}_n \equiv n^{-1}\sum_{i=1}^{n} X_i$, respectively. As noted in Chapter 4, by considering suitable transformations of the original observations, this formulation allows us to consider a wide class of estimators under the present framework. In particular, this includes the sample lag cross-covariance estimators, the sample autocorrelation estimators, and the Yule-Walker estimators for autoregressive processes. For the rest of this chapter, we suppose that there exists a smooth function $H : \mathbb{R}^d \to \mathbb{R}$, such that

$$\theta = H(\mu) \quad \text{and} \quad \hat{\theta}_n = H(\bar{X}_n).$$

Let φ_n denote the level-2 parameter of interest that is a functional of the distribution of $\hat{\theta}_n$. Also, let $\hat{\varphi}_n(j;\ell)$ be the bootstrap estimator of φ_n obtained by using the jth block bootstrap method with (expected) block length $\ell, j = 1,2,3,4$. As it turns out, the accuracy of the $\hat{\varphi}_n(\ell;\cdot)$ depends on the particular functional φ_n of the sampling distribution of $\hat{\theta}_n$ being estimated. For concreteness, we shall restrict attention to the following two level-2 parameters:

$$\varphi_{1n} \equiv \text{Bias}(\hat{\theta}_n) = E\hat{\theta}_n - \theta \tag{5.4}$$

and

$$\varphi_{2n} \equiv \text{Var}(\hat{\theta}_n) = E(\hat{\theta}_n - E\hat{\theta}_n)^2 , \tag{5.5}$$

and compare the performance of the block bootstrap methods for estimating these, using the MSE criterion. Similar results can be proved for the bootstrap estimators of the distribution function and for certain other functionals (e.g., the quantiles) of the sampling distribution of $\hat{\theta}_n$, although a different set of regularity conditions and arguments would be needed.

For the sake of completeness, we now briefly describe the specific versions of the bootstrap estimators of φ_{1n} and φ_{2n} considered here. Recall that, we index the methods MBB, NBB, CBB, and SB as method number 1,2,3, and 4, respectively, and denote the bootstrap samples generated by the jth method as $X^*_{j,1}, X^*_{j,2}, \ldots$, or as $X_1^{*(j)}, X_2^{*(j)}, \ldots$, as convenient, where $j = 1,2,3,4$. Let ℓ denote the block length for the first three methods and the expected block length for the SB method. For a given value of ℓ, we suppose that $b = \lfloor n/\ell \rfloor$ blocks are resampled for the MBB, the NBB, and the CBB, resulting in $n_1 \equiv b\ell$ bootstrap observations. Denote the corresponding sample mean by $\bar{X}^{*(j)}_{n,\ell}$, $j = 1,2,3$. Thus, the bootstrap version of the centered variable $T_n = \hat{\theta}_n - \theta$ under the MBB, the NBB, and the CBB are given by

$$T^{*(j)}_{n,\ell} = H(\bar{X}^{*(j)}_{n,\ell}) - H(E_* \bar{X}^{*(j)}_{n,\ell}),\ j = 1,2,3 , \tag{5.6}$$

where $\bar{X}^{*(j)}_{n,\ell} = n_1^{-1} \sum_{i=1}^{n_1} X^*_{j,i}$.

Next consider the SB method. Since we denote the expected block length by ℓ, the block length variables $L_1, L_2, \ldots$ of the SB method are now conditionally iid Geometric random variables with parameter $p = \ell^{-1}$. As a result, for the comparison here, we consider the SB estimators only corresponding to a finite set of values of the parameter $p \in (0,1)$, which are reciprocals of an integer ℓ in the interval $(1,n)$. Since the typical choice of p is such that $p \to 0$ as $n \to \infty$, it is possible to find an asymptotically equivalent choice, $p \sim \ell^{-1}$, for a suitable sequence $\ell \equiv \ell_n \to \infty$, and thus, this unified framework does not impose a serious restriction.

For a given value of ℓ, we suppose that under the SB method $K = \inf\{1 \le k \le n : L_1 + \ldots + L_k \ge n\}$ blocks are resampled, resulting in $N_1 = L_1 + \ldots + L_K$ bootstrap observations. Let $\bar{X}_{n,\ell}^{*(4)} \equiv n^{-1} \sum_{i=1}^{n} X_{4,i}^{*}$ denote the average of the first n SB observations. As noted earlier, $E_* \bar{X}_{n,\ell}^{*(4)} = \bar{X}_n$ for all ℓ. Hence, we define the bootstrap version of the centered variable $T_n = \hat{\theta}_n - \theta$ under the SB method by

$$T_{n,\ell}^{*(4)} = H(\bar{X}_{n,\ell}^{*(4)}) - H(\bar{X}_n) \,. \tag{5.7}$$

Note that the level-2 parameters of interest given in (5.4) and (5.5) are the first two moments of T_n, viz., $\varphi_{1n} = \text{Bias}(\hat{\theta}_n) = ET_n$ and $\varphi_{2n} = \text{Var}(\hat{\theta}_n) = \text{Var}(T_n)$. Hence, the bootstrap estimators of φ_{1n} and φ_{2n} are respectively defined as

$$\hat{\varphi}_{1n}(j;\ell) \equiv \widehat{\text{BIAS}}_j(\ell) = E_* T_{n,\ell}^{*(j)}, \; j = 1,2,3,4$$

and

$$\hat{\varphi}_{2n}(j;\ell) \equiv \widehat{\text{VAR}}_j(\ell) = \text{Var}_*(T_{n,\ell}^{*(j)}), \; j = 1,2,3,4 \,.$$

In the next section, we obtain expansions for the MSEs of the block bootstrap estimators $\widehat{\text{BIAS}}_j(\ell)$ and $\widehat{\text{VAR}}_j(\ell)$, $j = 1,2,3,4$.

5.4 Expansions for the MSEs

For deriving expansions for the MSEs of the bootstrap estimators, note that for any random variable Y, $\text{MSE}(Y) = [\text{Bias}(Y)]^2 + \text{Var}(Y)$, and, hence, an expansion of the MSE can be obtained from expansions for the bias part and the variance part. Consequently, we look at the bias and the variance of the bootstrap estimators separately and combine these to get a single measure of performance in terms of the MSE. Recall the notation $\Delta(r;\delta) = 1 + \sum_{n=1}^{\infty} n^{2r-1} \alpha(n)^{\delta/(2r+\delta)}$, where $\alpha(\cdot)$ denotes the strong mixing coefficient of the process $\{X_i\}_{i \in \mathbb{Z}}$. Also, recall that for $\alpha = (\alpha_1, \ldots, \alpha_d)' \in \mathbb{Z}_+^d$, we write $|\alpha| = \alpha_1 + \cdots + \alpha_d$, $\alpha! = \prod_{j=1}^{d} \alpha_j!$ and $D^\alpha = \frac{\partial^{|\alpha|}}{\partial x_1^{\alpha_1} \cdots \partial x_d^{\alpha_d}}$. We will use the following conditions for deriving the expansions in this section. Here r is a positive integer and its values are specified in the statements of the results below.

Condition D_r *The function* $H : \mathbb{R}^d \to \mathbb{R}$ *is* r*-times continuously differentiable and* $\max\{|D^\nu H(x)| : |\nu| = r\} \le C(1 + \|x\|^{a_0}), x \in \mathbb{R}^d$ *for some integer* $a_0 \ge 1$.

Condition M_r $E\|X_1\|^{2r+\delta} < \infty$ *and* $\Delta(r;\delta) < \infty$ *for some* $\delta > 0$.

Then, we have the following result on the bias part of the bootstrap estimators $\hat{\varphi}_{1n}(j;\ell)$ and $\hat{\varphi}_{2n}(j;\ell)$, $j = 1,2,3,4$.

Theorem 5.1 *Assume that* ℓ *is such that* $\ell^{-1} + n^{-1/2}\ell = o(1)$ *as* $n \to \infty$.

(a) Suppose that Condition D_r *holds with* $r = 3$ *and that Condition* M_r *holds with* $r = 3 + a_0$*, where* a_0 *is as specified by Condition* D_r*. Then,*

$$Bias(\widehat{BIAS}_j(\ell)) = n^{-1}\ell^{-1}A_1 + o(n^{-1}\ell^{-1}) \quad for \quad j = 1,2,3,4,$$

where $A_1 = -\sum_{|\alpha|=1}\sum_{|\beta|=1} c_{\alpha+\beta}(\sum_{j=-\infty}^{\infty} |j| E X_1^\alpha X_{1+j}^\beta)$ *and* $c_\alpha = D^\alpha H(\mu)/\alpha!, \alpha \in (\mathbb{Z}_+)^d$.

(b) Suppose that Condition D_r *holds with* $r = 2$ *and that Condition* M_r *holds with* $r = 4 + 2a_0$*, where* a_0 *is as specified by Condition* D_r*. Then,*

$$Bias(\widehat{VAR}_j(\ell)) = n^{-1}\ell^{-1}A_2 + o(n^{-1}\ell^{-1}) \quad for \quad j = 1,2,3,4,$$

where $A_2 = -\sum_{j=-\infty}^{\infty} |j| E\tilde{Z}_1\tilde{Z}_{1+j}$, $\tilde{Z}_i = \sum_{|\alpha|=1} c_\alpha (X_i - \mu)^\alpha$, $i \ge 1$, *and* c_α *is as defined in (a) above.*

Proof: See Section 5.7. □

Thus, it follows from Theorem 5.1, that the biases of the bootstrap estimators of φ_{1n} and φ_{2n} are identical up to the first-order terms *for* all four block bootstrap methods considered here. In particular, contrary to the common belief, the stationarity of the SB observations $X_{4,1}^*, X_{4,2}^*, \ldots$ does not contribute significantly toward reducing the bias of the resulting bootstrap estimators. Also, the use of either overlapping or nonoverlapping blocks results in the same amount of bias asymptotically. Since the bias of a block bootstrap estimator essentially results from replacing the original data sequence $X_1, \ldots, X_n$ by independent copies of *smaller* subsequences, all the methods perform similarly as long as the (expected) length ℓ of these subsequences are asymptotically equivalent.

Next we compare the variances of the block bootstrap estimators of φ_{1n} and φ_{2n}.

Theorem 5.2 *Assume that the conditions of Theorem 5.1 on the block length parameter* ℓ *and on the index* r *in Conditions* D_r *and* M_r *for the respective parts hold. Then, there exist symmetric, nonnegative real valued*

functions g_1, g_2 *such that*
(a)

$$
\begin{aligned}
Var(\widehat{BIAS}_j(\ell)) &= \{4\pi^2 g_1(0)/3\} n^{-3}\ell + o(n^{-3}\ell), \ j = 1,3 \ ; \\
Var(\widehat{BIAS}_j(\ell)) &= \{2\pi^2 g_1(0)\} n^{-3}\ell + o(n^{-3}\ell), \ j = 2 \ ; \\
Var(\widehat{BIAS}_j(\ell)) &= (2\pi)\left[2\pi g_1(0) + \int_{-\pi}^{\pi} (1+e^{iw}) g_1(w) dw\right](n^{-3}\ell) \\
&\quad + o(n^{-3}\ell), \ j = 4 \ ;
\end{aligned}
$$

(b)

$$
\begin{aligned}
Var(\widehat{VAR}_j(\ell)) &= \{(2\pi)^2 g_2(0)/3\} n^{-3}\ell + o(n^{-3}\ell), \ j = 1,3 \ ; \\
Var(\widehat{VAR}_j(\ell)) &= \{(2\pi)^2 g_2(0)/2\} n^{-3}\ell + o(n^{-3}\ell), \ j = 2 \ ; \\
Var(\widehat{VAR}_j(\ell)) &= (2\pi)\left[2\pi g_2(0) + \int_{-\pi}^{\pi} (1+e^{iw}) g_2(w) dw\right](n^{-3}\ell) \\
&\quad + o(n^{-3}\ell), \ j = 4 \ .
\end{aligned}
$$

Proof: See Section 5.7. □

The definitions of the functions $g_k(\cdot)$, $k = 1, 2$ are somewhat complicated and are given in Section 5.7 (cf. (5.8)). However, even without their exact definitions, we can compare the relative magnitudes of the variance parts of different block bootstrap estimators using Theorem 5.2. From parts (a) and (b) of Theorem 5.2, we see that the MBB and the CBB estimators (corresponding to $j = 1, 3$) of $\varphi_{1n} = \text{Bias}(\hat{\theta}_n)$ and $\varphi_{2n} = \text{Var}(\hat{\theta}_n)$ have 2/3-times smaller variances than the corresponding NBB estimators. Since the blocks in the MBB and the CBB are allowed to overlap, the amount of variability among the resampled blocks is lesser, leading to a smaller variance for these estimators. This advantage of the MBB over the NBB was first noted in Künsch (1989) (see Remark 3.3 of Künsch (1989)). It is interesting to note that in spite of all the differences in their resampling mechanisms, all four block bootstrap methods have the same *order of magnitude* for the variances of the resulting estimators. This is particularly surprising in the case of the SB method, since it introduces additional randomness in the resampled blocks. The effect of this additional randomness shows up in the constant of the leading term in the expansion for the variances of the SB estimators. Since $\Delta_k \equiv \int_{-\pi}^{\pi}(1 + e^{iw})g_k(w)dw \geq 0$ for $k = 1, 2$, it follows that the SB estimators have asymptotically the largest variances among all four block bootstrap methods for a given value of ℓ.

5.5 Theoretical Comparisons

5.5.1 Asymptotic Efficiency

In the following result, we summarize the implications of Theorems 5.1 and 5.2 for the asymptotic relative efficiency of different block bootstrap estimators in terms of their MSEs. For any two sequences of estimators $\{\hat{\psi}_{1n}\}_{n\geq 1}$ and $\{\hat{\psi}_{2n}\}_{n\geq 1}$, we define the *asymptotic relative efficiency* of the sequence $\{\hat{\psi}_{1n}\}_{n\geq 1}$ with respect to $\{\hat{\psi}_{2n}\}_{n\geq 1}$ as

$$\text{ARE}(\hat{\psi}_{1n};\hat{\psi}_{2n}) = \lim_{n\to\infty} \Big[\text{MSE}(\hat{\psi}_{2n})/\text{MSE}(\hat{\psi}_{1n})\Big] .$$

Thus, if $\text{ARE}(\hat{\psi}_{1n};\hat{\psi}_{2n}) < 1$, then the sequence of estimators $\{\hat{\psi}_{1n}\}_{n\geq 1}$ are less efficient than $\{\hat{\psi}_{2n}\}_{n\geq 1}$ in the sense that $\hat{\psi}_{1n}$'s have larger MSEs than the MSEs of the estimators $\hat{\psi}_{2n}$'s, for large n.

Theorem 5.3 *Assume that the conditions of Theorems 5.1 and 5.2 hold and that* $A_k \neq 0,\ g_k(0) \neq 0,\ k = 1,2$.

(i) *For* $\ell^{-1} + n^{-1/3}\ell = o(1)$, *for any* $i, j \in \{1,2,3,4\},\ k = 1,2,$

$$ARE\Big(\hat{\varphi}_{kn}(j;\ell);\hat{\varphi}_{kn}(i;\ell)\Big) = 1 .$$

(ii) *For* $\ell^{-1}n^{1/3} = o(1)$, *and for* $k = 1,2,$

$$ARE\Big(\hat{\varphi}_{kn}(2;\ell);\hat{\varphi}_{kn}(j;\ell)\Big) = 2/3 \quad \textit{for} \quad j = 1,3 ;$$

$$\begin{aligned} &ARE\Big(\hat{\varphi}_{kn}(4;\ell);\hat{\varphi}_{kn}(2;\ell)\Big) \\ &= \quad \left[2 + \pi^{-1}\int_{-\pi}^{\pi}(1+e^{iw})(g_k(w)/g_k(0))dw\right]^{-1} \in (0,1/2) . \end{aligned}$$

(iii) *For* $\ell = Cn^{1/3}(1+o(1))$, $C \in (0,\infty)$ *and for* $k = 1,2,$

$$ARE\Big(\hat{\varphi}_{kn}(2;\ell);\hat{\varphi}_{kn}(j;\ell)\Big) = \frac{3 + 4\pi^2C^3A_k^{-2}g_k(0)}{3 + 6\pi^2C^3A_k^{-2}g_k(0)} \in (2/3,1), j = 1,3;$$

$$\begin{aligned} &ARE\Big(\hat{\varphi}_{kn}(4;\ell);\hat{\varphi}_{kn}(2;\ell)\Big) \\ &= \frac{1 + 2\pi^2C^3A_k^{-2}g_k(0)}{1 + 2\pi C^3A_k^{-2}\left[2\pi g_k(0) + \int_{-\pi}^{\pi}(1+e^{iw})g_k(w)dw\right]} \in (0,1) . \end{aligned}$$

Proof: A direct consequence of Theorems 5.1 and 5.2. □

Theorems 5.1–5.3 are due to Lahiri (1999a). Note that asymptotic relative efficiency of the SB estimators with respect to the MBB and the CBB estimators in parts (ii) and (iii) of Theorem 5.3 can be found by the identity $\text{ARE}(\hat{\varphi}_{kn}(4;\ell); \hat{\varphi}_{kn}(j;\ell)) = \text{ARE}(\hat{\varphi}_{kn}(4;\ell); \hat{\varphi}_{kn}(2;\ell)) \cdot \text{ARE}(\hat{\varphi}_{kn}(2;\ell); \hat{\varphi}_{kn}(j;\ell))$, $j = 1, 3$ and, hence, are not stated separately. Also note that parts (i) and (ii) of Theorem 5.3 correspond to the cases where the leading terms of the MSEs of the block bootstrap estimators are determined solely by their biases and their variances, respectively. It follows that for smaller values of the block length parameter ℓ (i.e., under case (i)), all methods have an ARE of 1 with respect to one another. For large values of ℓ (i.e., under case (ii)), the ARE of the SB is less than 1/2 compared to the other block bootstrap methods based on nonrandom block lengths. In the intermediate case (i.e., under case (iii)), the MSE has nontrivial contributions from both the bias part and the variance part. In this case, the ARE of the NBB or the SB with respect to the MBB and the CBB lies between 1 and lower bound on the limits under case (iii). In particular, the ARE of the SB estimator $\hat{\varphi}_{kn}(4;\ell)$ with respect to the MBB estimator $\hat{\varphi}_{kn}(1;\ell)$ under case (iii) lies in the interval (0,1), depending on the value of the constant C and the function g_k.

5.5.2 *Comparison at Optimal Block Lengths*

From Theorems 5.1 and 5.2, we see that for *each* of the block bootstrap methods considered here, as the (expected) block length ℓ increases, the bias of a block bootstrap estimator decreases while its variance increases. As a result, for each block bootstrap method, there is a critical value of the block length parameter ℓ that minimizes the MSE. We call the value of ℓ minimizing the leading terms in the expansion of the MSE as the (first-order) MSE-optimal block length. Let

$$\ell_{1j}^0 = \text{argmin}\{\text{MSE}(\widehat{\text{BIAS}}_j(\ell)) : n^\epsilon \le \ell \le n^{(1-\epsilon)/2}\}$$

and

$$l_{2j}^0 = \text{argmin}\{\text{MSE}(\widehat{\text{VAR}}_j(\ell)) : n^\epsilon \le \ell \le n^{(1-\epsilon)/2}\},$$

$1 \le j \le 4$ denote the MSE-optimal block lengths for estimating the bias and the variance of $\hat{\theta}_n$, where $\epsilon \in (0, \frac{1}{3})$ is a given number. The following result gives the optimal block lengths ℓ_{kj}^0, $k = 1, 2$, $j = 1, 2, 3, 4$ for estimating φ_{1n}, φ_{2n} for the four block bootstrap methods considered in this chapter.

Theorem 5.4 *Suppose that the conditions of Theorem 5.3 hold. Then, for* $k = 1, 2$,

$$\ell_{kj}^0 \sim (3A_k^2/[2\pi^2 g_k(0)])^{1/3} \cdot n^{1/3}, \ j = 1, 3 ;$$

$$\ell_{kj}^0 \sim (A_k^2/[\pi^2 g_k(0)])^{1/3} \cdot n^{1/3}, \; j = 2 \; ;$$

$$\ell_{kj}^0 \sim \left(A_k^2 / \left[2\pi^2 g_k(0) + \pi \int_{-\pi}^{\pi} (1 + e^{iw}) g_k(w) dw \right] \right)^{1/3} \cdot n^{1/3}, \; j = 4 \; .$$

Proof: Follows from Theorems 5.1 and 5.2 and the fact that the function $h(x) = c_1 x + c_2 x^{-2}$, $x > 0$, with coefficients $c_1 > 0$, $c_2 > 0$ is minimized at

$$x^* = (2c_2/c_1)^{1/3}$$

and

$$h(x^*) = (3/\sqrt[3]{4}) c_1^{2/3} c_2^{1/3} \; .$$

□

The formulas in Theorem 5.4 for the MBB and the NBB were noted by Hall, Horowitz and Jing (1995). Note that the optimal block size for the MBB is larger than that of the NBB by a factor of $(3/2)^{1/3}$. For the SB variance estimator of the sample mean, Politis and Romano (1994b) show that the order of the MSE-optimal expected block length is $n^{1/3}$. The explicit formulas for the optimal block sizes for the SB bias and variance estimators under the Smooth Function Model are due to Lahiri (1999a).

It is clear from the definitions of ℓ_{kj}^0 that each block bootstrap method provides the most accurate estimator of the parameter φ_{kn} when it is used with the corresponding optimal block length. In the next result, we compare the block bootstrap methods at their *best* possible performances, i.e. when *each* method is used to estimate a given parameter with its *MSE-optimal block length.*

Theorem 5.5 *Suppose that the conditions of Theorem 5.3 hold.*

(a) Then, for $k = 1, 2$,

$$MSE(\hat{\varphi}_{kn}(j; \ell_{kj}^0)) = 3^{1/3}[2\pi^2 g_k(0) A_k]^{2/3} n^{-8/3} + o(n^{-8/3}), \quad j = 1, 3 \; ;$$

$$MSE(\hat{\varphi}_{kn}(j; \ell_{kj}^0)) = 3[\pi^2 g_k(0) A_k]^{2/3} n^{-8/3} + o(n^{-8/3}), \; j = 2 \; ;$$

$$MSE(\hat{\varphi}_{kn}(j; \ell_{kj}^0)) = 3\left[\left\{2\pi^2 g_k(0) + \pi \int_{-\pi}^{\pi} (1 + e^{iw}) g_k(w) dw\right\} A_k\right]^{2/3} \times n^{-8/3} + o(n^{-8/3}), \; j = 4 \; .$$

(b) For $k = 1, 2$,

$$ARE(\hat{\varphi}_{kn}(2; \ell_{k2}^0); \hat{\varphi}_{kn}(j; \ell_{kj}^0)) = \left(\tfrac{2}{3}\right)^{2/3}, \quad j = 1, 3 \; ;$$

$$ARE(\hat{\varphi}_{kn}(4; \ell_{k4}^0); \hat{\varphi}_{kn}(2; \ell_{k2}^0)) = \left\{2 + \pi^{-1} \int_{-\pi}^{\pi} (1 + e^{iw}) \tilde{g}_k(w) dw\right\}^{-2/3}$$

where $\tilde{g}_k(w) = g_k(w)/g_k(0)$.

Proof of Theorem 5.5: Follows from Theorems 5.1 and 5.2, and the proof of Theorem 5.4. □

Theorem 5.5 shows that when each method is used with the corresponding MSE-optimal value of ℓ, the MBB and the CBB has an optimal MSE that is $(2/3)^{2/3}$ times smaller than the optimal MSE for the NBB, and the MSE of the optimal NBB estimator is, in turn, at least $2^{-2/3}$-times smaller than that of the optimal SB estimator.

The following result shows that the ARE of the SB with respect to the NBB at the optimal block length admits a lower bound.

Theorem 5.6 *Assume that the conditions of Theorem 5.3 hold. Then, for* $k = 1, 2$

$$ARE\big(\hat{\varphi}_{kn}(4; \ell_{k4}^0); \hat{\varphi}_{kn}(2; \ell_{k2}^0)\big) \geq \left[\tfrac{7}{2}\right]^{-2/3} .$$

Proof of Theorem 5.6: By Cauchy-Schwarz inequality, the maximum value of the integral $\int_{-\pi}^{\pi}(1+\cos w)\tilde{g}_k(w)dw$ is attained if and only if $\tilde{g}_k(w) = C_0 \cdot (1 + \cos w)$ for almost all $w \in (-\pi, \pi)$ (with respect to the Lebesgue measure) for some $C_0 \in \mathbb{R}$. Since $\tilde{g}_k(\cdot)$ is continuous, letting $w \to 0$, we get $C_0 = 1/2$. Hence, from Theorem 5.5(b), for $k = 1, 2$,

$$\begin{aligned}
&\text{ARE}\big(\hat{\varphi}_{kn}(4; \ell_{k4}^0); \hat{\varphi}_{kn}(2; \ell_{k2}^0)\big) \\
&\geq \Big\{2 + (2\pi)^{-1} \int_{-\pi}^{\pi} (1 + \cos w)^2 dw\Big\}^{-2/3} \\
&= \big\{2 + (2\pi)^{-1}[3\pi]\big\}^{-2/3} .
\end{aligned}$$

□

Theorem 5.5 is due to Lahiri (1999a). The lower bound result in Theorem 5.6 is due to Politis and White (2003).

5.6 Concluding Remarks

The results of Sections 5.3-5.5 show that in terms of their MSEs, the MBB and the CBB estimators outperform the NBB estimators, which in turn outperform the SB estimators. This conclusion is valid as long as the corresponding (expected) block lengths grow at a rate not slower than the optimal block length. For estimating the bias and the variance functionals φ_{1n} and φ_{2n}, this optimal rate is $const.n^{1/3}$, where n denotes the sample of size. When the respective block lengths grow at a slower rate than the optimal rate, the main contribution to the MSE comes from the bias part. In this case, performance of all four methods are comparable with all the AREs being equal to 1. This is a simple consequence of the fact that, asymptotically, the biases of the bootstrap estimators derived from all four methods have the same leading term. The finite sample simulation example of Section 5.2 also supports this observation. When the block bootstrap methods are used with block lengths close to the corresponding optimal block lengths, the MBB and the CBB give the most accurate bootstrap estimators.

Going beyond the bias and the variance of $\hat{\theta}_n$, it is possible to carry out a comparison of the block bootstrap methods for more complicated functionals (e.g., quantiles) of the sampling distribution of (a suitably studentized version of) $\hat{\theta}_n$. For estimating the distribution function and quantiles of a studentized version of $\hat{\theta}_n$, the optimal block length is of the form $const.n^{1/4}$ for all four block bootstrap methods (cf. Hall, Horowitz and Jing (1995), Lahiri (1999c, 2003c)). In this case, the AREs of the block bootstrap distribution function estimators has an ordering that is exactly the same as that for the bias and the variance functionals. Indeed, for block lengths growing at a rate not slower than $const.n^{1/4}$, the MBB and the CBB are the most accurate among the four block bootstrap methods.

The results above show optimality of the MBB and the CBB only among the four methods considered above. Carlstein, Do, Hall, Hesterberg and Künsch (1998) have proposed a block bootstrap method, called the Matched Block Bootstrap (MaBB) where the bootstrap blocks are resampled using a Markov chain. Thus, unlike the four block bootstrap methods covered here, the resampled blocks under the MaBB are dependent. Under some structural assumptions on the underlying process (e.g., AR(p) or Markov), Carlstein et al. (1998) show that the MaBB estimator of the variance of the sample mean has a variance that is of a comparable order to the variance of the NBB estimator and has a bias that is of smaller order. Thus, the minimum MSE of the MaBB is of a smaller order than the minimum MSEs for the four methods considered here. Consequently, for processes with a Markovian structure, the MaBB outperforms the above methods at the respective optimal block sizes. For more general time series that may not necessarily have a Markovian structure, Paparoditis and Politis (2001, 2002) recently proposed a method, called the Tapered Block Bootstrap (TaBB) method, and showed that the TaBB method yields a more accurate estimator of the variance-type level-2 parameters than do the MBB and the CBB methods.

5.7 Proofs

Let $S(i;k) = \sum_{j=i}^{i+k-1} Y_{n,j}, i, k \in \mathbb{N}$ denote the partial sums of the periodically extended time series $\{Y_{n,i}\}_{i \geq 1}$. Recall that $\zeta_s = \left(E\|X_1\|^s\right)^{1/s}$, $s > 0$, and for $r \in \mathbb{Z}_+$ and $\delta \in (0, \infty)$, let $\Delta(r; \delta) = 1 + \sum_{n=1}^{\infty} n^{2r-1}[\alpha(n)]^{\frac{\delta}{2r+\delta}}$. Let $\hat{\mu}(j; \ell) = E_* \bar{X}_{n,\ell}^{*(j)}$, $\hat{c}_{\alpha,j} = D^\alpha H(\hat{\mu}(j; \ell))$, $1 \leq j \leq 4$, $\bar{c}_\alpha = D^\alpha H(\bar{X}_n)$, $\alpha \in \mathbb{Z}_+^d$, and $\Sigma_\infty = lim_{n \to \infty} \text{Cov}(\sqrt{n} \bar{X}_n)$.

Under Condition M_r for any $r \in \mathbb{N}$, $\{X_i\}_{i \in \mathbb{Z}}$ has a (continuous) spectral density matrix $f(\cdot)$, defined by the relation

$$\text{Cov}(X_1, X_{1+k}) = \int_{-\pi}^{\pi} e^{\iota k x} f(x) dx, \ k \in \mathbb{Z} .$$

We index the components of the $d \times d$ matrix-valued function $f(x)$ by unit vectors $\alpha; \beta \in \mathbb{Z}_+^d$ as $f(x; \alpha, \beta)$, $|\alpha| = 1 = |\beta|$. Next define

$$\begin{aligned} g_1(w) &= \sum_{|\alpha|=1} \sum_{|\beta|=1} \sum_{|\gamma|=1} \sum_{|\nu|=1} c_{\alpha+\beta} c_{\gamma+\nu} \Big\{ f(w; \alpha, \gamma) \bar{f}(w; \beta, \nu) \\ &\quad + f(w; \alpha, \nu) \bar{f}(w; \beta, \gamma) + f(w; \beta, \gamma) \bar{f}(w; \alpha, \nu) \\ &\quad + f(w; \beta, \nu) \bar{f}(w; \alpha, \gamma) \Big\} , \end{aligned} \tag{5.8}$$

$-\pi \leq w \leq \pi$, where for any complex number $z = u + \iota v$, $u, v \in \mathbb{R}$, $\bar{z} \equiv u - \iota v$ denotes its complex conjugate. Next let $g_2(w)$ be the function obtained by replacing $c_{\alpha+\beta}, c_{\gamma+\nu}$ in the definition of $g_1(w)$ by $c_\alpha c_\beta, c_\gamma c_\nu$, respectively. Note that, since $f(w; \alpha, \beta) = \bar{f}(w; \beta, \alpha) = f(-w; \beta, \alpha)$ for all α, β, the functions $g_1(w)$ and $g_2(w)$ are real valued and are symmetric about zero.

For clarity of exposition, we separately present the proofs of the main results for the block bootstrap methods based on nonrandom block lengths in Section 5.7.1 and for the SB in Section 5.7.2

5.7.1 Proofs of Theorems 5.1-5.2 for the MBB, the NBB, and the CBB

Lemma 5.1 *Assume that $\ell = O(n^{1-\epsilon})$ for some $0 < \epsilon < 1$, $E\|X_1\|^{2r+\delta} < \infty$ and $\Delta(r; \delta) < \infty$, for some positive integer r and for some $\delta > 0$. Then,*

(i) $E\{E_* \|S(I_{j,1}; \ell)\|^{2r}\} \leq C(r, d) \zeta_{2r+\delta}^{2r} \Delta(r; \delta) \cdot \ell^r$ *for* $j = 1, 2, 3$.

(ii) $E\|\hat{\mu}(j; \ell)\|^{2r} \leq C(r, d) \zeta_{2r+\delta}^{2r} \Delta(r; \delta) \cdot n^{-r}$ *for* $j = 1, 2, 3$.

(iii) $E\{E_* \|\bar{X}_{n,\ell}^{*(j)}\|^{2r}\} \leq C(r, d) \zeta_{2r+\delta}^{2r} \Delta(r; \delta) \cdot n^{-r}$, $j = 1, 2, 3$.

Proof: We prove part (i) first. Note that by Lemma 3.2, for $j = 1$, we get

$$\begin{aligned} E\left\{E_* \|S(I_{j,1}; \ell)\|^{2r}\right\} &= E\Big\{N^{-1} \sum_{i=1}^{N} \|S(i; \ell)\|^{2r}\Big\} = E\|S(i; \ell)\|^{2r} \\ &\leq C(d, r) \zeta_{2r+\delta}^{2r} \Delta(r; \delta) \ell^r . \end{aligned} \tag{5.9}$$

The proof is similar for $j = 2$. For $j = 3$, note that for any $\nu \in \mathbb{Z}_+^d$ and any $1 < m < n/2$,

$$
\begin{aligned}
& |E_*(S(I_{3,1}; m))^\nu - E_*(S(I_{1,1}; m))^\nu| \\
\leq\ & C(\nu)\Bigg[n^{-1} \sum_{i=1}^{m-1} \Big\{ \|S(n-i+1; i)\|^{|\nu|} + \|S(1; m-i)\|^{|\nu|} \Big\} \\
& + n^{-2} m \sum_{i=1}^{n-m+1} \|S(i; m)\|^{|\nu|} \Bigg] . \qquad (5.10)
\end{aligned}
$$

Hence, the bound for $j = 3$ follows from (5.9), (5.10), and Lemma 3.2. This proves part (i).

As for part (ii), note that for all $j \in \{1, 2, 3\}$, $\hat{\mu}(j; \ell) = \sum_{i=1}^n w_{ijn} X_i$ for some nonrandom weights w_{ijn} with $|w_{ijn}| \leq 1$ for all i, j, n. Hence, using cumulant expansion for moments and Lemma 3.2, we get (ii).

Part (iii) is a consequence of parts (i) and (ii) and the following result (cf. Lemma III.3.1 of Ibragimov and Hasminskii (1980)): *For zero mean independent random variables $W_1, \ldots, W_m$ and for any integer $r \geq 1$,*

$$
E\Big(\sum_{i=1}^m W_i \Big)^{2r} \leq C(r) \Big[m^{r-1} \sum_{i=1}^m E W_i^{2r} \Big]. \qquad (5.11)
$$

This completes the proof of Lemma 5.1. □

Proof of Theorem 5.1 for j=1,2,3: We prove the theorem only for the bias estimators $\hat{\varphi}_{1n}(j; \ell)$, $j = 1, 2, 3$. The proof for the variance estimators $\hat{\varphi}_{2n}(j; \ell)$, $j = 1, 2, 3$ is similar and, hence, is omitted. Without loss of generality, suppose that $\mu = 0$. (Otherwise, replace X_i's with $(X_i - \mu)$'s in every step below.) Note that by Taylor's expansion of $H(\bar{X}_{n,\ell}^{*(j)})$ around $\hat{\mu}(j; \ell)$, we have

$$
\begin{aligned}
& \hat{\varphi}_{1n}(j; \ell) \\
=\ & \sum_{|\alpha|=2} \hat{c}_{\alpha,j} \Big\{ E_* \Big(\bar{X}_{n,\ell}^{*(j)} - \hat{\mu}(j; \ell) \Big)^\alpha \Big\} \\
& + 3 \sum_{|\alpha|=3} (\alpha!)^{-1} E_* \Big(\bar{X}_{n,\ell}^{*(j)} - \hat{\mu}(j; \ell) \Big)^\alpha \\
& \times \int_0^1 (1-u)^2 D^\alpha H\Big(\hat{\mu}(j; \ell) + u \bar{X}_{n,\ell}^{*(j)} \Big) du . \qquad (5.12)
\end{aligned}
$$

For $j \in \{1, 2, 3\}$, conditional on $\mathcal{X}_n$, $\bar{X}_{n,\ell}^{*(j)}$ is the average of b iid random variables and, hence, we may further simplify (5.12) as

$$
\hat{\varphi}_{1n}(j; \ell) = b^{-1} \ell^{-2} \sum_{|\alpha|=2} c_\alpha E_* \big(S(I_{j,1}; \ell) \big)^\alpha + R_{1n}(j; \ell) , \qquad (5.13)
$$

where, after some lengthy and routine algebra, the remainder term $R_{1n}(j;\ell)$ can be shown to satisfy the inequality

$$\begin{aligned}
&|R_{1n}(j;\ell)| \\
&\quad\leq\ C(\zeta_2, a_0, d)\Big\{b^{-1}\|\hat{\mu}(j;\ell)\|^2 \\
&\qquad + (1+\|\hat{\mu}(j;\ell)\|^{a_0})\cdot\|\hat{\mu}(j;\ell)\|(b^{-1}E_*\|S(I_{j,1};\ell)/\ell\|^2) \\
&\qquad + E_*\Big(1+\|\hat{\mu}(j;\ell)\|^{a_0}+\|\bar{X}_{n,\ell}^{*(j)}\|^{a_0}\Big)\|\bar{X}_{n,\ell}^{*(j)}\|^3\Big\}\,. \qquad (5.14)
\end{aligned}$$

By Lemma 5.1 and Hölder's inequality

$$\begin{aligned}
&E(R_{1n}(j;\ell))^2 \\
&\leq\ C(\zeta_2, a_0, d)\Big[b^{-2}E\|\hat{\mu}(j;\ell)\|^4 \\
&\quad + \ell^{-4}b^{-2}(E\|\hat{\mu}(j;\ell)\|^4)^{1/2}\big(E(E_*\|S(I_{j1};\ell)\|^8)\big)^{1/2} \\
&\quad + \ell^{-4}b^{-2}\Big(E\|\hat{\mu}(j;\ell)\|^{6+2a_0}\Big)^{\frac{(2+2a_0)}{(6+2a_0)}}\Big(E\Big[E_*\|S(I_{j1};\ell)\|^{6+2a_0}\Big]\Big)^{\frac{4}{(6+2a_0)}} \\
&\quad + E\Big\{E_*\|\bar{X}_{n,\ell}^{*(j)}\|^6 + E_*\|\bar{X}_{n,\ell}^{*(j)}\|^{6+2a_0}\Big\} \\
&\quad + \Big(E\|\hat{\mu}(j;\ell)\|^{6+2a_0}\Big)^{\frac{2a_0}{(6+2a_0)}}\Big(E\Big(E_*\|\bar{X}_{n,\ell}^{*(j)}\|^{6+2a_0}\Big)^{\frac{6}{(6+2a_0)}}\Big)\Big] \\
&\leq\ C(\zeta_2, a_0, d)\big[n^{-3}+n^{-4}\ell^2\big]\,. \qquad (5.15)
\end{aligned}$$

Hence, Theorem 5.1 follows from (5.13) and (5.15) for $j \in \{1,2,3\}$, by noting that $\text{Bias}(\hat{\theta}_n) \equiv E\hat{\theta}_n - \theta = n^{-1}\sum_{|\alpha|=2} c_\alpha E Z_\infty^\alpha + O(n^{-2})$. □

For proving Theorem 5.2 for $j = 1,2,3$, we need a lemma.

Lemma 5.2 *Suppose that Condition M_r holds for some $r \geq 2$ and that $\mu = 0$. Then, for any integers $p \geq 1$, $q \geq 1$ with $p+q < 2r$ and for any $t_1,\ldots,t_p$, $s_1,\ldots,s_q \in \mathbb{R}^d$ with $\|t_i\| \leq 1$, $\|s_i\| \leq 1$,*

$$\begin{aligned}
&\lim_{\ell\to\infty} \ell^{-1}\sum_{j=1}^{\ell} E\Big\{\Big(\prod_{k=1}^{p} t_k' U_{11}\Big)\Big(\prod_{k=1}^{q} s_k' U_{1(1+j)}\Big)\Big\} \\
&\quad = \sum_I\sum_J\Big\{b(|I|+|J|;p+q)EZ_\infty(I^c)E\tilde{Z}_\infty(J^c)EZ_\infty(I)\tilde{Z}_\infty(J)\Big\}\,,
\end{aligned}$$

where $b(k;m) = \int_0^1 x^{k/2}(1-x)^{(m-k)/2}dx$, $m \geq k \geq 0$, $Z_\infty(I) = \prod_{i\in I} t_i' Z_\infty$, $\tilde{Z}_\infty(J) = \prod_{j\in J} s_j' Z_\infty$ for $I \subset \{1,\ldots,p\}$, $J \subset \{1,\ldots,q\}$, $Z_\infty \sim N(0,\Sigma_\infty)$, and where the summations $\sum_I$ and $\sum_J$ respectively extend over all subsets $I \subset \{1,\ldots,p\}$ and $J \subset \{1,\ldots,q\}$.

Proof: Let $\tilde{S}(k;m) = X_k + \cdots + X_m$, and for a set A, let

$$V(k,m;A) \ = \ \prod_{i\in A} t_i'\tilde{S}(k;m)\ ;$$

$$W(k,m;A) \equiv \prod_{i \in A} s_i' \tilde{S}(k;m) \ ,$$

$m \geq k \geq 1$. Then, for $\ell^{1/2} \leq j \leq \ell - \ell^{1/2}$, setting $m \equiv \lfloor \ell^{1/4} \rfloor$, we have

$$\begin{aligned}
\Delta_{j,\ell} &\equiv E\Big\{\Big(\prod_{k=1}^{p} t_k' U_{11}\Big)\Big(\prod_{k=1}^{q} s_k' U_{1(1+j)}\Big)\Big\} \\
&= \ell^{-(p+q)/2} E\bigg[\bigg\{\prod_{k=1}^{p} t_k'\Big(\tilde{S}(1;j-m) + \tilde{S}(j-m+1;j) \\
&\quad + \tilde{S}(j+1;\ell)\Big)\bigg\}\bigg\{\prod_{k=1}^{q} s_k'\Big(\tilde{S}(j+1;\ell) + \tilde{S}(\ell+1;\ell+m) \\
&\quad + \tilde{S}(\ell+m+1;\ell+j)\Big)\bigg\}\bigg] \\
&= \ell^{-(p+q)/2} \sum_I \sum_J E\Big\{V(1,j-m;I^c)V(j+1,\ell;I) \\
&\quad \times W(j+1,\ell;J)W(\ell+m+1,\ell+j;J^c)\Big\} \\
&\quad + Q_{n,j}
\end{aligned}$$

where, by Hölder's inequality and Lemma 3.2,

$$\begin{aligned}
&\max\Big\{|Q_{n,j}| : \ell^{1/2} \leq j \leq \ell - \ell^{1/2}\Big\} \\
&\leq C(p,q) \cdot \ell^{-\frac{p+q}{2}} \max\Big\{\Big(E\|\tilde{S}(1,k)\|^{p+q}\Big)^{\frac{a}{p+q}}\Big(E\|\tilde{S}(1,i)\|^{p+q}\Big)^{1-\frac{a}{p+q}} : \\
&\qquad\qquad 1 \leq k \leq m,\ 1 \leq i \leq \ell,\ 1 \leq a \leq p+q\Big\} \\
&\leq C(r,\delta)\zeta_{2r+\delta}^{p+q}\Delta(r;\delta) \cdot \ell^{-\frac{p+q}{2}} \max\Big\{\big(m^{\frac{p+q}{2}}\big)^{\frac{a}{p+q}}\big(\ell^{\frac{p+q}{2}}\big)^{1-\frac{a}{p+q}} : \\
&\qquad\qquad 1 \leq a \leq p+q\Big\} \\
&= O\Big([m/\ell]^{1/2}\Big) = o(1) \quad \text{as} \quad \ell \to \infty \ . \qquad (5.16)
\end{aligned}$$

Note that the variables $V(1,j-m;I^c)$, $V(j+1,\ell;I)W(j+1,\ell;I)$, and $W(\ell+m+1,\ell+j;J^c)$ are functions of disjoint sets of X_i-variables, which are separated by m many X_i-variables from one another. Hence, by Proposition 3.1, Lemma 3.2, and Hölder's inequality, with $\gamma = 2r - (p+q) > 0$, we get

$$\begin{aligned}
&\bigg|\Delta_{j,\ell} - \Big[\ell^{-\frac{p+q}{2}} \sum_I \sum_J EV(1,j-m;I^c) \\
&\qquad \times E\{V(j+1,\ell;I)W(j+1,\ell;I)\}EW(1,j-m;J^c)\Big]\bigg| \\
&\leq C(p,q)\ell^{-\frac{p+q}{2}} \sum_I \sum_J \alpha(m-1)^{\frac{\gamma}{p+q+\gamma}} \Big\| \|\tilde{S}(1;j-m)\| \Big\|_{p+q+\gamma}^{|I^c|}
\end{aligned}$$

$$\times \Big\| \|\tilde{S}(1;\ell-j)\| \Big\|_{p+q+\gamma}^{|I|+|J|} \Big\| \|\tilde{S}(1;j-m)\| \Big\|_{p+q+\gamma}^{|J^c|} + Q_{n,j}$$

$$\leq C(r,\delta)\zeta_{2r+\delta}^{p+q}[\Delta(r,\delta)] \cdot \alpha(m-1)^{\gamma/2r} + Q_{n,j} \tag{5.17}$$

uniformly in $\ell^{1/2} \leq j \leq \ell - \ell^{1/2}$.

Next note that by the Central Limit Theorem for strong mixing random variables (cf. Theorem A.8, Appendix A), $n^{-1/2}\tilde{S}(1;n) \longrightarrow^d Z_\infty$ as $n \to \infty$ and by Lemma 3.2, $[n^{-1/2}\tilde{S}(1,n)]^\nu$ is uniformly integrable for any $\nu \in \mathbb{Z}_+^d$ with $|\nu| < 2r$. Hence, it follows that

$$\lim_{i\to\infty} E\Big[i^{-1/2}\tilde{S}(1,i)\Big]^\nu = EZ_\infty^\nu \tag{5.18}$$

for any $\nu \in \mathbb{Z}_+^d$ with $|\nu| < 2r$. Hence, the lemma follows from (5.16)–(5.18) by observing that for any I, J,

$$\begin{aligned}
&\ell^{-(p+q+1)} \sum_{\ell^{1/2}\leq j\leq \ell-\ell^{1/2}} EV(1,j-m;I^c)E\Big\{V(j+1,\ell;I)W(j+1,\ell;J)\Big\} \\
&\qquad\qquad \times EW(1,j-m;J^c) \\
&= \ell^{-1} \sum_{\ell^{1/2}\leq j\leq \ell-\ell^{1/2}} \Big\{\frac{j-m}{\ell}\Big\}^{\frac{|I^c|+|J^c|}{2}} \Big\{\frac{\ell-j}{\ell}\Big\}^{\frac{|I|+|J|}{2}} u_{(j-m)}v_{(\ell-j)} \\
&= \Big[\int_0^1 x^{\frac{(|I^c|+|J^c|)}{2}}(1-x)^{\frac{|I|+|J|}{2}}dx\Big] u_\infty v_\infty(1+o(1)) \quad \text{as} \quad \ell\to\infty\,,
\end{aligned}$$

where

$$\begin{aligned}
u_j &= EV_j(I^c)EW_j(J^c)/j^{\frac{(|I^c|+|J^c|)}{2}}\,, \\
v_j &= E\{V_j(I)W_j(I)\}/j^{\frac{(|I|+|J|)}{2}}\,,\ j\geq 1\,,
\end{aligned}$$

and u_∞ and v_∞ denote the limits of the sequences $\{u_n\}_{n\geq 1}$ and $\{v_n\}_{n\geq 1}$ (cf. (5.18)), respectively. □

Proof of Theorem 5.2 for j=1,2,3: As in the proof of Theorem 5.1, we shall restrict attention to the bootstrap bias estimator $\hat{\varphi}_{1n}(j;\ell)$. Write $T_{n,j} = b^{-1}\ell^{-2}\sum_{|\alpha|=2} c_\alpha E_*\Big(S(I_{j,1};\ell)\Big)^\alpha$, $1 \leq j \leq 3$. Note that by (5.13) and the Cauchy-Schwarz inequality,

$$\begin{aligned}
&\Big|\mathrm{Var}\Big(\hat{\varphi}_{1n}(j;\ell)\Big) - \mathrm{Var}(T_{n,j})\Big| \\
&\leq \Big\|R_{1n}(j;\ell)\Big\|_2^2 + 2\Big[\mathrm{Var}(T_{n,j})\Big]^{1/2}\Big\|R_{1n}(j;\ell)\Big\|_2 .
\end{aligned}$$

Hence, in view of (5.15), it is enough to show that

$$\mathrm{Var}(T_{n,j}) = \Big[\frac{4\pi^2}{3}\, g_1(0)\Big] \cdot \Big(n^{-3}\ell\Big)(1+o(1)),\ j=1,3 \tag{5.19}$$

$$\mathrm{Var}(T_{n,j}) = \Big[2\pi^2 \; g_1(0)\Big] \cdot \Big(n^{-3}\ell\Big)\big(1+o(1)\big), \; j = 2 \; . \tag{5.20}$$

First consider (5.19) with $j = 1$. With $U_{1i} \equiv S(i;\ell)/\sqrt{\ell}$, $i \in \mathbb{Z}$, by Lemma 5.2, we get

$$\begin{aligned}
\mathrm{Var}(T_{n,1}) &= b^{-2}\ell^{-2} \; \mathrm{Var}\bigg(\sum_{|\alpha|=2} c_\alpha \Big[N^{-1} \sum_{i=1}^{N} U_{1i}^{\alpha}\Big]\bigg) \\
&= N^{-1}b^{-2}\ell^{-2} \sum_{|\alpha|=2} \sum_{|\beta|=2} c_\alpha c_\beta \Big[\sum_{i=-\ell}^{\ell} \mathrm{Cov}\Big(U_{11}^{\alpha}, U_{1,i+1}^{\beta}\Big)\Big] \\
&\quad + Q_{11n} \\
&= 2b^{-2}\ell^{-1}N^{-1} \sum_{|\alpha|=2} \sum_{|\beta|=2} c_\alpha c_\beta \Big[b(4;4) E Z_\infty^{\alpha+\beta} \\
&\quad + \{2 \cdot b(2,4) + b(0,4) - 1\} E Z_\infty^{\alpha} E Z_\infty^{\beta} \Big] \big(1 + o(1)\big) + Q_{11n} \\
&= \frac{2}{3} \, b^{-2}\ell^{-1}N^{-1} \sum_{|\alpha|=2} \sum_{|\beta|=2} c_\alpha c_\beta \mathrm{Cov}(Z_\infty^{\alpha}, Z_\infty^{\beta})[1 + o(1)] \\
&\quad + Q_{11n} \; ,
\end{aligned} \tag{5.21}$$

where the remainder term Q_{11n} is defined by subtraction and by Proposition 3.1, it satisfies the inequality

$$\begin{aligned}
|Q_{11n}| &\leq Cb^{-2}\ell^{-2}N^{-1} \bigg[\sum_{i=\ell+1}^{N-1} \alpha(i-\ell)^{(a_0+1)/(a_0+3)} \Big(E\|U_{11}\|^{6+2a_0}\Big)^{\frac{2}{a_0+3}} \\
&\quad + N^{-1}\ell^2 E\|U_{11}\|^4 \bigg] \\
&= o(n^{-3}\ell) \; .
\end{aligned}$$

Now (5.19) follows from (5.21) for the case $j = 1$.

Next consider (5.19) with $j = 3$. By (5.10) and Lemma 3.2, with $D_H = \max\{|D^\alpha H(\mu)| : 0 \leq |\alpha| \leq 3\}$, we have

$$\begin{aligned}
& E\bigg| b^{-1}\ell^{-2} \sum_{|\alpha|=2} c_\alpha \Big(E_*\big(S(I_{3,1};\ell)\big)^{\alpha} - E_*\big(S(I_{1,1};\ell)\big)^{\alpha}\Big)\bigg|^2 \\
& \leq \; C(D_H, d) n^{-2}\ell^{-2} \bigg[n^{-2} E\bigg\{ \sum_{i=1}^{\ell} \|S(1;i)\|^2 \bigg\}^2 \\
& \qquad + n^{-4}\ell^2 E\bigg(\sum_{j=1}^{N} \|S(i;\ell)\|^2 \bigg)^2 \bigg]
\end{aligned}$$

$$
\begin{aligned}
&= O\Big(n^{-2}\ell^{-2}\Big[n^{-2}\ell^4 + n^{-4}\ell^2 N^2 \ell^2\Big]\Big) \\
&= O(n^{-4}\ell^2) \,. \qquad (5.22)
\end{aligned}
$$

The expansion for $\mathrm{Var}(\hat{\varphi}_{1n}(j;\ell))$, $j = 3$ now follows from (5.22) and the result for the case $j = 1$.

To prove (5.20), note that for $j = 2$, with $U_{1i}^{(2)} = S\big((i-1)\ell + 1; \ell\big)/\sqrt{\ell}$ and $V_{2i} = \sum_{|\alpha|=2} c_\alpha [U_{1i}^{(2)}]^\alpha$, $i \in \mathbb{Z}$,

$$
\begin{aligned}
\mathrm{Var}(T_{n,j}) &= b^{-2}\ell^{-2}\mathrm{Var}\Big(\sum_{|\alpha|=2} c_\alpha \Big\{ b^{-1} \sum_{i=1}^{b} \big[U_{1i}^{(2)}\big]^\alpha \Big\}\Big) \\
&= b^{-2}\ell^{-2} b^{-1} \mathrm{Var}(V_{21}) + Q_{21n} \\
&= b^{-3}\ell^{-2} \sum_{|\alpha|=2} \sum_{|\beta|=2} c_\alpha c_\beta \mathrm{Cov}(Z_\infty^\alpha, Z_\infty^\beta)\big(1 + o(1)\big) + Q_{21n}
\end{aligned}
$$

where, by Proposition 3.1, with $a = a_0$,

$$
\begin{aligned}
|Q_{21n}| &\le Cb^{-3}\ell^{-2} \sum_{i=1}^{b-1} \Big|\mathrm{Cov}(V_{21}, V_{2(i+1)})\Big| \\
&\le Cb^{-3}\ell^{-2}\Big[\Big|\mathrm{Cov}(V_{21}, V_{22})\Big| \\
&\quad + \sum_{i=2}^{b-1} \alpha(i\ell - \ell)^{\frac{a+1}{a+3}} \Big(E\|U_{11}\|^{6+2a}\Big)^{\frac{2}{a+3}}\Big] \\
&\le Cb^{-3}\ell^{-2} \sum_{|\alpha|=2} \sum_{|\beta|=2} |c_\alpha c_\beta| \cdot \Big|\mathrm{Cov}\Big(U_{11}^\alpha, U_{1(\ell+1)}^\beta\Big)\Big| + o(b^{-3}\ell^{-2}) \\
&= o(n^{-3}\ell) \,,
\end{aligned}
$$

provided we show that $\Big|\mathrm{Cov}\Big(U_{11}^\alpha, U_{1(\ell+1)}^\beta\Big)\Big| = o(1)$ for all $|\alpha| = 2 = |\beta|$. This can be done using arguments similar to those used in the proof of Lemma 5.2. More precisely, writing $U_{1(\ell+1)}$ as the sum $U_{1(\ell+1)} = \ell^{-1/2}\big[\tilde{S}(\ell+1, \ell+m) + \tilde{S}(\ell+m+1, 2\ell)\big]$ with $m = \lfloor \ell^{1/4} \rfloor$, we have for any $|\alpha| = |\beta| = 2$,

$$
\begin{aligned}
&\Big|\mathrm{Cov}\Big(U_{11}^\alpha, U_{1(\ell+1)}^\beta\Big)\Big| \\
&\le \Big|\mathrm{Cov}\Big(U_{11}^\alpha, \ell^{-1}\tilde{S}(\ell+m+1; 2\ell)^\beta\Big)\Big| \\
&\quad + 4\ell^{-1}\Big(E\|U_{11}\|^4\Big)^{1/2}\Big(E\|\tilde{S}(1;m)\|^4\Big)^{1/4}\Big(E\|\tilde{S}(1;\ell-m)\|^4\Big)^{1/4} \\
&\quad + 2\ell^{-1}\Big(E\|U_{11}^4\|\Big)^{1/2}\Big(E\|\tilde{S}(1;m)\|^4\Big)^{1/2}
\end{aligned}
$$

$$
\begin{aligned}
&\leq C\Big(E\|U_{11}\|^{4+\delta}\Big)^{\frac{2}{4+\delta}}\Big(E\|\ell^{-1/2}\tilde{S}(1;\ell-m)\|^{4+\delta}\Big)^{\frac{2}{4+\delta}}[\alpha(m)]^{\frac{\delta}{4+\delta}} \\
&\quad + O(m^{1/2}\ell^{-1/2}) + O(m\ell^{-1}) \\
&= o(1)\ .
\end{aligned}
$$

This completes the proof of (5.20), and hence, of Theorem 5.2. □

5.7.2 Proofs of Theorems 5.1-5.2 for the SB

Lemma 5.3 *Assume that $\ell = O(n^{1-\epsilon})$ for some $0 < \epsilon < 1$, $E\|X_1\|^{2r+\delta} < \infty$ and $\Delta(r;\delta) < \infty$, for some positive integer r and for some $\delta > 0$. Then,*

(i) $E\{E_*\|S(I_{4,1};L_1)\|^{2r}\} \leq C(r,d)\zeta_{2r+\delta}^{2r}\Delta(r;\delta)\cdot\ell^r$.

(ii) $E\{E_*\|\bar{X}_{n,\ell}^{*(4)}\|^{2r}\} \leq C(r,d)\zeta_{2r+\delta}^{2r}\Delta(r;\delta)\cdot n^{-r}(1+\ell(np)^{-2r})$.

Proof: Note that the (conditional) distributions of $I_{3,1}$ and $I_{4,1}$ are the same and that $I_{4,1}$ and L_1 are independent. Hence, by part (i) of Lemma 5.2 with $j = 3$, by Lemma 3.4, and by the stationarity of the X_i's, we get

$$
\begin{aligned}
&E\Big(E_*\|S(I_{4,1};L_1)\|^{2r}\Big) \\
&= E\Big[\sum_{m=1}^{\infty}\Big\{E_*\|S(I_{4,1};m)\|^{2r}\cdot p(1-p)^{m-1}\Big\}\Big] \\
&\leq \sum_{m=1}^{2r\ell\,\log n} E\Big\{E_*\|S(I_{3,1};m)\|^{2r}p(1-p)^{m-1}\Big\} \\
&\quad + C(r)\cdot\max\Big\{E\|S(1;k)\|^{2r} : 1 \leq k \leq n\Big\}\exp(-2r\,\log n) \\
&\leq C(r,d)\Delta(r;d)\zeta_{2r+\delta}^{2r}\Big[\sum_{m=1}^{2r\ell\,\log n} m^r p(1-p)^{m-1} + n^r\exp(-2r\,\log n)\Big] \\
&\leq C(r,d)\Delta(r;d)\zeta_{2r+\delta}^{2r}\ell^r\ ,
\end{aligned}
$$

proving (i).

To prove (ii), write $R_{j,n}^* = \sum_{i=1}^{N_1} X_{j,i}^* - n\bar{X}_{n,\ell}^{*(j)}$, for the sum of the excess $(N_1 - n)$ many bootstrap data-values beyond $X_{j,1}^*,\ldots,X_{j,n}^*$, $j = 4$. Note that

$$
R_{4,n}^* = \sum_{i=1}^{L_K} w_i^* Y_{n,J(i)}\ ,
$$

where $J(i) = I_{4,K} + i - 1$ and $w_i^* = 0$ if $L_1 + \cdots + L_{K-1} + i \leq n$ and $= 1$, otherwise. Without loss of generality, assume that the bootstrap variables $I_{4,1},\ldots,I_{4,n}$ and $L_1,\ldots,L_n$ are also defined on the same probability space $(\Omega,\mathcal{F},P)$, supporting the sequence $\{X_i\}_{i\in\mathbb{Z}}$. Let $\mathcal{L}_n = \sigma\langle L_1,\ldots,L_n\rangle$, $\mathcal{X}_n =$

$\sigma\langle X_1, \ldots, X_n \rangle$ and $\mathcal{T}_n = \mathcal{L}_n \vee \mathcal{X}_n =$ the smallest σ-field containing both $\mathcal{L}_n$ and $\mathcal{X}_n$, $n \geq 1$, as in Chapter 3. Then, $R^*_{4,n}$ may be considered as a random vector on $(\Omega, \mathcal{F}, P)$ and

$$\begin{aligned} E\Big\{E_*\|R^*_{4,n}\|^{2r}\Big\} &\equiv E\Big\{E\Big(\|R^*_{4,n}\|^{2r}|\mathcal{X}_n\Big)\Big\} = E\Big(\|R^*_{4,n}\|^{2r}\Big) \\ &= E\Big[E\Big\{E\Big(\|R^*_{4,n}\|^{2r}|\mathcal{T}_n\Big)|\mathcal{L}_n\Big\}\Big] . \end{aligned}$$

Note that the random variables $\{I_{4,1}, \ldots, I_{4,n}\}$, $\{L_1, \ldots, L_n\}$, and $\{X_1, \ldots, X_n\}$ are all independent. Hence, it follows that conditional on $\mathcal{L}_n$, L_K and w_i^*'s may be treated as nonrandom quantities. Consequently, by Lemma 3.2,

$$\begin{aligned} & E\Big\{E\Big(\|R^*_{4,n}\|^{2r}|\mathcal{T}_n\Big)|\mathcal{L}_n\Big\} \\ &= E\bigg(n^{-1}\sum_{j=1}^{n}\Big\|\sum_{i=1}^{L_K} w_i^* Y_{n,j+i-1}\Big\|^{2r}\Big|\mathcal{L}_n\bigg) \\ &\leq C(r)\max\bigg\{E\Big\|\sum_{i=1}^{m} a_i X_i\Big\|^{2r} : 1 \leq m \leq L_K \wedge n,\ a_i \in \{0,1\}\bigg\} \\ &\leq C(r)\zeta^{2r}_{2r+\delta}\Delta(r;\delta)L_K^r . \end{aligned}$$

Therefore, by Lemma 3.4,

$$E\Big(E_*\|R^*_{4,n}\|^{2r}\Big) \leq C(r)\zeta^{2r}_{2r+\delta}\Delta(r;\delta)\ell^{r+1} . \tag{5.23}$$

Note that $\sum_{i=1}^{K} L_i \leq n + L_K$ and that conditional on $\mathcal{T}_n$, $S(I_{4,i}; L_i) - L_i\bar{X}_n$, $1 \leq i \leq K$ are zero mean independent (but not necessarily identically distributed) random vectors. By Lemma 3.4, (5.11), and the inequality above (5.23),

$$\begin{aligned} & E\Big(E_*\|\bar{X}^{*(4)}_{n,\ell}\|^{2r}\Big) \\ &= n^{-2r}E\bigg\{E_*\Big\|\sum_{i=1}^{K} S(I_{4,i};L_i) + R^*_{4,n}\Big\|^{2r}\bigg\} \\ &\leq C(r)n^{-2r}\bigg[E(E_*\|R^*_{4,n}\|^{2r}) + E\bigg\{\bigg(E_*\Big|\sum_{i=1}^{K} L_i\Big|^{2r}\bigg)\cdot\|\bar{X}_n\|^{2r}\bigg\} \\ &\quad + E\bigg\{E_*\Big\|\sum_{i=1}^{K}(S(I_{4,i};L_i) - L_i\bar{X}_n)\Big\|^{2r}\bigg\}\bigg] \\ &\leq C(r)n^{-2r}\zeta^{2r}_{2r+\delta}\Delta(r;\delta)\Big\{E(L_K^r) + (n^{2r} + EL_K^{2r})n^{-r}\Big\} \\ &\quad + C(r)n^{-2r}E\bigg(E\bigg\{E\bigg(\Big\|\sum_{i=1}^{K}(S(I_{4,i};L_i) - L_i\bar{X}_n)\Big\|^{2r}\Big|\mathcal{T}_n\bigg)\Big|\mathcal{X}_n\bigg\}\bigg) \end{aligned}$$

$$
\begin{aligned}
&\leq \; C(r)\zeta_{2r+\delta}^{2r}\Delta(r;\delta)n^{-r}\Big[1+n^{-2r}E(L_K)^{2r}\Big] \\
&\quad + C(r)n^{-2r}E\bigg(E\Big\{K^{r-1}\sum_{i=1}^{K}E(\|S(I_{4,i};L_i)\|^{2r}|\mathcal{T}_n)|\mathcal{X}_n\Big\}\bigg) \\
&\leq \; C(r)\zeta_{2r+\delta}^{2r}\Delta(r;\delta)n^{-r}\Big[1+n^{-2r}\ell^{2r+1}\Big] \\
&\quad + C(r)n^{-2r}E\Big\{K^{r-1}\sum_{i=1}^{K}E(\|S(I_{4,i};L_i)\|^{2r}|\mathcal{L}_n)\Big\} \\
&\leq \; C(r)\zeta_{2r+\delta}^{2r}\Delta(r;\delta)n^{-r}\Big[1+n^{-2r}\ell^{2r+1}\Big] \\
&\quad + C(r)n^{-2r}E\bigg[K^{r-1}\sum_{i=1}^{K}\Big\{(E\|S(1;L_i)\|^{2r}|\mathcal{L}_n)\mathbb{1}(L_i\leq 4r\ell\,\log n) \\
&\quad + \max\{E\|S(1;m)\|^{2r}:1\leq m\leq n\}\mathbb{1}(L_i>4r\ell\,\log n)\Big\}\bigg] \\
&\leq \; C(r)\zeta_{2r+\delta}^{2r}\Delta(r;\delta)\cdot n^{-r}\bigg[1+\ell(np)^{-2r}+n^{-r}\Big\{E\Big(K^{r-1}\sum_{i=1}^{K}L_i^r\Big) \\
&\quad + E\Big(n^{r-1}\sum_{i=1}^{n}\{n^r\cdot\mathbb{1}(L_i\geq 4r\ell\,\log n)\}\Big)\Big\}\bigg] \\
&\leq \; C(r)\zeta_{2r+\delta}^{2r}\Delta(r;\delta)n^{-r}\Big[1+\ell(np)^{-2r}\Big]\;.
\end{aligned}
$$

This completes the proof of Lemma 5.3. □

Lemma 5.4 *Let $g:(-\pi,\pi]\to[0,\infty)$ be a continuous function that is symmetric about zero. Then, with $p=\ell^{-1},\; q=1-p$,*

(i) $\displaystyle\lim_{n\to\infty}\int_{-\pi}^{\pi}\Big[e^{\iota w}/(1-qe^{\iota w})\Big]g(w)dw$
$\displaystyle = \pi g(0)+\int_{-\pi}^{\pi}\Big[2^{-1}\cos w+\big(\cos(w/2)\big)^2\Big]g(w)dw\;.$

(ii) $\displaystyle\lim_{n\to\infty}p\cdot\int_{-\pi}^{\pi}\Big[e^{\iota w}/(1-qe^{\iota w})\Big]\Big[e^{\iota 2w}/(1-q^2e^{\iota 2w})\Big]g(w)dw$
$\displaystyle = g(0)\bigg[2\int_{-\infty}^{\infty}(1+4y^2)^{-2}dy-\int_{-\infty}^{\infty}(1+4y^2)^{-1}dy\bigg]\;.$

(iii) $\displaystyle\int_{-\pi}^{\pi}\Big[e^{\iota 2w}/(1-q^2e^{\iota 2w})\Big]g(w)dw=O(1)$ *as* $n\to\infty$.

Proof: (i) Since g is real and symmetric, for any $M>1$, we have

$$
\int_{-\pi}^{\pi}e^{\iota w}(1-qe^{\iota w})^{-1}g(w)dw
$$

$$
\begin{aligned}
&= \int_{-\pi}^{\pi} (1+q^2-2q\cos w)^{-1}\Big[(1-q\cos w)\cos w + q(\sin w)^2\Big] g(w)dw \\
&= \Big(\int_{|w|\le Mp} + \int_{Mp<|w|<M^{-1}} + \int_{M^{-1}<|w|<\pi}\Big)\Big(p^2+4q\sin^2(w/2)\Big)^{-1} \\
&\quad \cdot \Big[\big(p+2q\sin^2(w/2)\big)\cos w + q(\sin w)^2\Big] g(w)dw \\
&\equiv I_1(M) + I_2(M) + I_3(M), \text{ say.} \qquad (5.24)
\end{aligned}
$$

Using the change of variable $y = w/2p$ and the Bounded Convergence Theorem (cf. Theorem 16.5, Billingsley (1995)), one can show that for any $M > 1$, as $n \to \infty$,

$$
\begin{aligned}
I_1(M) &= 2\int_{|y|<M/2} \frac{(1+2qp^{-1}\sin^2 yp)\cos 2py + qp^{-1}(\sin^2 2yp)}{1+4qy^2[(py)^{-2}\sin^2 py]} \\
&\qquad\qquad g(2py)dy \\
&= 2g(0)\cdot \int_{|y|<M/2} (1+4y^2)^{-1}dy + o(1)\ . \qquad (5.25)
\end{aligned}
$$

Since $x/3 < \sin x$ for all $x \in (0, \pi/2]$, for any $M > 1$, we have

$$
\begin{aligned}
I_2(M) &\le C\int_{Mp\le w<M^{-1}} \frac{(p+w^2)}{w^2} g(w)dw \\
&\le C\Big[\int_{M<y<(Mp)^{-1}} y^{-2}g(py)dy + \int_{0<w<M^{-1}} g(w)dw\Big] \\
&\le C\max\{g(w) : 0 < w < M^{-1}\}\cdot M^{-1}\ . \qquad (5.26)
\end{aligned}
$$

Finally, by the bounded convergence Theorem, for any $M > 1$,

$$
\begin{aligned}
&\lim_{n\to\infty} I_3(M) \\
&= \int_{M^{-1}<|w|<\pi} (4\sin^2\frac{w}{2})^{-1}\Big[(2\sin^2\frac{w}{2})\cos w + (\sin w)^2\Big] g(w)dw\ . \\
&\qquad (5.27)
\end{aligned}
$$

Part(i) follows from (5.24)–(5.27), by letting $M \to \infty$ and by noting that $\int_{-\infty}^{\infty}(1+4y^2)^{-1}dy = \pi/2$.

Next, consider part (ii). Let

$$
h_n(w) \equiv \frac{p^2(2-p)-2p(2-p)\sin^2(w/2)+2(\sin^2 w)(q-2\cos w)}{[p^2(1+q)^2+4q^2\sin^2 w][p^2+4q\sin^2(w/2)]} g(w), \qquad w \in (-\pi, \pi)\ .
$$

Then, using the symmetry of $g(w)$, it can be shown that the integral on the left side of (ii) equals $\int_{-\pi}^{\pi} h_n(w)dw$. As before, we split this integral into three parts, now ranging over the sets $[-Mp, Mp]$, $\{w : Mp < |w| < \pi/2\}$, and $\{w : \pi/2 \leq |w| < \pi\}$, where $M > 1$. Arguments similar to (5.25) and (5.26) yield,

$$\begin{aligned} &\lim_{n\to\infty} p \cdot \int_{|w|\leq Mp} h_n(w)dw \\ &= \left(2\int_{|y|\leq M/2} \frac{2-8y^2}{[4+16y^2][1+4y^2]}dy\right)g(0) \end{aligned} \tag{5.28}$$

and

$$\begin{aligned} &\int_{Mp<|w|<\pi/2} |h_n(w)|dw \\ &\leq Cq^{-3}\left[\int_{Mp<|w|<\pi/2} \frac{[p^2+w^2]}{w^4} g(w)dw\right] \\ &\leq Cq^{-3}p^{-1}\left[\int_M^\infty u^{-4}du + \int_M^\infty u^{-2}du\right] \cdot \|g\|_\infty , \end{aligned} \tag{5.29}$$

for any $M > 1$.

For the third region, note that for n large,

$$\begin{aligned} &\int_{\pi/2<|w|<\pi} |h_n(w)|dw \\ &\leq C\int_{\pi/2}^{\pi} \frac{(p+\sin^2 w)}{[p^2(1+q)^2+4q^2\sin^2 w]\cdot w^2} \cdot g(w)dw \\ &\leq C\int_0^{\pi/2} \frac{g(\pi-w)(p+\sin^2 w)}{[p^2+4\sin^2 w]}dw \\ &\leq C , \end{aligned} \tag{5.30}$$

by arguments similar to (5.25) and (5.27). Part (ii) now follows from (5.28)–(5.30).

Part (iii) also follows by similar arguments. We omit the details. □

Proof of Theorem 5.1, j=4: We prove the theorem only for φ_{1n}. Without loss of generality, let $\mu = 0$. Note that the SB observations $\{X^*_{4,i}\}_{i\geq 1}$ form a stationary *dependent* sequence. As in Section 5.7.1, using Taylor's expansion, for $j = 4$, we have (cf. (5.13)),

$$\hat{\varphi}_{1n}(j;\ell) = \sum_{|\alpha|=2} c_\alpha E_*(\bar{X}^{*(j)}_{n,\ell} - \bar{X}_n)^\alpha + R_{1n}(j;\ell) \tag{5.31}$$

where the remainder term $R_{1n}(j;\ell)$ now admits the bound

$$\begin{aligned} |R_{1n}(j;\ell)| \;\leq\; & C(a_0,d,\zeta_2)\Big\{\|\bar{X}_n\|(1+\|\bar{X}_n\|^{a_0})E_*\|\bar{X}^{*(j)}_{n,\ell}\|^2 \\ & + (1+\|\bar{X}_n\|^{a_0}+\|\bar{X}^{*(j)}_{n,\ell}\|^{a_0})E_*\|\bar{X}^{*(j)}_{n,\ell}\|^3\Big\} . \end{aligned}$$

Hence, using Hölder's inequality, Proposition 3.1, and Lemma 5.3 as in the derivation of (5.15), for the case $j = 4$ we have,

$$
\begin{aligned}
& E\big(R_{1n}(j;\ell)\big)^2 \\
\leq\ & C(a_0, d, \zeta_2)\Big[\Big\{E\|\bar{X}_n\|^4\Big\}^{1/2}\Big\{E(E_*\|\bar{X}_{n,\ell}^{*(j)}\|^8)\Big\}^{1/2} \\
& + \Big(E\|\bar{X}_n\|^{6+2a_0}\Big)^{\frac{(2+2a_0)}{(6+2a_0)}}\Big\{E(E_*\|\bar{X}_{n,\ell}^{*(j)}\|^{6+2a_0})\Big\}^{\frac{4}{(6+2a_0)}} \\
& + E\Big\{E_*(\|\bar{X}_{n,\ell}^{*(j)}\|^6 + \|\bar{X}_{n,\ell}^{*(j)}\|^{6+2a_0})\Big\}\Big] \\
\leq\ & Cn^{-3}\,.
\end{aligned}
\tag{5.32}
$$

Next for $0 \leq j \leq n-1$ and $\alpha, \beta \in \mathbb{Z}_+^d$ with $|\alpha| = 1 = |\beta|$, write $\hat{\sigma}(j;\alpha,\beta) = n^{-1}\sum_{i=1}^{n-j} X_i^\alpha X_{i+j}^\beta$. Then, using (3.24) from Chapter 3 and the stationarity of the $X_{4,i}^*$'s, we have

$$
\begin{aligned}
& \sum_{|\alpha|=1}\sum_{|\beta|=1} c_{\alpha+\beta} E_*(\bar{X}_{n,\ell}^{*(4)} - \bar{X}_n)^{\alpha+\beta} \\
=\ & n^{-2}\sum_{|\alpha|=1}\sum_{|\beta|=1} c_{\alpha+\beta}\Big[n\Big\{E_*(X_{4,1}^*)^{\alpha+\beta} - (\bar{X}_n)^{\alpha+\beta}\Big\} \\
& + \sum_{j=1}^{n-1}(n-j)\Big\{\Big(E_*(X_{4,1}^*)^\alpha(X_{4,(j+1)}^*)^\beta - (\bar{X}_n)^{\alpha+\beta}\Big) \\
& + \Big(E_*(X_{4,1}^*)^\beta(X_{4,(j+1)}^*)^\alpha - (\bar{X}_n)^{\alpha+\beta}\Big)\Big\}\Big] \\
=\ & n^{-1}\sum_{|\alpha|=1}\sum_{|\beta|=1} c_{\alpha+\beta}\Big[\Big\{\hat{\sigma}(0;\alpha,\beta) - (\bar{X}_n)^{\alpha+\beta}\Big\} \\
& + \sum_{j=1}^{n-1}(1-n^{-1}j)q^j\Big\{\Big(\hat{\sigma}(j;\alpha,\beta) + \hat{\sigma}(j;\beta,\alpha)\Big) \\
& + \Big(\hat{\sigma}(n-j;\alpha,\beta) + \hat{\sigma}(n-j;\beta,\alpha)\Big) - 2(\bar{X}_n)^{\alpha+\beta}\Big\}\Big] \\
=\ & n^{-1}\sum_{|\alpha|=1}\sum_{|\beta|=1} c_{\alpha+\beta}\Big[\sum_{j=0}^{n-1} q_{nj}\Big(\hat{\sigma}(j;\alpha,\beta) + \hat{\sigma}(j;\beta,\alpha)\Big) \\
& - \Big\{1 + 2\sum_{j=1}^{n-1}(1-n^{-1}j)q^j\Big\}(\bar{X}_n)^{\alpha+\beta}\Big]\,,
\end{aligned}
\tag{5.33}
$$

where $q_{nj} = (1-n^{-1}j)q^j + (n^{-1}j)q^{(n-j)}$, $1 \leq j \leq n-1$; and $q_{n0} = 1/2$. Therefore, by (5.31), (5.32), and (5.33), it follows that

$$
\begin{aligned}
& \text{Bias}(\hat{\varphi}_{1n}(4;\ell)) \\
=\ & E\hat{\varphi}_{1n}(4;\ell) - \text{Bias}(\hat{\theta}_n)
\end{aligned}
$$

$$
\begin{aligned}
&= n^{-1} \sum_{|\alpha|=1} \sum_{|\beta|=1} c_{\alpha+\beta} \Bigg[\sum_{j=1}^{n-1} (q_{nj} - 1)(1 - n^{-1}j) \\
&\qquad\qquad \times \Big\{ EX_1^{\alpha} X_{j+1}^{\beta} + EX_1^{\beta} X_{j+1}^{\alpha} \Big\} \Bigg] \\
&\quad + O\bigg(n^{-1} \Big(\sum_{j=1}^{\infty} q^j \Big) E\|\bar{X}_n\|^2 \bigg) + O(n^{-3/2}) . \qquad (5.34)
\end{aligned}
$$

Note that by Taylor's expansion, $|1 - q^j - jp| \leq j^2 p^2/2$ for all $j \geq 1$ and all $0 < p < 1$, and that

$$
p^{-1}(1 - q_{nj})(1 - n^{-1}j) \to j \quad \text{as} \quad n \to \infty \quad \text{for all} \quad j \geq 1 .
$$

Also, by the mixing and moment condition, $\sum_{j=1}^{\infty} j^2 |EX_1^{\alpha} X_{1+j}^{\beta}| < \infty$. Hence, using the Dominated Convergence Theorem (DCT), from (5.34), we get

$$
\begin{aligned}
&\text{Bias}\Big(\hat{\varphi}_{1n}(4;\ell) \Big) \\
&= -n^{-1} p \sum_{|\alpha|=1} \sum_{|\beta|=1} c_{\alpha+\beta} \sum_{j=1}^{\infty} j (EX_1^{\alpha} X_{1+j}^{\beta} + EX_1^{\beta} X_{1+j}^{\alpha}) \big(1 + o(1)\big) \\
&\quad + O(n^{-2}\ell) + O(n^{-3/2}) .
\end{aligned}
$$

This completes the proof of Theorem 5.1 for $j = 4$. □

Proof of Theorem 5.2 for j=4: We restrict attention to the bootstrap estimator $\hat{\varphi}_{1n}(4;\ell)$ and without loss of generality, set $\mu = 0$. Since $E\{n^{-1}(1 + 2\sum_{j=1}^{n-1}(1 - n^{-1}j)q^j)\|\bar{X}_n\|^2\}^2 = O(n^{-4}\ell^2)$, in view of (5.31), (5.32), and (5.33), it is enough to show that

$$
\begin{aligned}
&\text{Var}\bigg(n^{-1} \sum_{|\alpha|=1} \sum_{|\beta|=1} c_{\alpha+\beta} \Big[\sum_{j=0}^{n-1} q_{nj} \Big(\hat{\sigma}(j;\alpha,\beta) + \hat{\sigma}(j;\beta,\alpha) \Big) \Big] \bigg) \\
&= (2\pi) \bigg[2\pi g_1(0) + \int_{-\pi}^{\pi} (1 + e^{iw}) g_1(w) dw \bigg] (n^{-3}\ell) + o(n^{-3}\ell) \quad (5.35)
\end{aligned}
$$

To this end, we make use of Lemma 3.3 of Chapter 3. Note that under the conditions of Theorem 5.2 for $j = 4$, the remainder terms $R_n(j,k;\alpha,\beta,\gamma,\nu)$'s satisfy the bound (cf. (3.36) of Chapter 3)

$$
\max_{|\alpha|=|\beta|=|\gamma|=|\nu|=1} \sum_{j=0}^{n-1} \sum_{k=0}^{n-1} q_{nj} q_{nk} |R_n(j,k;\alpha,\beta,\gamma,\nu)| = O(n^{-1}) . \qquad (5.36)
$$

Let $s = s_n = \lfloor \ell(\log n)^2 + n^{1/3} \rfloor$. Write $\sigma(k;\alpha,\beta) = EX_1^{\alpha} X_{1+k}^{\beta}$, $k \in \mathbb{Z}$, $|\alpha| = 1 = |\beta|$ and $\tilde{\eta}_{jvm} = [\eta_{jv}(m) + j + v]$. Then,

$$
\max\{q_{nj} : s \leq j \leq n - s\} \quad \leq \quad \max\{q^j \vee q^{n-j} : s \leq j \leq n - s\}
$$

$$\begin{aligned} &\le q^s = O\Big(\exp(-s/l)\Big) \\ &= O\Big(\exp(-(\log n)^2)\Big) ; \end{aligned} \tag{5.37}$$

$$|q_{nj}q_{n(j+v)} - q^j q^{j+v}| \le 2n^{-1}\big(jq^j + (j+v)q^{j+v}\big), \ 1 \le j, v \le 2s ; \tag{5.38}$$

$$\begin{aligned} & n^{-3} \sum_{j=n-s}^{n-2} \sum_{v=1}^{n-1-j} q_{nj}q_{n(j+v)} \sum_{m=-(n-j)+1}^{(n-j)-v-1} \\ & \qquad \Big|(1 - n^{-1}\tilde{\eta}_{jvm})\sigma(m;\alpha,\gamma)\sigma(v+m;\beta,\nu)\Big| \\ \le\ & n^{-3} \sum_{j=n-s}^{n-2} \sum_{v=1}^{s} \sum_{m=-s+1}^{s-v-1} \\ & \qquad \big\{n^{-1}(|n-j| + |m| + v)\big\}|\sigma(m;\alpha,\gamma)||\sigma(v+m;\beta,\nu)| \\ \le\ & 3n^{-4}s^2 \bigg(\sum_{m=-\infty}^{\infty} |\sigma(m;\alpha,\gamma)| \bigg) \bigg(\sum_{u=-\infty}^{\infty} |\sigma(u;\beta,\nu)| \bigg) , \end{aligned} \tag{5.39}$$

and by similar arguments,

$$\begin{aligned} & n^{-3} \sum_{j=0}^{s} \sum_{v=n-2s}^{(n-j)-1} q_{nj}q_{n(j+v)} \sum_{m=-(n-j)+1}^{n-j-v-1} \\ & \qquad \Big\{1 - \frac{\tilde{\eta}_{jvm}}{n}\Big\}|\sigma(m;\alpha,\gamma)||\sigma(v+m;\beta,\nu)| \\ =\ & O(n^{-4}s^2) . \end{aligned} \tag{5.40}$$

Next note that the functions $\psi_m(x) \equiv (2\pi)^{-1/2}\exp(\iota m x)$, $x \in (-\pi,\pi]$, $m \in \mathbb{Z}$ form an orthonormal basis of the Hilbert space $L^2(-\pi,\pi]$ with respect to the inner product $\langle f_1, f_2\rangle = \int_{-\pi}^{\pi} f_1(x)\bar{f}_2(x)dx$, $f_1, f_2 \in L^2(-\pi,\pi]$, and, hence, $\sum_{m=-\infty}^{\infty}\langle f_1,\psi_m\rangle\overline{\langle f_2,\psi_m\rangle} = \langle f_1, f_2\rangle$ for any $f_1, f_2 \in L^2(-\pi,\pi]$. Now using (5.37)–(5.40) and Condition M_r, it can be shown that for any unit vectors $\alpha,\beta,\gamma,\nu \in \mathbb{Z}_+^d$,

$$\begin{aligned} & n^{-3} \sum_{j=0}^{n-2} \sum_{v=1}^{n-1-j} q_{nj}q_{n(j+v)} \bigg\{ \sum_{m=-(n-j)+1}^{(n-j)-v-1} \Big(1 - \frac{\tilde{\eta}_{jvm}}{n}\Big) \\ & \qquad \times \sigma(m;\alpha,\gamma)\sigma(v+m;\beta,\nu)\bigg\} \\ =\ & n^{-3}\bigg(\sum_{v=1}^{s} 2^{-1}q^v + \sum_{j=1}^{s}\sum_{v=1}^{s} q^{2j+v}\bigg) \\ & \qquad \times \bigg\{ \sum_{m=-(n-j)+1}^{(n-j)-v-1} \sigma(m;\alpha,\gamma)\sigma(v+m;\beta,\nu)\bigg\} \end{aligned}$$

$$
\begin{aligned}
&+ O\Big(n^{-4}\Big(\sum_{j=0}^{s} jq^j\Big)\Big(\sum_{u=-\infty}^{\infty}(1+|u|)\|EX_1X'_{1+u}\|\Big) \\
&\qquad \times \Big(\sum_{m=-\infty}^{\infty}\|EX_1X'_{1+m}\|\Big)\Big) \\
&+ O\Big(n^{-4}s^2 + n^{-2}\exp(-(\log n)^2)\Big) \\
= \; & n^{-3}\Big(2^{-1}\sum_{v=1}^{\infty} q^v + \sum_{j=1}^{\infty}\sum_{v=1}^{\infty} q^{2j+v}\Big)\Big[\sum_{m=-\infty}^{\infty}\Big(\int_{-\pi}^{\pi} e^{\iota m w} f(w;\alpha,\gamma)dw\Big) \\
&\qquad \times \Big(\int_{-\pi}^{\pi} e^{-\iota(v+m)w}\bar{f}(w;\beta,\nu)dw\Big)\Big] \\
&+ O(n^{-4}s^2) \\
= \; & n^{-3}\Big(2^{-1} + (1-q^2)^{-1}q^2\Big)\sum_{v=1}^{\infty} q^v \\
&\qquad \times \Big(2\pi\int_{-\pi}^{\pi} f(w;\alpha,\gamma)e^{-ivw}\bar{f}(w;\beta,\nu)dw\Big) \\
&+ O(n^{-4}s^2) \\
= \; & n^{-3}\Big(2^{-1} + q^2/(1-q^2)\Big) \\
&\qquad \times 2\pi\int_{-\pi}^{\pi} qe^{iw}(1-qe^{iw})^{-1} f(-w;\alpha,\gamma)\bar{f}(-w;\beta,\nu)dw \\
&+ O(n^{-4}s^2)\,. \qquad (5.41)
\end{aligned}
$$

By similar arguments, it follows that for any $\alpha,\beta,\gamma,\nu \in \mathbb{Z}_+^d$ with $|\alpha| = 1 = |\beta| = |\gamma| = |\nu|$,

$$
\begin{aligned}
& n^{-3}\sum_{j=0}^{n-2}\sum_{v=1}^{n-1-j} q_{nj}q_{n(j+v)}\Big\{\sum_{m=-(n-j)+1}^{(n-j)-v-1} \\
&\qquad (1-n^{-1}\tilde{\eta}_{jvm})\sigma(m+j+v;\alpha,\nu)\sigma(m-j;\beta,\gamma)\Big\} \\
= \; & n^{-3}2\pi\int_{-\pi}^{\pi}\Big\{2^{-1} + (qe^{\iota w})^2(1-(qe^{\iota w})^2)^{-1}\Big\}(1-qe^{\iota w})^{-1} \\
&\qquad qe^{\iota w} f(w;\beta,\gamma)\bar{f}(w;\alpha,\nu)dw \\
&+ O(n^{-4}s^2) \qquad (5.42)
\end{aligned}
$$

and

$$
n^{-3}\sum_{j=0}^{n-1} q_{nj}^2 \sum_{m=-(n-j)+1}^{(n-j)-1}\big(1-n^{-1}(|m|+j)\big)\times
$$

$$
\begin{aligned}
&\Big\{\sigma(m;\alpha,\gamma)\sigma(m;\beta,\nu)+\sigma(m+j;\alpha,\nu)\sigma(m-j;\beta,\gamma)\Big\}\\
= \;& n^{-3}2\pi[4^{-1}+q^2(1-q^2)^{-1}]\int_{-\pi}^{\pi} f(w;\alpha,\gamma)\bar{f}(w;\beta,\nu)dw\\
&+n^{-3}2\pi\int_{-\pi}^{\pi}\Big[4^{-1}+(qe^{\iota w})^2\Big(1-(qe^{\iota w})^2\Big)^{-1}\Big]\\
&\qquad\qquad\qquad\times f(-w;\beta,\gamma)\bar{f}(-w;\alpha,\nu)dw\\
&+O(n^{-4}s^2)\;. \qquad (5.43)
\end{aligned}
$$

Let $\hat{\sigma}_{1n}(j)=\sum_{|\alpha|=1}\sum_{|\beta|=1}c_{\alpha+\beta}\Big(\hat{\sigma}(j;\alpha,\beta)+\hat{\sigma}(j;\beta,\alpha)\Big)$, $0\le j\le n-1$. Then, by (5.36), (5.41)–(5.43) and Lemmas 3.3 and 5.4, we have

$$
\begin{aligned}
&\mathrm{Var}\Bigg(n^{-1}\sum_{|\alpha|=1}\sum_{|\beta|=1}c_{\alpha+\beta}\Bigg[\sum_{j=0}^{n-1}q_{nj}\Big(\hat{\sigma}(j;\alpha,\beta)+\hat{\sigma}(j;\beta,\alpha)\Big)\Bigg]\Bigg)\\
= \;& n^{-2}\Bigg[\sum_{j=0}^{n-1}q_{nj}^2\,\mathrm{Var}\Big(\hat{\sigma}_{1n}(j)\Big)\\
&+2\sum_{j=0}^{n-1}\sum_{v=1}^{n-1-j}q_{nj}q_{n(j+v)}\mathrm{Cov}\Big(\hat{\sigma}_{1n}(j),\hat{\sigma}_{1n}(j+v)\Big)\Bigg]\\
= \;& 2\pi n^{-3}\Bigg[\int_{-\pi}^{\pi}\Big\{2^{-1}+q^2(1-q^2)^{-1}\\
&\qquad\qquad +(qe^{\iota w})^2\Big(1-(qe^{\iota w})^2\Big)^{-1}\Big\}g_1(w)dw\\
&+2\Big\{\int_{-\pi}^{\pi}\Big[2^{-1}+q^2(1-q^2)^{-1}\Big](1-qe^{\iota w})^{-1}qe^{\iota w}g_1(w)dw\\
&+\int_{-\pi}^{\pi}\Big[2^{-1}+(qe^{\iota w})^2\Big(1-(qe^{\iota w})^2\Big)^{-1}\Big](1-qe^{\iota w})^{-1}qe^{\iota w}g_1(w)dw\Big\}\Bigg]\\
&+O(n^{-4}s^2)\\
= \;& 2\pi n^{-3}\ell\Bigg[\int_{-\pi}^{\pi}g_1(w)dw+2\pi g_1(0)+\int_{-\pi}^{\pi}\cos w\,g_1(w)dw\Bigg]\\
&+o(n^{-3}l)\;.
\end{aligned}
$$

This completes the proof of Theorem 5.2 for $j=4$. □

6

Second-Order Properties

6.1 Introduction

In this chapter, we consider second-order properties of block bootstrap estimators for estimating the sampling distribution of a statistic of interest. The basic tool for studying second-order properties of block bootstrap distribution function estimators is based on the theory of Edgeworth expansions. Let $\hat{\theta}_n$ be an estimator of a level-1 parameter θ and $T_n = \sqrt{n}(\hat{\theta}_n - \theta)/s_n$ be a scaled version of $\hat{\theta}_n$ such that $T_n \longrightarrow^d N(0,1)$. If we set s_n to be the (asymptotic) standard deviation of $\sqrt{n}(\hat{\theta}_n - \theta)$, then T_n is called a *normalized* or *standardized* version of $\hat{\theta}_n$. If s_n is an estimator of the asymptotic standard deviation of $\sqrt{n}(\hat{\theta}_n - \theta)$, then T_n is called a *studentized* version of $\hat{\theta}_n$. In many instances, it is possible to expand the distribution function of T_n in a series of the form

$$P(T_n \leq x) = \Phi(x) + n^{-1/2}p_1(x;\gamma)\phi(x) + o(n^{-1/2}) \tag{6.1}$$

uniformly in $x \in \mathbb{R}$, where Φ and ϕ, respectively, denote the distribution function and the density (with respect to the Lebesgue measure) of the standard normal distribution on $\mathbb{R}$ and where $p_1(\cdot;\gamma)$ is a polynomial such that its coefficients are (smooth) functions of some population parameters γ. The right side of (6.1) is called a first-order Edgeworth expansion for the distribution function of T_n. Next, let T_n^* denote the bootstrap version of T_n based on one of the several block bootstrap methods presented in Chapter 2. Under suitable regularity conditions on T_n, on the resampling mechanism, and on the underlying time series, we can often expand the conditional distribution function of T_n^* in an Edgeworth expansion of the form

$$P_*(T_n^* \leq x) = \Phi(x) + n^{-1/2} p_1(x; \hat{\gamma}_n)\phi(x) + o_p(n^{-1/2}) \tag{6.2}$$

uniformly in $x \in \mathbb{R}$, where $p_1(\cdot;\cdot)$ is the same function as in (6.1) and where $\hat{\gamma}_n$ is a data-based version of the population parameter γ, generated by the particular resampling method in use. Relations (6.1) and (6.2) may be readily combined to assess the rate of approximation of the bootstrap distribution function estimator $P_*(T_n^* \leq x)$. Indeed, by (6.1) and (6.2), it follows that

$$\begin{aligned}
&\sup_{x \in \mathbb{R}} \Big| P_*(T_n^* \leq x) - P(T_n \leq x) \Big| \\
&\quad = n^{-1/2} \sup_{x \in \mathbb{R}} \Big| p_1(x; \hat{\gamma}_n)\phi(x) - p_1(x; \gamma)\phi(x) \Big| + o_p(n^{-1/2}) \\
&\quad = o_p(n^{-1/2}) \, , \qquad (6.3)
\end{aligned}$$

provided $\hat{\gamma}_n$ is a consistent estimator of γ and the coefficients of the polynomial $p_1(\cdot; t)$ is continuous in the second argument t. In this case, the bootstrap approximation $P_*(T_n^* \leq x)$ to $P(T_n \leq x)$ has a smaller order of error than the normal approximation to $P(T_n \leq x)$, which is only of the order $O(n^{-1/2})$ (cf. (6.1)). This property is referred to as the *second-order correctness* of the bootstrap approximation, as it captures the second-order term (i.e., the term of order $n^{-1/2}$) asymptotically. This line of arguments for studying second-order properties of bootstrap methods was pioneered by Singh (1981), who established second-order correctness of the IID bootstrap method of Efron (1979) for the normalized sample mean of iid random variables and provided the first theoretical confirmation of the superiority of the bootstrap approximation over the classical normal approximation.

In this chapter, we consider second-order properties of different block bootstrap methods for normalized and studentized versions of the sample mean and of (more general) estimators that satisfy the requirements of the Smooth Function Model. In Section 6.2, we introduce the basic theory of Edgeworth expansion for the sample mean under independence. Since the bootstrap blocks are drawn by independent resampling, Edgeworth expansions for the conditional distribution of the bootstrap sample mean can be derived using these techniques. In Section 6.3, we describe the framework and the fundamental results of Götze and Hipp (1983) on Edgeworth expansions for the sample mean of weakly dependent random vectors. We next discuss extensions of the Edgeworth expansion theory for the sample mean to the Smooth Function Model in Section 6.4. In Section 6.5, we describe the results on second-order properties of various block bootstrap methods based on independent resampling, including the MBB and the SB, for normalized and studentized statistics under the Smooth Function Model.

6.2 Edgeworth Expansions for the Mean Under Independence

Let $X_1, \ldots, X_n$ be independent but not necessarily identically distributed $\mathbb{R}^d$-valued random vectors with $EX_i = 0$ and $E\|X_i\|^s < \infty$ for all $1 \leq i \leq n$, where $s \geq 3$ is an integer. An excellent account of rigorous Edgeworth expansion theory for the scaled sample mean $S_n \equiv \sqrt{n}\bar{X}_n = n^{-1/2}\sum_{i=1}^n X_i$ is given in Bhattacharya and Rao (1986) for the multivariate case ($d \geq 1$), and in Petrov (1975) for the univariate case ($d = 1$). In this section, we recast some of these results for the multivariate case, which serve as a basis for deriving Edgeworth expansions for bootstrapped statistics.

For easy reference later on, here we recall some relevant notation and definitions. For a smooth function $h : \mathbb{R}^d \to \mathbb{R}$, we write $D_j h$ to denote the partial derivative of $h(x)$ with respect to the jth coordinate of x, $1 \leq j \leq d$. For $d \times 1$ vectors $\nu = (\nu_1, \ldots, \nu_d)' \in \mathbb{Z}_+^d$ and $x = (x_1, \ldots, x_d)' \in \mathbb{R}^d$, let $|\nu| = \nu_1 + \cdots + \nu_d$, $\nu! = \nu_1! \cdots \nu_d!$, $x^\nu = \prod_{i=1}^d (x_i)^{\nu_i}$, and $\|x\| = (x_1^2 + \cdots + x_d^2)^{1/2}$. Also, let D^ν denote the differential operator $D_1^{\nu_1} \cdots D_d^{\nu_d}$. For a nonzero complex number $z = r\exp(\iota w)$, $r > 0$, $-\pi < w \leq \pi$, we define

$$\log z = (\log r) + \iota w \ ,$$

where $\iota^2 = -1$. Then, $\log z$ is the so-called principal branch of the logarithm, and it is analytic in the domain $\mathbb{C} \setminus (-\infty, 0]$.

Let X be a $\mathbb{R}^d$-valued random vector and let $\xi(t) = E\exp(\iota t'X)$, $t \in \mathbb{R}^d$ denote the characteristic function of X. Note that under the moment condition $E\|X\|^s < \infty$, $\log \xi(t)$ is s-times differentiable in a neighborhood of zero. For $\nu \in \mathbb{Z}_+^d$ with $1 \leq |\nu| \leq s$, let χ_ν denote the νth cumulant of X, defined by

$$\chi_\nu = (-\iota)^{|\nu|} D^\nu \log \xi(0) \ . \tag{6.4}$$

Let μ_ν denote the νth moment of X, i.e., $\mu_\nu = E(X^\nu)$, $1 \leq |\nu| \leq s$. Then, it is possible to express the cumulants of X in terms of the moments of X, and vice versa. Note that the exponential and the logarithm functions admit the power series expansions

$$e^z = \sum_{k=0}^\infty \frac{z^k}{k!}, \ z \in \mathbb{C}$$

and

$$\log(1+z) = \sum_{k=1}^\infty (-1)^{k-1}\frac{z^k}{k}, \ z \in \mathbb{C}, \ |z| < 1 \ .$$

Using these expansions, we get the formal identity

$$\sum_{|\nu|=1}^\infty (\iota t)^\nu \chi_\nu/\nu! \quad = \quad \log\big(\xi(t)\big) = \log\Big(1 + \sum_{|\nu|=1}^\infty (\iota t)^\nu \mu_\nu/\nu!\Big)$$

$$= \sum_{k=1}^{\infty}(-1)^{k-1}\Big(\sum_{|\nu|=1}^{\infty}(\iota t)^{\nu}\mu_{\nu}/\nu!\Big)^{k}/k\,, \tag{6.5}$$

$t \in \mathbb{R}^d$. Equating the coefficients of t^ν in (6.5), we can obtain an expression for χ_ν's in terms of μ_ν's. In the multi-dimensional case, writing down an exact expression for χ_ν is rather cumbersome. For most applications, a working formula is given by (see page 46, Bhattacharya and Rao (1986))

$$\chi_\nu = \sum_{k=1}^{|\nu|}\sum_{(k)} c(\nu_1,\ldots,\nu_k; i_1,\ldots,i_k)\mu_{\nu}^{i_1},\ldots,\mu_{\nu_k}^{i_k} \tag{6.6}$$

where $c(\nu_1,\ldots,\nu_k; i_1,\ldots,i_k)$ are combinatorial constants, depending only on their arguments, and where the summation $\sum_{(k)}$ extends over all $\nu_1,\ldots,\nu_k \in \mathbb{Z}_+^d$ and $i_1,\ldots,i_k \in \mathbb{N}$ satisfying $\sum_{m=1}^{k} i_m\nu_m = \nu$. In the one-dimensional case, we have the following relations (cf. page 46, Bhattacharya and Rao (1986)):

$$\begin{aligned}
\chi_1 &= \mu_1\\
\chi_2 &= \mu_2 - \mu_1^2\\
\chi_3 &= \mu_3 - 3\mu_2\mu_1 + 2\mu_1^2\\
\chi_4 &= \mu_4 - 4\mu_3\mu_1 - 3\mu_2^2 + 12\mu_2\mu_1^2 - 6\mu_1^4\\
\chi_5 &= \mu_5 - 5\mu_4\mu_1 - 10\mu_3\mu_2 + 20\mu_3\mu_1^2 + 30\mu_2^2\mu_1\\
&\quad - 60\mu_2\mu_1^3 + 24\mu_1^5\,.
\end{aligned} \tag{6.7}$$

For expressions for higher order cumulants, see Petrov (1975) and Kendall and Stuart (1977).

Cumulants play an important role in the development of Edgeworth expansions for sums of independent random vectors. To gain some insight into the derivation of the Edgeworth expansions for sums of independent random vectors, first we consider the simpler situation involving *iid* random vectors, at a heuristic level. Suppose that X is a $\mathbb{R}^d$-valued random vector with $EX = 0$ and $E\|X\|^s < \infty$ for some integer $s \geq 3$. Then, in a neighborhood of $t = 0$, by Taylor's expansion, we have

$$E\exp(\iota t'X) = \exp\Big(\sum_{1\leq|\nu|\leq s}(\iota t)^{\nu}\chi_{\nu}/\nu! + R(t)\Big)\,, \tag{6.8}$$

where $R(t) = o(\|t\|^s)$ as $\|t\| \to 0$. Let $\tilde{T}_n$ denote the sum of n independent copies of X. Then, noting that $\chi_\nu = EX^\nu = 0$ for all $|\nu| = 1$, by (6.8), for any *given* $t \in \mathbb{R}^d$, we have

$$\begin{aligned}
&E\exp\Big(\iota t'(n^{-1/2}\tilde{T}_n)\Big)\\
&= \Big[E\exp(\iota t'X/\sqrt{n})\Big]^n
\end{aligned}$$

$$
\begin{aligned}
&= \exp\Big(n\Big[\sum_{2\le|\nu|\le s} (\iota t/\sqrt{n})^{\nu}\chi_\nu/\nu! + R(t/\sqrt{n})\Big]\Big) \\
&= \exp(-t'\Sigma t/2)\exp\Big(\sum_{r=1}^{s-2} n^{-r/2}\Big[\sum_{|\nu|=r+2}(\iota t)^{\nu}\chi_\nu/\nu!\Big] + nR_n(t/\sqrt{n})\Big) \\
&= \exp(-t'\Sigma t/2)\Big[1 + \\
&\sum_{m=1}^{\infty}\Big(\sum_{r=1}^{s-2} n^{-r/2}\Big\{\sum_{|\nu|=r+2}(\iota t)^{\nu}\chi_\nu/\nu!\Big\} + o\big(n^{-(s-2)/2}\big)\Big)^m/m!\Big] \qquad (6.9)
\end{aligned}
$$

as $n \to \infty$, where $\Sigma = EXX'$ is the covariance matrix of X. If the terms of order $n^{-r/2}$ on the last line of (6.9) are grouped together, then we may *formally* expand the characteristic function of the scaled sum $\tilde{T}_n$ as

$$
\begin{aligned}
E\exp\big(\iota t'(n^{-1/2}\tilde{T}_n)\big) &= \exp(-t'\Sigma t/2)\Big[1 + \sum_{r=1}^{s-2} n^{-r/2}\tilde{p}_r(\iota t)\Big] \\
&\quad + o\big(n^{-(s-2)/2}\big) \quad \text{as} \quad n \to \infty\,, \qquad (6.10)
\end{aligned}
$$

for each *fixed* $t \in \mathbb{R}^d$, where $\tilde{p}_r(\cdot)$'s are polynomials. The Edgeworth expansions for $n^{-1/2}\tilde{T}_n$ (or, equivalently, for the scaled sample mean) is obtained by *inverting* the expansion for the characteristic function in (6.10). Theoretical justification for this inversion step and for the formal approximations in (6.9) and (6.10) are quite involved, and will not be presented here. We refer the interested reader to the monograph of Bhattacharya and Rao (1986) for details.

Using the heuristic arguments above, we now describe the Edgeworth expansion theory for the normalized sum of independent (but not necessarily identically distributed) random vectors. Let $X_1, \ldots, X_n$ be a collection of independent $\mathbb{R}^d$-valued random vectors with $EX_j = 0$ for $1 \le j \le n$. Let $\chi_{\nu,j}$ denote the νth cumulant of X_j, $1 \le j \le n$ and let $\bar{\chi}_\nu = n^{-1}\sum_{j=1}^{n}\chi_{\nu,j}$. Then, define the polynomials $\tilde{p}_j(z) \equiv \tilde{p}_j(z;\cdot)$ in $z \in \mathbb{C}^d$ by the *formal* identity in $u \in \mathbb{R}$ (cf. (6.9) and (6.10))

$$
1 + \sum_{j=1}^{\infty} u^j\tilde{p}_j(z;\{\bar{\chi}_\nu\}) = \exp\Big(\sum_{j=1}^{\infty} u^j\Big[\sum_{|\nu|=j+2}\bar{\chi}_\nu z^{\nu}/\nu!\Big]\Big)\,. \qquad (6.11)
$$

It can be shown (cf. Lemma 7.1, Bhattacharya and Rao (1986)) that for each $j \ge 1$, $\tilde{p}_j(z;\{\bar{\chi}_\nu\})$ is a polynomial of degree $3j$ in z and its coefficients are smooth functions of the cumulants $\bar{\chi}_\nu$ of order $|\nu| \le j+2$. The density $\psi_{n,s}$ of the $(s-2)$-th order Edgeworth expansion of the scaled sample mean $S_n \equiv n^{-1/2}\sum_{j=1}^{n} X_j$ is defined via its Fourier transform

$$
\psi^{\dagger}_{n,s}(t) \equiv \int \exp(\iota t'x)\psi_{n,s}(x)dx
$$

$$= \left[1 + \sum_{j=1}^{s-2} n^{-j/2} \tilde{p}_j(\iota t; \{\bar{\chi}_\nu\})\right] \exp(-t'\bar{\Sigma}t/2) , \qquad (6.12)$$

$t \in \mathbb{R}^d$, where $\bar{\Sigma} = n^{-1}\sum_{i=1}^n EX_jX_j'$. Note that $\psi_{n,s}$ depends on the cumulants $\bar{\chi}_\nu$ for $|\nu| \leq s$.

The density $\psi_{n,s}$ can be recovered from its Fourier transform in (6.12) by using the inversion formula:

$$\psi_{n,s}(x) = (2\pi)^{-d} \int_{\mathbb{R}^d} \psi^\dagger_{n,s}(t) \exp(-\iota t'x)dt, \; x \in \mathbb{R}^d . \qquad (6.13)$$

Next we evaluate the integral on the right side of (6.13). Note that the function

$$\tilde{f}_j(t) \equiv \tilde{p}_j(\iota t; \{\bar{\chi}_\nu\}) \exp(-t'\bar{\Sigma}t/2), \; t \in \mathbb{R}^d \qquad (6.14)$$

is the Fourier transform of the function

$$f_j(x) = \tilde{p}_j(-D; \{\bar{\chi}_\nu\})\phi_{\bar{\Sigma}}(x), \; x \in \mathbb{R}^d , \qquad (6.15)$$

where $\tilde{p}_j(-D; \{\bar{\chi}_\nu\})$ is a differential operator obtained by formally substituting $-D \equiv (-D_1, \ldots, -D_d)'$ in place of $z = (z_1, \ldots, z_d)'$ in $\tilde{p}_j(z; \{\bar{\chi}_\nu\})$ and where $\phi_{\bar{\Sigma}}$ denotes the density of the $N(0, \bar{\Sigma})$ distribution on $\mathbb{R}^d$. For example, if $\tilde{p}_j(z; \{\bar{\chi}_\nu\}) = \sum_{k=0}^{3j} \sum_{|\nu|=k} a_\nu z^\nu$ for some constants a_ν's depending on $\{\bar{\chi}_\nu\}$, then $f_j(x) = \sum_{k=0}^{3j} \sum_{|\nu|=k} a_\nu (-1)^{|\nu|} D^\nu \phi_{\bar{\Sigma}}(x)$, $x \in \mathbb{R}^d$. Since the νth order partial derivative of $\phi_{\bar{\Sigma}}(x)$ is of the form $p_\nu(x)\phi_{\bar{\Sigma}}(x)$ for a polynomial $p_\nu(x)$ (with coefficients depending on $\bar{\Sigma}$), from (6.12)–(6.15), it follows that

$$\psi_{n,s}(x) = \left[1 + \sum_{j=1}^{s-2} n^{-j/2} p_j(x; \{\bar{\chi}_\nu\})\right] \phi_{\bar{\Sigma}}(x), \; x \in \mathbb{R}^d , \qquad (6.16)$$

where $p_j(x; \{\bar{\chi}_\nu\})$ are polynomials determined by the relation

$$p_j(x; \{\bar{\chi}_\nu\})\phi_{\bar{\Sigma}}(x) = \tilde{p}_j(-D; \{\bar{\chi}_\nu\})\phi_{\bar{\Sigma}}(x), \; x \in \mathbb{R}^d \qquad (6.17)$$

for $j = 1, \ldots, s-2$. The exact forms of the polynomials $p_j(\cdot;\cdot)$ are difficult to write down explicitly for a general $d \geq 1$. See Bhattacharya and Rao (1986), Chapter 7, for an illustrative example. Here we list the first two polynomials for $d = 1$. For simplicity, suppose that the X_i's are iid with mean zero and variance 1. Then, $\bar{\Sigma} = 1$. Furthermore, in the one-dimensional case, the derivatives of the standard normal density function $\phi(x) = (2\pi)^{-1/2} \exp(-x^2/2)$, $x \in \mathbb{R}$, may be expressed in terms of the Hermite polynomials $H_k(x)$'s, defined by the relation

$$H_k(x)\phi(x) = (-1)^k \frac{d^k}{dx^k}(\phi(x)), \; x \in \mathbb{R} , \qquad (6.18)$$

$k \geq 0$. The first few Hermite polynomials are given by

$$H_0(x) \equiv 1,\ H_1(x) = x,\ H_2(x) = x^2 - 1,\ H_3(x) = x^3 - 3x,$$
$$H_4(x) = x^4 - 6x^2 + 3,\ H_5(x) = x^5 - 10x^3 + 15x, \text{ etc.} \tag{6.19}$$

See, for example, Hall (1992), page 44 and Petrov (1975), page 137. The polynomials $p_j(\cdot;\cdot)$, $j = 1, 2$ are given by

$$p_1(x; \{\bar{\chi}_\nu\}) = \frac{1}{6} H_3(x)\mu_3,\ x \in \mathbb{R}\ , \tag{6.20}$$

$$p_2(x; \{\bar{\chi}_\nu\}) = \frac{1}{72} H_6(x)\mu_3^2 + \frac{1}{24} H_4(x)[\mu_4 - 3],\ x \in \mathbb{R}\ , \tag{6.21}$$

where $\mu_k = EX_1^k$, $k \geq 1$. For explicit expressions of the "distribution function" $\Psi_{n,s}((-\infty, x]) \equiv \int_{-\infty}^{x} \psi_{n,s}(y)dy$, $x \in \mathbb{R}$ for the first few s values in the case when X_j's are independent (and not necessarily identically distributed) random variables, see Petrov (1975), page 138.

Next, we return to the general set up of $\mathbb{R}^d$-valued independent random vectors. Let $\Psi_{n,s}$ be a signed measure on $(\mathbb{R}^d, \mathcal{B}(\mathbb{R}^d))$ with density $\psi_{n,s}$ (with respect to the Lebesgue measure) where $\psi_{n,s}$ is as defined in (6.16). The main result of this section says that under suitable regularity conditions, asymptotic approximations for the probabilities $P(S_n \in \cdot)$, or, more generally, for the expected values $Ef(S_n)$ for Borel-measurable functions $f : \mathbb{R}^d \to \mathbb{R}$, with an error of the order $o(n^{-(s-2)/2})$ are given by $\int f d\Psi_{n,s}$. For $\epsilon > 0$ and $s \in \mathbb{N}$, let $a_n(s, \epsilon) = n^{-1} \sum_{j=1}^{n} E\|X_j\|^s \mathbb{1}(\|X_j\| > \epsilon\sqrt{n})$ and let $a_n(s) = a_n(s, 1)$. Also, for a function $f : \mathbb{R}^d \to \mathbb{R}$ and a measure μ on $(\mathbb{R}^d, \mathcal{B}(\mathbb{R}^d))$, define the integrated modulus of continuity of f with respect to μ as

$$w(\epsilon; f, \mu) = \int \sup\{|f(x) - f(y)| : \|x - y\| \leq \epsilon\}\mu(dy)\ , \tag{6.22}$$

$\epsilon > 0$. With $\beta(s) = 2\lfloor s/2 \rfloor$, let

$$M_s(f) = \sup\left\{\left(1 + \|x\|^{\beta(s)}\right)^{-1} |f(x)| : x \in \mathbb{R}^d\right\}. \tag{6.23}$$

Then, we have the following result.

Theorem 6.1 *Let $X_1, \ldots, X_n$ $(n \in \mathbb{N})$ be a collection of $\mathbb{R}^d$-valued independent random vectors with $EX_j = 0$ for $1 \leq j \leq n$, $n^{-1} \sum_{j=1}^{n} EX_jX_j' = \mathbb{I}_d$, and $\bar{\rho}_{n,s} \equiv n^{-1} \sum_{i=1}^{n} E\|X_j\|^s < \infty$ for some integer $s \geq 3$. Let $f : \mathbb{R}^d \to \mathbb{R}$ be a Borel-measurable function with $M_s(f) < \infty$. Then, for any $\epsilon \in (0, 1)$, there exists a constant $C = C(d, s) \in (0, \infty)$ (not depending on f and n) such that, with $u_n \equiv n^{-(s-2)/2}$,*

$$\left|Ef(S_n) - \int f d\Psi_{n,s}\right|$$

$$\begin{aligned}
&\le C \cdot M_s(f)\Big[u_n\tilde{\rho}_{n,s} + (1+\tilde{\rho}_{n,s})\Big\{(a_n(s)+u_n)u_n \\
&\quad + v_n^{-2(s+d+1)} + n^{(s+2d)/2}v_n^{-(s+d+1)(d+2s)}\Big\} \\
&\quad + \Big\{\gamma_n(\epsilon)\epsilon^{-2d} + n^{s+d+1}\epsilon^{-8d}\exp(-\epsilon^{-1})\Big\}\Big] \\
&\quad + C\cdot(1+\bar{\rho}_{n,s})\cdot w(2\epsilon; f,\Phi) \qquad (6.24)
\end{aligned}$$

whenever

$$a_n(s;2/3) \le (2^{s+4}d)^{-1}n^{(s-2)/2}\ . \qquad (6.25)$$

Here $v_n = (\bar{\rho}_{n,s}u_n)^{-1/(s+d+1)}$, $\tilde{\rho}_{n,s} = n^{-3/2}\sum_{j=1}^{n} E\|X_j\|^{s+1}\mathbb{1}(\|X_j\| \le n^{1/2})$, *and* $\gamma_n(\epsilon) = \sum_{1\le j_1,\ldots,j_{s+d+1}\le n} \sup\{\Pi_{j\neq j_1,\ldots,j_{s+d+1}}|\theta_{n,j}(t)| : (16\bar{\rho}_{n,3})^{-1} \le \|t\| \le \epsilon^{-4}\}$, *with* $\theta_{n,j}(t) = |E\exp(\iota t'X_j)| + 2P(\|X_j\| > \sqrt{n})$, $1 \le j \le n$, $t \in \mathbb{R}^d$.

Proof: See Appendix B. □

Theorem 6.1 gives an upper bound on the difference between $Ef(S_n)$ and the expansion $\int f d\Psi_{n,s}$ for any given collection of random vectors $X_1,\ldots,X_n$ $(n \ge 1)$ for which (6.25) and the other conditions of Theorem 6.1 hold. This form of the Edgeworth expansion is most useful for establishing validity of Edgeworth expansions for the conditional expectation $E_*f(S_n^*)$ in the (block) bootstrap context, where S_n^* corresponds to a sum of independent random vectors from a *triangular array.* Standard results available in the literature are often stated for a sequence of random vectors and, hence, they are not directly applicable in the bootstrap case. Theorem 6.1 can be proved by careful modification of the main steps associated with the proof in the "sequence" case. An outline of the proof of Theorem 6.1 is given in Appendix B. Note that the error of approximation in (6.24) depends on the quantities $\gamma_n(\epsilon)$ and $w(2\epsilon; f,\Phi)$. For $\gamma_n(\epsilon)$ to be small, the distribution of X_j's must satisfy some smoothness condition (such as Cramer's condition; see (6.28) and (6.31) below), while for $w(2\epsilon; f,\Phi)$ to be small, the function f cannot have too much oscillation. In most applications, we are interested in expansions for $P(S_n \in B)$ for B ranging over some classes of sets in $\mathcal{B}(\mathbb{R}^d)$. In this case, $f = \mathbb{1}_B$ and the last term is small under additional smoothness conditions on the boundary, ∂B, of the set B.

We now state a version of Theorem 6.1 for probabilities of events involving the scaled row sum of a triangular array of independent random vectors. In the statement of Theorem 6.2, for $\epsilon \in (0,\infty)$ and $B \subset \mathbb{R}^d$, let ∂B denote the boundary of B and let

$$(\partial B)^\epsilon = \{x \in \mathbb{R}^d : \|x-y\| < \epsilon \quad \text{for some} \quad y \in \partial B\}\ .$$

Theorem 6.2 *Let* $\{X_{n,j} : 1 \le j \le n\}_{n\ge1}$ *be a triangular array of row wise independent* $\mathbb{R}^d$*-valued random vectors* $X_{n,1},\ldots,X_{n,n}$ *with* $EX_{n,j} =$

0, $1 \leq j \leq n$ and $n^{-1}\sum_{j=1}^{n} EX_{n,j}X'_{n,j} = \mathbb{I}_d$ for each $n \geq 1$. Suppose that for some integer $s \geq 3$ and some $\delta \in (0, 1/2)$,

$$\lim_{n\to\infty} n^{-1}\sum_{j=1}^{n} E\|X_{n,j}\|^{s}\mathbb{1}\Big(\|X_{n,j}\| > n^{\frac{1}{2}-\delta}\Big) = 0 , \tag{6.26}$$

$$\limsup_{n\to\infty} n^{-1}\sum_{j=1}^{n} E\|X_{n,j}\|^{s} < \infty , \tag{6.27}$$

and for some sequence $\{\eta_n\}_{n\geq 1} \subset (0,\infty)$ with $\eta_n = o(n^{-(s-2)/2})$,

$$\limsup_{n\to\infty} \sup\Big\{|E\exp(\iota t'X_{n,j})| : (16\bar{\rho}_{n,3})^{-1} \leq \|t\| < \eta_n^{-4},\ 1 \leq j \leq n\Big\} < 1 , \tag{6.28}$$

where $\bar{\rho}_{n,r} = n^{-1}\sum_{j=1}^{n} E\|X_{n,j}\|^{r}$, $r \in \mathbb{N}$. Then,

$$\sup_{B\in\mathcal{B}} \big|P(S_n \in B) - \Psi_{n,s}(B)\big| = o\big(n^{-(s-2)/2}\big) \tag{6.29}$$

for any collection $\mathcal{B}$ of Borel sets in $\mathbb{R}^d$ satisfying

$$\sup_{B\in\mathcal{B}} \Phi\big((\partial B)^{\epsilon}\big) = O(\epsilon) \quad as \quad \epsilon \downarrow 0 . \tag{6.30}$$

Proof: See Appendix B for an outline of the proof. □

Let $\mathcal{C}$ denote the collection of all measurable convex subsets of $\mathbb{R}^d$. Then, (6.30) holds with $\mathcal{B} = \mathcal{C}$. For $d = 1$, if we set $\mathcal{B} = \{(-\infty, x] : x \in \mathbb{R}\}$, then also (6.30) holds and Theorem 6.2 yields a $(s-2)$-th order Edgeworth expansion for the distribution function of S_n.

Next we consider the important special case where the triangular array $\{X_{n,j} : 1 \leq j \leq n\}_{n\in\mathbb{N}}$ derives from a sequence of iid random vectors $\{X_n\}_{n\geq 1}$, i.e., $X_{n,j} = X_j$ for all $1 \leq j \leq n$, $n \geq 1$. Then, (6.26) and (6.27) holds if and only if $E\|X_1\|^s < \infty$. And condition (6.28) holds if and only if

$$\limsup_{\|t\|\to\infty} \big|E\exp(\iota t'X_1)\big| < 1 . \tag{6.31}$$

Inequality (6.31) is a smoothness condition on the distribution of X_1 and is known as *Cramer's condition.* A sufficient condition for (6.31) is that the probability distribution of X_1 has an absolutely continuous component with respect to the Lebesgue measure on $\mathbb{R}^d$. This is an immediate consequence of the Riemann-Lebesgue Theorem (cf. Theorem 26.1, Billingsley (1995)). In general, (6.31) does not hold when X_1 has a purely discrete distribution.

6.3 Edgeworth Expansions for the Mean Under Dependence

Let $\{X_i\}_{i\in\mathbb{Z}}$ be a sequence of $\mathbb{R}^d$-valued random vectors with $EX_i = 0$ for all $i \in \mathbb{Z}$. The process $\{X_i\}_{i\in\mathbb{Z}}$ need *not* be stationary. In this section, we state an Edgeworth expansion result for the scaled sample mean $S_n = \sqrt{n}\bar{X}_n = n^{-1/2}\sum_{i=1}^n X_i$, when the X_i's are weakly dependent. Derivation of Edgeworth expansions for dependent random vectors is technically difficult primarily due to the fact that unlike the independent case, the characteristic function of the scaled sum S_n does not factorize into the product of marginal characteristic functions. Extensions of the Edgeworth expansion theory to dependent variables arising from a Markov chain have been established by Statulevicius (1969a, 1969b, 1970), Hipp (1985), Malinovskii (1986), and Jensen (1989). For weakly dependent processes $\{X_i\}_{i\in\mathbb{Z}}$ that do not necessarily have a Markovian structure, Edgeworth expansions for the scaled sum S_n under a very general framework have been obtained by Götze and Hipp (1983). In this section, we state some basic Edgeworth expansion results for S_n under the Götze and Hipp (1983) framework. Suppose that the process $\{X_i\}_{i\in\mathbb{Z}}$ is defined on a probability space $(\Omega, \mathcal{F}, P)$ and that $\{\mathcal{D}_i\}_{i\in\mathbb{Z}}$ is a collection of sub-σ-fields of $\mathcal{F}$. A key feature of the Götze and Hipp (1983) framework is the introduction of the auxiliary set of σ-fields $\{\mathcal{D}_i\}_{i\in\mathbb{Z}}$ that allows one to treat various classes of weakly dependent processes under a common framework, by suitable choices of the sequence $\{\mathcal{D}_i\}_{i\in\mathbb{Z}}$. In the following, we first state and discuss the regularity conditions that specify the role played by the $\mathcal{D}_i$'s, and then give some examples of processes $\{X_i\}_{i\in\mathbb{Z}}$ and the corresponding choices of the σ-fields $\{\mathcal{D}_i\}_{i\in\mathbb{Z}}$ to illustrate the generality of the framework. For $-\infty \le a \le b \le \infty$, write $\mathcal{D}_a^b = \sigma\langle\{\mathcal{D}_i : i \in \mathbb{Z},\ a \le i \le b\}\rangle$. We will make use of the following conditions:

(C.1) For some integer $s \ge 3$ and a real number $\alpha(s) > s^2$,

$$\sup\Big\{E\|X_j\|^s\Big[\log(1+\|X_j\|)\Big]^{\alpha(s)} : j \ge 1\Big\} < \infty\ .$$

(C.2) (i) $EX_j = 0$ for all $j \ge 1$ and

$$\Sigma = \lim_{n\to\infty} n^{-1}\mathrm{Var}\Big(\sum_{j=1}^n X_j\Big) \tag{6.32}$$

exists and is nonsingular.

(ii) There exists $\delta \in (0,1)$ such that for all $n > \delta^{-1}$, $m > \delta^{-1}$,

$$\inf\Big\{t'\mathrm{Var}\Big(\sum_{i=n+1}^{n+m} X_i\Big)t : \|t\| = 1\Big\} > \delta m\ .$$

(C.3) There exists $\delta \in (0,1)$ such that for all $n, m = 1, 2, \ldots$ with $m > \delta^{-1}$, there exists a $\mathcal{D}_{n-m}^{n+m}$-measurable random vector $X_{n,m}^{\ddagger}$ satisfying

$$E\|X_n - X_{n,m}^{\ddagger}\| \leq \delta^{-1} \exp(-\delta m) .$$

(C.4) There exists $\delta \in (0,1)$ such that for all $i \in \mathbb{Z}$, $m \in \mathbb{N}$, $A \in \mathcal{D}_{-\infty}^{i}$, and $B \in \mathcal{D}_{i+m}^{\infty}$,

$$\big|P(A \cap B) - P(A)P(B)\big| \leq \delta^{-1} \exp(-\delta m) .$$

(C.5) There exists $\delta \in (0,1)$ such that for all $m, n, k = 1, 2, \ldots$, and $A \in \mathcal{D}_{n-k}^{n+k}$

$$E\big|P(A \mid \mathcal{D}_j : j \neq n) - P(A \mid \mathcal{D}_j : 0 < |j-n| \leq m+k)\big| \\ \leq \delta^{-1} \exp(-\delta m).$$

(C.6) There exists $\delta \in (0,1)$ such that for all $m, n = 1, 2, \ldots$ with $\delta^{-1} < m < n$ and for all $t \in \mathbb{R}^d$ with $\|t\| \geq \delta$,

$$E\Big|E\Big\{\exp(\iota t'[X_{n-m} + \cdots + X_{n+m}]) \mid \mathcal{D}_j : j \neq n\Big\}\Big| \leq \exp(-\delta) .$$

Now we briefly discuss the Conditions (C.1)–(C.6) stated above. Condition (C.1) is a moment condition used by Lahiri (1993a) to derive an $(s-2)$-th order Edgeworth expansion for the normalized sample mean. It is slightly weaker than the corresponding moment condition imposed by Götze and Hipp (1983), which requires existence of the $(s+1)$-th order moments of the X_j's. When the sequence $\{X_i\}_{i\in\mathbb{Z}}$ is m-dependent for some $m \in \mathbb{N}$, Lahiri (1993a) also shows that an $(s-2)$-th order expansion for the distribution of S_n remains valid under the following reduced moment condition:

$$\sup\big\{E\|X_j\|^s : j \in \mathbb{Z}\big\} < \infty , \tag{6.33}$$

as in the case of *independent* random vectors. The nonsingularity of Σ in Condition (C.2)(i) is required for a nondegenerate normal limit distribution of the scaled mean S_n. When the process $\{X_i\}_{i\in\mathbb{Z}}$ is second-order stationary, Condition (C.2)(ii) automatically follows from (C.2)(i). Condition (C.4) is a strong-mixing condition on the underlying auxiliary sequence of σ-fields $\mathcal{D}_j$'s. Condition (C.4) requires the σ-fields $\mathcal{D}_j$'s to be strongly mixing at an exponential rate. For Edgeworth expansions for the normalized sample mean under polynomial mixing rates, see Lahiri (1996b). Condition (C.3) connects the strong mixing condition on the σ-fields $\mathcal{D}_j$'s to the weak-dependence structure of the random vectors X_j's. If, for all $j \in \mathbb{Z}$, we set $\mathcal{D}_j = \sigma\langle X_j\rangle$, the σ-field generated by X_j, then Condition (C.3) is trivially satisfied with $X_{n,m}^{\ddagger} = X_n$ for all m. However, this choice of $\mathcal{D}_j$ is not always the most useful one for the verification of the rest of the conditions. See the examples given below, illustrating various choices of the σ-fields $\mathcal{D}_j$'s in different problems.

Condition (C.5) is an approximate Markov-type property, which says that the conditional probability of an event $A \in \mathcal{D}_{n-k}^{n+k}$, given the larger σ-field $\vee\{\mathcal{D}_j : j \neq n\}$, can be approximated with increasing accuracy when the conditioning σ-field $\vee\{\mathcal{D}_j : 0 < |j-n| \leq m+k\}$ grows with m. This condition trivially holds if X_j is $\mathcal{D}_j$-measurable and $\{X_i\}_{i\in\mathbb{Z}}$ is itself a Markov chain of a finite order. Finally, we consider (C.6). It is a version of the Cramer condition in the weakly dependent case. Note that if X_j's are iid and the σ-fields $\mathcal{D}_j$'s are chosen as $\mathcal{D}_j = \sigma\langle X_j\rangle$, $j \in \mathbb{Z}$, then Condition (C.6) is equivalent to requiring that for some $\delta \in (0,1)$,

$$\begin{aligned} 1 > e^{-\delta} &\geq E\big|E\{\exp(\iota t' X_n) \mid X_j : j \neq n\}\big| \\ &= \big|E\exp(\iota t' X_1)\big| \quad \text{for all} \quad \|t\| \geq \delta \,. \end{aligned}$$

It is easy to check that this is equivalent to the standard Cramer condition (cf. (6.31))

$$\limsup_{\|t\|\to\infty} |E\exp(\iota t' X_1)| < 1 \,.$$

However, for weakly dependent stationary X_j's, the standard Cramer condition on the *marginal* distribution of X_1 is not enough to ensure a "regular" Edgeworth expansion for the normalized sample mean, as shown by Götze and Hipp (1983). Here, by a regular Edgeworth expansion, we mean an Edgeworth expansion with a density of the form

$$\xi_{n,s}(x) = [1 + \sum_{r=1}^{s-2} n^{-r/2} p_r(x)]\phi_V(x), \ x \in \mathbb{R}^d$$

for some polynomials $p_1(\cdot), \ldots, p_r(\cdot)$ and for some positive definite matrix V, where ϕ_V is the density of the $N(0,V)$ distribution on $\mathbb{R}^d$. The sequence $\{X_i\}_{i\in\mathbb{Z}}$ in the example of Götze and Hipp (1983) is stationary and m-dependent with $m = 1$. Furthermore, X_1 has finite moments of all orders and it satisfies the standard Cramer condition (6.31). However, a "regular" Edgeworth expansion for the sum of the X_j's does not hold.

Next, we give examples of some important classes of weakly dependent processes that fit into the above framework and we indicate the choices of the σ-fields $\mathcal{D}_j$'s and the variables $X_{n,m}^{\ddagger}$'s for the verification of Conditions (C.3)–(C.6).

Example 6.1: Suppose that $\{X_i\}_{i\in\mathbb{Z}}$ is a linear process, given by

$$X_i = \sum_{j\in\mathbb{Z}} a_j \epsilon_{i-j}, \ i \in \mathbb{Z} \,, \tag{6.34}$$

where $\{a_i\}_{i\in\mathbb{Z}}$ is a sequence of real numbers and $\{\epsilon_i\}_{i\in\mathbb{Z}}$ is a sequence of iid random variables with $E\epsilon_1 = 0$, $E\epsilon_1^2 = 1$. Furthermore, suppose that $\sum_{i\in\mathbb{Z}} a_i \neq 0$ and for some $\delta \in (0,1)$,

$$|a_j| = O\big(\exp(-\delta|j|)\big) \quad \text{as} \quad |j| \to \infty. \tag{6.35}$$

If, in addition, ϵ_1 satisfies the standard Cramer condition,

$$\limsup_{|t|\to\infty} |E\exp(\iota t\epsilon_1)| < 1 \ , \tag{6.36}$$

then Conditions (C.3)–(C.6) hold with $\mathcal{D}_j = \sigma\langle\epsilon_j\rangle$, $j \in \mathbb{Z}$. In this case, we may take $X_{n,m}^{\ddagger} = \sum_{|j|\leq m} a_j\epsilon_{n-j}$.

A special case of (6.34) is the ARMA(p,q)-model

$$X_i = \alpha_1 X_{i-1} + \cdots + \alpha_p X_{i-p} + \epsilon_i + \beta_1\epsilon_{i-1} + \cdots + \beta_q\epsilon_{i-q} \ , \tag{6.37}$$

where $\alpha_1, \ldots, \alpha_p$, $\beta_1, \ldots, \beta_q$ ($p \in \mathbb{N}, q \in \mathbb{N}$) are real numbers and $\{\epsilon_i\}_{i\in\mathbb{Z}}$ is a sequence of iid random variables as in (6.34). We also suppose that the polynomials $\alpha(z) \equiv 1 - (\alpha_1 z + \cdots + \alpha_p z^p)$, and $\beta(z) \equiv 1 + \beta_1 z + \cdots + \beta_q z^q$, $z \in \mathbb{C}$ have no common zeros in $\mathbb{C}$ and $\alpha(z) \neq 0$ for all z in the closed unit disc $\{z \in \mathbb{C} : |z| \leq 1\}$. Then, it can be shown that there exists a sequence of constants $\{a_i\}_{i\in\mathbb{Z}} \subset \mathbb{R}$ satisfying (6.35) such that representation (6.34) holds (see, for example, Chapter 3, Brockwell and Davis (1991)). If, in addition, $\beta(1)/\alpha(1) \neq 0$, then $\sum_{i\in\mathbb{Z}} a_i \neq 0$. Thus, Conditions (C.3)–(C.6) hold for the ARMA(p,q) model (6.37), provided ϵ_1 satisfies the standard Cramer's condition (6.36), and the polynomials $\alpha(z)$ and $\beta(z)$ satisfy the regularity conditions pointed out above. □

Example 6.2: Let $\{\epsilon_i\}_{i\in\mathbb{Z}}$ be a sequence of iid random variables and let

$$X_i = h(\epsilon_{i+1}, \ldots, \epsilon_{i+m_0}), \ i \in \mathbb{Z} \tag{6.38}$$

for some continuously differentiable function $h : \mathbb{R}^{m_0} \to \mathbb{R}$, where $m_0 \in \mathbb{N}$. The sequence $\{X_i\}_{i\in\mathbb{Z}}$ of (6.38) is known as an m_0-dependent shift. Note that according to our definition (cf. Section 2.3), $\{X_i\}_{i\in\mathbb{Z}}$ is an m-dependent sequence with $m = m_0 - 1$. In this case, we set $\mathcal{D}_j = \sigma\langle\epsilon_j\rangle$, $j \in \mathbb{Z}$ and take $X_{n,m}^{\ddagger} = X_n$ for all $m, n \in \mathbb{N}$, with $m \geq m_0$. Then, it is easy to see that Conditions (C.3)–(C.5) hold with these choices of $\mathcal{D}_j$ and $X_{n,m}^{\ddagger}$. A set of sufficient conditions for (C.6) with $\mathcal{D}_j = \sigma\langle\epsilon_j\rangle$, $j \in \mathbb{Z}$ is that ϵ_1 has a density g with respect to the Lebesgue measure on $\mathbb{R}$, and, that there exist $y_1, \ldots, y_{2m_0-1} \in \mathbb{R}$ and an open subset U containing $y_1, \ldots, y_{2m_0-1}$ such that g is (everywhere) positive on U and

$$\sum_{j=1}^{m_0} \frac{\partial}{\partial x_j} h(x_1, \ldots, x_{m_0})\Big|_{(x_1,\ldots,x_{m_0})=(y_j,\ldots,y_{j+m_0-1})} \neq 0 \ . \tag{6.39}$$

See Götze and Hipp (1983), pages 218–219, for the details of the verification of (C.6). As mentioned earlier, in this case Condition (C.1) may be replaced by the weaker moment condition (6.33) for a valid $(s-2)$-th order Edgeworth expansion for the probability distribution of S_n. See Theorem 2.2, Lahiri (1993a). □

Example 6.3: Let $\{X_i\}_{i\in\mathbb{Z}}$ be a stationary homogeneous Markov chain with transition kernel $q(x; A)$, $x \in \mathbb{R}^d$, $A \in \mathcal{B}(\mathbb{R}^d)$. Suppose that X_1 satisfies the standard Cramer condition (6.31) and that

$$\sup\Big\{|q(x,A) - q(y,A)| : x, y \in \mathbb{R}^d,\ A \in \mathcal{B}(\mathbb{R}^d)\Big\} < 1\ . \tag{6.40}$$

Then, $\{X_i\}_{i\in\mathbb{Z}}$ satisfies Conditions (C.3)–(C.6) with $\mathcal{D}_i = \sigma\langle X_i\rangle$, $i \in \mathbb{Z}$ and $X^{\ddagger}_{n,m} = X_n$ for $n, m \in \mathbb{N}$. See Götze and Hipp (1983), page 219 for more details. □

Example 6.4: Let $\{Y_i\}_{i\in\mathbb{Z}}$ be a stationary Gaussian process with a positive analytic density and let

$$X_i = f(Y_i),\ i \in \mathbb{Z}$$

for some continuously differentiable nonconstant function $f : \mathbb{R} \to \mathbb{R}$. Then, Conditions (C.3)–(C.6) hold with $\mathcal{D}_i = \sigma\langle Y_i\rangle$, $i \in \mathbb{Z}$, and $X^{\ddagger}_{n,m} = X_n$ for $m, n \in \mathbb{N}$. See pages 219–220, Götze and Hipp (1983). □

For verification of Conditions (C.3)–(C.6) in other problems, see Bose (1988), Janas (1993), and Götze and Hipp (1994).

Next, we describe the form of the Edgeworth expansion for S_n in the dependent case. Like Section 6.2, we define a set of polynomials $\tilde{q}_{r,n}(t)$, $1 \le r \le s-2$, appearing in the Fourier transform of the Edgeworth expansion, by the identity (in $u \in \mathbb{R}$),

$$\begin{aligned}\exp&\left(\sum_{r=3}^{s} u^{r-2}\Big[n^{(r-2)/2}\chi_{r,n}(t)\Big]/r!\right)\\ &= \ 1 + \sum_{r=1}^{\infty} u^r \tilde{q}_{r,n}(t)\ ,\end{aligned} \tag{6.41}$$

where $\chi_{r,n}(t)$ is the rth cumulant of the random variable $t'S_n$, $t \in \mathbb{R}^d$. For $1 \le r \le s-2$, this definition of $\tilde{q}_{r,n}(t)$ is essentially equivalent to the definition of the polynomials $\tilde{p}_r(\cdot; \{\bar{\chi}_\nu\})$ given in (6.11) in the independent case. To appreciate why, note that we may replace the sum on the left side of (6.41) by $\sum_{r=3}^{\infty}$ and *formally* define the polynomials $\tilde{q}_{r,n}(\cdot)$ using the resulting identity. But this modification does not affect the first $(s-2)$

polynomials, because the cumulants of order $r \geq s+1$ do not appear in these polynomials. Hence, both identities yield the same polynomials $\tilde{q}_{r,n}$'s for $1 \leq r \leq s-2$.

It can be shown that under Conditions (C.1)–(C.6), for any given $t \in \mathbb{R}^d$, $\chi_{r,n}(t) = O(n^{-(r-2)/2})$ as $n \to \infty$, and, hence, the coefficients of the polynomials $\tilde{q}_{r,n}(t)$ are *bounded* sequences in n. However, for a sequence of nonstationary random vectors $\{X_i\}_{i \in \mathbb{Z}}$, the coefficients of $\tilde{q}_{r,n}(\cdot)$ typically depend on n. If $\{X_i\}_{i \in \mathbb{Z}}$ is stationary, $n^{(r-2)/2}\chi_{r,n}(t)$, $2 \leq r \leq s$ may be expanded further into a sum of the form

$$\begin{aligned} n^{(r-2)/2}\chi_{r,n}(t) &= \tilde{\chi}_{r,1}(t) + n^{-1/2}\tilde{\chi}_{r,2}(t) + \cdots \\ &\quad + n^{-(s-2)/2}\tilde{\chi}_{r,s-1} + o\big(n^{-(s-2)/2}\big) \end{aligned} \tag{6.42}$$

(for $t \in \mathbb{R}^d$ fixed) for some polynomials $\tilde{\chi}_{r,1}(t), \ldots, \tilde{\chi}_{r,k}(t)$, not depending on n. As a result, for a stationary sequence $\{X_i\}_{i \in \mathbb{Z}}$, the Edgeworth expansion for S_n may be written in terms of a set of polynomials that do not depend on n. See Remark 2.12, Götze and Hipp (1983) for more details.

Next, with $\tilde{q}_{r,n}(t)$ given by (6.41), we define density $\xi_{n,s}$ of the Edgeworth expansion for S_n in terms of its Fourier transform, by the relation

$$\begin{aligned} \xi^{\dagger}_{n,s}(t) &\equiv \int e^{\iota t' x}\xi_{n,s}(x)dx \\ &= \exp(-\chi_{2,n}(t)/2)\Big[1 + \sum_{r=1}^{s-2} n^{-r/2}\tilde{q}_{r,n}(\iota t)\Big], \quad t \in \mathbb{R}^d . \end{aligned} \tag{6.43}$$

The $(s-2)$-th order Edgeworth expansion $\Upsilon_{n,s}$ for S_n is defined as the signed measure on $(\mathbb{R}^d, \mathcal{B}(\mathbb{R}^d))$ having density $\xi_{n,s}$ with respect to the Lebesgue measure on $\mathbb{R}^d$. As in Section 6.2, let $\beta(s) = 2\lfloor s/2 \rfloor$ and for a Borel measurable function $f : \mathbb{R}^d \to \mathbb{R}$, define $w(\epsilon; f, \Phi_\Sigma)$ by (6.22), i.e., by the relation $w(\epsilon; f, \Phi_\Sigma) = \int \sup\{|f(x+y) - f(x)| : \|y\| \leq \epsilon\}\Phi_\Sigma(dx)$, $\epsilon > 0$, where Σ is as in (C.2). Then, we have the following result on asymptotic expansion for $Ef(S_n)$ in the dependent case.

Theorem 6.3 *Suppose that Conditions (C.1)–(C.6) hold. Let $f : \mathbb{R}^d \to \mathbb{R}$ be a Borel measurable function satisfying $\sup\{(1 + \|x\|^{\beta(s)})^{-1}|f(x)| : x \in \mathbb{R}^d\} \equiv M_s(f) < \infty$. Then, for any real number $a \in (0, \infty)$, there exists a constant $C = C(a) \in (0, \infty)$ such that*

$$\begin{aligned} &\Big|Ef(S_n) - \int f d\Upsilon_{n,s}\Big| \\ &\quad \leq C \cdot w(n^{-a}; f, \Phi_\Sigma) + o(n^{-(s-2)/2}) \quad \text{as} \quad n \to \infty . \end{aligned} \tag{6.44}$$

Further, the term $o(n^{-(s-2)/2})$ in (6.44) depends on f only through the constant $M_s(f)$.

Proof: See Theorem 2.1, Lahiri (1993a). □

Theorem 6.3 readily yields an $(s-2)$-th order Edgeworth expansion for the probability distribution of S_n uniformly over classes of Borel sets satisfying an analog of the boundary condition (6.30). We note this in the following result.

Theorem 6.4 *Suppose Conditions (C.1)–(C.6) hold. Then,*

$$\sup_{B\in\mathcal{B}} \Big| P(S_n \in B) - \Upsilon_{n,s}(B) \Big| = o(n^{-(s-2)/2}) \tag{6.45}$$

for any class $\mathcal{B}$ of Borel sets in $\mathbb{R}^d$ satisfying

$$\sup_{B\in\mathcal{B}} \Phi_\Sigma\big((\partial B)^\epsilon\big) = O(\epsilon) \quad as \quad \epsilon \downarrow 0 \ , \tag{6.46}$$

where $(\partial B)^\epsilon = \{x \in \mathbb{R}^d : \|x-y\| < \epsilon$ for some $y \in \partial B\}$, $\epsilon > 0$ and ∂B denotes the boundary of B.

6.4 Expansions for Functions of Sample Means

6.4.1 Expansions Under the Smooth Function Model Under Independence

Edgeworth expansion theory presented in the previous two sections deal with the sample mean. We now discuss some extensions of the theory to statistics that can be represented as smooth functions of sample means. First, we consider the case where $\{X_i\}_{i\in\mathbb{Z}}$ is a sequence of iid $\mathbb{R}^d$-valued random vectors with a finite mean vector $EX_1 = \mu \in \mathbb{R}^d$. Suppose that the statistic of interest $\hat{\theta}_n$ and its target parameter θ obey the Smooth Function Model of Chapter 4, i.e.,

$$\hat{\theta}_n = H(\bar{X}_n) \quad \text{and} \quad \theta = H(\mu) \tag{6.47}$$

for some (smooth) function $H : \mathbb{R}^d \to \mathbb{R}$, where $\bar{X}_n = n^{-1}\sum_{i=1}^n X_i$.

Let $W_{1n} = \sqrt{n}(\hat{\theta}_n - \theta)$. If $E\|X_1\|^2 < \infty$ and H is differentiable at μ and the vector of first-order partial derivatives of H at μ is nonzero, then a first-order Taylor's expansion of H around μ shows that

$$W_{1n} = \sum_{|\alpha|=1} D^\alpha H(\mu)[\sqrt{n}(\bar{X}_n - \mu)]^\alpha + o_p(1) \ ,$$

and, hence,

$$W_{1n} \longrightarrow^d N(0, \tau^2) \ , \tag{6.48}$$

where $\tau^2 = \sum_{|\beta|=1}\sum_{|\alpha|=1} D^\alpha H(\mu) D^\beta H(\mu) \mathrm{Cov}(X_1^\alpha, X_1^\beta)$. This is often referred to as the *Delta method.* Edgeworth expansions for W_n may be derived by considering higher-order Taylor's expansions of the function H around μ. Suppose that for some integer $s \geq 3$, H is s-times continuously differentiable in a neighborhood of μ. Then, we may express W_n as

$$\begin{aligned} W_{1n} &= \sum_{|\alpha|=1}^{s-1} n^{-(|\alpha|-1)/2} D^\alpha H(\mu) \big[\sqrt{n}(\bar{X}_n - \mu)\big]^\alpha / \alpha! + R_{n,s} \\ &\equiv V_{n,s} + R_{n,s}, \text{ say }, \end{aligned} \tag{6.49}$$

where $R_{n,s}$ is a remainder term that, under the moment condition $E\|X_1\|^s < \infty$, satisfies

$$P\big(|R_{n,s}| > \delta_{n,s}\big) = \delta_{n,s} \tag{6.50}$$

for some sequence $\delta_{n,s} = o(n^{-(s-2)/2})$. Here the random variable $V_{n,s}$ is called a $(s-2)$-th order *stochastic approximation* to W_{1n}. Under (6.50), the $(s-2)$-th order Edgeworth expansions for W_{1n} and $V_{n,s}$ coincide. It is customary to describe the $(s-2)$-th order Edgeworth expansion for W_{1n} using that for $V_{n,s}$. Supposing (for the time being) that X_1 has sufficiently many finite moments, the rth cumulant $\chi_r(V_{n,s})$ of $V_{n,s}$ can be expressed as

$$\chi_r(V_{n,s}) = \tilde{\chi}_{r,n,s} + o(n^{-(s-2)/2}) \tag{6.51}$$

for $1 \leq r \leq s$, where

$$\tilde{\chi}_{r,n,s} = \begin{cases} \sum_{j=1}^{s-2} n^{-j/2} \tilde{\chi}_{r,j} & \text{if } \ 1 \leq r \leq s, \ r \neq 2 \\ \tau^2 + \sum_{j=1}^{s-2} n^{-j/2} \tilde{\chi}_{2,j} & \text{if } \ r = 2 \end{cases} \tag{6.52}$$

for some constants $\tilde{\chi}_{r,j}$, not depending on n. It can be shown that the constants $\tilde{\chi}_{r,j}$, $1 \leq j \leq s-2$, $1 \leq r \leq s$ depend only on the moments EX_1^ν for $1 \leq |\nu| \leq s$ and on the partial derivatives $D^\nu H(\mu)$ for $|\nu| \leq s-1$. The $\tilde{\chi}_{r,n,s}$'s in (6.51) are called the *approximate cumulants* of $V_{n,s}$. Thus, when $E\|X_1\|^s < \infty$, we may *formally* expand $\chi_r(V_{n,s})$ (pretending that all moments of X_1 are finite) and then extract the approximate cumulants $\tilde{\chi}_{r,n,s}$ for $1 \leq r \leq s$, which involves only moments of order s or less. The Fourier transform of (the density of) the Edgeworth expansion for $V_{n,s}$ (and, hence, for W_{1n}) is given by

$$\psi_{n,s}^{[1]\dagger}(t) = \exp(-t^2\tau^2/2)\left[1 + \sum_{r=1}^{s-2} n^{-r/2} \tilde{p}_r^{[1]}(\iota t)\right], \tag{6.53}$$

$t \in \mathbb{R}$, where $\tilde{p}_1^{[1]}(\cdot), \ldots, \tilde{p}_{s-2}^{[1]}(\cdot)$ are polynomials defined by the identity (in $u \in \mathbb{R}$)

$$1 + \sum_{m=1}^{\infty} \left[\sum_{r=1}^{s} (r!)^{-1} \left(\sum_{j=1}^{s-2} u^j \tilde{\chi}_{r,j}\right) (\iota t)^r\right]^m \bigg/ m!$$

$$= 1 + \sum_{j=1}^{\infty} u^j \tilde{p}_j^{[1]}(\iota t) , \tag{6.54}$$

for $t \in \mathbb{R}$. As in Section 6.2, the Edgeworth expansion $\Psi_{n,s}^{[1]}$ for $V_{n,s}$ is the signed measure having the density (with respect to the Lebesgue measure on $\mathbb{R}$)

$$\begin{aligned} \psi_{n,s}^{[1]}(x) &= (2\pi)^{-1} \int \exp(-\iota t x) \psi_{n,s}^{[1]\dagger}(t) dt \\ &= \Big[1 + \sum_{r=1}^{s-2} n^{-r/2} \tilde{p}_r^{[1]} \Big(-\frac{d}{dx} \Big) \Big] \phi_{\tau^2}(x) , \\ &\equiv \Big(1 + \sum_{r=1}^{s-2} n^{-r/2} p_r^{[1]}(x) \Big) \phi_{\tau^2}(x), \ x \in \mathbb{R} , \end{aligned} \tag{6.55}$$

say, where $\tilde{p}_r^{[1]}(-\frac{d}{dx})$ is defined by replacing $(\iota t)^j$ in the definition of the polynomial $\tilde{p}_r^{[1]}(\iota t)$ with the differential operator $(-1)^j \frac{d^j}{dx^j}$, $j \geq 1$, and where $\phi_{\tau^2}(x) = (2\pi\tau^2)^{-1/2} \exp(-x^2/2\tau^2)$, $x \in \mathbb{R}$. The following result of Bhattacharya and Ghosh (1978) shows that $\Psi_{n,s}^{[1]}$ is a *valid* $(s-2)$-th order expansion for W_{1n}, i.e., the error of approximating the probability distribution of W_{1n} by the signed measure $\Psi_{n,s}^{[1]}$ is of the order $o(n^{-(s-2)/2})$ uniformly over classes of sets satisfying an analog of (6.46).

Theorem 6.5 *Suppose that $\{X_i\}_{i \in \mathbb{Z}}$ is a sequence of iid $\mathbb{R}^d$-valued random vectors with $E\|X_1\|^s < \infty$ and that H is s-times continuously differentiable in a neighborhood of $\mu = EX_1$, where $s \geq 3$ is an integer. If, in addition, X_1 satisfies the standard Cramer condition (6.31),then*

$$\sup_{B \in \mathcal{B}} \Big| P(W_{1n} \in B) - \Psi_{n,s}^{[1]}(B) \Big| = o(n^{-(s-2)/2})$$

for any collection $\mathcal{B}$ of Borel subsets of $\mathbb{R}$ satisfying (6.46) with $d = 1$ and $\Sigma = \tau^2$.

Proof: See Theorem 2(b) of Bhattacharya and Ghosh (1978). □

In the literature, the expansion $\Psi_{n,s}^{[1]}$, defined in terms of the "approximate cumulants" $\tilde{\chi}_{r,n,s}$ of (6.51) and (6.52), is often referred to as the *formal* Edgeworth expansion of W_{1n}. The seminal work of Bhattacharya and Ghosh (1978) established *validity* of this approach of deriving an Edgeworth expansion for W_{1n}, settling a conjecture of Wallace (1958). They developed a transformation technique that yielded an alternative valid expansion for W_{1n} and then showed that the formal expansion coincided with the alternative expansion up to terms of order $O(n^{-(s-2)/2})$. As a result, $\Psi_{n,s}^{[1]}$ gives a valid $(s-2)$-th order Edgeworth expansion for W_{1n}. For related work on this problem, see Chibishov (1972), Skovgaard (1981), Bhattacharya and Ghosh (1988), Bai and Rao (1991), and the references therein.

6.4.2 Expansions for Normalized and Studentized Statistics Under Independence

Note that Theorem 6.5 readily yields an $(s-2)$-th order Edgeworth expansion for the distribution of the normalized (or standardized) version of $\hat{\theta}_n$, defined by

$$W_{2n} \equiv \sqrt{n}(\hat{\theta}_n - \theta)/\tau \; . \tag{6.56}$$

Indeed, $P(W_{2n} \leq x) = P(W_{1n} \leq \tau x) = \Psi_{n,s}^{[1]}((-\infty, \tau x]) + o(n^{-(s-2)/2})$ uniformly in $x \in \mathbb{R}$. Hence, a valid $(s-2)$-th order Edgeworth expansion for the distribution function of W_{2n} is given by $\Psi_{n,s}^{[2]}$, with

$$\begin{aligned}\Psi_{n,s}^{[2]}((-\infty, x]) &= \Psi_{n,s}^{[1]}((-\infty, \tau x]) \\ &= \int_{-\infty}^{x} \Big(1 + \sum_{r=1}^{s-2} n^{-r/2} p_r^{[2]}(y)\Big)\phi(y)dy, \; x \in \mathbb{R} \; ,\end{aligned}$$

for polynomials $p_1^{[2]}, \ldots, p_{s-2}^{[2]}$, where by (6.55) and a change of variables, it easily follows that $p_r^{[2]}(x) = p_r^{[1]}(\tau x)$, $x \in \mathbb{R}$.

Next consider the case of studentized statistics. It turns out that in the independent case, we can also apply Theorem 6.5 with a "suitable H" to obtain an Edgeworth expansion for the studentized version of $\hat{\theta}_n$, given by

$$W_{3n} \equiv \sqrt{n}(\hat{\theta}_n - \theta)/\hat{\tau}_n \; , \tag{6.57}$$

where $\hat{\tau}_n^2 \equiv \sum_{|\alpha|=1} \sum_{|\beta|=1} D^\alpha H(\bar{X}_n) D^\beta H(\bar{X}_n)[n^{-1}\sum_{i=1}^n (X_i - \bar{X}_n)^\alpha (X_i - \bar{X}_n)^\beta]$ is an estimator of the asymptotic variance τ^2 of $\sqrt{n}(\hat{\theta}_n - \theta)$ (cf. (6.48)). To appreciate why, note that we may express W_{3n} as a smooth function of the sample mean of the $d + d(d+1)/2$-dimensional iid random vectors Y_i, $i = 1, \ldots, n$, where the first d components of Y_i are given by X_i and the last $d(d+1)/2$ components are given by the diagonal and the above-the-diagonal elements of the $d \times d$ matrix $X_i X_i'$. If the function H (defining $\hat{\theta}_n$) in (6.47) is s-times continuously differentiable in a neighborhood of $\mu = EX_1$, if $E\|Y_1\|^s < \infty$, and if Y_1 satisfies the standard Cramer's condition, then by Theorem 6.4, W_{3n} has an $(s-2)$-th order Edgeworth expansion of the form

$$\sup_{B \in \mathcal{B}} \left| P(W_{3n} \in B) - \int_B \Big[1 + \sum_{j=1}^{s-2} n^{-j/2} p_j^{[3]}(x)\Big] \phi(x)dx \right| = o(n^{-(s-2)/2}) \tag{6.58}$$

for any collection $\mathcal{B}$ of Borel subsets of $\mathbb{R}$ satisfying (6.30) with $d = 1$, where $p_1^{[3]}, \ldots, p_{s-2}^{[3]}$ are polynomials and where, $\phi(x) = (2\pi)^{-1/2}\exp(-x^2/2)$,

$x \in \mathbb{R}$ is the density of a standard normal random variable. Without additional parametric distributional assumptions on the X_i's, the polynomials $p_1^{[3]}, \ldots, p_{s-2}^{[3]}$ are typically different from the polynomials $p_1^{[2]}, \ldots, p_{s-2}^{[2]}$ that appear in the expansion for the normalized version W_{2n} of $\hat{\theta}_n$. For an example, consider the case when $\hat{\theta}_n = \bar{X}_n$, the sample mean of a set of n iid random variables (with $d = 1$). Then, $\tau^2 = \sigma^2 = \mathrm{Var}(X_1)$, and by (6.16) and (6.20), a first-order Edgeworth expansion (with $s = 3$) for W_{2n} is given by

$$P(W_{2n} \leq x) = \Phi(x) - \frac{1}{6\sqrt{n}} \frac{\mu_3}{\sigma^3}(x^2 - 1)\phi(x) + o(n^{-1/2}) \tag{6.59}$$

uniformly in $x \in \mathbb{R}$, where $\mu_3 = E(X_1 - \mu)^3$, $\sigma^2 = \mathrm{Var}(X_1)$, and $\Phi(x) = \int_{-\infty}^{x} \phi(y)dy$, $x \in \mathbb{R}$ is the distribution function of a standard normal random variable. The corresponding first-order Edgeworth expansion for the studentized version W_{3n} of (6.57) for $\hat{\theta}_n = \bar{X}_n$ is given by (cf. Hall (1992), page 71–72),

$$P(W_{3n} \leq x) = \Phi(x) + \frac{1}{6\sqrt{n}} \frac{\mu_3}{\sigma^3}(2x^2 + 1)\phi(x) + o(n^{-1/2}) \;, \tag{6.60}$$

uniformly in $x \in \mathbb{R}$. Of course, the regularity conditions required for the validity of the two expansions are different, with the studentized case requiring stronger moment and/or distributional smoothness conditions. The key observation here is that in the independent case, Edgeworth expansions for the studentized statistics can be obtained using the same techniques as those employed for the normalized statistics under the Smooth Function Model. However, the same is no longer true in the dependent case, as explained below. For various alternative approaches to deriving expansions for studentized estimators under independence, see Hall (1987), Götze (1987), Helmers (1991), Lahiri (1994), Hall and Wang (2003), and the references therein.

6.4.3 Expansions for Normalized Statistics Under Dependence

Next we turn our attention to the case of dependent random vectors. Let $\{X_i\}_{i \in \mathbb{Z}}$ be a sequence of stationary $\mathbb{R}^d$-valued random vectors with $EX_1 = \mu$ and let $\hat{\theta}_n$ be an estimator of a parameter of interest θ based on $X_1, \ldots, X_n$, where θ and $\hat{\theta}_n$ satisfy the Smooth Function Model formulation (6.47). If the function H is continuously differentiable at μ and $\bar{X}_n$ satisfies the Central Limit Theorem (cf. Theorem A.8, Appendix A), then

$$\sqrt{n}(\hat{\theta}_n - \theta) \to^d N(0, \tau_\infty^2) \;, \tag{6.61}$$

where $\tau_\infty^2 = \sum_{|\alpha|=1} \sum_{|\beta|=1} c_\alpha c_\beta \Sigma_\infty(\alpha, \beta)$, $c_\alpha = D^\alpha H(\mu)/\alpha!$ and for $|\alpha| = |\beta| = 1$, $\alpha, \beta \in \mathbb{Z}_+^d$, $\Sigma_\infty(\alpha, \beta) \equiv \Sigma(\alpha, \beta) = \lim_{n\to\infty} E[\sqrt{n}(\bar{X}_n - \mu)]^{\alpha+\beta} =$

$\sum_{j\in\mathbb{Z}} E(X_1-\mu)^\alpha(X_{1+j}-\mu)^\beta$. In the dependent case, a valid $(s-2)$-th order Edgeworth expansion ($s \geq 3$) can be derived for the normalized version

$$\tilde{W}_{2n} \equiv \sqrt{n}(\hat{\theta}_n - \theta)/\tau_\infty \tag{6.62}$$

of the estimator $\hat{\theta}_n$ by applying the transformation technique of Bhattacharya and Ghosh (1978) to the $(s-2)$-th order Edgeworth expansion for the centered and scaled mean $S_n \equiv \sqrt{n}(\bar{X}_n - \mu)$. Indeed, if the conditions of Theorem 6.3 hold and the function H is s-times continuously differentiable in a neighborhood of μ, then there exist polynomials $q_r^{[2]}$, $r = 1, \ldots, s-2$ such that

$$\sup_{x\in\mathbb{R}} \left| P(\tilde{W}_{2n} \leq x) - \Upsilon_{n,s}^{[2]}\big((-\infty, x]\big) \right| = o(n^{-(s-2)/2}) , \tag{6.63}$$

where $\Upsilon_{n,s}^{[2]}$ is the signed measure with the Lebesgue density

$$\xi_{n,s}^{[2]}(x) = \Phi(x) + \sum_{r=1}^{n-2} n^{-r/2} q_r^{[2]}(x)\phi(x), \ x \in \mathbb{R} .$$

As mentioned in Section 6.3, under the stationarity of the process $\{X_i\}_{i\in\mathbb{Z}}$, the νth cumulant $\chi_{\nu,n}$ of S_n for $\nu \in \mathbb{Z}_+^d$, $2 \leq |\nu| \leq s$, may be expressed in the form (cf. (6.42))

$$\begin{aligned} n^{(|\nu|-2)/2}\chi_{\nu,n} &= \tilde{\chi}_{\nu,1,\infty} + n^{-1/2}\tilde{\chi}_{\nu,2,\infty} + \cdots + n^{-(s-2)/2}\tilde{\chi}_{\nu,s-1,\infty} \\ &\quad + o\big(n^{-(s-2)/2}\big) \quad \text{as} \quad n \to \infty \end{aligned} \tag{6.64}$$

for some $\tilde{\chi}_{\nu,j,\infty} \in \mathbb{R}$. The coefficients of the polynomials $q_1^{[2]}, \ldots, q_{s-2}^{[2]}$ are smooth functions of the partial derivatives $D^\nu H(\mu)$, $|\nu| \leq s-1$, and of the constants $\tilde{\chi}_{\nu,j,\infty}$, $1 \leq j \leq s-1$, $2 \leq |\nu| \leq s$, appearing in (6.64).

Although under the stationarity assumption on the process $\{X_i\}_{i\in\mathbb{Z}}$, it is possible to describe the Edgeworth expansion of $\tilde{W}_{2n}$ in terms of the polynomials $q_r^{[2]}$ that do not depend on n, in practice one may group some of these terms together to describe the Edgeworth expansion in terms of the moments (or cumulants) of the centered and scaled sample mean S_n directly. For example, a first-order Edgeworth expansion for $P(\tilde{W}_{2n} \leq x)$ (with $s = 3$) is given by

$$\Upsilon_{n,3}^{[2]}\big((-\infty, x]\big) = \Phi(x) - n^{-1/2}\big[\mathcal{K}_{31} + \mathcal{K}_{32}(x^2 - 1)\big]\phi(x) , \tag{6.65}$$

$x \in \mathbb{R}$, where the constants $\mathcal{K}_{31}$ and $\mathcal{K}_{32}$ are given by $\mathcal{K}_{31} \equiv \mathcal{K}_{31n} = \sum_{|\alpha|=2} c_\alpha E S_n^\alpha/\tau_n$ and $\mathcal{K}_{32} \equiv \mathcal{K}_{32n} = [\sqrt{n}E(\sum_{|\alpha|=1} c_\alpha S_n^\alpha)^3 - 3\tau_n^3\mathcal{K}_{31} + 3E\{(\sum_{|\alpha|=1} c_\alpha S_n^\alpha)^2(\sum_{|\alpha|=2} c_\alpha S_n^\alpha)\}]/(6\tau_n^3)$. Here, $\tau_n^2 = \text{Var}(\sum_{|\alpha|=1} c_\alpha S_n^\alpha)$ and $c_\alpha = D^\alpha H(\mu)/\alpha!$, $\alpha \in \mathbb{Z}_+^d$.

The expansion $\Upsilon_{n,3}^{[2]}$ of (6.65) may be further simplified and rewritten in the form (6.63). We also point out that $\Upsilon_{n,3}^{[2]}$ also gives the first-order Edgeworth expansion of the alternative normalized version of $\hat{\theta}_n$,

$$\check{W}_{2n} = \sqrt{n}(\hat{\theta}_n - \theta)/\tau_n \ ,$$

where the limiting standard deviation τ_∞ is replaced by τ_n. This follows by noting that, under the condition of Theorem 6.3 with $s = 3$, $\tau_\infty^2 - \tau_n^2 = O(n^{-1})$, and hence, the effect of replacing τ_∞ by τ_n is only $O(n^{-1})$, which is negligible for a first-order Edgeworth expansion.

6.4.4 Expansions for Studentized Statistics Under Dependence

Next, we consider the studentized case. Under weak dependence, the asymptotic variance of $\sqrt{n}(\hat{\theta}_n - \theta)$ is given by (cf. (6.61))

$$\tau_\infty^2 = \sum_{j \in \mathbb{Z}} \mathrm{Cov}(Y_1, Y_{j+1}) \ ,$$

where we write $Y_j = \sum_{|\alpha|=1} c_\alpha (X_j - \mu)^\alpha$, $j \in \mathbb{Z}$. Since τ_∞^2 is an infinite sum of lag covariances, a studentizing factor must estimate an *unbounded* number of lag-covariances, as the sample size n increases. A class of estimators of τ_∞^2 (cf. Götze and Künsch (1996)) is given by

$$\hat{\tau}_n^2 = \sum_{k=0}^{(\ell-1)} w_{kn} \Big[h(\bar{X}_n)' \hat{\Gamma}_n(k) h(\bar{X}_n) \Big] \ , \tag{6.66}$$

where $\hat{\Gamma}_n(k) = n^{-1} \sum_{j=1}^{n-\ell} (X_j - \bar{X}_n)(X_{j+k} - \bar{X}_n)'$, h is the $d \times 1$ vector of first-order partial derivatives of H, and w_{kn}'s are lag weights, with $w_{0n} = 1$ and $w_{kn} = 2w(k/\ell)$, $1 \leq k \leq \ell - 1$ for some continuous function $w : [0, 1) \to [0, 1]$ with $w(0) = 1$. If $\ell \to \infty$ and $n/\ell \to \infty$ as $n \to \infty$, then $\hat{\tau}_n^2$ is consistent for τ_∞^2. We define the studentized version of $\hat{\theta}_n$ as

$$\tilde{W}_{3n} \equiv \sqrt{n}(\hat{\theta}_n - \theta)/\hat{\tau}_n \ , \tag{6.67}$$

which has a standard normal distribution, asymptotically. In contrast to the case of studentized statistics under independence, Edgeworth expansions for $\tilde{W}_{3n}$ cannot be directly obtained from the Edgeworth expansion theory described above. This is because $\tilde{W}_{3n}$ is a (smooth) function of an *unbounded* number of sample means, while the classical theory deals mainly with sample means of a fixed finite dimension. Recently, first-order Edgeworth expansions for studentized statistics of the form $\tilde{W}_{3n}$ have been independently derived by Götze and Künsch (1996) and Lahiri (1996a). While Götze and Künsch (1996) considered studentized statistics under the Smooth Function Model (6.47), Lahiri (1996a) considered studentized versions of M-estimators of the regression parameters in a multiple linear regression model. Here we follow Götze and Künsch (1996) to describe the Edgeworth expansion result for $\tilde{W}_{3n}$. Recall the notation $\bar{Y}_n = \frac{1}{\sqrt{n}}\sum_{j=1}^n Y_j = \sum_{|\alpha|=1} c_\alpha S_n^\alpha$, $S_n = \frac{1}{\sqrt{n}}\sum_{j=1}^n (X_j - \mu)$, and $\tau_n^2 = n^{-1}\text{Var}(\sum_{i=1}^n Y_i)$. Let $\tau_{1n}^2 = \sum_{k=0}^{\ell-1} w_{kn} EY_1Y_{1+k}$, $\pi_n = n^{-1}\sum_{i=1}^n \sum_{j=1}^n \sum_{k=0}^{\ell-1} w_{kn} E(Y_iY_jY_{j+k})$, and $\mu_{3,n} = n^2 E(\bar{Y}_n)^3$. Also, let Ξ_n denote the variance matrix of the $(d+1)\times 1$ dimensional vector $\tilde{W}_{4n} \equiv (\sqrt{n}\bar{Y}_n; S_n')'$ and let a_γ's be constants defined by the identity

$$\begin{aligned}\sum_{|\gamma|=2}^{*} a_\gamma \tilde{W}_{4n}^\gamma &= (2\tau_n)^{-1}\sum_{|\alpha|=2} D^\alpha H(\mu) S_n^\alpha \\ &\quad - \tau_n^{-3}\{\sqrt{n}\,\bar{Y}_n\} S_n' \Big[D^2 H(\mu)\Sigma_\infty h(\mu)\Big] ,\end{aligned}$$

where $D^2H(\mu)$ is the $d\times d$ matrix of second-order partial derivatives of H at μ, $\Sigma_\infty \equiv \Sigma = \lim_{n\to\infty}\text{Var}(S_n)$ (cf. Condition (C.2)), and $h(\mu)$ is the $d\times 1$ vector of first-order partial derivatives of H at μ. Note that in the left-hand side of the identity, the index $\gamma \in \mathbb{Z}_+^{d+1}$, while on the right-hand side, the index $\alpha \in \mathbb{Z}_+^d$. With this, we define the first-order Edgeworth expansion $\Upsilon_{n,3}^{[3]}$ of $\tilde{W}_{3n}$ in terms of its Fourier transform

$$\begin{aligned}\xi_{n,3}^{[3]\dagger}(t) &= \int \exp(\iota t'x) d\Upsilon_{n,3}^{[3]}(x) \\ &= 1 + \frac{1}{\sqrt{n}}\cdot\frac{1}{\tau_n^3}\left[\left(\frac{\mu_{3n}}{6} - \frac{\pi_n}{2}\right)(\iota t)^3 - \frac{(\iota t)\pi_n}{2}\right]\exp(-t^2/2) \\ &\quad + \frac{1}{\sqrt{n}}(\iota t)\sum_{\gamma}^{*} a_\gamma(-1)^{|\gamma|} D^\gamma \exp(-w'\Xi_n w/2)\Big|_{w=(t,0,\ldots,0)} .\end{aligned} \tag{6.68}$$

Then, we have the following result due to Götze and Künsch (1996) on Edgeworth expansion for the studentized statistic $\tilde{W}_{3n}$ under dependence.

Theorem 6.6 *Suppose that Condition (5.D_r) of Section 5.4 on the function H holds with $r = 3$, $\sum_{|\alpha|=1}|D^\alpha H(\mu)| \neq 0$, and that $E\|X_1\|^{p+\delta} < \infty$ for some $\delta > 0$ and $p \geq 8$, $p \in \mathbb{N}$. Furthermore, suppose that*

$$\log n \ll \ell \leq n^{1/3} \tag{6.69}$$

and that Conditions (C.2)–(C.6) of Section 6.3 hold. Then,

$$\sup_{x\in\mathbb{R}}\left|P(\tilde{W}_{3n} \leq x) - \Upsilon_{n,3}^{[3]}((-\infty, x])\right| = O\Big(\ell n^{-1+[2/p]} + \left|\tau_n^2 - \tau_{1n}^2\right|\Big) . \tag{6.70}$$

Proof: See relations (6) and (7) and Theorem 4.1 of Götze and Künsch (1996). □

Note that under the conditions of Theorem 6.6, the second term $|\tau_n^2 - \tau_{1n}^2|$ on the right side of (6.70) is $o(n^{-1/2})$ if the weight function $w(x) \equiv 1$ for all $x \in [0, 1)$. A drawback of this choice of the weight function is that it does not guarantee that the estimator $\hat{\tau}_n^2$ of the asymptotic variance τ_∞^2 is always nonnegative. However, under the regularity conditions of Theorem 6.6, the event $\{\hat{\tau}_n^2 \leq 0\}$ has a negligible probability and it does not affect the rate of approximation $O(\ell n^{-1+2/p})$ of the first-order Edgeworth expansion $\Upsilon_{n,3}^{[3]}((-\infty, x])$ to $P(\tilde{W}_{3n} \leq x)$. Another class of popular weights are given by functions $w(\cdot)$ that satisfy $w(x) = 1 + O(x^2)$ as $x \to 0+$. For such weights, $|\tau_n^2 - \tau_{1n}^2| = O(\ell^{-2})$ and thus, in such cases, ℓ must grow at a faster rate than $n^{1/4}$ to yield an error of $o(n^{-1/2})$ in (6.70).

6.5 Second-Order Properties of Block Bootstrap Methods

In this section, we establish second-order correctness of block bootstrap methods under the Smooth Function Model (6.47). Accordingly, let $\{X_j\}_{j \in \mathbb{Z}}$ be a sequence of $\mathbb{R}^d$-valued stationary random vectors and let θ and $\hat{\theta}_n$ be as given by $\theta = H(\mu)$, $\hat{\theta}_n = H(\bar{X}_n)$, where $\mu = EX_1$, $\bar{X}_n = n^{-1} \sum_{i=1}^n X_i$, and $H : \mathbb{R}^d \to \mathbb{R}$ is a smooth function. Also, let $\tilde{W}_{2n}$ be the normalized version of $\hat{\theta}_n$ and $\tilde{W}_{3n}$ be the studentized version of $\hat{\theta}_n$, given by (6.62) and (6.67), respectively. Then, $\tilde{W}_{2n}$ and $\tilde{W}_{3n}$ are asymptotically *pivotal* quantities for the parameter θ, in the sense that the limit distributions of $\tilde{W}_{2n}$ and $\tilde{W}_{3n}$ are free of parameters. Block bootstrap methods applied to these pivotal quantities are second-order correct. The bootstrap estimators of the distribution functions of $\tilde{W}_{2n}$ and $\tilde{W}_{3n}$ not only capture the limiting standard normal distribution function, but also capture the next smaller order terms (viz., terms of order $n^{-1/2}$) in the Edgeworth expansions of $\tilde{W}_{2n}$ and $\tilde{W}_{3n}$. As a result, for such pivotal quantities, the bootstrap distribution function estimators outperform the normal approximation and are *second-order correct.* As indicated in Section 6.1, this can be easily shown by comparing the Edgeworth expansions of $\tilde{W}_{kn}$'s and their bootstrap versions W_{kn}^*, $k = 2, 3$. First we consider the normalized statistic $\tilde{W}_{2n}$ and the bootstrap approximation generated by the MBB method. Let $\bar{X}_n^*$ denote the MBB sample mean based on a random sample of $b = \lfloor n/\ell \rfloor$ blocks from the collection of overlapping blocks $\{\mathcal{B}_i : 1 \leq i \leq N\}$ of length ℓ, where, recall that, $\mathcal{B}_i = (X_i, \ldots, X_{i+\ell-1})$, $1 \leq i \leq N$, and $N = n - \ell + 1$. Then, the MBB version of $\tilde{W}_{2n}$ is given by

$$\tilde{W}_{2n}^* = \sqrt{n_1}(\theta_n^* - \tilde{\theta}_n)/\tilde{\tau}_n , \tag{6.71}$$

where $n_1 = b\ell$, $\theta_n^* = H(\bar{X}_n^*)$, and, with $\hat{\mu}_n \equiv E_*(\bar{X}_n^*)$, $\tilde{\theta}_n = H(\hat{\mu}_n)$ and $\tilde{\tau}_n^2 = n_1 \cdot \mathrm{Var}_*(\sum_{|\alpha|=1} D^\alpha H(\hat{\mu}_n)(\bar{X}_n^*)^\alpha)$. Note that conditional on $X_1, \ldots, X_n$, $\bar{X}_n^*$ is the average of a collection of b iid random vectors. Hence, an expansion for $\tilde{W}_{2n}^*$ may be derived using the Edgeworth expansion theory of Sections 6.2 and 6.4 for independent random vectors. The exact form of the first-order Edgeworth expansion for $\tilde{W}_{2n}^*$ is given by

$$\hat{\Upsilon}_{n,3}^{[2]}\big((-\infty, x]\big) = \Phi(x) - \frac{1}{\sqrt{n}}\Big(\hat{\mathcal{K}}_{31} + (x^2-1)\hat{\mathcal{K}}_{32}\Big)\phi(x), \; x \in \mathbb{R}\,, \tag{6.72}$$

where, with $\hat{c}_\alpha = D^\alpha H(\hat{\mu}_n)/\alpha!$, $\alpha \in \mathbb{Z}_+^d$, and $S_n^* = \sqrt{n_1}(\bar{X}_n^* - \hat{\mu}_n)$, the coefficients $\hat{\mathcal{K}}_{31}$ and $\hat{\mathcal{K}}_{32}$ are defined as

$$\hat{\mathcal{K}}_{31} \equiv \hat{\mathcal{K}}_{31n}(\ell) = \sum_{|\alpha|=2} \hat{c}_\alpha E_*(S_n^*)^\alpha / \tilde{\tau}_n\ ,$$

and

$$\begin{aligned}\hat{\mathcal{K}}_{32} \equiv \hat{\mathcal{K}}_{32n}(\ell) \quad &= \quad \Big[\sqrt{n}E_*\Big(\sum_{|\alpha|=1} \hat{c}_\alpha (S_n^*)^\alpha\Big)^3 \\ &\quad + 3E_*\Big\{\Big(\sum_{|\alpha|=1} \hat{c}_\alpha (S_n^*)^\alpha\Big)^2\Big(\sum_{|\alpha|=2} \hat{c}_\alpha (S_n^*)^\alpha\Big)\Big\} \\ &\quad - 3\tilde{\tau}_n \hat{\mathcal{K}}_{31}\Big]\Big/(6\tilde{\tau}_n^3)\ .\end{aligned}$$

The following result establishes second-order correctness of the MBB for the normalized statistic $\tilde{W}_{2n}$.

Theorem 6.7 *Suppose that $\{X_i\}_{i\in\mathbb{Z}}$ is stationary, Conditions (C.2)–(C.6) hold and $E\|X_1\|^{35+\delta} < \infty$ for some $\delta > 0$. Furthermore, suppose that Condition (5.D$_r$) of Section 5.4 on the function H holds with $r = 4$ and that the block length ℓ satisfies*

$$\epsilon n^\epsilon \le \ell \le \epsilon^{-1} n^{1/3} \tag{6.73}$$

for all $n \ge \epsilon^{-1}$, for some $\epsilon \in (0,1)$. Then,
(a) as $n \to \infty$,

$$\sup_{x\in\mathbb{R}} \Big| P_*(W_{2n}^* \le x) - \hat{\Upsilon}_{n,3}^{[2]}\big((-\infty, x]\big)\Big| = O_p(n^{-1}\ell)\ ,$$

(b) as $n \to \infty$,

$$\sup_{x\in\mathbb{R}} \Big| P_*(W_{2n}^* \le x) - P(\tilde{W}_{2n} \le x)\Big| = O_p(n^{-1}\ell + n^{-1/2}\ell^{-1})\ . \tag{6.74}$$

Proof: Part (a) is an easy consequence of Lemma 5.6 of Lahiri (1996d), who also obtains a bound on the MSE of the MBB distribution function estimator $P_*(W_{2n}^* \leq \cdot)$. As for part (b), note that under the regularity conditions of Theorem 6.7,

$$\sup_{x \in \mathbb{R}} \left| P(\tilde{W}_{2n} \leq x) - \Upsilon_{n,3}^{[2]}((-\infty, x]) \right| = O(n^{-1}) .$$

Hence, by part (a) and (6.65),

$$\begin{aligned} & \sup_{x \in \mathbb{R}} \left| P_*(\tilde{W}_{2n}^* \leq x) - P(\tilde{W}_{2n} \leq x) \right| \\ &= \sup_{x \in \mathbb{R}} \left| \hat{\Upsilon}_{n,3}^{[2]}((-\infty, x]) - \Upsilon_{n,3}^{[2]}((-\infty, x]) \right| + O_p(n^{-1}\ell) \\ &= O_p(n^{-1/2}|\hat{\mathcal{K}}_{31} - \mathcal{K}_{31}| + n^{-1/2}|\hat{\mathcal{K}}_{32} - \mathcal{K}_{32}|) . \end{aligned} \tag{6.75}$$

To complete the proof of part (b), without loss of generality, we set $\mu = 0$. Then, it is easy to check that $\hat{\mathcal{K}}_{31} - \mathcal{K}_{31}$ is a smooth function of the centered bootstrap moments $\{(E_*(U_{11}^*)^\nu - E(U_{11})^\nu) : |\nu| = 1, 2\}$ and $(\hat{\mathcal{K}}_{32} - \hat{\mathcal{K}}_{32})$ is a smooth function of $\{\sqrt{\ell}(E_*(U_{11}^*)^\nu - E(U_{11}^\nu)) : |\nu| = 3\} \cup \{(E_*(U_{11}^*)^\nu - E(U_{11}^\nu)) : |\nu| = 1, 2\}$, where $U_{11} = (X_1 + \cdots + X_\ell)/\sqrt{\ell}$ and $U_{11}^* = (X_1^* + \cdots + X_\ell^*)/\sqrt{\ell}$. The rate of error in (6.75) is determined by the first set of terms $\{\sqrt{\ell}(E_*(U_{11}^*)^\nu - EU_{11}^\nu) : |\nu| = 3\}$, whose root-mean-squared-error is bounded by

$$\max_{|\nu|=3} \left\{ \ell E \left| E_*(U_{11}^*)^\nu - E(U_{11}^\nu) \right|^2 \right\}^{1/2} + \max_{|\nu|=3} \left| \sqrt{\ell} E(U_{11}^\nu) - \sqrt{n} E(S_n^\nu) \right| .$$

The first term is of the order $O(n^{-1/2}\ell)$, by Lemma 3.1. It is easy to check that the second term is of the order $O(\ell^{-1})$. This completes the proof of Theorem 6.7. □

Theorem 6.7 shows that the MBB approximation to the distribution of the normalized statistic $\tilde{W}_{2n}$ is more accurate than the normal approximation, which has an error of $O(n^{-1/2})$. Thus, like the IID bootstrap for independent data, the MBB also outperforms the normal approximation under dependence.

A proof of this fact, with the right side of (6.74) replaced by "$o(n^{-1/2})$ a.s.," was first given in Lahiri (1991, 1992a). The second-order analysis of Lahiri (1991, 1992a) also show that for the MBB, the correct centering for the bootstrapped estimator $\theta_n^* = H(\bar{X}_n^*)$ is $\tilde{\theta}_n = H(\hat{\mu}_n)$, *not* the more naive choice $\hat{\theta}_n = H(\bar{X}_n)$. Indeed, if θ_n^* is centered at $\hat{\theta}_n$ and we define the bootstrap version of $\tilde{W}_{2n}$ as

$$W_{2n}^{**} = \sqrt{n}(\theta_n^* - \hat{\theta}_n)/\tilde{\tau}_n , \tag{6.76}$$

then, the error of approximation, $\sup_x |P_*(W_{2n}^{**} \leq x) - P(\tilde{W}_{2n} \leq x)|$ goes to zero precisely at the rate $n^{-1/2}\ell^{1/2}$, in probability. As a result, centering

θ_n^* at $\hat{\theta}_n$ yields an approximation that is *worse* than the normal approximation. This problem does not occur with the IID bootstrap method for independent data as the conditional expected value of $\bar{X}_n^*$ is $\bar{X}_n$.

A second and more important difference of the MBB with the IID bootstrap is that the rate of MBB approximation depends on the block length and is typically worse than $O_p(n^{-1})$. Indeed, compared to the IID bootstrap of Efron (1979) for independent data, where the error of approximation is of the order $O_p(n^{-1})$ (cf. Section 2.2), the best possible rate of MBB approximation for distribution function estimation is only $O(n^{-3/4})$, which is attained by blocks of length ℓ of the order $n^{1/4}$.

Next, we consider the MBB approximation to the distribution of the studentized statistic $\tilde{W}_{3n}$. Here, we follow Götze and Künsch (1996) to define the bootstrap version of $\tilde{W}_{3n}$, although other alternative definitions of the bootstrap version of $\tilde{W}_{3n}$ are possible (cf. Lahiri (1996a)). Recall that $U_{1i}^* = (X_{(i-1)\ell+1}^* + \cdots + X_{i\ell}^*)/\sqrt{\ell}$ denotes the sum of the ith resampled MBB block scaled by $\ell^{-1/2}$, $i = 1, \ldots, b$ and that $U_{11}^*, \ldots, U_{1b}^*$ are conditionally iid with the common distribution

$$P_*(U_{11}^* = U_{1i}) = \frac{1}{N}, \; 1 \le i \le N \; ,$$

where $U_{1i} = (X_i + \cdots + X_{i+\ell-1})/\sqrt{\ell}$ and $b = \lfloor n/\ell \rfloor$. To define the bootstrap version of the studentizing factor for $\sqrt{n_1}(\theta_n^* - \tilde{\theta}_n)$, note that by Taylor's approximation, the linear part of $\sqrt{n_1}(\theta_n^* - \tilde{\theta}_n)$ is

$$\begin{aligned} L_n^* &\equiv \sum_{|\alpha|=1} \hat{c}_\alpha \Big[\sqrt{n_1} (\bar{X}_n^* - \hat{\mu}_n)^\alpha \Big] \\ &= b^{-1/2} \sum_{i=1}^{b} \Big\{ \sum_{|\alpha|=1} \hat{c}_\alpha \big(U_{1i}^* - \hat{\mu}_n \sqrt{\ell} \big)^\alpha \Big\} \\ &\equiv b^{-1/2} \sum_{i=1}^{b} Y_{1i}^*, \text{ say }, \end{aligned}$$

where $\hat{c}_\alpha = D^\alpha H(\hat{\mu}_n)/\alpha!$, $\alpha \in \mathbb{Z}_+^d$. Hence, $\mathrm{Var}_*(L_n^*) = \mathrm{Var}_*(Y_{11}^*)$. This suggests that an estimator of the conditional variance $\mathrm{Var}_*(Y_{11}^*)$ is given by the "sample variance" of the iid random variables $Y_{11}^*, \ldots, Y_{1b}^*$. Hence, with $\bar{Y}_{1b}^* = b^{-1} \sum_{i=1}^b Y_{1i}^*$, we define

$$\tau_n^{*2} = b^{-1} \sum_{i=1}^{b} (Y_{1i}^* - \bar{Y}_{1b}^*)^2 \; , \tag{6.77}$$

as an "estimator" of $\mathrm{Var}_*(Y_{11}^*)$ and define the bootstrap version of the studentized statistic $\tilde{W}_{3n}$ as

$$W_{3n}^* = \sqrt{n_1}(\theta_n^* - \tilde{\theta}_n)/\tau_n^* \; .$$

Götze and Künsch (1996) suggested setting the MBB block length ℓ to be equal to the smoothing parameter ℓ in the definition of the studentizing factor $\hat{\tau}_n^2$ (cf. (6.66)). However, as they pointed out, second-order correctness of the MBB approximation holds for other choices of the block length ℓ satisfying (6.69). See the last paragraph on page 1217 or Götze and Künsch (1996). For notational simplicity, we suppose that the block size parameter ℓ and the lag-window parameter ℓ in (6.66) are equal. With this, we now define the first-order Edgeworth expansion $\hat{\Upsilon}_{n,3}^{[3]}$ of W_{3n}^* in terms of its Fourier transformation (cf. (6.68))

$$\begin{aligned}\hat{\xi}_{n,3}^{[3]\dagger}(t) &\equiv \int \exp(\iota t'x)d\hat{\Upsilon}_{n,3}^{[3]}(x) \\ &= \left[1 + \frac{\hat{\mu}_{3,n}}{\sqrt{n}\tilde{\tau}_n^3}\left\{-\frac{1}{3}(\iota t)^3 - \frac{1}{2}(\iota t)\right\}\right]\exp(-t^2/2) \\ &\quad + \frac{1}{\sqrt{n}}(\iota t)\sum_{\gamma}^{*} \hat{a}_\gamma(-1)^{|\gamma|}D^\gamma \exp(-w'\hat{\Xi}_n w/2)\Big|_{w=(t,0,\ldots,0)},\end{aligned} \tag{6.78}$$

where $\hat{\mu}_{3,n} = \ell^{1/2}E_*(Y_{11}^*)^3$, $\tilde{\tau}_n^2 = E_*(Y_{11}^*)^2$, $\hat{\Xi}_n = \text{Var}_*((Y_{11}^*, U_{11}^{*'})')$, and $\hat{a}_\alpha$'s are defined in analogy to the a_α's of (6.68), with μ replaced by $\hat{\mu}_n$.

The following result establishes second-order correctness of the MBB in the studentized case.

Theorem 6.8 *Suppose that $\{X_i\}_{i\in\mathbb{Z}}$ is stationary, Conditions (C.2)–(C.6) hold, and $E\|X_1\|^{qp+\delta} < \infty$ for some $\delta > 0$, and for some integers $q \geq 3$, $p \geq 8$. Also, suppose that Condition (5.D$_r$) of Section 5.4 holds with $r = 3$ and that ℓ satisfies (6.73). Then,*

(a) as $n \to \infty$,

$$\sup_{x\in\mathbb{R}}\left|P_*(W_{3n}^* \leq x) - \hat{\Upsilon}_{n,3}^{[3]}((-\infty, x])\right| = O_p\left(n^{-1+2/p}\ell + n^{-1/2}\ell^{-1}\right),$$

(b) as $n \to \infty$,

$$\sup_{x\in\mathbb{R}}\left|P_*(W_{3n}^* \leq x) - P(W_{3n} \leq x)\right| = O_p\left(n^{-1+2/p}\ell + n^{-1/2}\ell^{-1} + |\tau_n^2 - \tau_{1n}^2|\right).$$

Proof: See Theorems 4.1 and 4.2 of Götze and Künsch (1996). □

As in the case of the normalized statistic $\tilde{W}_{2n}$, under additional moment conditions, the rate of approximation in part (a) of Theorem 6.8 can be shown to be $O_p(n^{-1}\ell + n^{-1/2}\ell^{-1})$ (cf. Lahiri (2003c)). In particular, the rate of MBB approximation in the studentized case also depends on the block length. For second-order correctness, not only is the choice of ℓ (which now represents the block length and, also, the smoothing parameter appearing in the definition of the studentizing factor $\hat{\tau}_n^2$) important, but also is the choice of the weight function $w(\cdot)$. Lahiri (1996a) considers the case where the weight function $w(\cdot) \equiv 1$ and employs a different definition of the bootstrap studentized statistic to establish second-order correctness of the MBB for M-estimators in a multiple linear regression model. Relative merits of the two approaches are not clear at this stage.

Second-order correctness of the NBB and the CBB, which are also based on independent resampling of blocks of a *nonrandom* length, can be established using arguments similar to those used in the proofs of Theorems 6.7 and 6.8. See Hall, Horowitz and Jing (1995) and Politis and Romano (1992a) for a proof in the normalized case for the NBB and the CBB, respectively. As for the SB, Lahiri (1999c) developed some iterated conditioning argument to deal with the random block lengths in the SB method and established second-order correctness of the SB method for studentized statistics. For second and higher order investigations into the properties of bootstrap methods for some popular classes of estimators in Econometrics (e.g., the "Generalized Method of Moments" estimators), see Hall and Horowitz (1996), Inoue and Shintani (2001), Andrews (2002), , and the references therein.

7

Empirical Choice of the Block Size

7.1 Introduction

As we have seen in the earlier chapters, performance of block bootstrap methods critically depends on the block size. In this chapter, we describe the theoretical optimal block lengths for the estimation of various level-2 parameters and discuss the problem of choosing the optimal block sizes empirically. For definiteness, we restrict attention to the MBB method. Analogs of the block size estimation methods presented here can be defined for other block bootstrap methods. In Section 7.2, we describe the forms of the MSE-optimal block lengths for estimating the variance and the distribution function. In Section 7.3, we present a data-based method for choosing the optimal block length based on the subsampling method. This is based on the work of Hall, Horowitz and Jing (1995). A second method based on the Jackknife-After-Bootstrap (JAB) method is presented in Section 7.4. Numerical results on finite sample performance of these optimal block length selection rules are also given in the respective sections.

7.2 Theoretical Optimal Block Lengths

Let $(X_1, \ldots, X_n) = \mathcal{X}_n$ denote a finite stretch of random variables, observed from a stationary weakly dependent process $\{X_i\}_{i \in \mathbb{Z}}$ in $\mathbb{R}^d$. Let $\hat{\theta}_n$ be an estimator of a level-1 parameter of interest $\theta \in \mathbb{R}$, based on $\mathcal{X}_n$. In this section, we obtain expansions for the MSEs of block bootstrap estimators for various characteristics of the distribution of $\hat{\theta}_n$. Let G_n denote the distribution of the centered estimator $(\hat{\theta}_n - \theta)$, i.e.,

$$G_n(x) = P(\hat{\theta}_n - \theta \leq x), \; x \in \mathbb{R} \; . \tag{7.1}$$

The level-2 parameters of interest here are given by

$$\varphi_{1n} = \text{Bias}(\hat{\theta}_n) = \int x dG_n(x) \tag{7.2}$$

$$\varphi_{2n} = \text{Var}(\hat{\theta}_n) = \int x^2 dG_n(x) - \Big(\int x dG_n(x)\Big)^2 \tag{7.3}$$

$$\varphi_{3n} \equiv \varphi_{3n}(x_0) = P\bigg(\frac{\sqrt{n}(\hat{\theta}_n - \theta)}{\tau_n} \leq x_0\bigg) = G_n\Big(\frac{x_0\tau_n}{\sqrt{n}}\Big) \tag{7.4}$$

$$\varphi_{4n} = \varphi_{4n}(y_0) \equiv P\bigg(\bigg|\frac{\sqrt{n}(\hat{\theta}_n - \theta)}{\tau_n}\bigg| \leq y_0\bigg) = G_n\Big(\frac{y_0\tau_n}{\sqrt{n}}\Big) - G_n\Big(\frac{-y_0\tau_n}{\sqrt{n}}\Big) \; , \tag{7.5}$$

where $x_0 \in \mathbb{R}$ and $y_0 \in (0, \infty)$ are given real numbers and where τ_n^2 is the asymptotic variance of $\sqrt{n}(\hat{\theta}_n - \theta)$. Here, φ_{1n} and φ_{2n} are, respectively, the bias and the variance of the estimator $\hat{\theta}_n$, φ_{3n} denotes the (one-sided) distribution function of $\sqrt{n}(\hat{\theta}_n - \theta)$ at a given point $x_0 \in \mathbb{R}$, and φ_{4n} denotes the two-sided distribution function of $\sqrt{n}(\hat{\theta}_n - \theta)$ at $y_0 \in (0, \infty)$. The latter is useful for constructing symmetric confidence intervals for θ (cf. Hall (1992)). Next, for $k = 1, 2, 3, 4$, let $\hat{\varphi}_{kn}(\ell)$ denote the MBB estimators of the level-2 parameter φ_{kn} based on blocks of length ℓ. We define the theoretical optimal block length ℓ_{kn}^0 as the minimizer of the MSE of $\hat{\varphi}_{kn}(\ell)$ over a set of values of the block size ℓ, depending on $k = 1, 2, 3, 4$. Specifically, we define

$$\ell_{kn}^0 = \text{argmin}\Big\{\text{MSE}\big(\hat{\varphi}_{kn}(\ell)\big) : \epsilon n^\epsilon < \ell < \epsilon^{-1} n^{1/2-\epsilon}\Big\}, \; k = 1, 2 \tag{7.6}$$

$$\ell_{kn}^0 = \text{argmin}\Big\{\text{MSE}\big(\hat{\varphi}_{kn}(\ell)\big) : \epsilon n^\epsilon \leq \ell \leq \epsilon^{-1} n^{1/3-\epsilon}\Big\}, \; k = 3, 4 \tag{7.7}$$

for some small $\epsilon > 0$. It will follow from the arguments and results below that the theoretical optimal block length ℓ_{kn}^0 is of the order $n^{1/3}$ for the bias and the variance functionals (with $k = 1, 2$), while the order of ℓ_{kn}^0 for the one- and the two-sided distribution functions, with $k = 3$ and $k = 4$, are of the orders $n^{1/4}$ and $n^{1/5}$, respectively. Thus, the ranges $[\epsilon n^\epsilon, \epsilon^{-1} n^{1/2-\epsilon}]$ and $[\epsilon n^\epsilon, \epsilon^{-1} n^{1/3-\epsilon}]$ of block lengths ℓ in (7.6) and (7.7), respectively, contain the optimal block lengths ℓ_{kn}^0 for all $k = 1, 2, 3, 4$. Indeed, it can be shown that under some additional regularity conditions, the theoretical optimal block lengths ℓ_{kn}^0 have the same order even when the ranges of ℓ values in (7.6) and (7.7) are replaced by the larger interval $[\epsilon n^\epsilon, \epsilon^{-1} n^{1-\epsilon}]$ for an arbitrarily small $\epsilon \in (0, 1)$. However, we will restrict attention to the range of ℓ values specified by (7.6) and (7.7) and will not pursue such generalizations here.

For deriving expansions for the MSEs of the block bootstrap estimators $\hat{\varphi}_{kn}(\ell)$'s, $k = 1, 2, 3, 4$, we shall suppose that the level-1 parameter θ and its estimator $\hat{\theta}_n$ satisfy the requirements of the Smooth Function Model (cf. Section 4.2). Thus, there exists a function $H : \mathbb{R}^d \to \mathbb{R}$ such that

$$\hat{\theta}_n = H(\bar{X}_n), \ \theta = H(\mu) \tag{7.8}$$

and the function H is "smooth" in a neighborhood of μ, where $\mu = EX_1$ and $\bar{X}_n = n^{-1}\sum_{i=1}^n X_i$. Recall that we write $c_\alpha = D^\alpha H(\mu)/\alpha!$, D^α for the differential operator $\frac{\partial^{\alpha_1+\cdots+\alpha_d}}{\partial x_1^{\alpha_1}\cdots\partial x_d^{\alpha_d}}$ and $\alpha! = \prod_{i=1}^d \alpha_i!$ for $\alpha = (\alpha_1, \ldots, \alpha_d)' \in \mathbb{Z}_+^d$.

7.2.1 Optimal Block Lengths for Bias and Variance Estimation

Expansions of the MSEs of the MBB estimators of the bias and the variance of the estimator $\hat{\theta}_n$ under the Smooth Function Model (7.8) was given in Chapter 5. Here, we recast the relevant results in a slightly different form by expressing relevant population quantities in the time domain. Let Z_∞ be a d-dimensional Gaussian random vector with mean zero and covariance matrix $\Sigma_\infty = \sum_{j=-\infty}^{\infty} E\{(X_1 - \mu)(X_{1+j} - \mu)'\}$.

Theorem 7.1 *Suppose that $\ell^{-1} + n^{-1/2}\ell = o(1)$ as $n \to \infty$.*

(a) Suppose that Conditions (5.D_r) and (5.M_r) of Section 5.4 hold with $r = 3$ and $r = 3 + a_0$, respectively, where a_0 is as specified by (5.D_r). Then

$$\begin{aligned} MSE\Big(n \cdot \hat{\varphi}_{1n}(\ell)\Big) &= \Big[(n^{-1}\ell)\frac{2}{3}\mathrm{Var}\Big(\sum_{|\alpha|=2} c_\alpha Z_\infty^\alpha\Big) + \ell^{-2}A_1^2\Big] \\ &\quad + o(n^{-1}\ell + \ell^{-2}) \,, \end{aligned} \tag{7.9}$$

where

$$A_1 = -\sum_{|\alpha|=1}\sum_{|\beta|=1} c_{\alpha+\beta}\Big[\sum_{j=-\infty}^{\infty} |j| E(X_1 - \mu)^\alpha (X_{1+j} - \mu)^\beta\Big] \,.$$

(b) Suppose that Conditions (5.D_r) and (5.M_r) of Section 5.4 hold with $r = 2$ and $r = 4 + 2a_0$, respectively, where a_0 is as specified by Condition (5.D_r). Then,

$$\begin{aligned} MSE\Big(n \cdot \hat{\varphi}_{2n}(\ell)\Big) &= \Big[(n^{-1}\ell)\frac{2}{3}\mathrm{Var}\Big(\Big(\sum_{|\alpha|=1} c_\alpha Z_\infty^\alpha\Big)^2\Big) + \ell^{-2}A_2^2\Big] \\ &\quad + o(n^{-1}\ell + \ell^{-2}) \,, \end{aligned} \tag{7.10}$$

where

$$A_2 = -\sum_{|\alpha|=1}\sum_{|\beta|=1} c_\alpha c_\beta \Big[\sum_{j=-\infty}^{\infty} |j| E(X_1-\mu)^\alpha (X_{1+j}-\mu)^\beta\Big] .$$

Proof: Follows from the proofs of Theorems 5.1 and 5.2 for the case '$j = 1$' (corresponding to the MBB estimators). □

Note that under the regularity conditions of Theorem 7.1, both the bias and the variance of the estimator $\hat{\theta}_n$ are of the order $O(n^{-1})$. Hence, we state the MSEs of the scaled bootstrap bias estimator $n \cdot \hat{\varphi}_{1n}(\ell)$ and of the scaled bootstrap variance estimator $n \cdot \hat{\varphi}_{2n}(\ell)$, in Theorem 7.1. Alternatively, we may think of the scaled bootstrap estimators $n \cdot \hat{\varphi}_{kn}(\ell)$ as estimators of the limiting level-2 parameters $\varphi_{k,\infty} \equiv \lim_{n\to\infty} n \cdot \varphi_{kn}$, $k = 1, 2$, given by

$$\varphi_{1,\infty} = \sum_{|\alpha|=1}\sum_{|\beta|=1} c_{\alpha+\beta} \Big[\sum_{j=-\infty}^{\infty} E(X_1-\mu)^\alpha (X_{1+j}-\mu)^\beta\Big]$$

and

$$\varphi_{2,\infty} = \sum_{|\alpha|=1}\sum_{|\beta|=1} c_\alpha c_\beta \Big[\sum_{j=-\infty}^{\infty} E(X_1-\mu)^\alpha (X_{1+j}-\mu)^\beta\Big] .$$

Theorem 7.1 immediately yields expressions for the leading terms of the theoretical optimal block lengths for bias and variance estimation. We note these down in the following corollary.

Corollary 7.1 *Suppose that the respective set of conditions of Theorem 7.1 hold for the bias functional ($k = 1$) and the variance functional ($k = 2$), and that the constants A_1 and A_2 are nonzero. Then, for $k = 1, 2$,*

$$\ell^0_{kn} = n^{1/3}(2A_k^2/v_k^2)^{1/3} + o(n^{1/3}) , \tag{7.11}$$

where $v_1^2 = \frac{2}{3}\mathrm{Var}(\sum_{|\alpha|=2} c_\alpha Z_\infty^\alpha)$ and $v_2^2 = \frac{2}{3}\mathrm{Var}([\sum_{|\alpha|=1} c_\alpha Z_\infty^\alpha]^2)$.

Künsch (1989) derived the leading term of the theoretical optional block length for the variance functional while Hall, Horowitz and Jing (1995) derived the leading terms for both the bias and the variance functionals φ_{1n} and φ_{2n}. The conclusions of Corollary 7.1 can be strengthened to some extent. A more detailed analysis of the remainder term in the proof of Theorem 7.1 can be used to show that under some additional smoothness and moment conditions, the $o(n^{1/3})$ term on the right side (7.11) is indeed $O(1)$ as $n \to \infty$, for both $k = 1$ and $k = 2$. Thus, the fluctuations of the true optimal block length from its leading term is bounded for both bias and variance functionals. In the next section, we consider theoretical optimal block lengths for the estimation of distribution functions.

7.2.2 Optimal Block Lengths for Distribution Function Estimation

First we consider the one-sided distribution function φ_{3n} of (7.4), given by

$$\varphi_{3n} = P(\sqrt{n}(\hat{\theta}_n - \theta_0)/\tau_n \leq x_0)$$

for a given value $x_0 \in \mathbb{R}$. Hall, Horowitz and Jing (1995) consider both the NBB and the MBB estimators of φ_{3n} and derive expansions for the MSEs in the case of the sample mean, i.e., in the case where $\hat{\theta}_n = \bar{X}_n$ and $\theta = EX_1$. An expansion for the MSE of the MBB estimator $\hat{\varphi}_{3n}(\ell)$ (say) of φ_{3n} is obtained by Lahiri (1996d) under the Smooth Function Model (7.8). Here we follow the exposition of Lahiri (1996d) and describe an expansion for MSE $(\hat{\varphi}_{3n}(\ell))$ under the framework of Götze and Hipp (1983), introduced in Chapter 6. Suppose that $\{X_i\}_{i\in\mathbb{Z}}$ is defined on a probability space $(\Omega, \mathcal{F}, P)$, $\{X_i\}_{i\in\mathbb{Z}}$ is stationary, and that $\{\mathcal{D}_i\}_{i\in\mathbb{Z}}$ is a given sequence of sub-σ-fields of $\mathcal{F}$. For $-\infty \leq a \leq b \leq \infty$, let $\mathcal{D}_a^b$ denote the smallest σ-field containing $\{\mathcal{D}_i : i \in [a, b] \cap \mathbb{Z}\}$. For easy reference, we now restate some of the conditions from Section 6.3, under the stationarity assumption on the process $\{X_i\}_{i\in\mathbb{Z}}$.

(C.1) There exists $\delta \in (0, 1)$ such that for all $n, m = 1, 2, \ldots$ with $m > \delta^{-1}$, there exists a $\mathcal{D}_{n-m}^{n+m}$-measurable random vector $X_{n,m}^{\ddagger}$ satisfying

$$E\|X_n - X_{n,m}^{\ddagger}\| \leq \delta^{-1} \exp(-\delta m) .$$

(C.2) There exists $\delta \in (0, 1)$ such that for all $i \in \mathbb{Z}$, $m \in \mathbb{N}$, $A \in \mathcal{D}_{-\infty}^i$, and $B \in \mathcal{D}_{i+m}^{\infty}$,

$$\big|P(A \cap B) - P(A)P(B)\big| \leq \delta^{-1} \exp(-\delta m) .$$

(C.3) There exists $\delta \in (0, 1)$ such that for all $m, n, k = 1, 2, \ldots$, and $A \in \mathcal{D}_{n-k}^{n+k}$

$$E\big|P(A \mid \mathcal{D}_j : j \neq n) - P(A \mid \mathcal{D}_j : 0 < |j - n| \leq m + k)\big| \leq \delta^{-1} \exp(-\delta m) .$$

(C.4) There exists $\delta \in (0, 1)$ such that for all $m, n = 1, 2, \ldots$ with $\delta^{-1} < m < n$, and for all $t \in \mathbb{R}^d$ with $\|t\| \geq \delta$,

$$E\Big|E\Big\{\exp(\iota t'[X_{n-m} + \cdots + X_{n+m}]) \mid \mathcal{D}_j : j \neq n\Big\}\Big| \leq \delta^{-1} \exp(-\delta m) .$$

(C.5) $E\|X_1\|^{35+\delta} < \infty$ for some $\delta \in (0, 1)$.

Conditions (C.1)–(C.4) are restatements of Conditions (6.C.3)–(6.C.6) from Chapter 6, respectively. For a discussion of these conditions, see Chapter 6. We do not state Condition (6.C.2) separately here, as it follows from the conditional Cramer Condition (C.4) and the stationarity of $\{X_i\}_{i\in\mathbb{Z}}$. The moment Condition (C.5) is rather stringent. Lahiri (1996b) used this condition to prove negligibility of the remainder terms in the second-order Edgeworth expansion of the bootstrap distribution function estimator $\hat{\varphi}_{3n}(\ell)$ in the L^2-norm.

The following result gives an expansion for the MSE of $\hat{\varphi}_{3n}(\ell) \equiv \hat{\varphi}_{3n}(x_0;\ell)$ for a given $x_0 \in \mathbb{R}$.

Theorem 7.2 *Assume that Conditions (C.1)–(C.5) hold and that the smoothness Condition (5.D$_r$) of Section 5.4 on the function H holds with $r = 4$. Also, suppose that for some $\epsilon \in (0,3)$,*

$$\epsilon n^{\epsilon} \le \ell \le \epsilon^{-1} n^{1/3} \quad \textit{for all} \quad n > \epsilon^{-1} . \tag{7.12}$$

Then, there exist constants $v_{31}, v_{32} \in (0,\infty)$ and $B_{31}, B_{32} \in \mathbb{R}$ such that for $|x_0| \ne 1$,

$$\begin{aligned} MSE(\hat{\varphi}_{3n}(x_0;\ell)) &= \Big[(x_0^2-1)\phi(x_0)\Big]^2 v_{31}^2 \cdot n^{-2}\ell^2 \\ &\quad + \Big[\phi(x_0)\{B_{31} + B_{32}(x_0^2-1)\}\Big]^2 n^{-1}\ell^{-2} \\ &\quad + o\Big(n^{-2}\ell^2 + n^{-1}\ell^{-2}\Big) , \end{aligned} \tag{7.13}$$

and for $|x_0| = 1$,

$$\begin{aligned} MSE(\hat{\varphi}_{3n}(x_0;\ell)) &= [\phi(x_0)]^2 v_{32}^2 \cdot n^{-2}\ell + [\phi(x_0)]^2 B_{31}^2 n^{-1}\ell^{-2} \\ &\quad + o\Big(n^{-2}\ell^2 + n^{-1}\ell^{-2}\Big) . \end{aligned} \tag{7.14}$$

Proof: See Lahiri (1996d). □

From the Edgeworth expansion results of Chapter 6 (cf. Theorem 6.7), it follows that

$$\begin{aligned} \hat{\varphi}_{3n}(x_0;\ell) &= \Phi(x_0) - n^{-1/2}\Big\{\hat{\mathcal{K}}_{31}(\ell) + (x_0^2-1)\hat{\mathcal{K}}_{32}(\ell)\Big\}\phi(x_0) \\ &\quad + O_p(n^{-1}) , \end{aligned}$$

where $\hat{\mathcal{K}}_{3i}(\ell) \equiv \hat{\mathcal{K}}_{3in}(\ell)$, $i = 1,2$ are smooth functions of certain bootstrap moments. For $|x_0| \ne 1$, the leading term of the variance of $\hat{\varphi}_{3n}(x_0;\ell)$ comes from the variance of the dominant term $n^{-1/2}(x_0^2-1)\hat{\mathcal{K}}_{32}(\ell)$, which is of the order $(n^{-1/2})^2 \cdot n^{-1}\ell^2$. In contrast, for $|x_0| = 1$, the term $n^{-1/2}(x_0^2-1)\hat{\mathcal{K}}_{31}(\ell)$ is zero and in this case, the leading term in the variance of $\hat{\varphi}_{3n}(x_0;\ell)$ is given by the variance of $n^{-1/2}\hat{\mathcal{K}}_{31}(\ell)$, which is of the order $(n^{-1/2})^2 \cdot n^{-1}\ell$.

On the other hand, the contribution to the bias of $\hat{\varphi}_{3n}(x_0;\ell)$ comes from both $\hat{\mathcal{K}}_{31}(\ell)$ and $\hat{\mathcal{K}}_{32}(\ell)$, each having a bias of the order ℓ^{-1}. This explains the sources of the various terms in the expansions for $\text{MSE}(\hat{\varphi}_{3n}(x_0;\ell))$ in (7.13) and (7.14). The exact forms of the population quantities v_{31}, v_{32}, B_{31}, and B_{32} are very complicated, and hence are not presented here. Interested readers are referred to Lahiri (1996d) for explicit expressions for these parameters. Interestingly, neither of the two empirical methods, that we describe in Sections 7.3 and 7.4 below for data-based selection of the optimal block sizes, requires explicit definitions of these parameters.

Theorem 7.2 readily yields the following asymptotic expressions for the optimal block lengths for estimating $\varphi_{3n}(x_0)$.

Corollary 7.2 *Assume that the conditions of Theorem 7.2 hold. Then, for* $|x_0| \neq 1$,

$$\ell_{3n}^0 \equiv \ell_{3n}^0(x_0) = n^{1/4}\Big[\Big\{B_{31} + (x_0^2-1)B_{32}\Big\}^2 \Big/ \Big\{(x_0^2-1)v_{31}\Big\}^2\Big]^{1/4} + o(n^{1/4}) \tag{7.15}$$

and for $|x_0| = 1$,

$$\ell_{3n}^0 \equiv \ell_{3n}^0(x_0) = n^{1/3}\Big[2B_{31}^2/v_{32}^2\Big]^{1/3} + o(n^{1/3})\ . \tag{7.16}$$

Thus, the optimal block length for estimating the distribution function of the normalized version of $\hat{\theta}_n$ is of the order $n^{1/4}$ at any given point $x_0 \in \mathbb{R}$, $|x_0| \neq 1$. For $|x_0| = 1$, the optimal order is $n^{1/3}$, the same as that for estimating the bias and variance parameters φ_{1n} and φ_{2n} (cf. (7.11)). Relations (7.15) and (7.16) give optional block lengths for *local* estimation of the distribution function of the pivotal quantity $\sqrt{n}(\hat{\theta}_n - \theta)/\tau_n$. The optimal block length for *global* estimation of the distribution function $\varphi_{3n}(\cdot) \equiv P(\sqrt{n}(\hat{\theta}_n - \theta)/\tau_n \leq \cdot)$ can be obtained by minimizing an expansion for the (weighted) mean integrated squared error (MISE) of $\hat{\varphi}_{3n}(\cdot)$. An integration of the expansions (7.13) and (7.14) yields

$$\begin{aligned}\text{MISE}\Big(\hat{\varphi}_{3n}(\cdot;\ell)\Big) &\equiv E\int\Big[\hat{\varphi}_{3n}(x;\ell) - \varphi_{3n}(x;\ell)\Big]^2 w_0(x)dx \\ &= v_{33}^2 n^{-2}\ell^2 + B_{33}^2 n^{-1}\ell^{-2} + o(n^{-2}\ell^2 + n^{-1}\ell^{-2})\ ,\end{aligned} \tag{7.17}$$

where $w_0(\cdot) : \mathbb{R} \to (0,\infty)$ is a nonnegative weight function with $\int w_0(x)dx \in (0,\infty)$ and where $v_{33}^2 = v_{31}^2 \int (x^2-1)^2\phi(x)^2 w_0(x)dx$ and $B_{33}^2 = \int \phi(x)^2[B_{31} + B_{32}(x^2-1)]^2 w_0(x)dx$. Hence, the global optimal block length, defined as

$$\ell_{3n,\text{global}}^0 \equiv \text{argmin}\Big\{\text{MISE}\Big(\hat{\varphi}_{3n}(\cdot;\ell)\Big) : \epsilon n^{\epsilon} \leq \ell \leq \epsilon^{-1}n^{1/3}\Big\} \tag{7.18}$$

for a given $\epsilon \in (0, \frac{1}{3})$, is given by

$$\ell^0_{3n,\text{global}} = n^{1/4}\Big[B^2_{33}/v^2_{33}\Big]^{1/4} + o(n^{1/4}) \,. \tag{7.19}$$

Next, consider the two-sided distribution function $\varphi_{4n}(x_0) \equiv P(|\sqrt{n}(\hat{\theta}_n - \theta)/\tau_n| \leq x_0)$, $x_0 \in (0, \infty)$. Hall, Horowitz and Jing(1995) give an expansion of the MSE of the MBB estimator $\hat{\varphi}_{4n}(x_0; \ell)$ for the case where $\hat{\theta}_n = \bar{X}_n$ and $\theta = EX_1$. In this case, they show that the optimal block length for estimating the level-2 parameter $\varphi_{4n}(x_0)$ is of the form

$$\ell^0_{4n} \equiv \ell^0_{4n}(x_0) = n^{1/5} C_0(x_0) + o(n^{+1/5}) \tag{7.20}$$

for some constant $C_0(x_0) \in (0, \infty)$. We refer the interested reader to Hall, Horowitz and Jing (1995) for further details in the two-sided case. Thus, one needs to use blocks of a smaller order (viz., $n^{1/5}$) for optimal estimation of probabilities assigned to symmetric intervals than those in the asymmetric case.

As pointed out in Chapter 1, the MSE and the optimal block length ℓ^0 are population-parameters that are determined by the sampling distributions of the bootstrap estimators of a level-2 parameter, and therefore, may be regarded as level-3 parameters. Thus, a general approach to the estimation of MSE$(\hat{\varphi}_n(\ell))$ and ℓ^0 is to apply two rounds of resampling methods iteratively. In Sections 7.3 and 7.4, we describe two such general methods, proposed by Hall, Horowitz and Jing (1995) and Lahiri, Furukawa and Lee (2003), respectively. The method proposed by Hall, Horowitz and Jing (1995) uses a combination of subsampling and bootstrapping the data. The other method, proposed by Lahiri, Furukawa and Lee (2003), is based on the Jackknife-After-Bootstrap method and it uses a combination of jackknifing and bootstrapping the data. In the same vein, one may use two rounds of block bootstrapping to estimate the level-3 parameters MSE$(\hat{\varphi}_n(\ell))$ and ℓ^0, although properties of this third alternative remain unexplored at this time. Estimation methods tailored to estimate the optimal block size for a specific functional are also known. For the case of the variance functional, Bühlmann and Künsch (1999b) propose some novel plug-in estimators of the optional block length for block bootstrap variance estimation and establish their convergence rates. For a more direct plug-in method in the variance functional case, see Politis and White (2003). They employ the "flat-top" kernel method of Politis and Romano (1995) to estimate the relevant population parameters in the leading term of the optimal block size given by Corollary 7.1 above.

7.3 A Method Based on Subsampling

In this section, we describe the Hall, Horowitz and Jing (1995) method for choosing the theoretical optimal block size. For concreteness, suppose that

$\hat{\varphi}_n(\ell)$ denotes the MBB estimator of the level-2 parameter φ_n, based on blocks of length ℓ, where n is the sample size. Furthermore, suppose that the MSE of $\hat{\varphi}_n(\ell)$ admits an expansion of the form

$$\text{MSE}\big(\hat{\varphi}_n(\ell)\big) = a_n\Big[C_1 n^{-1}\ell^r + C_2\ell^{-2}\Big](1+o(1)) \quad \text{as} \quad n \to \infty \tag{7.21}$$

for some constants $C_1, C_2 \in (0,\infty)$, $r \in \mathbb{N}$, and for some sequence $\{a_n\}_{n\geq 1}$ of positive real numbers, over a suitable set $\mathcal{J}_n \subset \mathbb{N}$ of block sizes. We shall assume that the set $\mathcal{J}_n$ contains the set $[n^{\frac{1}{r+2}-\epsilon}, n^{\frac{1}{r+2}+\epsilon}]$ for some small $\epsilon \in (0,1)$. Next, define the optimal block length ℓ_n^0 by

$$\ell_n^0 \equiv \text{argmin}\Big\{\text{MSE}\big(\hat{\varphi}_n(\ell)\big) : l \in \mathcal{J}_n\Big\} . \tag{7.22}$$

Note that by (7.21) and (7.22), the optimal block length ℓ_n^0 is of the order $n^{\frac{1}{r+2}}$. To define the Hall, Horowitz and Jing (1995) estimator of the theoretical optimal block length ℓ_n^0, we proceed as follows. Let $m \equiv m_n$ be a sequence of real numbers satisfying

$$m^{-1} + n^{-1}m = o(1) \quad \text{as} \quad n \to \infty . \tag{7.23}$$

Consider the subsamples $\mathcal{X}_{i,m} \equiv (X_i, \ldots, X_{i+m-1})$, $i = 1, \ldots, n-m+1$ of length m. Let φ_m denote the level-2 parameter φ_n at $n = m$. For each $i = 1, \ldots, n-m+1$, let $\hat{\varphi}_{m,i}(\ell)$ be the MBB estimator of φ_m obtained by resampling blocks of length ℓ from the m observations $\mathcal{X}_{i,m}$. Next define the subsampling estimator of $\text{MSE}(\hat{\varphi}_m(\ell))$, the mean squared error of the MBB estimator of the level-2 parameter φ_m based on a sample of size m, as

$$\widehat{\text{MSE}}_m(\ell) = (n-m+1)^{-1}\sum_{i=1}^{n-m+1}\Big[\hat{\varphi}_{m,i}(\ell) - \hat{\varphi}_n(\ell_n^*)\Big]^2 , \tag{7.24}$$

where ℓ_n^* is a plausible pilot block size. Let

$$\hat{\ell}_m^0 = \text{argmin}\Big\{\widehat{\text{MSE}}_m(\ell) : \ell \in \mathcal{J}_m\Big\} , \tag{7.25}$$

where we employ the set $\mathcal{J}_m$ (not $\mathcal{J}_n$) to define $\hat{\ell}_m^0$. Then, $\hat{\ell}_m^0$ is an estimator of the theoretical optimal block length when the sample size is m. We need to rescale this initial estimator to get an estimator of ℓ_n^0 of (7.22). Since the optimal block length ℓ_n^0 in (7.22) is of the order $n^{\frac{1}{r+2}}$, the right scaling factor here is $[n/m]^{\frac{1}{r+2}}$. The Hall, Horowitz and Jing (1995) estimator of ℓ_n^0 is given by

$$\hat{\ell}_n^0 = (\hat{\ell}_m^0) \cdot [n/m]^{\frac{1}{r+2}} . \tag{7.26}$$

Note that the Hall, Horowitz and Jing (1995) estimation method is applicable quite generally, requiring only that the MSE of the bootstrap estimator has (an expansion of) the form (7.21) for some $r \geq 1$ and that the subsampling estimator $\widehat{\mathrm{MSE}}_m(\ell)$ of $\mathrm{MSE}(\hat{\varphi}_m(\ell))$ converges in some suitable sense, say, in probability. In particular, the method is applicable even without an explicit expression for the constants C_1 and C_2 in (7.21). Similarly, the method can be applied when a block bootstrap method other than the MBB is employed. A set of sufficient conditions for the consistency of the subsampling estimator $\widehat{\mathrm{MSE}}_m(\ell)$ are that the series $\{X_i\}_{i\in\mathbb{Z}}$ is stationary and has an absolutely summable strong mixing coefficient.

From the description of the method, it is clear that accuracy of the Hall, Horowitz and Jing (1995) method depends on the choices of the subsampling parameter m and the pilot block size ℓ_n^*. The optimal order of m is unknown at this stage. However, for the other smoothing parameter, viz., the pilot block size ℓ_n^*, Hall, Horowitz and Jing (1995) suggest a way to reduce the effect of ℓ_n^* on the optimal block length estimator ℓ_n^0 of (7.26). To reduce the effect of ℓ_n^*, they suggest iterating the main steps of the algorithm, by replacing the pilot block size ℓ_n^* with the estimated value $\hat{\ell}_n^0$ for the second iteration, and repeating this process until convergence. However, convergence of this iterative scheme is not guaranteed (see the numerical example below).

We now describe the results of a small simulation study on finite sample properties of the Hall, Horowitz and Jing (1995) method. We consider the time series model

$$X_i = (\epsilon_i + \epsilon_{i-1})/\sqrt{2}, \ i \in \mathbb{Z} \tag{7.27}$$

where $\{\epsilon_i\}_{i\in\mathbb{Z}}$ is a sequence of iid random variables with common distribution $(\chi^2(1) - 1)$, the centered Chi-squared distribution with one degree of freedom. Thus, $E\epsilon_1 = 0$ and $E\epsilon_1^2 = 2$. We took the level-1 parameter θ as EX_1, and the estimator $\hat{\theta}_n$ as $\hat{\theta}_n = \bar{X}_n$, the sample mean with sample size $n = 125$. The level-2 parameters of interest are given by (cf. (7.3) and (7.4))

$$\varphi_{2n} = n.\mathrm{Var}(\bar{X}_n) \tag{7.28}$$

and

$$\begin{aligned} \varphi_{3n} &= P\Big(\frac{\sqrt{n}(\hat{\theta}_n - \theta)}{\tau_n} \leq 0\Big) \\ &= P(\hat{\theta}_n \leq \theta) \ . \end{aligned} \tag{7.29}$$

True values of φ_{2n} and φ_{3n} were found by 20,000 simulation runs. These are given by $\varphi_{2n} = 3.984$ and $\varphi_{3n} = .5226$.

To find the theoretical optimal block lengths for φ_{2n} and φ_{3n} , we applied the MBB method to generate block bootstrap estimators of the level-2 parameters φ_{2n} and φ_{3n} with several values of the block length ℓ. Table 7.1 below gives the expected value (Mean), the bias, the standard deviation (SD) and the MSE's of the MBB estimators based on 1000 simulation runs. From the table, it is evident that the optimal block lengths for estimating φ_{2n} and φ_{3n} are respectively given by $\ell_{2n}^0 = 3$ and $\ell_{3n}^0 = 2$. Next the subsampling-based method of Hall, Horowitz and Jing (1995) was applied to select an optional block size empirically. We chose the subsample size $m = 30$, and the pilot block size parameter $\ell_{kn}^* = 5$ for both φ_{2n} and φ_{3n}. Thus, for $k = 1, 2$, the MSE estimator $\widehat{\text{MSE}}_m(\ell)$ of (7.24) for the level-2 parameter φ_{kn} is now evaluated by resampling overlapping blocks of size ℓ from each of the 96 $(= 125 - 30 + 1)$ overlapping subsamples of size $m = 30$ and then computing the MBB estimators $\hat{\varphi}_{km,i}(\ell)$ (say) of φ_{kn} for the ith subsample, for $i = 1, \ldots, 96$. The centering value $\hat{\varphi}_{kn}(\ell_{kn}^*)$ in $\widehat{\text{MSE}}_m(\ell)$ is computed using the full sample of size $n = 125$, with $\ell_{kn}^* = 5$ for both k. All bootstrap estimates (including those related to Table 7.2 below) were evaluated using 800 Monte-Carlo replicates.

TABLE 7.1. Determination of the true optimal block sizes for MBB estimation of the level-2 parameters φ_{2n} and φ_{3n} of (7.28) and (7.29) for model (7.27). The results are based on 1000 simulation runs. An asterisk(*) denotes the minimun MSE value for a functional.

		(a) Variance Estimation		
L	Mean	Bias	SD	MSE
1	1.947	−2.037	0.705	4.645
2	2.902	−1.082	1.089	2.358
3	3.204	−0.780	1.244	2.157*
4	3.320	−0.664	1.334	2.221
5	3.394	−0.590	1.412	2.341
6	3.437	−0.547	1.482	2.497
7	3.452	−0.532	1.542	2.660
8	3.460	−0.524	1.594	2.814
9	3.460	−0.524	1.648	2.990
10	3.469	−0.515	1.713	3.198

	(b) Distribution Function Estimation			
L	E.phi	Bias	SD	MSE
1	0.5099	−0.0126	0.0136	0.000345
2	0.5132	−0.0094	0.0132	0.000262*
3	0.5127	−0.0099	0.0142	0.000299
4	0.5136	−0.0089	0.0139	0.000272
5	0.5123	−0.0103	0.0144	0.000313
6	0.5125	−0.0100	0.0149	0.000322
7	0.5125	−0.0100	0.0150	0.000324
8	0.5121	−0.0105	0.0154	0.000347
9	0.5123	−0.0103	0.0157	0.000352
10	0.5103	−0.0122	0.0164	0.000419

Table 7.2 gives the frequency distribution of the optimal block size estimator $\hat{\ell}_{kn}^0$ for φ_{kn}, computed by formula (7.26) using 500 simulation runs. As in Hall, Horowitz and Jing (1995), in this simulation study also, the optimal block size estimators converged after a couple of iterations in a majority of the cases. However, in some instances, there was a circular behavior of the estimated optimal block size in successive iterations (e.g., the initial value 5 led to 8 which led to 3 and then, 3 led back to 5). The frequency of such cases is given under the value -1. This problem appeared to be more prevalent for distribution function estimation (i.e., for φ_{3n} of (7.29)) than for variance estimation (i.e., for φ_{2n} of (7.28)). In such a situation, one may pick a value of $\hat{\ell}_{kn}^0$ (from the set of all optimal block lengths in different iterations) that corresponds to the minimum estimated $\widehat{\text{MSE}}_m(\ell)$.

Parts (a) and (b) of Table 7.2 show that for both level-2 parameters φ_{2n} and φ_{3n}, the estimated optimal block sizes have a pronounced mode at the true optional block sizes, i.e., at $\ell_{2n}^0 = 3$ for φ_{2n} and at $\ell_{3n}^0 = 2$ for φ_{3n}. Furthermore, the distribution of the estimated optimal block size for variance estimation has a longer right tail compared to that for the distribution function estimation. However, the performance of this method improves as the sample size n increases. See Hall, Horowitz and Jing (1995) for further numerical examples and discussions.

TABLE 7.2. Frequency distribution of the optimal block sizes selected by the Hall, Horowitz and Jing (1995) method for model (7.27) with $n = 125$, $m = 30$, and initial block size $\ell_{kn}^* = 5$, $k = 1, 2$. Results are based on 500 simulation runs. The value -1 of $\hat{\ell}_{kn}^0$, $k = 1, 2$, corresponds to the cases where the iterations of the method failed to converge.

(a) Variance Estimation

$\hat{\ell}_{2n}^0$	-1	2	3	5	7	9	10	12	14	15	17	19	21
Freq.	35	137	200	63	18	12	13	8	6	2	3	1	2

(b) Distribution Function Estimation

$\hat{\ell}_{3n}^0$	-1	2	3	4	6	7	9	10	12	13
Freq.	137	276	50	5	7	13	8	1	2	1

7.4 A Nonparametric Plug-in Method

In this section, we describe a plug-in method for selecting the optimal block length based on a recent work of Lahiri, Furukawa and Lee (2003). The plug-in method estimates the leading term in the first-order expansion of the optimal block length using a resampling method, and does not require an explicit expression for the level-3 population parameters.

In Section 7.4.1, we describe the motivation and basic construction of the plug-in estimator and in Section 7.4.2, we describe estimation of the level-3 parameter associated with the bias part of the block bootstrap estimators. Estimation of the level-3 parameter associated with the variance part employ the Jackknife-After-Bootstrap (JAB) method of Efron (1992) and Lahiri (2002a). In Section 7.4.3, we describe the JAB method for dependent data. The nonparametric plug-in estimators of the optimal block lengths are presented in Section 7.4.4. Some finite sample results are given in Section 7.4.5. In all of Section 7.4, we restrict attention to the optimal block lengths for the MBB method.

7.4.1 Motivation

Let φ_n be a level-2 parameter of interest and let $\hat{\varphi}_n(\ell)$ be a block bootstrap estimator of φ_n based on blocks of length ℓ. From the discussion of Section 7.2, it follows that under suitable regularity conditions, the variance of $\hat{\varphi}_n(\ell)$ and the bias of $\hat{\varphi}_n(\ell)$ admit expansions of the form

$$n^{2a} \cdot \text{Var}\big(\hat{\varphi}_n(\ell)\big) = vn^{-1}\ell^r + o(n^{-1}\ell^r) \tag{7.30}$$

and

$$n^a \cdot \text{Bias}\big(\hat{\varphi}_n(\ell)\big) = B\ell^{-1} + o(\ell^{-1}) \tag{7.31}$$

for some population parameters $B \in \mathbb{R}$, $v \in (0, \infty)$ and for some known constants $a \in (0, \infty)$, $r \in \mathbb{N}$. For example, for the bias and variance functionals $\varphi_n = \varphi_{1n}, \varphi_{2n}$, $r = 1$, and $a = 1$, while for the distribution function (at a given point x_0) $\varphi_n = \varphi_{3n}(x_0)$ with $|x_0| \neq 1$, $r = 2$ and $a = 1/2$. In this case, the MSE-optimal block size $\ell_n^0 \equiv \ell_n^0(\varphi_n)$ is given by

$$\ell_n^0 = \left(\frac{2B^2}{rv}\right)^{\frac{1}{r+2}} n^{\frac{1}{r+2}} \big(1 + o(1)\big) . \tag{7.32}$$

Like any other plug-in method, the nonparametric plug-in method focuses on the leading term $\left(\frac{2B^2}{rv}\right)^{\frac{1}{r+2}} n^{\frac{1}{r+2}}$ but estimates the level-3 parameters B and v nonparametrically, as follows. Note that from (7.30) and (7.31), we have

$$\lim_{n\to\infty} (n^{-1}\ell^r)^{-1} n^{2a} \,\text{Var}\big(\hat{\varphi}_n(\ell)\big) = v \tag{7.33}$$

and

$$\lim_{n\to\infty} \ell \cdot n^a \,\text{Bias}\big(\hat{\varphi}_n(\ell)\big) = B . \tag{7.34}$$

This suggests that consistent estimators of v and B may be derived if we can estimate $\text{Var}(\hat{\varphi}_n(\ell))$ and $\text{Bias}(\hat{\varphi}_n(\ell))$ consistently. Let $\widehat{\text{VAR}}_n$ and $\widehat{\text{BIAS}}_n$ be nonparametric estimators of $\text{Var}(\hat{\varphi}_n(\ell))$ and $\text{Bias}(\hat{\varphi}_n(\ell))$, respectively, that are consistent in the following sense:

$$\frac{\widehat{\text{VAR}}_n}{\text{Var}\big(\hat{\varphi}_n(\ell_1)\big)} \longrightarrow_p 1 \quad \text{as} \quad n \to \infty \tag{7.35}$$

and

$$\frac{\widehat{\text{BIAS}}_n}{\text{Bias}(\hat{\varphi}_n(\ell_1))} \longrightarrow_p 1 \quad \text{as} \quad n \to \infty \tag{7.36}$$

along some suitable sequence $\{\ell_1\} \equiv \{\ell_{1n}\}_{n\geq 1}$.

Then, using (7.33) and (7.34), we define estimators of the parameters v and B as

$$\hat{v}_n \equiv \hat{v}_n(\ell_1) = n^{2a}\, \widehat{\text{VAR}}_n \cdot (n^{-1}\ell_1^r)^{-1}\ , \tag{7.37}$$

and

$$\hat{B}_n = \hat{B}_n(\ell_1) = n^a\, \widehat{\text{BIAS}}_n \cdot \ell_1\ . \tag{7.38}$$

The nonparametric plug-in estimator $\hat{\ell}_n^0$ of the optimal block length ℓ_n^0 is then given by replacing the level-3 parameters v and B in the leading term in (7.32) by the above estimators, i.e., by

$$\hat{\ell}_n^0 = \left[\frac{2\hat{B}_n^2}{r\hat{v}_n}\right]^{\frac{1}{r+2}} n^{\frac{1}{r+2}}\ . \tag{7.39}$$

It is clear that the performance of the estimator $\hat{\ell}_n^0$ depends on the sequence $\{\ell_{1n}\}_{n\geq 1}$, on the level-2 parameter φ_n, and on the basic estimators $\widehat{\text{VAR}}_n$ and $\widehat{\text{BIAS}}_n$ employed in the construction of $\hat{v}_n$ and $\hat{B}_n$ in (7.37) and (7.38), respectively. In the next section we describe the plug-in method of Lahiri, Furukawa and Lee (2003) who used the JAB method for estimating $\text{Var}(\hat{\varphi}_n(\ell))$ and constructed an estimator of $\text{Bias}(\hat{\varphi}_n(\ell))$ by combining two block bootstrap estimators suitably. The use of these basic estimators were prompted by considerations regarding computational efficacy and accuracy of the proposed plug-in method. As explained below, the JAB variance estimator has some computational advantage over other common resampling methods in that the JAB variance estimator can be computed by *reusing* the block bootstrap resamples used in the Monte-Carlo evaluation of $\hat{\varphi}_n(\ell_1)$, and thus, do not involve iterated levels of resampling. Similarly, the bias estimator proposed in Lahiri, Furukawa and Lee (2003) also involves a single level of resampling. In the section below, we describe further details of the construction.

7.4.2 The Bias Estimator

For constructing the bias estimator, we begin with relation (7.31), which gives an asymptotic representation for the bias part of the bootstrap estimator $\hat{\varphi}_n(\ell)$ and may be rewritten as

$$E\hat{\varphi}_n(\ell) = \varphi_n + \frac{B}{n^a \ell} + o(n^{-a}\ell^{-1}) \quad \text{as} \quad n \to \infty\ . \tag{7.40}$$

If (7.40) holds for the sequences $\{\ell_1\} \equiv \{\ell_{1n}\}_{n\geq 1}$ and $\{2\ell_1\} \equiv \{2\ell_{1n}\}_{n\geq 1}$, then we may combine the corresponding expansions to conclude that

$$\begin{aligned} & E\Big[\hat{\varphi}_n(\ell_1) - \hat{\varphi}_n(2\ell_1)\Big] \\ = & \Big[\Big\{\varphi_n + \frac{B}{n^a \ell_1} + o(n^{-a}\ell_1^{-1})\Big\} - \Big\{\varphi_n + \frac{B}{2n^a \ell_1} + o(n^{-a}\ell_1^{-1})\Big\}\Big] \\ = & \frac{B}{2n^a\ell_1} + o(n^{-a}\ell_1^{-1}) \quad \text{as} \quad n \to \infty \,. \end{aligned} \tag{7.41}$$

This suggests that a consistent estimator of $\text{Bias}(\hat{\varphi}_n(\ell_1))$ satisfying (7.36) may be constructed as

$$\widehat{\text{BIAS}}_n \equiv \widehat{\text{BIAS}}_n(\ell_1) = 2\big(\hat{\varphi}_n(\ell_1) - \hat{\varphi}_n(2\ell_1)\big) \,. \tag{7.42}$$

Indeed, if the optimal order of the block length for estimating φ_n is $n^{\frac{1}{r+2}}$ (cf. (7.32)), then by Cauchy-Schwarz inequality, it follows that for any sequence $\{\ell_1\} = \{\ell_{1n}\}_{n\geq 1}$ satisfying the requirement

$$1 \ll \ell_1 \ll n^{\frac{1}{r+2}} \quad \text{as} \quad n \to \infty \,, \tag{7.43}$$

$\widehat{\text{BIAS}}_n$ is consistent. A specific choice of $\{\ell_{1n}\}_{n\geq 1}$ will be suggested in Section 7.4.4 for the plug-in estimator $\hat{\ell}_n^0$ of (7.39). Note that, as pointed out earlier, the estimator $\widehat{\text{BIAS}}_n$ is based on only two block bootstrap estimator of φ_n and may be computed using only one level of resampling.

In the next section, we describe the JAB method for dependent data. Readers familiar with the method may skip this section and proceed to Section 7.4.4.

7.4.3 The JAB Variance Estimator

The JAB method was proposed by Efron (1992) for assessing accuracy of bootstrap estimators based on the IID bootstrap method for independent data. A modified version of the method for block bootstrap estimators in the case of dependent data was proposed by Lahiri (2002a). The JAB method for dependent data applies a version of the block jackknife method to a block bootstrap estimator. For the sake of completeness, first we briefly describe the block jackknife method.

Let $\mathcal{X}_n = \{X_1, \ldots, X_n\}$ be the observations and let $\hat{\gamma}_n \equiv t_n(\mathcal{X}_n)$ be an estimator of a level-1 parameter of interest γ. The block jackknife method systematically deletes blocks of consecutive observations to define the jackknife copies (called the *block jackknife point values*) of $\hat{\gamma}_n$ and combines these to produce estimators of the bias and the variance of $\hat{\gamma}_n$. Like the block bootstrap methods, different versions of the block jackknife method, such as, overlapping, nonoverlapping, and weighted block jackknife methods have been proposed in the literature (cf. Künsch (1989), Liu and Singh

(1992)). Here we describe the overlapping version of the block jackknife or the moving blocks jackknife (MBJ) of Künsch (1989) and Liu and Singh (1992). (Like the term "MBB," the term MBJ was also introduced by Liu and Singh (1992)). Let $m \equiv m_n$ be a sequence of integers such that m goes to infinity but at a rate slower than n, i.e.,

$$m^{-1} + n^{-1}m = o(1) \quad \text{as} \quad n \to \infty \,. \tag{7.44}$$

Here m denotes the number of observations (or the size of the block) to be deleted for defining the MBJ point values. For $i = 1, \ldots, n-m+1$, let $\mathcal{X}_{n,i} = \mathcal{X}_n \backslash \{X_i, \ldots, X_{i+m-1}\}$ denote the set of observations after the block $\{X_i, \ldots, X_{i+m-1}\}$ of size m has been deleted from $\mathcal{X}_n$. Then, the ith *MBJ point value* $\hat{\gamma}_n^{(i)}$ is defined as

$$\hat{\gamma}_n^{(i)} = t_{n-m}(\mathcal{X}_{n,i}), \; i = 1, \ldots, n-m+1 \,. \tag{7.45}$$

The MBJ estimator of the variance of $\hat{\gamma}_n$ is given by

$$\widehat{\text{VAR}}_{\text{MBJ}}(\hat{\gamma}_n) = \frac{m}{(n-m)} \cdot \frac{1}{n-m+1} \sum_{i=1}^{n-m+1} \left(\tilde{\gamma}_n^{(i)} - \hat{\gamma}_n\right)^2 , \tag{7.46}$$

where $\tilde{\gamma}_n^{(i)} \equiv m^{-1}\left(n\hat{\gamma}_n - (n-m)\hat{\gamma}_n^{(i)}\right)$ is the ith *MBJ pseudo-value* corresponding to $\hat{\gamma}_n$. For consistency and finite sample properties of the MBJ and its other variants, we refer the reader to Künsch (1989), Liu and Singh (1992), Shao and Tu (1995), Davison and Hinkley (1997), and the references therein. Note that, if we set $m \equiv 1$, i.e., if we delete a single observation at a time, then the MBJ variance estimator in (7.46) reduces to the classical delete-1 jackknife variance estimator for independent data

$$\widehat{\text{VAR}}_J(\hat{\gamma}_n) = \frac{1}{n(n-1)} \sum_{i=1}^{n} \left(\tilde{\gamma}_n^{(i)} - \hat{\gamma}_n\right)^2 . \tag{7.47}$$

For properties of the jackknife method for independent data, see Miller (1974), Efron (1982), Wu (1990), Liu and Singh (1992), Efron and Tibshirani (1993), Shao and Tu (1995), Davison and Hinkley (1997), and the references therein.

Next we describe the JAB method for dependent data. Let $\hat{\varphi}_n \equiv \hat{\varphi}_n(\ell)$ be the MBB estimator of a level-2 parameter φ_n based on (overlapping) blocks of size ℓ from $\mathcal{X}_n = \{X_1, \ldots, X_n\}$. Let $\mathcal{B}_i = \{X_i, \ldots, X_{i+\ell-1}\}, i = 1, \ldots, N$ (with $N = n - \ell + 1$) denote the collection of all overlapping blocks contained in $\mathcal{X}_n$ that are used for defining the MBB estimator $\hat{\varphi}_n$. Also, let m be an integer such that (7.44) holds. Note that the MBB estimator $\hat{\varphi}_n(\ell)$ is defined in terms of the "basic building blocks" $\mathcal{B}_i$'s. Hence, instead of deleting blocks of original observations $\{X_i, \ldots, X_{i+m-1}\}$, as done in the MBJ method described above, the JAB method of Lahiri (2002a) defines the jackknife point-values by deleting blocks of $\mathcal{B}_i$'s. Later in this section, we will discuss how this simple modification plays an important role in ensuring computational efficacy of the JAB method.

Since there are N observed blocks of length ℓ, we can define $M \equiv N - m + 1$ many JAB point-values corresponding to the bootstrap estimator $\hat{\varphi}_n$, by deleting the *overlapping* "blocks of blocks" $\{\mathcal{B}_i, \ldots, \mathcal{B}_{i+m-1}\}$ of size m for $i = 1, \ldots, M$. Let $I_i^0 = \{1, \ldots, N\} \backslash \{i, \ldots, i+m-1\}$, $i = 1, \ldots, M$. To define the ith JAB point-value $\hat{\varphi}_n^{(i)} \equiv \hat{\varphi}_n^{(i)}(\ell)$, we need to resample $b = \lfloor n/\ell \rfloor$ blocks randomly and with replacement from the reduced collection $\{\mathcal{B}_j : j \in I_i^0\}$ and construct the MBB estimator of φ_n using these resampled blocks. More precisely, suppose that $T_n = t_n(\mathcal{X}_n; \theta)$ be a random variable with probability distribution G_n and let $\varphi_n = \varphi(G_n)$ for some functional φ. Let $J_{i1}, \ldots, J_{ib}$ be a collection of b random variables such that, conditional on $\mathcal{X}_n$, these are iid with common distribution

$$P_*(J_{i1} = j) = (N-m)^{-1} \quad \text{for all} \quad j \in I_i^0 \; . \tag{7.48}$$

Then, the resampled blocks to be used for defining the JAB point-value $\hat{\varphi}_n^{(i)}$ are given by

$$\left\{\mathcal{B}_j^{*(i)} \equiv \mathcal{B}_{J_{ij}} : j = 1, \ldots, b\right\} . \tag{7.49}$$

Let $\mathcal{X}_n^{*(i)}$ denote the resampled data obtained by concatenating $\{\mathcal{B}_j^{*(i)}, j = 1, \ldots, b\}$. Also, let $T_n^{*(i)} \equiv t_{n_1}(\mathcal{X}_n^{*(i)}; \tilde{\theta}_{n,i})$ be the MBB version of T_n, defined using the resampled data $\mathcal{X}_n^{*(i)}$ and using a suitable estimator $\tilde{\theta}_{n,i}$ of θ. Then, the JAB point-value $\hat{\varphi}_n^{(i)}$ is given by applying the functional φ to the conditional distribution $\hat{G}_{n,i}$ (say) of $T_n^{*(i)}$ as

$$\hat{\varphi}_n^{(i)} = \varphi(\hat{G}_{n,i}) \; . \tag{7.50}$$

For an example illustrating the definition of $T_n^{*(i)}$, suppose that

$$T_n = \sqrt{n}(\hat{\theta}_n - \theta) \tag{7.51}$$

with $\hat{\theta}_n = H(\bar{X}_n)$ and $\theta = H(\mu)$ for some (smooth) function $H : \mathbb{R}^d \to \mathbb{R}$, where $\{X_i\}_{i \in \mathbb{Z}}$ is a stationary sequence of $\mathbb{R}^d$-valued random vectors, $\bar{X}_n = n^{-1} \sum_{i=1}^n X_i$ and $\mu = EX_1$. Let $\bar{X}_n^{*(i)}$ denote the MBB sample mean based on the $n_1 = b.\ell$ resampled values in $\{\mathcal{B}_j^{*(i)}, j = 1, \ldots, b\}$. Then, the MBB version $T_n^{*(i)}$ for the ith JAB point-value is defined as

$$T_n^{*(i)} = \sqrt{n_1}\left(\theta_n^{*(i)} - \tilde{\theta}_{n,i}\right) , \tag{7.52}$$

where $\theta_n^{*(i)} = H(\bar{X}_n^{*(i)})$ and where we set $\tilde{\theta}_{n,i} = H(\hat{\mu}_{n,i})$ with $\hat{\mu}_{n,i} = E_* \bar{X}_n^{*(i)}$, $i = 1, \ldots, M$.

Next we return to the general case of $T_n \equiv t_n(\mathcal{X}_n;\theta)$ and define the JAB variance estimator of $\hat{\varphi}_n$ as (cf. (7.46))

$$\widehat{\mathrm{VAR}}_{\mathrm{JAB}}(\hat{\varphi}_n) = \frac{m}{(N-m)} \cdot \frac{1}{M}\sum_{i=1}^{M}\left(\tilde{\varphi}_n^{(i)} - \hat{\varphi}_n\right)^2 , \qquad (7.53)$$

where $\tilde{\varphi}_n^{(i)} = m^{-1}(N\hat{\varphi}_n - (N-m)\hat{\varphi}_n^{(i)})$ denotes the ith JAB *pseudo-value* corresponding to $\hat{\varphi}_n$ and where $\hat{\varphi}_n^{(i)}$'s are defined by (7.50). As with the given MBB estimator $\hat{\varphi}_n$, computation of the point-values $\hat{\varphi}_n^{(i)}$'s and hence, of the pseudo-values $\tilde{\varphi}_n^{(i)}$ are typically done using the Monte-Carlo method. A simple representation result, initially noted by Efron (1992) in the context of IID bootstrap, makes it possible to compute the JAB variance estimator by reusing the resampled blocks used in the computation of the given bootstrap estimator $\hat{\varphi}_n$. We now give a statement of this result below.

Proposition 7.1 *Let $J_1, \ldots, J_b$ be iid random variables with the Discrete Uniform Distribution on $\{1, \ldots, N\}$ and let $J_{i1}, \ldots, J_{ib}$ be iid random variables with the Discrete Uniform Distribution on I_i^0, $1 \leq i \leq M$. Let $\hat{p}_i = b^{-1}\sum_{j=1}^{b} \mathbb{1}(J_j \in I_i^0)$, $1 \leq i \leq M$. Then, for any $i = 1, \ldots, M$, the conditional distribution of $(J_1, \ldots, J_b)$ given $\hat{p}_i = 1$ is the same as the unconditional distribution of $(J_{i1}, \ldots, J_{ib})$.*

Proof: For any $j_1, \ldots, j_b \in I_i^0$,

$$\begin{aligned} & P(J_1 = j_1, \ldots, J_b = j_b \mid \hat{p}_i = 1) \\ = \;& P(J_1 = j_1, \ldots, J_b \in j_b)/P(\hat{p}_i = 1) \\ = \;& [N^{-b}]/[(N-m)/N]^b \\ = \;& (N-m)^{-b} = P(J_{i1} = j_1, \ldots, J_{ib} = j_b) \;. \end{aligned}$$

This completes the proof of the proposition. □

To appreciate the relevance of this result, suppose that $\{{}_k\mathcal{B}_1^*, \ldots, {}_k\mathcal{B}_b^*\}$, $k = 1, \ldots, K$ denote the set of blocks drawn randomly, with replacement from the collection $\{\mathcal{B}_1, \ldots, \mathcal{B}_N\}$ for the Monte-Carlo evaluation of the given block bootstrap estimator $\hat{\varphi}_n$. Let $\{{}_kJ_1, \ldots, {}_kJ_b\}$ denote the random indices corresponding to $\{{}_k\mathcal{B}_1^*, \ldots, {}_k\mathcal{B}_b^*\}$, i.e., ${}_k\mathcal{B}_j^* = {}_k\mathcal{B}_{{}_kJ_j}$, $1 \leq j \leq b$, $k = 1, \ldots, K$. Then for any k, if all b indices ${}_kJ_1, \ldots, {}_kJ_b$ lie in I_i^0, by Proposition 7.2, we may consider $({}_kJ_1, \ldots, {}_kJ_b)$ as a random sample of size b from the reduced index set $I_i^0 = \{1, \ldots, N\}\backslash\{i, \ldots, i+m-1\}$. Let

$$I_i^* = \{k : 1 \leq k \leq K,\; {}_kJ_j \in I_i^0 \quad \text{for all} \quad j = 1, \ldots, b\}$$

denote the index set of all such random vectors $({}_kJ_1, \ldots, {}_kJ_b)$. Then, $\{({}_kJ_1, \ldots, {}_kJ_b) : k \in I_i^*\}$ gives us an iid collection of random vectors (of possibly different sizes for different $i \in \{1, \ldots, M\}$), each having the same distribution as $(J_{i1}, \ldots, J_{ib})$ of the Proposition. Thus, the resamples for computing the ith JAB point-value $\hat{\varphi}_n^{(i)}$ may be obtained by *extracting* the subcollection $\{({}_k\mathcal{B}_1^*, \ldots, {}_k\mathcal{B}_b^*) : k \in I_i^*\}$ from the original resamples $\{({}_k\mathcal{B}_1^*, \ldots, {}_k\mathcal{B}_b^*) : 1 \le k \le K\}$, and no additional resampling is needed. The Monte-Carlo approximations generated by this method are close to the true values of $\hat{\varphi}_n^{(i)}$'s, provided K is large.

As an illustration, consider the random variable T_n of (7.51) and suppose that the level-2 parameter of interest is $\varphi_n = \varphi(G_n)$ for some functional φ where G_n is the sampling distribution of T_n. Figures 7.1 and 7.2 give a schematic description of the main steps involved in the computations of the MBB estimator $\hat{\varphi}_n$ and its JAB point-values $\hat{\varphi}_n^{(i)}$, $i = 1, \ldots, M$. For computing $\hat{\varphi}_n$, we generate K iid sets of b many blocks $\{{}_k\mathcal{B}_1^*, \ldots, {}_k\mathcal{B}_b^*\}$ for $k = 1, \ldots, K$, compute the bootstrap sample mean ${}_k\bar{X}_n^*$ and the bootstrap version ${}_kT_n^* = \sqrt{n}({}_k\theta_n^* - \tilde{\theta}_n)$ for each set with ${}_k\theta_n^* = H({}_k\bar{X}_n^*)$ and $\tilde{\theta}_n = H(\hat{\mu}_n)$. Then, the Monte-Carlo approximation to $\hat{\varphi}_n$ is given by $\varphi(G_n^*)$ where G_n^* denotes the empirical distribution of the bootstrap replicates $\{{}_kT_n^* : k = 1, \ldots, K\}$. For computing $\hat{\varphi}_n^{(i)}$, we scan the K sets of resampled blocks $\{{}_k\mathcal{B}_1^*, \ldots, {}_k\mathcal{B}_b^*\}$, $k = 1, \ldots, K$ and extract the ${}_k\theta_n^*$-values corresponding to the block-sets $\{{}_k\mathcal{B}_1^*, \ldots, {}_k\mathcal{B}_b^*\}$ that do not contain any of the blocks $\mathcal{B}_i, \ldots, \mathcal{B}_{i+m-1}$. Next, the bootstrap version of $T_n^{*(i)}$ are computed by employing these ${}_k\theta_n^*$'s in the formula ${}_kT_n^{*(i)} = \sqrt{n_1}({}_k\theta_n^* - \tilde{\theta}_{n,i})$ where $\tilde{\theta}_{n,i} \equiv H(\hat{\mu}_{n,i})$. Note that $\hat{\mu}_{n,i}$ is given by the average of block-averages in the reduced collection $\{\mathcal{B}_j : j \in I_i^0\}$ and can be computed without any resampling. The copies ${}_kT_n^{*(i)}$'s are now combined to generate the Monte-Carlo approximation to $\hat{\varphi}_n^{(i)}$, just in the same way the ${}_kT_n^*$'s are used for computing the original bootstrap estimate $\hat{\varphi}_n$.

7.4.4 The Optimal Block Length Estimator

We now return to the problem of choosing the optimal block length for block bootstrap methods using the nonparametric plug-in method. Let $\hat{\varphi}_n \equiv \hat{\varphi}_n(\ell)$ be an MBB estimator of a level-2 parameter φ_n with an MSE of the form (cf. (7.30),(7.31))

$$n^{2a} \cdot \text{MSE}\big(\hat{\varphi}_n(\ell)\big) = vn^{-1}\ell^r + B^2\ell^{-2} + o(n^{-1}\ell^r + \ell^{-2}) \,, \qquad (7.54)$$

where $v \in (0, \infty)$, $B \in \mathbb{R}$, $B \neq 0$ are unknown parameters and where $r \in \mathbb{N}$, $a \in (0, \infty)$. Then, the theoretical optimal block length ℓ_n^0 is given by (cf. (7.32))

$$\ell_n^0 = \left(\frac{2B^2}{rv}\right)^{\frac{1}{r+2}} n^{\frac{1}{r+2}} \big(1 + o(1)\big) \,. \qquad (7.55)$$

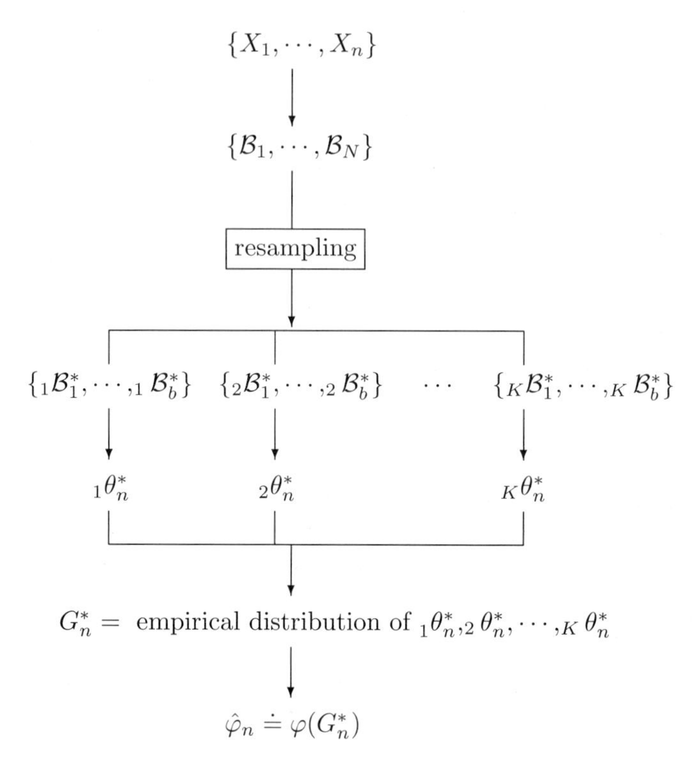

FIGURE 7.1. Monte-Carlo computation of the MBB estimator of $\varphi_n = \varphi(G_n)$ where G_n is the probability distribution of T_n of (7.51).

The nonparametric plug-in method, described in Section 7.4.1, suggests (cf. (7.39))

$$\hat{\ell}_n^0 = \left[\frac{2\hat{B}_n^2}{r\hat{v}_n}\right]^{\frac{1}{r+2}} n^{\frac{1}{r+2}} \tag{7.56}$$

as an estimator of the optional block length, where $\hat{B}_n = n^a \, \ell_1 \, \widehat{\mathrm{BIAS}}_n$ and $\hat{v}_n = [n^{-1}\ell_1^r]^{-1} n^{2a} \, \widehat{\mathrm{VAR}}_n$ are estimators of the level-3 parameters B and v, and $\widehat{\mathrm{BIAS}}_n \equiv \widehat{\mathrm{BIAS}}_n(\ell_1)$ and $\widehat{\mathrm{VAR}}_n \equiv \widehat{\mathrm{VAR}}_n(\ell_1)$ are some consistent estimators of the bias and the variance parts of the block bootstrap estimator $\hat{\varphi}_n(\ell_1)$ based on some suitable initial block length ℓ_1 (cf. (7.35), (7.36)). Lahiri, Furukawa and Lee (2003) suggest using the bias estimator $\widehat{\mathrm{BIAS}}_n$ of (7.42) to define $\hat{B}_n$ and using the JAB variance estimator $\widehat{\mathrm{VAR}}_{\mathrm{JAB}}(\hat{\varphi}_n(\ell_1))$ for defining $\hat{v}_n$. With these choices, the plug-in estimator of the optimal block length ℓ_n^0 is given by

$$\tilde{\ell}_n^0 = \left[\frac{2\tilde{B}_n^2}{r\tilde{v}_n}\right]^{\frac{1}{r+2}} n^{\frac{1}{r+2}}\ , \tag{7.57}$$

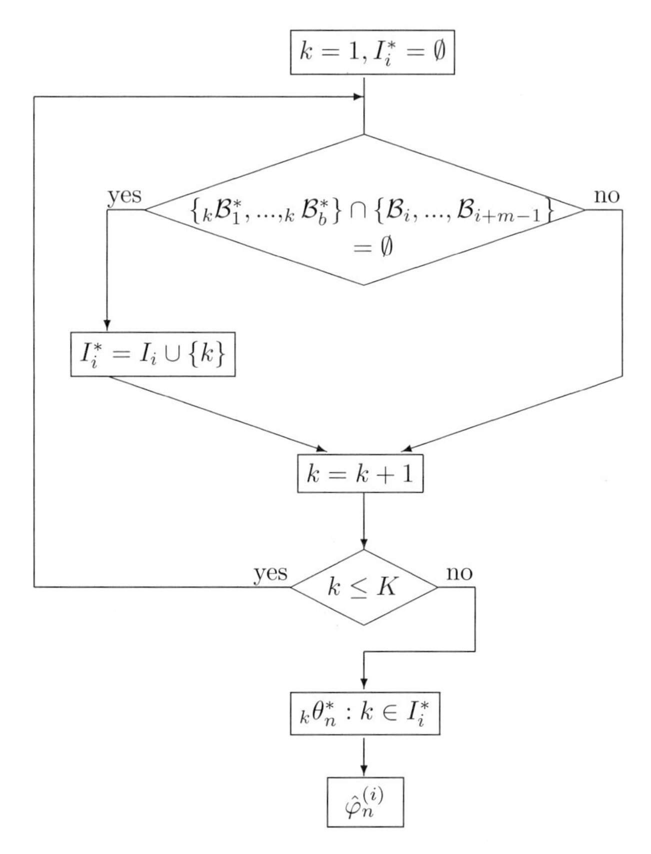

FIGURE 7.2. Computation of the ith JAB point value $\hat{\varphi}_n^{(i)}$ starting with the resampled blocks $\{{}_1\mathcal{B}_1^*, \ldots, {}_1\mathcal{B}_b^*\}, \ldots, \{{}_k\mathcal{B}_1^*, \ldots, {}_k\mathcal{B}_b^*\}$ generated for the Monte-Carlo computation of the block bootstrap estimator $\hat{\varphi}_n$ of Figure 7.1.

where $\tilde{B}_n = 2\ell_1[\hat{\varphi}_n(\ell_1) - \hat{\varphi}_n(2\ell_1)]$ and $\tilde{v}_n = (n\ell_1^{-r}) \cdot \widehat{\text{VAR}}_{\text{JAB}}(\hat{\varphi}_n(\ell_1))$, and $\widehat{\text{VAR}}_{\text{JAB}}(\hat{\varphi}_n(\ell_1))$ is defined by (7.53) with $\ell = \ell_1$. The n^a and n^{2a} factors in the definitions of $\hat{B}_n$ and $\hat{v}_n$ are left out as they cancel from the numerator and the denominator of (7.56).

We now show that this naive construction yields consistent estimators of ℓ_n^0 for various functionals φ_n without explicit form of the constants B and v in (7.54). For this, we suppose that $\{X_i\}_{i\in\mathbb{Z}}$ is a sequence of stationary random vectors with values in $\mathbb{R}^d$ and the level-1 parameter θ and its estimator $\hat{\theta}_n$ satisfy the requirements of the Smooth Function Model (7.8), i.e., $\hat{\theta}_n = H(\bar{X}_n)$ and $\theta = H(\mu)$ for some smooth function $H : \mathbb{R}^d \to \mathbb{R}$, $\bar{X}_n = n^{-1}\sum_{i=1}^n X_i$ and $\mu = EX_1$. First, we consider the bias and the variance functionals (cf. (7.2),(7.3))

$$\varphi_{1n} \equiv \text{Bias}(\hat{\theta}_n) = E(\hat{\theta}_n) - \theta \tag{7.58}$$

$$\varphi_{2n} = \text{Var}(\hat{\theta}_n) . \tag{7.59}$$

For $k = 1, 2$, let ℓ_{kn}^0 denote the optional block length for estimating φ_{kn}, defined by (7.6). Then, we have the following result.

Theorem 7.3 *Suppose that Condition (5.D$_r$) of Section 5.4 holds with $r = 4$,*

$$\ell_1^{-1} + n^{-1/3}\ell_1 + m^{-1}\ell_1 = o(1) , \tag{7.60}$$

and

$$n^{-1}\ell_1^{-2}m^3 = O(1) . \tag{7.61}$$

Also, suppose that Condition (5.M$_r$) of Section 5.4 holds with $r = 2+k(2+a_0)$ where a_0 is as in the statement of Condition (5.D$_r$). Then

$$\tilde{\ell}_{kn}^0 / \ell_{kn}^0 \longrightarrow_p 1 \quad as \quad n \to \infty , \tag{7.62}$$

for $k = 1, 2$.

Proof: See Lahiri, Furukawa and Lee (2003). □

Under suitable regularity conditions, Lahiri, Furukawa and Lee (2003) also prove consistency of the plug-in estimator $\tilde{\ell}_n^0$ for bootstrap distribution function estimation and for bootstrap quantile estimation for certain studentized versions of $\hat{\theta}_n$. For these functionals, expansion (7.54) for the corresponding MSEs hold with $r = 2$ and $a = 1/2$, as in the case of the distribution function $\varphi_{3n}(x_0)$ of the normalized version of $\hat{\theta}_n$ for $|x_0| \neq 1$. Thus, the optimal block sizes for these functionals in the studentized case are of the order $n^{1/4}$ and the corresponding plug-in estimators $\tilde{\ell}_n^0$ are defined with $r = 2$ in such cases. For the *normalized* version of $\hat{\theta}_n$, consistency of $\tilde{\ell}_n^0$ for the distribution function estimator φ_{3n} of (7.4) also holds (cf. Lahiri (1996d)), provided we set $r = 2$ for $|x_0| \neq 1$, and $r = 1$ for $|x_0| = 1$. Thus, the plug-in estimator provides a consistent and computationally efficacious method for estimating the optimal block length for a variety of level-2 parameters.

Although the nonparametric plug-in method produces a valid (i.e., consistent) estimator of the optimal block length, finite sample performance of the estimator depends on the choice of the smoothing parameter ℓ_1, and on the JAB "blocks of blocks" deletion parameter m. It turns out that a reasonable choice of ℓ_1 in (7.57) depends on the functional φ_n. A careful analysis of the MSE of $\tilde{B}_n$ shows that the optimal choice of ℓ_1 is of the form

$$\ell_1 = C_3 n^{\frac{1}{r+4}} , \tag{7.63}$$

where r is as in (7.54), and C_3 is a population parameter. As for the other smoothing parameter, an heuristic argument in Lahiri (2002a) suggests that a reasonable choice of the JAB parameter m is given by

$$m = C_4 n^{1/3} \ell_1^{2/3} \tag{7.64}$$

for some constant C_4. Numerical results of Lahiri, Furukawa and Lee (2003) show that the choice $C_3 = 1$ in (7.63) for the initial block size ℓ_1 yields good results for both the variance and the distribution function estimation problems, while the corresponding values for C_4 in (7.64) are given by $C_4 = 1.0$ for the variance functional and $C_4 = 0.1$ for the distribution function. Below we report the results from a small simulation study with the above choices of C_3 and C_4. For more simulation results, see Lahiri, Furukawa and Lee (2003).

We consider the moving average model of Section 7.3, given by (cf. (7.27)) $X_i = (\epsilon_i + \epsilon_{i-1})/\sqrt{2}$, $i \in \mathbb{Z}$, where $\{\epsilon_i\}_{i\in\mathbb{Z}}$ is a sequence of iid random variables having the centered Chi-squared distribution with one degree of freedom. As in Section 7.3, we also set the level-1 parameter to be $\theta = EX_1$, the estimator $\hat{\theta}_n$ to be the sample mean $\bar{X}_n$, and the level-2 parameters as $\varphi_{2n} = n.\text{Var}(\bar{X}_n)$ and $\varphi_{3n} = P(\sqrt{n}(\hat{\theta}_n - \theta)/\tau_n \le 0)$. The true value of θ is zero. Also, we take the sample size n to be 125. As stated in Section 7.3, the true values of φ_{2n} and φ_{3n} are $\varphi_{2n} = 3.984$ and $\varphi_{3n} = 0.5226$. Furthermore, the theoretical optimal block sizes for estimating φ_{2n} and φ_{3n} by the MBB are $\ell_{2n}^0 = 3$ and $\ell_{3n}^0 = 2$, as shown in Table 7.1.

Next we applied the nonparametric plug-in method to estimate the target values ℓ_{2n}^0 and ℓ_{3n}^0. Table 7.3 gives the frequency distribution of the estimated optimal block sizes based on 500 simulation runs. The block boostrap estimators in each case were evaluated using 1000 Monte-Carlo replicates. Table 7.3 shows that more than 80% of the mass of the estimated block size $\hat{\ell}_{2n}^0$ for variance estimation lies in the interval [2,5] (the true value being $\ell_{2n}^0 = 3$). The method also produces very good results for distribution function estimation, with a pronounced mode at the true value $\ell_{3n}^0 = 2$, and a small support set $\{1, 2, 3\}$.

TABLE 7.3. Frequency distribution of the optimal block sizes selected by the nonparametric plug-in method for model (7.27) with $n = 125$.

	(a) Variance Estimation									
$\hat{\ell}_{2n}^0$	1	2	3	4	5	6	7	8	9	10
Frequency	50	114	125	94	71	29	10	3	2	2

	(b) Distribution Function Estimation		
$\hat{\ell}_{3n}^0$	1	2	3
Frequency	172	268	60

8
Model-Based Bootstrap

8.1 Introduction

In this chapter, we consider bootstrap methods for some popular time series models, such as the autoregressive processes, that are driven by iid random variables through a structural equation. As indicated in Chapter 2, for such models, it is often possible to adapt the basic ideas behind bootstrapping a linear regression model with iid error variables (cf. Freedman (1981)). In Section 8.2, we consider stationary autoregressive processes of a general order and describe a version of the autoregressive bootstrap (ARB) method. Like Efron's (1979) IID resampling scheme, the ARB also resamples a single value at a time. We describe theoretical and empirical properties of the ARB for the stationary case in Section 8.2. In Section 8.3, we consider the explosive autoregressive processes. In the explosive case, the initial variables defining the model have nontrivial effects on the limit distributions of the least squares estimators of the autoregression (AR) parameters. As a result, the validity of the ARB critically depends on the initial values. In Section 8.3, we describe the relevant issues and provide conditions for the validity of the ARB method in the explosive case.

The unstable autoregressive processes are considered in Section 8.4. Here, the ARB with the natural choice of the resample size fails. A remedy to this problem is given by the "m out of n" ARB, where the resample size m grows to infinity at a rate slower than the sample size n. In the unstable case, we describe the theoretical and numerical aspects of the ARB for the first-order AR-models only. In Section 8.5, we present some results on bootstrapping autoregressive and moving average (ARMA) processes of a general (finite) order, in the stationary case.

8.2 Bootstrapping Stationary Autoregressive Processes

Let $\{X_i\}_{i\in\mathbb{Z}}$ be a stationary autoregressive process of order p (AR(p)), satisfying the linear difference equation

$$X_i = \beta_1 X_{i-1} + \cdots + \beta_p X_{i-p} + \epsilon_i,\ i \in \mathbb{Z}\ , \tag{8.1}$$

where $p \in \mathbb{N}$, $\beta_1, \ldots, \beta_p$ are the autoregression parameters and $\{\epsilon_i\}_{i\in\mathbb{Z}}$ is a sequence of zero mean iid random variables with a common distribution F. In the sequel, we shall often assume that the autoregression parameters $\beta_1, \ldots, \beta_p$ are such that

$$\beta(z) \equiv 1 - \sum_{j=1}^{p} \beta_j z^j \neq 0 \quad \text{for all} \quad z \in \mathbb{C} \quad \text{with} \quad |z| \leq 1\ . \tag{8.2}$$

It is well known (cf. Brockwell and Davis (1991), Chapter 3) that under (8.2), the AR(p) process $\{X_i\}_{i\in\mathbb{Z}}$ of (8.1) admits an infinite-order moving-average representation

$$X_i = \sum_{j=0}^{\infty} b_j \epsilon_{i-j}\ , \tag{8.3}$$

where $\{b_j\}_{j=0}^{\infty}$ are constants, determined by the power series expansion of the function $b(z) \equiv [\beta(z)]^{-1}$, given by

$$b(z) = \sum_{j=0}^{\infty} b_j z^j\ , \quad |z| \leq 1\ .$$

Although the random variables X_i's under the AR(p) model (8.1) are dependent, here we can use the model structure to generate valid bootstrap approximations without any block resampling. As described in Chapter 2, the basic idea is to consider the "residuals" from the fitted model, which turn out to be "approximately independent," and then resample the residuals (with a suitable centering adjustment) to define the bootstrap observations through an estimated version of the structural equation (8.1). Suppose that a finite segment $X_1, \ldots, X_n$ of the process $\{X_i\}_{i\in\mathbb{Z}}$ is observed. Let $\hat\beta_{1n}, \ldots, \hat\beta_{pn}$ denote the least squares estimators of $\beta_1, \ldots, \beta_p$ based on $X_1, \ldots, X_n$. Thus, $\hat\beta_{1n}, \ldots, \hat\beta_{pn}$ are given by the relation

$$(\hat\beta_{1n}, \ldots, \hat\beta_{pn})' = (V_n' V_n)^{-1} V_n' (X_{p+1}, \ldots, X_n)'\ , \tag{8.4}$$

where V_n is a $(n-p)\times p$ matrix with ith row $(X_{i+p-1},\ldots,X_i)$, $i=1,\ldots,n-p$. Let, $\hat{\epsilon}_i = X_i - \hat{\beta}_{1n}X_{i-1} - \ldots - \hat{\beta}_{pn}X_{i-p}$, $i=p+1,\ldots,n$ denote the residuals. Note that by using (8.1), we may express the residuals as

$$\hat{\epsilon}_i = \epsilon_i - \sum_{j=1}^{p}(\hat{\beta}_{jp} - \beta_j)X_{i-j},\ p+1 \le i \le n\ .$$

As a consequence, when $\beta_{jn} \longrightarrow_p \beta_j$ for $j=1,\ldots,p$, the second term is small for large values of n and thus, the residuals are approximately independent. This suggests that we may resample the residuals, a single value at a time as in Efron's (1979) IID bootstrap, to define the bootstrap version of a random variable $T_n \equiv t_n(X_1,\ldots,X_n;\beta_1,\ldots,\beta_p,F)$. However, to generate a valid approximation, we need to center the residuals ϵ_i's first and resample from the collection of the *centered* residuals, defined by

$$\tilde{\epsilon}_i = \hat{\epsilon}_i - \bar{\epsilon}_n\ ,\quad i=p+1,\ldots,n\ , \tag{8.5}$$

where $\bar{\epsilon}_n = (n-p)^{-1}\sum_{i=p+1}^{n}\hat{\epsilon}_i$. Next, generate the bootstrap error variables ϵ_i^*, $i\in\mathbb{Z}$ by sampling randomly with replacement from $\{\tilde{\epsilon}_{p+1},\ldots,\tilde{\epsilon}_n\}$. Thus, the random variables ϵ_i^*, $i\in\mathbb{Z}$ are conditionally iid (given $X_1,\ldots,X_n$) with common distribution

$$P_*(\epsilon_1^* = \tilde{\epsilon}_i) = \frac{1}{n-p}\ ,\quad p+1 \le i \le n\ . \tag{8.6}$$

Note that by (8.5) and (8.6), $E_*\epsilon_1^* = (n-p)^{-1}\sum_{i=p+1}^{n}\tilde{\epsilon}_i = 0$. Thus, the bootstrap error variables ϵ_i^*'s satisfy an analog of the model condition $E\epsilon_1 = 0$ at the bootstrap level. Now, define the bootstrap version of equation (8.1) by

$$X_i^* = \hat{\beta}_{1n}X_{i-1}^* + \cdots + \hat{\beta}_{pn}X_{i-p}^* + \epsilon_i^*\ ,\quad i\in\mathbb{Z}\ , \tag{8.7}$$

and let $\{X_i^*\}_{i\in\mathbb{Z}}$ be a stationary solution of (8.7). If $\hat{\beta}_{jn} \longrightarrow_p \beta_j$ as $n\to\infty$ for $j=1,\ldots,p$, then such a solution exists on a set of X_i's that has probability close to one for n large. In practice, one makes use of the recursion relation (8.7) for $i \ge p+1$ to generate the bootstrap "observations" by setting the initial p variables (arbitrarily) equal to $X_1,\ldots,X_p$ or equal to zeros. When the polynomial $\hat{\beta}(z) \equiv 1 - \sum_{j=1}^{p}\hat{\beta}_{jn}z^j$ does not vanish in the region $\{|z| \le 1\}$ (cf. (8.2)), the coefficients of the p initial values die out geometrically fast and, therefore, have a negligible effect in the long run. As a consequence, one may generate a long chain using the recursion relation (8.7) until stationarity is reached (i.e., the effect of the initial values become inappreciable) and may take the next m-values as the desired "resample" of size m. The autoregressive bootstrap (ARB) version of a random variable $T_n = t_n(X_1,\ldots,X_n;\beta_1,\ldots,\beta_p,F)$ based on a resample of size $m > p$ is given by

$$T_{m,n}^* = t_m(X_1^*,\ldots,X_m^*;\hat{\beta}_{1n},\ldots,\hat{\beta}_{pn},\hat{F}_n)\ , \tag{8.8}$$

where $\hat{F}_n$ is the empirical distribution of the centered residuals $\tilde{\epsilon}_i$, $p+1 \leq i \leq n$. Typically, the resample size m is chosen equal to the original sample size n. However, in some cases, a smaller value of m may be desirable; see, for example, Section 8.4. Furthermore, depending on the values of the parameters $\beta_1, \ldots, \beta_p$, it may be desirable to use two different sets of estimators of the parameters $\beta_1, \ldots, \beta_p$, in the formulation of the ARB method, employing one set of estimators to define the residuals $\tilde{\epsilon}_i$'s and the other set to define the bootstrap version (8.7) of the AR model; see Datta and Sriram (1997) for such an alternative approach.

Properties of the ARB have been investigated by many authors. The idea of exploiting the structural equation (8.1) to adapt Efron's (1979) IID resampling scheme to the AR-processes has been noted and formalized in different problems by Freedman and Peters (1984), Efron and Tibshirani (1986), and Swanepoel and van Wyk (1986), among others. The first paper applied the bootstrap to estimate the mean square forecasting error for a class of autoregressive-type time series models and obtained some empirical results. Theoretical analysis of their method has been subsequently carried out by Findley (1986). Efron and Tibshirani (1986) developed a bootstrap estimate of the standard errors of autoregression parameter estimators. Properties of certain bootstrap confidence bands for the autoregression spectral density have been studied by Swanepoel and van Wyk (1986).

Bose (1988) investigated higher-order properties of the ARB for the normalized least squares estimators of the autoregression parameters. Here we state a result of Bose (1988) that shows that the ARB approximation is indeed second-order correct. Suppose that $\{X_i^*\}_{i\in\mathbb{Z}}$ is the stationary solution to (8.7). Let Σ and $\hat{\Sigma}_n$ be the $p \times p$ matrices with (i,j)-th elements $\text{Cov}(X_1, X_{1+|i-j|})$ and $\text{Cov}_*(X_1^*, X_{1+|i-j|}^*)$, respectively, $1 \leq i$, $j \leq p$, where Cov_* denotes conditional covariance given $\{X_i\}_{i\in\mathbb{Z}}$. Also, write $\hat{\beta}_n = (\hat{\beta}_{1n}, \ldots, \hat{\beta}_{pn})'$ for the vector of least squares estimators of $\beta_1, \ldots, \beta_p$, given by (8.4). Let $\beta_n^* = (\beta_{1n}^*, \ldots, \beta_{pn}^*)'$ denote the bootstrap version of $\hat{\beta}_n$, obtained by replacing $\{X_1, \ldots, X_n\}$ in the definition of $\hat{\beta}_n$ by $\{X_1^*, \ldots, X_n^*\}$. For a nonnegative definite matrix A of order p, let $A^{1/2}$ denote a $p \times p$ symmetric matrix satisfying $A = A^{1/2} \cdot A^{1/2}$. Also, recall that $\iota = \sqrt{-1}$. Then, we have the following theorem.

Theorem 8.1 *Assume that $\{\epsilon_i\}_{i\in\mathbb{Z}}$ is a sequence of iid random variables such that*

(i) $E\epsilon_1 = 0$, $E\epsilon_1^2 = 1$, $E\epsilon_1^8 < \infty$ and

(ii) $(\epsilon_1, \epsilon_1^2)$ satisfies Cramer's condition, i.e.,

$$\limsup_{\|t\|\to\infty} \left| E\exp(\iota(\epsilon_1, \epsilon_1^2)t) \right| < 1 .$$

Also, suppose that all roots of the characteristic polynomial

$$\Psi_p(z) = z^p - \beta_1 z^{p-1} - \cdots - \beta_p \tag{8.9}$$

lie inside the unit circle. Then,

$$\sup_{x \in \mathbb{R}^p} \left| P_* \left(n^{1/2} \hat{\Sigma}_n^{1/2} (\beta_n^* - \hat{\beta}_n) \le x \right) - P\left(\sqrt{n} \Sigma^{1/2} (\hat{\beta}_n - \beta) \le x \right) \right|$$
$$= o(n^{-1/2}) \quad a.s.$$

Proof: See Bose (1988). □

Theorem 8.1 shows that the ARB approximation is second-order correct and, thus, it outperforms the normal approximation, which has an error of order $O(n^{-1/2})$. Under additional regularity conditions, the $o(n^{-1/2})$ bound in Theorem 8.1 can be sharpened further and the ARB can attain a level of accuracy similar to that of the IID bootstrap under independence (cf. Choi and Hall (2000)). As a result, the accuracy of the ARB approximation is higher than the MBB approximation.

It is important to note that Theorem 8.1 asserts second-order correctness of the ARB method for the normalized least squares estimator when the observations X_i's (are *known* to) have zero mean. For stationary autoregressive processes with an unknown mean, the least squares estimators are defined through (8.4) by centering the X_i's at the sample mean $\bar{X}_n$. Second-order accuracy of the ARB continues to hold in this case (cf. Remark 2, page 1710, Bose (1988)).

Next we briefly consider finite sample performance of the ARB method and compare it with the MBB method. Indeed, the superiority of the ARB method over the MBB for stationary AR(p) processes shows up in numerical studies even in samples of moderate sizes. As an example, we consider model (8.1) with $\beta_1 = 0.5$, $p = 1$, and $\epsilon_1 \sim N(0, 1)$. Figure 8.1 shows a data set of size $n = 100$ from this model. We applied the ARB method described above to approximate the sampling distribution of

$$T_n = \left[\sum_{t=1}^{n-1} X_t^2 \right]^{1/2} (\hat{\beta}_{1n} - \beta_1) \ .$$

The histogram of the ARB version T_n^* of T_n based on $B = 500$ bootstrap replicates is given in the left panel of Figure 8.2, which looks reasonably symmetric. We also obtained the true distribution of T_n using 10,000 simulation runs. A plot of the cdf of T_n^* against the true cdf of T_n is given in the right panel of Figure 8.2. Even for a sample size of $n = 100$, the approximation appears to be very good.

In the absence of the knowledge of the model (8.1), we may apply a block bootstrap method to derive alternative estimates of the sampling distribution of T_n. Left panels of Figure 8.3 show the histograms generated by the

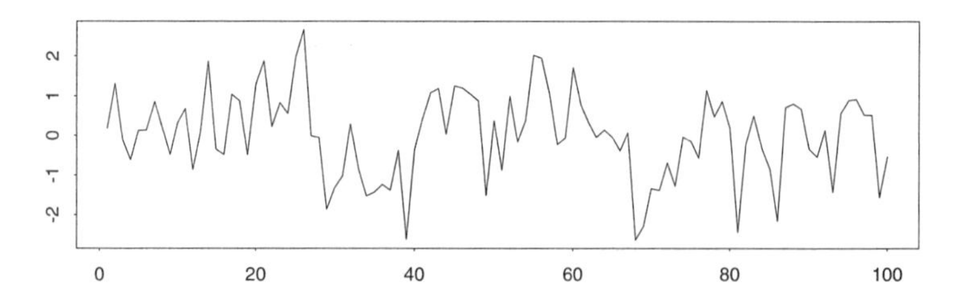

FIGURE 8.1. A data set of size $n = 100$ simulated from the AR(1) model $X_i = 0.5X_{i-1} + \epsilon_i$, $i \in \mathbb{Z}$ where ϵ_i's are iid $N(0,1)$ variables.

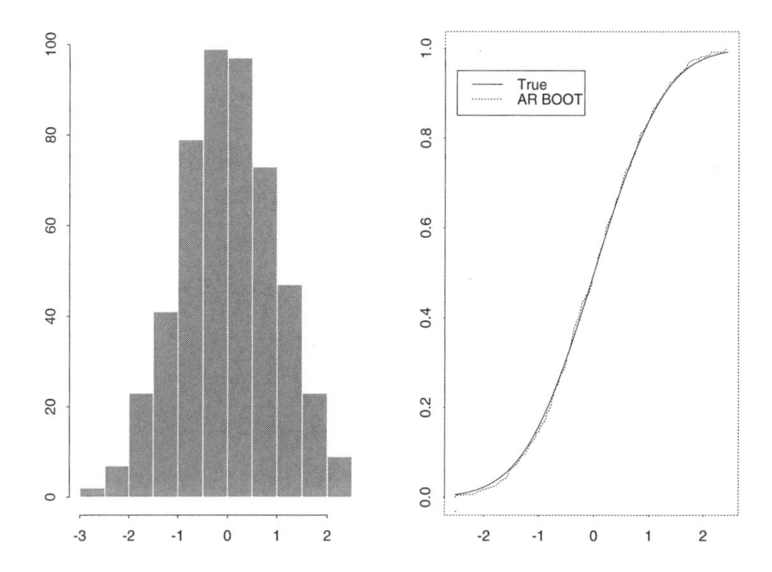

FIGURE 8.2. The ARB estimate of the sampling distribution of T_n for the data set of Figure 8.1 is given as a histogram on the left. The corresponding cdf (denoted by the dotted curve) and the true cdf of T_n (denoted by the solid curve) are given on the right.

MBB method with block sizes $\ell = 5, 10$. Here, in each case, $B = 500$ bootstrap replicates were used. It appears that the MBB distribution function estimates are both skewed, with $\ell = 5$ leading to a higher level of skewness compared to $\ell = 10$. The overall errors of the resulting bootstrap approximations are effectively captured by the plots of the MBB cdf estimates against the true sampling distribution of T_n, as given by the right panels of Figure 8.3.

Next consider bootstrap CIs for β_1 at nominal coverage levels 80% and 90%. Using the bootstrap quantiles from the above computations, we derived two-sided equal tailed CIs for β_1 based on the ARB method and based on the MBB method with block lengths $\ell = 2, 5, 10, 20$. The upper and the lower end points of the resulting CIs are given in Table 8.1. CIs for

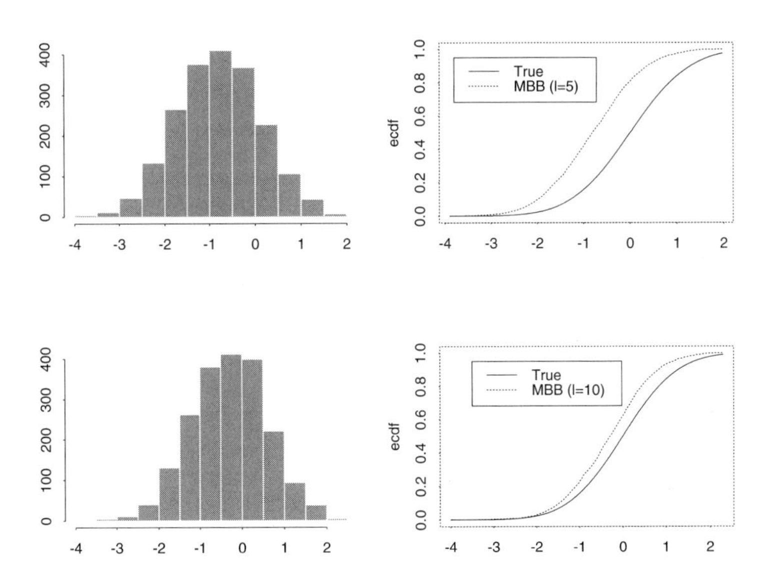

FIGURE 8.3. MBB estimates of the sampling distribution of the normalized least square estimator $T_n = (\sum_{i=1}^{n-1} X_i^2)^{1/2}(\hat{\beta}_{1n} - \beta_1)$ with block lengths $\ell = 5$ (top row) and $\ell = 10$ (bottom row) for the data set of Figure 8.1. The left panels give the MBB estimates as histograms, while the right panels give the corresponding MBB cdfs (denoted by dotted curves) against the true cdf of T_n (denoted by solid curves).

β_1 based on the true distribution of T_n are also included for comparison. Except for the 80% MBB CI with $\ell = 20$, all CIs contain the target value $\beta_1 = 0.5$. Note that the CIs given by the ARB have end points that are closer to those of the exact CIs, than the end points of the MBB CIs.

The ARB approximation tends to be more accurate than the MBB approximation because it explicitly makes use of the structure of the model (8.1). The quality of ARB approximation becomes poor when the model assumptions are violated. In particular, if the order of the autoregressive process is misspecified, the ARB method may be invalid, but the MBB would still give a valid approximation. Also, the standard ARB method is not robust against the values of the autoregressive parameters, particularly when some of these parameters lie on the boundary, e.g., when $\beta_1 \in \{-1, 1\}$ for the AR(1) case (cf. Section 8.4).

8.3 Bootstrapping Explosive Autoregressive Processes

Let $\{X_i\}_{i \geq 1}$ be an autoregressive process defined by the recursion relation

$$X_i = \beta_1 X_{i-1} + \cdots + \beta_p X_{i-p} + \epsilon_i \ , \quad i \geq p+1 \ , \tag{8.10}$$

TABLE 8.1. Two-sided equal tailed CIs for β_1 for the data set of Figure 8.1 at 80% and 90% nominal levels, obtained using the true distribution of T_n, the ARB method, and the MBB method with block sizes $\ell = 2, 5, 10, 20$.

	90% CI		80% CI	
	Lower	Upper	Lower	Upper
TRUE	0.253	0.559	0.287	0.525
ARB	0.255	0.551	0.288	0.520
MBB ($\ell = 2$)	0.464	0.752	0.494	0.715
MBB ($\ell = 5$)	0.340	0.629	0.368	0.594
MBB ($\ell = 10$)	0.300	0.576	0.325	0.546
MBB ($\ell = 20$)	0.262	0.524	0.292	0.495

where $\{X_1, \ldots, X_p\}$ is a given initial set of random variables, $\{\epsilon_i\}_{i \geq p+1}$ is a sequence of iid random variables that are independent of $\{X_1, \ldots, X_p\}$, and the autoregression parameters $\beta_1, \ldots, \beta_p$ are such that the roots of the characteristic polynomial $\Psi_p(z) = z^p - \beta_1 z^{p-1} - \cdots - \beta_p$ all lie in the region $\{z \in \mathbb{C} : |z| > 1\}$ of the complex plane. The model given by (8.10) is known as an explosive autoregressive model of order (p). Note that unlike the stationary case, the error variables ϵ_i's in (8.10) are not required to have zero mean. This is because, in the following, we allow the ϵ_i's to be heavy-tailed so that the expectation of ϵ_1 may not even exist. Another notable difference of the explosive AR(p) model with the stationary case is that the initial random variables $\{X_1, \ldots, X_p\}$ have nontrivial effects on the subsequent X_i's. To gain some insight, consider the case $p = 1$. Then, the ith random variable X_i generated by the recursion relation (8.10) involves the term $\beta_1^{i-1} X_1$. For the stationary case, $|\beta_1| < 1$ and the effect of the initial random variable X_1 on X_i becomes small at a geometric rate as $i \to \infty$. In contrast, for the explosive case, $|\beta_1| > 1$ and hence, the contribution from this term "explodes" in the long run, unless $X_1 = 0$ a.s. Furthermore, as we shall shortly see, the initial set of random variables have a nontrivial effect on the limit distribution of the least squares estimators.

Suppose that $\{X_1, \ldots, X_n\}$ denote the observations from model (8.10). Define the least square estimator $\hat{\beta}_n = (\hat{\beta}_{1n}, \ldots, \hat{\beta}_{pn})'$ of the autoregression parameters $\beta = (\beta_1, \ldots, \beta_p)'$ by relation (8.4) and let

$$A = \begin{bmatrix} \beta' & \\ \mathbb{I}_{p-1} & 0 \end{bmatrix} \tag{8.11}$$

be a $p \times p$ matrix with its first row equal to $\beta' \equiv (\beta_1, \ldots, \beta_p)$, where 0 denotes the $(p-1) \times 1$ vector of zeros and where, recall that, for $k \in \mathbb{N}$, $\mathbb{I}_k$ denotes the identity matrix of order k. Also, let

$$U = \sum_{j=1}^{\infty} A^{-j} \epsilon_{j+p} \tag{8.12}$$

and write $U^{(1)}$ for the first column of U. The following result, due to Datta (1995), gives the limit distribution of the least square estimator vector $\hat{\beta}_n$ in the explosive case.

Theorem 8.2 *Suppose that the error variables $\{\epsilon_i\}_{i \geq p+1}$ are iid nondegenerate random variables with $E \log\big(1 + |\epsilon_{p+1}|\big) < \infty$. Then,*

$$T_n \equiv (A')^n(\hat{\beta}_n - \beta) \longrightarrow^d Q^{-1}\Lambda W \ , \tag{8.13}$$

where $Q = \sum_{i=p+1}^{\infty} A^{-i}(WW')(A^{-i})'$, $W = (X_1, \ldots, X_p)' + U^{(1)}$, and where Λ is independent of W and Λ has the same distribution as $A^{-p}U$.

Proof: See Theorem 2.2 of Datta (1995). □

Note that the limit distribution of the normalized least square estimator T_n is nonnormal and depends on the initial variables $(X_1, \ldots, X_p)'$ through W. As a result, in the explosive case, any bootstrap method must use a consistent estimator of the joint distribution of $(X_1, \ldots, X_p)'$ to produce a valid approximation to the sampling distribution of T_n. This requires one to impose further restrictions on the joint distribution of $X_1, \ldots, X_p$, e.g., $X_1, \ldots, X_p$ are degenerate or $(X_1, \ldots, X_p)'$ follows a "known" p-dimensional distribution. Alternatively, one may consider the conditional distribution of T_n given $(X_1, \ldots, X_p)'$. In view of the independence assumption on $(X_1, \ldots, X_p)'$ and the sequence $\{\epsilon_i\}_{i \geq p+1}$ of error variables, the conditional (limit) distribution of T_n is determined by the joint distribution of $\{\epsilon_i\}_{i \geq p+1}$, with $(X_1, \ldots, X_p)'$ held fixed at its observed value. The bootstrap method described here follows this latter approach and generates the bootstrap observation using the "bootstrap" recursion relation

$$X_i^* = \hat{\beta}_{1n}X_{i-1}^* + \cdots + \hat{\beta}_{pn}X_{i-p}^* + \epsilon_i^* \ , \quad i \geq p+1 \tag{8.14}$$

by setting $(X_1^*, \ldots, X_p^*)' \equiv (X_1, \ldots, X_p)'$. Here, the bootstrap error variables ϵ_i^*'s are generated by random, with replacement sampling of the residuals $\{\hat{\epsilon}_i \equiv X_i - \sum_{j=1}^{p} \hat{\beta}_{jn}X_{i-j} : p+1 \leq i \leq n\}$. Unlike the stationary case, because the expectation of the ϵ_i's may not be finite, centering of the residuals is not carried out in the explosive case. However, in case $E|\epsilon_{p+1}| < \infty$ and $E\epsilon_{p+1} = 0$ in (8.10), one may center the residuals $\hat{\epsilon}_i$ and resample from $\{\tilde{\epsilon}_i \equiv \hat{\epsilon}_i - \bar{\epsilon}_n : i = p+1, \ldots, n\}$, where $\bar{\epsilon}_n = (n-p)^{-1}\sum_{i=p+1}^{n} \hat{\epsilon}_i$. The resulting bootstrap approximation also yields consistent estimators of the sampling distribution of T_n (conditional on $X_1, \ldots, X_p$) (cf. Theorem 3.1, Datta (1995)).

Let β_n^* denote the bootstrap version of $\hat{\beta}_n$, obtained by replacing the X_i's in (8.4) with X_i^*'s of (8.14). Also, define $\hat{A}_n$ and A_n^* by replacing β in (8.11) by $\hat{\beta}_n$ and β_n^*, respectively. Then, a studentized version of $\hat{\beta}_n$ is given by

$$T_{1n} = (\hat{A}_n')^n(\hat{\beta}_n - \beta) \ .$$

The ARB versions of T_n and T_{1n} are respectively given by $T_n^* = (\hat{A}_n')^n(\beta_n^* - \hat{\beta}_n)$ and $T_{1n}^* = ([A_n^*]')^n(\beta_n^* - \hat{\beta}_n)$.

To state the main results on the ARB method in the explosive case, write $Y = (X_1, \ldots, X_p)'$. Also, for any $y \in \mathbb{R}^p$ and for a random vector $R = r(\epsilon_{p+1}, \epsilon_{p+2}, \ldots; Y)$, depending on $\{\epsilon_i\}_{i \geq p+1}$ and Y, let $P_y(R \in \cdot)$ denote the conditional distribution of R given Y at $Y = y$. Then, we have the following result.

Theorem 8.3 *Suppose that the conditions of Theorem 8.2 hold and that the ARB samples $X_1^*, \ldots, X_n^*$, $n \geq p+1$ are generated by (8.14) with the initial values $X_i^* = X_i$ for $i = 1, \ldots, p$. Then, for any $y \in \mathbb{R}^p$,*

$$(a)\ \sup_{x \in \mathbb{R}^p} \left| P_y(T_n \leq x) - P_*(T_n^* \leq x) \right| = o_p(1) \quad as \quad n \to \infty ,$$

$$(b)\ \sup_{x \in \mathbb{R}^p} \left| P_y(T_{1n} \leq x) - P_*(T_{1n}^* \leq x) \right| = o_p(1) .$$

Proof: See Theorem 3.1 and Theorem 4.1 of Datta (1995). □

Thus, the ARB approximations to the (conditional) distributions of the normalized as well as the studentized least square estimator of β are consistent under the very mild moment condition $E\log(1 + |\epsilon_{p+1}|) < \infty$, which, in particular, allows the error variables ϵ_i's to be heavy tailed. We shall show in Chapter 11 that for iid heavy-tailed random variables and also for stationary heavy-tailed processes satisfying certain weak dependence conditions, the IID bootstrap of Efron (1979) and the MBB of Künsch (1989) fail when the bootstrap resample sizes are equal to the sample size. In view of these results, the consistency of the ARB under the same choice of the resample size is surprising and may be attributed to the specific nonstationary structure of the explosive AR(p) process. For the resample size equal to the sample size, Datta (1995) also proves almost sure validity of the ARB in the explosive case, requiring $E|\epsilon_{p+1}| < \infty$.

Theorem 8.3 also shows that the ARB provides a valid approximation if the random variables $\{X_i\}_{i \geq p+1}$ in the explosive case were generated by (8.10), starting with any initial set of p (nonrandom) real numbers. In this case, the bootstrap observations must also be generated starting with the same initial values. On the other hand, to get a valid approximation to the *unconditional* distribution of T_n or T_{1n}, the initial p random variables for the bootstrap recursion should be generated from the (joint) distribution of $(X_1, \ldots, X_p)'$, which therefore, must be estimable consistently.

8.4 Bootstrapping Unstable Autoregressive Processes

In this section, we consider properties of the ARB method for the unstable autoregressive processes. We call an autoregressive process of order p unstable if one or more of the roots of the characteristic polynomial $\Psi_p(z)$ (cf. (8.9)) lie on the unit circle $\{z \in \mathbb{C} : |z| = 1\}$. The standard ARB method is known to perform poorly for such processes. To describe the major issues involved in this case, we consider the first-order autoregressive process $\{X_i\}_{i\geq 1}$ given by

$$X_i = \beta_1 X_{i-1} + \epsilon_i \ , \quad i \geq 1 \tag{8.15}$$

with $X_0 = 0$ and $\beta_1 \in \{-1, 1\}$, where $\{\epsilon_i\}_{i\geq 1}$ is a sequence of iid random variables with $E\epsilon_1 = 0$ and $E\epsilon_1^2 = \sigma^2 \in (0, \infty)$. The AR(1) process in (8.15) is unstable because the root of the polynomial $\Psi_1(z) \equiv z - \beta_1$ lies on the unit circle. The least square estimator $\hat{\beta}_{1n}$, defined by (8.4), is still a consistent estimator of β_1, but has a different convergence rate and a different limit law from the stationary and the explosive cases. Indeed, for $|\beta_1| = 1$, $\hat{\beta}_{1n} - \beta_1 = O_p(n^{-1})$ and

$$\begin{aligned} T_n &\equiv \Big(\sum_{i=1}^{n-1} X_i^2\Big)^{1/2} \Big(\hat{\beta}_{1n} - \beta_1\Big) \\ &\longrightarrow^d \pm\frac{\sigma}{2}\big(W^2(1) - 1\big)\Big/\Big[\int_0^1 W^2(t)dt\Big]^{1/2} \quad \text{as} \quad n \to \infty \end{aligned} \tag{8.16}$$

under $\beta = \pm 1$, where $W(\cdot)$ denotes the standard Brownian motion on $[0, 1]$ (see, for example, Fuller (1996)). In particular, the limit distribution of T_n is nonnormal. In comparison, for the stationary case (viz., $|\beta_1| < 1$), $\hat{\beta}_{1n} - \beta_1 = O_p(n^{-1/2})$ and $T_n \longrightarrow^d N(0, \sigma^2)$ as $n \to \infty$, while for the explosive case (viz., $|\beta_1| > 1$), $\hat{\beta}_{1n} - \beta_1 = O_p(\beta_1^{-n})$ and the limit distribution of T_n is also nonnormal and is given by (8.13) with $p = 1$.

For bootstrapping the unstable AR(1) process, we combine the recipes for the stationary and the explosive cases as follows. Define the centered residuals $\tilde{\epsilon}_i = \hat{\epsilon}_i - n^{-1}\sum_{i=1}^n \hat{\epsilon}_i$, $1 \leq i \leq n$, where $\hat{\epsilon}_i = X_i - \hat{\beta}_{1n} X_{i-1}$, $1 \leq i \leq n$. Starting with $X_0^* = 0$, generate the ARB sample $X_1^*, \ldots, X_m^*$, $m \geq 1$, using the bootstrap version of the relation (8.15)

$$X_i^* = \hat{\beta}_{1n} \cdot X_{i-1}^* + \epsilon_i^* \ , \quad i \geq 1 \ , \tag{8.17}$$

where ϵ_i^*'s are obtained by simple random sampling from the collection of centered residuals $\{\tilde{\epsilon}_i : 1 \leq i \leq n\}$, with replacement. Unlike the stationary case treated in Section 8.2, here the stationarity of $\{X_1^*, \ldots, X_m^*\}$ is not of paramount interest, as the AR(1) process (8.15) is itself nonstationary in the unstable case.

The ARB version of T_n based on a resample of size m is given by

$$T^*_{m,n} = \Big(\sum_{i=1}^{m-1} X_i^* \Big)^{1/2} \Big(\beta^*_{1m} - \hat{\beta}_{1n} \Big) , \tag{8.18}$$

where $\beta^*_{1m} = \Big[\sum_{i=1}^{m-1} X_i^{*2} \Big]^{-1} \Big(\sum_{i=1}^{m-1} X_i^* X_{i+1}^* \Big)$ is obtained by replacing the X_i's in the definition of $\hat{\beta}_{1m}$ by X_i^*'s, $1 \le i \le m$ (cf. (8.4)).

An important result of Datta (1996) shows that the ARB method fails in the unstable case for the natural choice of the resample size $m = n$. Indeed, the bootstrap distribution of $T^*_{n,n}$ has a random limit for $|\beta_1| = 1$ and thus, it does not converge to the (nonrandom) limit distribution of T_n, given by (8.16). A precursor of this result was obtained by Basawa et al. (1991). They considered the model (8.15) with $N(0,1)$ error variables and used a parametric variant of the ARB method where the bootstrap error variables ϵ_i^*'s were generated from the standard normal distribution. See Inoue and Kilian (2002) for some recent advances on the problem.

To describe Datta's (1996) result, let $G(\cdot; \gamma)$ denote the probability distribution of the random variable

$$Z_\gamma \equiv \frac{\sigma \Big[\int_0^1 (1 - t + te^{2\gamma})^{-1} W(t) dW(t) \Big]}{\Big[\int_0^1 (1 - t + te^{2\gamma})^{-2} W^2(t) dt \Big]^{1/2}}, \ \gamma \in \mathbb{R} ,$$

where $W(\cdot)$ is the standard Brownian motion on $[0,1]$. Also, let $\Gamma = \big(W^2(1) - 1 \big) \Big/ \Big[2 \int_0^1 W^2(t) dt \Big]$. Define a random probability measure on $\big(\mathbb{R}, \mathcal{B}(\mathbb{R}) \big)$ by

$$\hat{G}_\infty(A) = \int_A G(dx; -\Gamma) , \quad A \in \mathcal{B}(\mathbb{R}) , \tag{8.19}$$

where the randomness in $\hat{G}_\infty$ is engendered through the random variable Γ. This $\hat{G}_\infty(\cdot)$ turns out to be the limit of the bootstrap distribution function estimator $\hat{G}_n(\cdot) \equiv P_*(T^*_{n,n} \in \cdot)$ in a suitable sense. Let $\mathbb{P}$ denote the collection of all probability measures on $\big(\mathbb{R}, \mathcal{B}(\mathbb{R}) \big)$, equipped with the topology of weak convergence. Then, $\mathbb{P}$ is metricizable (cf. Parthasarathi (1967)) and $\hat{G}_n$ and $\hat{G}_\infty$ can be viewed as $\mathbb{P}$-valued (Borel-measurable) random elements. The following result asserts convergence in distribution of $\hat{G}_n$ to $\hat{G}_\infty$ as $\mathbb{P}$-valued random elements for the case $\beta_1 = 1$. (See the discussion following Theorem A.1 in Appendix A or see Chapter 1 of Billingsley (1968).) An analogous result holds also for the case $\beta_1 = -1$.

Theorem 8.4 *Suppose that $\beta_1 = 1$ and that $E|\epsilon_1|^{2+\delta} < \infty$ for some $\delta > 0$. Let $\hat{G}_n(A) = P_*(T^*_{n,n} \in A)$ denote the bootstrap distribution of $T^*_{n,n}$ with resample size $m = n$, where $T^*_{n,n}$ is as defined in (8.18) with $m = n$. Then,*

$$\hat{G}_n \longrightarrow^d \hat{G}_\infty ,$$

where $\longrightarrow^d$ *denotes convergence in distribution on the metric space* $\mathbb{P}$.

Proof: See Datta (1996). □

Thus, Theorem 8.4 shows that for any $x \in \mathbb{R}$, if n is large, the ARB estimator $\hat{G}_n((-\infty, x]) \equiv P_*(T^*_{n,n} \le x)$ of the target probability $P(T_n \le x)$ behaves like the random variable $\hat{G}_\infty\big((-\infty, x]\big)$, which has a nondegenerate distribution on $\mathbb{R}$. As a result, there exists an η_0, $0 < \eta_0 < 1$, such that

$$\begin{aligned} &P\Big(\Big|P_*(T^*_{n,n} \le x) - P(T_n \le x)\Big| > \eta_0\Big) \\ &\longrightarrow P\Big(\Big|\hat{G}_\infty\big((-\infty, x]\big) - P(T_\infty \le x)\Big| > \eta_0\Big) > \eta_0 \end{aligned} \tag{8.20}$$

as $n \to \infty$, where T_∞ denotes the random variable appearing on the right side of $\longrightarrow^d$ in (8.16). Thus, (8.20) shows that with a positive probability, the ARB estimator $P_*(T^*_{n,n} \le x)$ takes values that are at least η_0-distance away from the target $P(T_n \le x)$ for large n. In practical applications, this means that for a nontrivial part of the sample space, the bootstrap estimator $P_*(T^*_{n,n} \le x)$ will fail to come to within η_0-distance of the true value even for an arbitrarily large sample size.

In the literature, similar inconsistency of bootstrap estimators have been noted in other problems. For sums of heavy-tailed random variables, inconsistency of the IID bootstrap of Efron (1979) has been established by Athreya (1987) under independence. A similar result for the MBB has been proved by Lahiri (1995) in the weakly dependent case (cf. Chapter 11). See also Fukuchi (1994) and Bretagnolle (1983) for other examples. The main reason for the failure of the ARB method in the unstable case seems to be different from the failure of the bootstrap methods in the other situations mentioned above. The ARB method fails here apparently because of the fact that the least square estimator $\hat{\beta}_{1n}$ of β_1, which we have used here to define the residuals for ARB resampling, does not converge at a "fast enough" rate when $|\beta_1| = 1$. Datta and Sriram (1997) propose a modified ARB where they replace the least square estimator $\hat{\beta}_{1n}$ in the resampling stage by a shrinkage estimator of β_1 that converges at a faster rate for $|\beta_1| = 1$. With this, they show that the modified ARB method produces a valid approximation to the normalized statistics T_n for *all* possible values of $\beta_1 \in \mathbb{R}$.

A second modification that is known to have worked in the other examples mentioned earlier, including the heavy-tail case and the sample extremes, is to use a resample size m that grows to infinity at a rate slower than the sample size n. On some occasions, this has been called the "m out of n" bootstrap (cf. Bickel et al. (1997)) in the literature. We shall refer to the ARB method based on a smaller resample size m as the "m out of n" ARB method. Validity of the "m out of n" ARB method for the unstable case (as well as for the other two cases) has been independently established by Datta (1996) and Heimann and Kreiss (1996).

The following result of Datta (1996) provides conditions on the resample size m for the "m out of n" ARB that ensure validity of the bootstrap approximation to the distribution of T_n "almost surely" and also "in probability". A version of the "in probability" convergence result was proved by Heimann and Kreiss (1996) under slightly weaker conditions, assuming finiteness of the second moment of ϵ_1 only.

Theorem 8.5 *Suppose that $E|\epsilon_1|^{2+\delta} < \infty$ for some $\delta > 0$, that the AR parameter $\beta_1 \in \mathbb{R}$, and that $m \uparrow \infty$ as $n \to \infty$. Also, suppose that $T^*_{m,n}$ is as defined in (8.18).*

(a) If $m/n \to 0$ as $n \to \infty$, then

$$\Delta_n \equiv \sup_{x\in\mathbb{R}} \Big| P(T_n \leq x) - P_*(T^*_{m,n} \leq x) \Big| \longrightarrow_p 0 \quad as \quad n \to \infty \,.$$

(b) If $m(\log\log n)^2/n \to 0$ as $n \to \infty$, then $\Delta_n = o(1)$ an $n \to \infty$, a.s.

Proof: See Theorem 2.1, Datta (1996). □

Theorem 8.5 shows that, for a wide range of choices of the resample size m, the "m out of n" ARB approximation adapts itself to the different shapes of the sampling distribution $\mathcal{L}(T_n)$ of T_n in all three cases, viz., in the stationary case ($|\beta_1| < 1$), to $\mathcal{L}(T_n)$ that has a normal limit, and in the explosive ($|\beta_1| > 1$) and the unstable ($|\beta_1| = 1$) cases, where $\mathcal{L}(T_n)$ has distinct nonnormal limits. An optimal choice of m seems to be unknown at this stage and it is expected to depend on the value of β_1. In a related problem Datta and McCormick (1995) have used a version of the Jackknife-After-Bootstrap method of Efron (1992) to choose m empirically. The Jackknife-After-Bootstrap method seems to be a reasonable approach for data-based choice of m in the present set up as well. Also, see Sakov and Bickel (1999) for a related work on the choice of m.

An important implication of Theorem 8.5 is that the "m out of n" ARB can be effectively used to construct valid CIs for the AR parameter β_1 under all three cases. Indeed, as the scaling factor $(\sum_{i=1}^{n-1} X_i^2)^{1/2}$ in the definition of T_n is the same in all three cases, this provides a unified way of constructing CIs for β_1 that attain the nominal coverage probability asymptotically for all $\beta_1 \in \mathbb{R}$. For $\alpha \in (0,1)$, let $\hat{t}_{m,n}(\alpha)$ denote the αth quantile of $T^*_{m,n}$, defined by $\hat{t}_{m,n}(\alpha) = \inf\{t \in \mathbb{R} : P_*(T^*_{m,n} \leq t) \geq \alpha\}$. Then, for $0 < \alpha < 1/2$, a $100(1-2\alpha)\%$ equal tailed "m out of n" bootstrap CI for β_1 is given by

$$I_{m,n}(\alpha) = \Big(\hat{\beta}_{1n} - \hat{t}_{m,n}(1-\alpha)\cdot s_n^{-1}, \;\; \hat{\beta}_{1n} - \hat{t}_{m,n}(\alpha)\cdot s_n^{-1}\Big)\,, \tag{8.21}$$

where $s_n^2 = (\sum_{i=1}^{n-1} X_i^2)$, $n \geq 2$. By Theorem 8.5, if $m = o(n)$, then

$$P_{\beta_1}\Big(\beta_1 \in I_{m,n}(\alpha)\Big) \to 1 - 2\alpha \quad \text{as} \quad n \to \infty \tag{8.22}$$

for all $\beta_1 \in \mathbb{R}$, where P_{β_1} denotes the joint distribution of $\{X_i\}_{i\geq 1}$ under a given value β_1. Thus, the CI $I_{m,n}(\alpha)$ enjoys a "robustness" property over the values of the parameter β_1 in the sense that it gives an asymptotically valid CI for all $\beta_1 \in \mathbb{R}$. However, the price paid for this remarkable property is that in the stationary case, the "m out of n" CI $I_{m,n}(\alpha)$ has a larger coverage error than the usual CI $I_{n,n}(\alpha)$ where the resample size m equals n. Thus, if there is enough evidence in the data to suggest that $\beta_1 \in (-1, 1)$, then $m = n$ is a better choice.

We now describe a numerical example to illustrate finite sample properties of the ARB in the unstable case. We considered model (8.15) with $\epsilon_i \sim N(0, 1)$ and $\beta_1 = 1$, and compared the accuracy of the usual "$m = n$" and the "m out of n" ARB approximations to the distribution function of the normalized statistic T_n when the sample of size $n = 100$. The choice of m in the "m out of n" bootstrap was taken as $m = 30$, which was close to the choice $m = n^{3/4}$, considered in Datta (1996). Figure 8.4 shows the usual ARB distribution function estimators with $m = n = 100$ and the "m out of n" ARB distribution function estimators with $m = 30$ for four data sets of size $n = 100$, generated from the AR(1) model (8.15) with the above specifications. In each case, $B = 500$ bootstrap replicates have been used to compute the bootstrap estimator $P_*(T^*_{m,n} \leq \cdot)$. The true distribution of T_n, found by 10,000 simulation runs is shown by a solid curve, while the "$m = n$" and the "$m = o(n)$" ARB distribution function estimators are denoted by dotted and dashed curves, respectively. Notice that for all four data sets, the modified ARB produced a better fit to the true distribution function of T_n. A more quantitative comparison is carried out in Table 8.2, which gives the values of the Kolmogorov-Smirnov goodness-of-fit statistic for the four data sets. For all four data sets, the distance of the "$m = n$" ARB from the true distribution function of T_n is at least 34% larger than that of the "m out of n" ARB, as measured by the Kolmogorov-Smirnov statistic.

TABLE 8.2. Values of the Kolmogorov-Smirnov goodness-of-fit statistic comparing the usual "$m = n$" ARB (column 2) and the "m out of n" ARB (column 3) distribution function estimators for four data sets of size $n = 100$ from model (8.15) in the unstable case ($\beta_1 = 1$). Column 4 is the ratio column 2/column 3.

Data Set	$m = 100$	$m = 30$	Relative Discrepancy
1	0.159	0.119	1.34
2	0.137	0.077	1.78
3	0.165	0.123	1.34
4	0.075	0.02	3.75

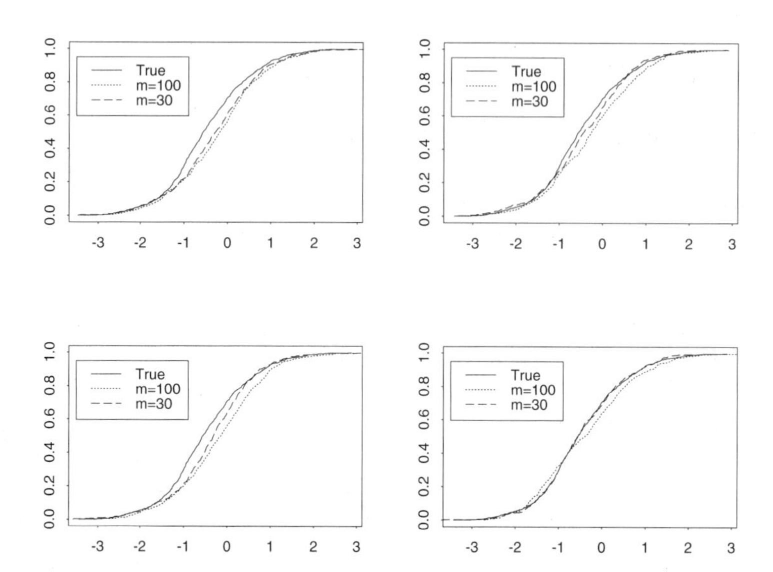

FIGURE 8.4. Bootstrap distribution function estimates and the sampling distribution of the normalized least square estimator $T_n = [\sum_{i=1}^{n-1} X_i^2]^{1/2}(\hat{\beta}_{1n} - \beta_1)$ for four data sets of size $n = 100$ from model (8.15) with $\beta_1 = 1$, $\epsilon_i \sim N(0, 1)$. The solid line is for the true distribution function, while the dashed and the dotted lines respectively denote the usual "$m = n$" approximation and the "m out of n" ARB approximation with $m = 30$.

8.5 Bootstrapping a Stationary ARMA Process

The idea of bootstrapping a stationary AR(p) model can be easily adapted to the more general class of stationary autoregressive and moving average (ARMA) processes. The key observation here is that a stationary ARMA process may be expressed both as an infinite order autoregressive process as well as an infinite-order moving average process, when the invertibility conditions hold. The autoregressive representation allows one to identify the "residuals" in terms of the observable random variables, which are then resampled to generate the "bootstrap error variables". These, in turn, are used to generate the bootstrap observations by employing the ARMA recursive relation. The details of the method are notationally awkward. As a result, we will first look at some auxiliary properties of the ARMA process itself that will help us understand the main steps of the ARMA bootstrap method better.

Let $\{X_i\}_{i\in\mathbb{Z}}$ be a stationary ARMA (p, q) process satisfying the difference equation

$$X_i = \sum_{j=1}^{p} \beta_i X_{i-j} + \sum_{j=1}^{q} \alpha_j \epsilon_{i-j} + \epsilon_i \ , \quad i \in \mathbb{Z} \ , \tag{8.23}$$

where $p, q \in \mathbb{Z}_+$ with $p+q \in \mathbb{N}$, $\{\epsilon_i\}_{i\in\mathbb{Z}}$ is a sequence of iid random variables with $E\epsilon_1 = 0$ and where $\beta_1, \ldots, \beta_p, \alpha_1, \ldots, \alpha_q \in \mathbb{R}$ are parameters. Let $\beta(z) = 1 - \sum_{j=1}^{p} \beta_j z^j$ and $\alpha(z) = 1 + \sum_{j=1}^{q} \alpha_j z^j$, $z \in \mathbb{C}$ denote the characteristic polynomials associated with the autoregressive part and the moving average part, respectively. We shall suppose that the parameters $\beta_1, \ldots, \beta_p, \alpha_1, \ldots, \alpha_q$ satisfy the causality and invertibility conditions that

$$\begin{cases} \beta(z) \neq 0 & \text{for all } |z| \leq 1 \\ \alpha(z) \neq 0 & \text{for all } |z| \leq 1 \end{cases} \tag{8.24}$$

and $\alpha(z)$ and $\beta(z)$ have no common zero. Furthermore, we suppose that $\alpha_q \neq 0$, and $\beta_p \neq 0$. Then there exists a $\eta_0 > 1$ (depending on the values of $\beta_1, \ldots, \beta_p$ and $\alpha_1, \ldots, \alpha_q$) such that in the disc $|z| \leq \eta_0$, we have the power series expansions

$$\begin{aligned} \big[\beta(z)\big]^{-1} &= \sum_{j=0}^{\infty} b_j z^j, \\ \big[\alpha(z)\big]^{-1} &= \sum_{j=0}^{\infty} a_j z^j, \quad \text{and} \\ \big[\beta(z)\big]^{-1}\alpha(z) &= \sum_{j=0}^{\infty} \rho_j z^j = \Big(\sum_{j=0}^{\infty} \gamma_j z^j\Big)^{-1}. \end{aligned} \tag{8.25}$$

As a consequence of this, we may express the X_i's as an infinite order AR process and also as an infinite-order moving average (MA) process (cf. Chapter 3, Brockwell and Davis (1991)). Indeed, the following are true:

$$\begin{aligned} X_i &= \sum_{j=0}^{\infty} \rho_j \epsilon_{i-j}\,, \quad i \in \mathbb{Z} \\ \epsilon_i &= \sum_{j=0}^{\infty} \gamma_j X_{i-j}\,, \quad i \in \mathbb{Z}, \quad \text{and} \\ \epsilon_i &= \sum_{j=0}^{\infty} a_j (X_{i-j} - \beta_1 X_{i-j-1} - \cdots - \beta_p X_{i-j-p})\,, \quad i \in \mathbb{Z}\,, \end{aligned} \tag{8.26}$$

where the constants ρ_i's, γ_i's, and a_i's are given by (8.25). From (8.25), it follows that $b_0 = a_0 = \rho_0 = \gamma_0 = 1$. We adopt the convention that for $i < 0$, $b_i = a_i = \rho_i = \gamma_i = 0$. Then, using the identity

$$[\alpha(z)]\Big[\sum_{j=0}^{\infty} a_j z^j\Big] = 1 \quad \text{for all} \quad |z| < \eta_0\,,$$

and, for all $k \geq 1$, equating the coefficients of z^k in the product on the left side to zero, we have

$$a_k + \alpha_1 a_{k-1} + \cdots + \alpha_q a_{k-q} = 0 \quad \text{for all} \quad k \geq 1\,. \tag{8.27}$$

Now, setting $\alpha_0 = 1$ and $\beta_0 = -1$, interchanging the summations three times and using (8.23), (8.25)–(8.27), we have, for all $i \geq 1 - q$ (cf. (2.3), Kreiss (1987)),

$$\begin{aligned}
\epsilon_i &= \sum_{j=0}^{\infty} a_j \Big(X_{i-j} - \beta_1 X_{i-j-1} - \cdots - \beta_p X_{i-j-p} \Big) \\
&= \sum_{j=1}^{i} a_{j-1} \Big(-\sum_{k=0}^{p} \beta_k X_{i+1-j-k} \Big) + \sum_{j=i+1}^{\infty} a_{j-1} \Big(\sum_{k=0}^{q} \alpha_k \epsilon_{i+1-j-k} \Big) \\
&= \sum_{j=1}^{i} a_{j-1} \Big(-\sum_{k=0}^{p} \beta_k X_{i+1-j-k} \Big) + \sum_{k=0}^{q} \sum_{s=k}^{\infty} \alpha_k a_{i+1+s-k} \epsilon_{-s} \\
&= \sum_{j=1}^{i} a_{j-1} \Big(-\sum_{k=0}^{p} \beta_k X_{i+1-j-k} \Big) + \sum_{s=0}^{q-1} \epsilon_{-s} \Big(\sum_{k=0}^{s} a_{i+1+s-k} \alpha_k \Big) .
\end{aligned} \tag{8.28}$$

Note that by (8.25), $a_i = O(\eta_0^{-i})$ as $i \to \infty$. Hence, for large i's, the contribution of the second term in (8.28) is small. Thus, we may concentrate on the first term only and define an "approximation" to ϵ_i by estimating the coefficients a_{j-1}'s and β_k's above. This observation forms the basis for defining a residual-based resampling method for a stationary ARMA (p, q) process, which we describe next.

Suppose that a finite segment $\mathcal{X}_{n+p} = \{X_{1-p}, \ldots, X_n\}$ of the ARMA (p, q) process $\{X_i\}_{i \in \mathbb{Z}}$ of (8.23) is observed. Let $(\tilde{\beta}_{1n}, \ldots, \tilde{\beta}_{pn})'$ and $(\tilde{\alpha}_{1n}, \ldots, \tilde{\alpha}_{qn})'$ respectively denote some estimators of the parameter vectors $(\beta_1, \ldots, \beta_p)'$ and $(\alpha_1, \ldots, \alpha_q)'$ based on $\mathcal{X}_{n+p}$ such that

$$\sum_{j=1}^{p} |\tilde{\beta}_{jn} - \beta_j| + \sum_{j=1}^{q} |\tilde{\alpha}_{jn} - \alpha_j| \longrightarrow_p 0 \quad \text{as} \quad n \to \infty . \tag{8.29}$$

Then, there exists $1 < \eta_1 < \eta_0$ such that, with high probability, the reciprocal of the function $\tilde{\alpha}(z) \equiv 1 + \sum_{j=1}^{q} \tilde{\alpha}_{jn} z^j$ admits the power series expansion

$$\big[\tilde{\alpha}(z)\big]^{-1} = \sum_{j=0}^{\infty} \tilde{a}_{jn} z^j , \quad |z| \leq \eta_1 \tag{8.30}$$

for large values of n (cf. Lemma 2.2, Kreiss and Franke (1992)). Then, using (8.28)–(8.30), we define the "residuals" $\hat{\epsilon}_{in}$'s by the relation

$$\hat{\epsilon}_{in} = \sum_{j=1}^{i} \tilde{a}_{j-1,n} \Big(-\sum_{k=0}^{p} \tilde{\beta}_{kn} X_{i+1-j-k} \Big) , \quad i = 1, \ldots, n , \tag{8.31}$$

where $\tilde{\beta}_{0n} = -1$. Note that for a purely AR(p) process, if we set the moving average parameters $\alpha_1, \ldots, \alpha_q$ equal to zero and also take $\tilde{\alpha}_{jn} = 0$, $1 \leq j \leq$

q, then it follows from (8.30) that $\tilde{a}_{0n} = 1$ and $\tilde{a}_{jn} = 0$ for all $j \geq 1$. In this case, the "residual" $\hat{\epsilon}_{in}$ of (8.31) reduces to $\hat{\epsilon}_{in} = X_i - \sum_{k=1}^{p} \tilde{\beta}_{kn} X_{i-k}$, which corresponds to the residuals defined in Section 8.2 for the ARB method with $\tilde{\beta}_{kn} = \hat{\beta}_{kn}$, the least square estimator of β_k, $1 \leq k \leq p$.

The remaining steps of the bootstrap procedure for the stationary ARMA process (we will call it the ARMA bootstrap or the ARMAB, in short) parallel the steps in the ARB. Starting with $\hat{\epsilon}_{in}$, $1 \leq i \leq n$, we form the centered residuals $\tilde{\epsilon}_{in} = \hat{\epsilon}_{in} - \bar{\epsilon}_n$, $1 \leq i \leq n$, where $\bar{\epsilon}_n = n^{-1} \sum_{i=1}^{n} \hat{\epsilon}_{in}$. Next, we generate iid bootstrap error variables ϵ_i^*, $i \geq 1 - \max\{p, q\}$ by sampling at random, with replacement from $\{\tilde{\epsilon}_{in} : 1 \leq i \leq n\}$. Then, we define the bootstrap observations by using the recursion relation

$$X_i^* = \sum_{j=1}^{p} \tilde{\beta}_{jn} X_{i-j}^* + \sum_{j=1}^{q} \tilde{\alpha}_{jn} \epsilon_{i-j}^* + \epsilon_i^* \tag{8.32}$$

for $i \geq 1 - \max\{p, q\}$, where, for $i \leq -\max\{p, q\}$, we set $X_i^* = 0$ and $\epsilon_i^* = 0$. The ARMA-bootstrap version of a random variable $T_n \equiv t_n(\mathcal{X}_{n+p}; \beta_1, \ldots, \beta_p, \alpha_1, \ldots, \alpha_q; F)$ is now defined by

$$T_n^* = t_n\left(\mathcal{X}_{n+p}^*; \tilde{\beta}_{1n}, \ldots, \tilde{\beta}_{pn}, \tilde{\alpha}_{1n}, \ldots, \tilde{\alpha}_{qn}; \tilde{F}_n\right) , \tag{8.33}$$

where $\mathcal{X}_{n+p}^* = (X_{1-p}^*, \ldots, X_n^*)$, F is the unknown distribution function of ϵ_1 and $\tilde{F}_n(x) = n^{-1} \sum_{i=1}^{n} \mathbb{1}(\tilde{\epsilon}_{in} \leq x)$, $x \in \mathbb{R}$ denotes the distribution function of ϵ_1^*. One may, if desired, use a resample size $m + p$ at the bootstrap stage instead of the original sample size $n + p$; it is a relatively simple task to modify the definition of the bootstrap variable T_n^* in this case.

As in the case of the ARB, the variables X_i^*'s generated by (8.32) (starting with $X_1^* = 0 = \epsilon_i^*$ for $i \leq -\max\{p, q\}$) are not stationary. However, as noted earlier, the effect of the initial values under (8.29) dies out exponentially fast with high probability as $n \to \infty$. As a result, the nonstationarity of the X_i^*'s typically does not have an effect on the limit. Alternatively, one may proceed as in the ARB by generating a long enough chain and discarding a set of beginning values to obtain the bootstrap observations.

Bose (1990) considers bootstrapping the pure moving average model ARMA(0,1) and establishes second-order correctness of a version of the ARMAB for the least square estimator of the MA parameter. The general version of the ARMA bootstrap method presented here is due to Kreiss and Franke (1992), who establish the validity of the ARMAB for a class of M-estimators. Allen and Datta (1999) propose a modification to Kreiss and Franke's (1992) proposal for bootstrapping M-estimators and show superiority of the modified version using some simulation results.

In the following, we describe the results of Kreiss and Franke (1992) and Allen and Datta (1999) in more details. Let F denote the distribution function of the error variable ϵ_1 and let $\psi : \mathbb{R} \to \mathbb{R}$ be a function such that

$$E\psi(\epsilon_1) = 0 . \tag{8.34}$$

Then, an M-estimator $\hat{\theta}_n$ of the parameter $\theta_0 = (\beta_1, \ldots, \beta_p, \alpha_1, \ldots, \alpha_q)'$ based on the function ψ is given by a measurable solution to the equation (in $\theta \in \mathbb{R}^{p+q}$)

$$n^{-1} \sum_{j=1}^{n} \psi\big(\epsilon_j(\theta)\big) Z(j-1;\theta) = 0 , \tag{8.35}$$

where, with $\theta = (\theta_1, \ldots, \theta_{p+q})'$, the variables $\epsilon_j(\theta)$ and $Z(j-1;\theta)$ are defined as $\epsilon_j(\theta) \equiv \sum_{k=0}^{j-1} [a_k(\theta)] \ (X_{j-k} - \sum_{i=1}^{p} \theta_i X_{j-k-i})$ and $Z(j-1;\theta) = \sum_{k=0}^{j-1} [a_k(\theta)] \ \big(X_{j-k-1}, \ldots, X_{j-k-p}; \epsilon_{j-k-1}(\theta), \ldots, \epsilon_{j-k-q}(\theta)\big)'$, $1 \leq j \leq n$. Here, the factors $a_k(\theta)$'s are formally defined by the relation (cf. (8.25),(8.30))

$$\sum_{k=0}^{\infty} [a_k(\theta)] z^k = \left(1 + \sum_{j=1}^{q} \theta_{p+j} z^j\right)^{-1} , \ |z| \leq 1 . \tag{8.36}$$

Furthermore, in the definition of $Z(j-1;\theta)$'s, we set $\epsilon_i(\theta) = 0$ if $i \leq 0$. Note that at the true parameter value, $\theta = \theta_0$, $\theta_{p+j} = \alpha_j$, $1 \leq j \leq q$ and hence, by (8.25) and (8.36), $a_k(\theta) = a_k$, $k \geq 0$. Thus, the variables $\epsilon_j(\theta)$'s, $1 \leq j \leq n$ play the role of "residuals" when θ_0 is "estimated" by θ. In particular, from (8.31) and (8.36), for $\theta = (\tilde{\beta}_{1n}, \ldots, \tilde{\beta}_{pn}; \tilde{\alpha}_{1n}, \ldots, \tilde{\alpha}_{qn})'$, we get $\epsilon_j(\theta) = \hat{\epsilon}_{jn}$, $1 \leq j \leq n$.

Next we describe a way of studentizing the multivariate M-estimator $\hat{\theta}_n$. Under some regularity conditions on ψ and F, it can be shown (cf. Kreiss and Franke (1992)) that

$$\sqrt{n}(\hat{\theta}_n - \theta_0) = \Gamma_n^{-1} \bar{\Psi}_n + o_p(1) , \tag{8.37}$$

where $\Gamma_n = n^{-1} \sum_{j=1}^{n} \psi'(\epsilon_j) Z(j-1) Z(j-1)'$, $\bar{\Psi}_n = n^{-1/2} \sum_{j=1}^{n} \psi(\epsilon_j) Z(j-1)$ and $Z(j-1) = Z(j-1;\theta_0)$, $1 \leq j \leq n$. In the definition of Γ_n and elsewhere in this section, ψ' and ψ'', respectively, denote the first and the second derivatives of ψ. From (8.37), now it follows that we may studentize $\hat{\theta}_n$ using the variance matrix estimator

$$\hat{\Sigma}_n = \hat{\Gamma}_n^{-1} \hat{\Lambda}_n \hat{\Gamma}_n^{-1} , \tag{8.38}$$

where, with $\hat{Z}_n(j-1) = Z(j-1;\hat{\theta}_n)$, $\hat{\Gamma}_n = n^{-1} \sum_{j=1}^{n} \psi'(\hat{\epsilon}_{jn}) \hat{Z}_n(j-1) \hat{Z}_n(j-1)'$, and $\hat{\Lambda}_n = n^{-1} \sum_{j=1}^{n} \psi^2(\hat{\epsilon}_{jn}) \hat{Z}_n(j-1) \hat{Z}_n(j-1)'$. For definiteness, here and in the rest of this section, we assume that the generic estimator $\tilde{\theta}_n \equiv (\tilde{\beta}_{1n}, \ldots, \tilde{\beta}_{pn}, \tilde{\alpha}_{1n}, \ldots, \tilde{\alpha}_{qn})'$ of the ARMA model parameters is chosen to be the M-estimator $\hat{\theta}_n = (\hat{\theta}_{1n}, \ldots, \hat{\theta}_{p+q,n})'$. In particular, the residuals $\hat{\epsilon}_{in}$'s in (8.31) are defined with $\tilde{\theta}_n = \hat{\theta}_n$. With this, the studentized version of the M-estimator is given by

$$T_n = \sqrt{n} \hat{\Sigma}_n^{-1/2} (\hat{\theta}_n - \theta_0) , \tag{8.39}$$

where $\hat{\Sigma}_n^{-1/2}$ is a $(p+q) \times (p+q)$ matrix (not necessarily symmetric) such that $\left(\hat{\Sigma}_n^{-1/2}\right)\left(\hat{\Sigma}_n^{-1/2}\right)' = \hat{\Sigma}_n^{-1}$. For example, we may use the Cholesky decomposition of $\hat{\Sigma}_n$ to find $\hat{\Sigma}_n^{-1/2}$.

Next, we define the bootstrap version of the studentized M-estimator T_n of (8.39). With the ARMAB "observations" X_i^*'s generated by (8.32), define the bootstrapped M-estimator θ_n^* as a solution of the equation (cf. (8.35))

$$n^{-1} \sum_{j=1}^{n} \psi_1\big(\epsilon_j^*(\theta)\big) Z^*(j-1;\theta) = 0 \ , \tag{8.40}$$

where $\epsilon_j^*(\theta)$ and $Z^*(j-1;\theta)$ are defined by replacing X_i's in the definitions of $\epsilon_j(\theta)$ and $Z(j-1;\theta)$, respectively, by X_i^*'s, and where $\psi_1(x) \equiv \psi(x) - E_*\psi(\epsilon_1^*)$. As in Section 4.3, we center the function ψ in the definition of the bootstrapped M-estimator θ_n^* to ensure

$$E_*\psi_1(\epsilon_1^*) = 0 \ ,$$

the bootstrap analog of the structural equation (8.34). It is worth noting that for the IID bootstrap of Efron (1979), such an explicit centering is important in linear regression models (cf. Freedman (1981), Shorack (1982), Lahiri (1992b)), but not so for identically distributed observations (cf. Lahiri (1992c, 1994)), as the centering factor is automatically negligible, by the definition of the M-estimator $\hat{\theta}_n$ itself. For the ARMA bootstrap, the centering has a negligible effect asymptotically (cf. Allen and Datta (1999), p. 368), but it improves finite sample accuracy of the ARMAB, particularly for models with a nontrivial MA-component.

Next, define Σ_n^* by replacing the X_i's and $\hat{\theta}_n$ in the definition of $\hat{\Sigma}_n$ with X_i^*'s and θ_n^*, respectively. Then, the bootstrap version of T_n is given by

$$T_n^* = \sqrt{n}\big[\Sigma_n^*\big]^{-1/2}(\theta_n^* - \hat{\theta}_n) \ , \tag{8.41}$$

where $[\Sigma_n^*]^{-1/2}$ is a matrix satisfying $([\Sigma_n^*]^{-1/2})([\Sigma_n^*]^{-1/2})' = [\Sigma_n^*]^{-1}$. The following result asserts the validity of the ARMAB for T_n.

Theorem 8.6 *Suppose that ψ is twice continuously differentiable with bounded derivatives ψ' and ψ'' such that $E\psi(\epsilon_1) = 0$, $E(\psi(\epsilon_1))^2 \in (0,\infty)$ and $E\psi'(\epsilon_1) \neq 0$. Also, suppose that $\{\hat{\theta}_n\}_{n\geq 1}$ is a sequence of measurable solutions of (8.35) that is strongly consistent for θ_0. If, in addition, $E|\epsilon_1|^3 < \infty$, then*

(a) $P_*\Big($there exists a solution θ_n^* of (8.40) such that $|\theta_n^* - \hat{\theta}_n| \leq Cn^{-1/2}\log n\Big) \geq 1 - C(\log n)^{-2}$ a.s. ;

(b) for the sequence of solutions $\{\theta_n^*\}_{n\geq 1}$ of part (a),

$$\sup_{x \in \mathbb{R}^{p+q}} \Big| P_*(T_n^* \leq x) - P(T_n \leq x) \Big| \to 0 \quad as \quad n \to \infty \ , \qquad a.s.$$

Proof: Part (a) follows by using arguments similar to the proof of Theorem 4.2, Chapter 4. Allen and Datta (1999) gives a proof of part (b) assuming $\sqrt{n}(\theta_n^* - \hat{\theta}_n) = O_{P_*}(1)$, a.s. Essentially, the same proof works in this case. We leave the details of the modification of their proof to the reader. □

Theorem 8.6 shows that the ARMAB provides a valid approximation to the distribution of the studentized M-estimator T_n for a large class of score functions ψ. Asymptotically valid approximations to the distribution of T_n may also be obtained by applying the ARMAB method to the studentized version of the "linear approximation" $\Gamma_n^{-1}\bar{\Psi}_n$ to $\sqrt{n}(\hat{\theta}_n - \theta_0)$, given by

$$T_{1n} = \sqrt{n}\hat{\Sigma}_n^{-1/2}(\Gamma_n^{-1}\bar{\Psi}_n) \ .$$

Indeed, Kreiss and Franke (1992) establish validity of the ARMAB to the distribution of $\sqrt{n}(\hat{\theta}_n - \theta_0)$ by applying the bootstrap to $\sqrt{n}\Gamma_n^{-1}\bar{\Psi}_n$. As the effect of ignoring the higher-order terms in the stochastic expansion of $\sqrt{n}(\hat{\theta}_n - \theta_0)$ would show up only in the second- and higher-order terms in the corresponding Edgeworth expansions, a first-order analysis fails to distinguish between the two versions. A finite sample simulation study in Allen and Datta (1999) points out the superiority of applying the ARMAB to T_n. A study of the second-order properties of the ARMAB and theoretical analysis of the two approaches seem to be nonexistent in the literature at this point.

9
Frequency Domain Bootstrap

9.1 Introduction

In this chapter, we describe a special type of transformation based-bootstrap, known as the frequency domain bootstrap (FDB). Given a finite stretch of observations from a stationary time series, here we consider the discrete Fourier transforms (DFTs) of the data and use the transformed values in the frequency domain to derive bootstrap approximations (hence, the name FDB). In Section 9.2, we describe the FDB for a class of estimators, called the *ratio statistics*. Dahlhaus and Janas's (1996) results show that under suitable regularity conditions, the FDB is second-order accurate for approximating the sampling distributions of ratio statistics. In Section 9.3, we describe the FDB method and its properties in the context of spectral density estimation. Material covered in Section 9.3 is based on the work of Franke and Härdle (1992). In Section 9.4, we describe a modified version of the FDB due to Kreiss and Paparoditis (2003) that, under suitable regularity conditions, removes some of the limitations of the standard FDB and yields valid approximations to the distributions of a larger class of statistics than the class of ratio statistics. It is worth pointing out that the results presented in this chapter on the FDB are valid only for linear processes.

9.2 Bootstrapping Ratio Statistics

9.2.1 Spectral Means and Ratio Statistics

Let $\{X_i\}_{i \in \mathbb{Z}}$ be a stationary time series with $EX_0 = 0$ and with spectral density f. Let $\Pi = [-\pi, \pi]$. Let $d_n(w) = \sum_{t=1}^{n} X_t \exp(-\iota wt)$, $w \in \Pi$ denote the finite Fourier transform of $X_1, \ldots, X_n$ and let

$$I_n(w) \equiv (2\pi n)^{-1} d_n(w) d_n(-w), \; w \in \Pi \tag{9.1}$$

denote the periodogram, where $\iota = \sqrt{-1}$. Statistical analysis of $\{X_i\}_{i \in \mathbb{Z}}$ in the frequency domain is carried out in terms of the transformed values $d_n(w)$'s. A class of level-1 parameters of interest that are commonly considered in frequency domain analysis of the time series are given by

$$A(\xi; f) = \int \xi f = \Big(\int_0^\pi \xi_1(w) f(w) dw, \ldots, \int_0^\pi \xi_p(w) f(w) dw \Big)', \tag{9.2}$$

where $\xi = (\xi_1, \ldots, \xi_p)'$ and where, for $i = 1, \ldots, p$, $\xi_i : [0, \pi] \to \mathbb{R}$ is a function of bounded variation. The parameter $A(\xi, f)$ in (9.2) is called a *spectral mean.* A canonical estimator of $A(\xi, f)$ is given by

$$A(\xi; I_n) = \int \xi I_n = \Big(\int_0^\pi \xi_1(w) I_n(w) dw, \ldots, \int_0^\pi \xi_p(w) I_n(w) dw \Big)'. \tag{9.3}$$

The following are examples of some common spectral means and their canonical estimators.

Example 9.1: (*Autocovariance estimator*). Let $\xi(w) = 2\cos kw$, $w \in [0, \pi]$ for some integer k. Then,

$$\begin{aligned}
A(\xi; I_n) &= 2\int_0^\pi (\cos kw) I_n(w) dw \\
&= \int_{-\pi}^{\pi} I_n(w) \exp(-\iota kw) dw \\
&= (2\pi n)^{-1} \sum_{t=1}^{n} \sum_{j=1}^{n} \int_{-\pi}^{\pi} X_t X_j \exp(-\iota wt) \exp(\iota wj) \exp(-\iota kw) dw \\
&= n^{-1} \sum_{t=1}^{n-k} X_t X_{t+k} \,,
\end{aligned}$$

as $\int_{-\pi}^{\pi} \exp(\iota mw) dw = 0$ for any nonzero integer m. By similar arguments,

$$A(\xi; f) = EX_1 X_{1+k} \,.$$

Thus, $A(\xi; I_n)$ is the usual moment estimator of the autocovariance $EX_1 X_{1+k}$, for any $k \in \mathbb{Z}$.

Example 9.2: (*Spectral distribution function estimator*). Let $\xi(w) = \mathbb{1}_{[0,\lambda]}(w)$, $w \in [0, \pi]$ for some $0 \leq \lambda \leq \pi$. Then,

$$A(\xi; f) = \int_0^\lambda f(w)dw \equiv F(\lambda) \tag{9.4}$$

and

$$A(\xi; I_n) = \int_0^\lambda I_n(w)dw \equiv F_n(\lambda) \tag{9.5}$$

are, respectively, the spectral distribution function and its periodogram based estimator. □

Suppose that $\{X_i\}_{i \in \mathbb{Z}}$ is a linear process, i.e., $\{X_i\}_{i \in \mathbb{Z}}$ has a representation of the form

$$X_i = \sum_{j \in \mathbb{Z}} a_j \zeta_{i-j}, \ i \in \mathbb{Z}$$

where $\{a_j\}_{j \in \mathbb{Z}} \subset \mathbb{R}$ are constants satisfying $\sum_{j \in \mathbb{Z}} j^2 |a_j| < \infty$ and where $\{\zeta_j\}_{j \in \mathbb{Z}}$ is a collection of iid random variables with $E\zeta_1 = 0$ and $\sigma^2 \equiv E\zeta_1^2 = 1$. Then, the results of Dahlhaus (1983, 1985) imply that under some suitable regularity conditions, for a real-valued ξ (for simplicity of discussion),

$$\begin{aligned} R_n &\equiv \sqrt{n}\big(A(\xi; I_n) - A(\xi; f)\big) \\ &\longrightarrow^d N\Big(0, \Big[2\pi \int \xi^2 f^2 + \frac{\kappa_4}{\sigma^4}\Big(\int \xi f\Big)^2\Big]\Big) , \end{aligned} \tag{9.6}$$

where κ_4 is the fourth cumulant of ζ_1. Dahlhaus and Janas (1996) showed that under some regularity conditions, the FDB version of R_n (to be described below) converges in distribution to $N(0, 2\pi \int \xi^2 f^2)$, with probability 1. As a consequence, the FDB yields a valid approximation if either $\kappa_4 = 0$ or $\int \xi f = 0$. The first condition is restrictive, as it specifies the fourth cumulant of the innovations *exactly* and it holds, for example, if ζ_1 is Gaussian. In comparison, the second condition is less restrictive on the distribution of the innovations but it limits the collection of $\xi(\cdot)$ functions. Dahlhaus and Janas (1996) identified a large class of spectral mean estimators, called the *ratio statistics*, for which the $\xi(\cdot)$ functions satisfy the second condition $\int \xi f = 0$ for any given spectral density f. We now describe the ratio statistics.

Let

$$g(w) = f(w)/F(\pi), \ w \in [0, \pi] \tag{9.7}$$

denote the normalized spectral density of the process $\{X_i\}_{i \in \mathbb{Z}}$, where F is the spectral distribution function, given by (9.4). Then, $A(\xi; g) \equiv \int \xi g$ is a *normalized spectral mean* parameter with kernel $\xi : [0, \pi] \to \mathbb{R}^p$. The corresponding canonical estimator is defined as

$$A(\xi; J_n) \equiv \int_0^\pi \xi(w) J_n(w) dw \ , \tag{9.8}$$

where $J_n(w) = I_n(w)/F_n(\pi)$ is the normalized periodogram and $F_n(\cdot)$ is the spectral distribution estimator of (9.5). Note that the normalized spectral mean estimator $A(\xi; J_n)$ can be written as the ratio of two spectral mean estimators as

$$A(\xi; J_n) = \int_0^\pi \xi(w) I_n(w) dw \Big/ \int_0^\pi I_n(w) dw \tag{9.9}$$

and, hence, is called a *ratio statistic* or a *ratio estimator*. For $\xi(w) = 2\cos kw$, $w \in [0, \pi]$ of Example 9.1, we get $A(\xi; g) = EX_1X_{1+k}/E(X_1X_1)$, the lag k autocorrelation of the process $\{X_i\}_{i\in\mathbb{Z}}$ and the corresponding ratio estimator $A(\xi; J_n) = \sum_{i=1}^{n-k} X_iX_{i+k}/\sum_{i=1}^{n} X_i^2$ is the lag k sample autocorrelation. Similarly, $\xi = \mathbb{1}_{[0,\lambda]}$ of Example 9.2 yields the normalized spectral distribution function and its canonical estimator. Although the class of ratio statistics is more restricted than the class of spectral mean estimators, ratio statistics play an important role in many inference problems in the frequency domain. For example, the Yule-Walker estimators of autoregressive parameters are based on estimators of autocorrelation, Bartlett's U_p-statistic for a goodness-of-fit test is based on the normalized distribution function estimator, etc. See Dahlhaus and Janas (1996) for further discussion on ratio statistics.

Next we show that for the ratio statistics, the second term in the asymptotic variance vanishes (cf. (9.6)). Indeed, for a given kernel $\xi : [0, \pi] \to \mathbb{R}^p$, we have

$$\begin{aligned} \sqrt{n}\big(A(\xi; J_n) - A(\xi; g)\big) &= \sqrt{n}\Big(\frac{\int \xi I_n}{\int I_n} - \frac{\int \xi f}{\int f}\Big) \\ &= \frac{\sqrt{n}}{\int I_n \cdot \int f}\Big[\int f \cdot \int \xi I_n - \int I_n \int \xi f\Big] \\ &= \frac{\sqrt{n}}{\int I_n \cdot \int f} \cdot \int \psi I_n \ , \end{aligned} \tag{9.10}$$

where $\psi(w) \equiv \xi(w)\int f - \int \xi f$, $w \in [0, \pi]$. Because $\int \psi f = 0$, ψ satisfies the second requirement for the validity of the FDB. In Section 9.2.3, we shall show that under suitable conditions, the FDB outperforms the normal approximation when applied to the ratio statistics.

9.2.2 Frequency Domain Bootstrap for Ratio Statistics

It is well known (cf. Chapters 4 and 5, Brillinger (1981), Lahiri (2003a)) that under some mild conditions on the process $\{X_i\}_{i\in\mathbb{Z}}$, the finite Fourier transforms $d_n(w_1), \ldots, d_n(w_k)$ are asymptotically independent for any given set of distinct frequencies $w_1, \ldots, w_k \in [0, \pi]$. In an unpublished article, Hurvich and Zeger (1987) proposed the FDB based on this observation. Since the DFTs at distinct frequencies are approximately independent, they suggested resampling suitably studentized DFT values *one at a time* as in Efron's (1979) IID bootstrap. Since its inception, properties of the FDB have been studied by many authors. Nordgaard (1992) considered the FDB for Gaussian processes, while Franke and Härdle (1992) applied the FDB to the problem of spectral density estimation (see Section 9.3 for more details). In this section, we describe the FDB for the ratio statistics of Section 9.2.1, based on the work of Dahlhaus and Janas (1996).

Let $I_{jn} = I_n(2\pi j/n)$, $j = 1, \ldots, n_0$ denote the periodogram at discrete ordinates $2\pi j/n$, where $n_0 = \lfloor n/2 \rfloor$. Note that for any given set of frequencies $0 < w_1 < \cdots < w_k < \pi$, the scaled periodogram values $I_n(w_j)/f(w_j)$, $j = 1, \ldots, k$ are asymptotically pivotal in the sense that they are asymptotically distributed as iid Exponential (1) random variables (cf. Theorem 5.2.6, Brillinger (1981)) and, hence, have limit distributions free of unknown parameters. This suggests that we may use an estimator $\hat{f}_n$ of f (say, a kernel density estimator) to studentize I_{jn}'s. The main steps of the FDB for the ratio statistics are as follows:

Step 1: Form the studentized periodogram ordinates $\hat{\epsilon}_{jn} = I_{jn}/\hat{f}_{jn}$, $j = 1, 2, \ldots, n_0$, where $\hat{f}_{jn} = \hat{f}_n(\lambda_{jn})$ and $\lambda_{jn} \equiv 2\pi j/n$.

Step 2: Define the rescaled variables $\tilde{\epsilon}_{jn} = \hat{\epsilon}_{jn}/\hat{\epsilon}_{\cdot n}$, $1 \leq j \leq n_0$, where $\hat{\epsilon}_{\cdot n} = n_0^{-1} \sum_{i=1}^{n_0} \hat{\epsilon}_{in}$.

Step 3: Generate the bootstrap variables ϵ_{jn}^*, $j = 1, \ldots, n_0$ by sampling randomly with replacement from the collection $\{\tilde{\epsilon}_{jn} : j = 1, \ldots, n_0\}$.

Step 4: Define the bootstrap periodogram values by $I_{jn}^* = \hat{f}_{jn} \cdot \epsilon_{jn}^*$, $1 \leq j \leq n_0$.

Define the bootstrap versions of $J_n(w)$ and $g(w)$ at $w = \lambda_{jn}$ by $J_{jn}^* = I_{jn}^*/[\frac{2\pi}{n} \sum_{i=1}^{n_0} I_{in}^*]$ and $\hat{g}_{jn} = \hat{f}_{jn}/[\frac{2\pi}{n} \sum_{i=1}^{n_0} \hat{f}_{in}]$, $j = 1, \ldots, n_0$. Then, the FDB version of the centered and scaled ratio estimator $A(\xi, J_n)$, i.e., of

$$T_n \equiv \sqrt{n}\big(A(\xi; J_n) - A(\xi; g)\big) \tag{9.11}$$

is given by

$$T_n^* = \sqrt{n}\big(B(\xi; J_n^*) - B(\xi; \hat{g}_n)\big) , \tag{9.12}$$

where

$$B(\xi; J_n^*) = \frac{2\pi}{n} \sum_{j=1}^{n_0} \xi(\lambda_{jn}) J_{jn}^*$$

and

$$B(\xi; \hat{g}_n) = \frac{2\pi}{n} \sum_{j=1}^{n_0} \xi(\lambda_{jn}) \hat{g}_{jn} \ .$$

The summations in $B(\xi; J_n^*)$ and $B(\xi; \hat{g}_n)$ above are approximations to the corresponding integrals over the interval $[0, \pi]$, where the approximating step functions are constant over subintervals of length $2\pi/n$. This results in the factor $2\pi/n$, which is comparable to the factor (π/n_0) appearing in Dahlhaus and Janas (1996). However, the effect of this scalar multiplier vanishes for ratio statistics, as the constants from the numerator and the denominator cancel out each other.

The rescaling in Step 2 plays a role similar to the centering of the estimating equations in the context of bootstrapping the M-estimators (cf. Section 4.3). Without the rescaling, the FDB approximation may fail to be consistent. Although the given variables X_i's are dependent, the studentized periodogram variables $\hat{\epsilon}_{jn}$'s are approximately iid and, hence, the resampling scheme in Step 3 above resamples a *single* value at a time as in Efron's (1979) iid resampling scheme (cf. Section 2.2). An alternative version of the FDB can be defined by replacing the bootstrap variables ϵ_j^*'s by iid standard exponentially distributed variables E_j^*'s (say) in Step 3 and using $I_{jn}^{**} \equiv \hat{f}_{jn} \cdot E_j^*$, $1 \leq j \leq n_0$ as the bootstrap periodogram values, instead of I_{jn}^*'s of Step 4. Both versions of the FDB are known to have a similar accuracy up to the second-order (cf. Remark 6, Dahlhaus and Janas (1996)).

9.2.3 *Second-Order Correctness of the FDB*

In this section, we consider second-order properties of the FDB approximation to the distribution of T_n. The following conditions will be used.

Conditions:

(C.1) $\{X_i\}_{i \in \mathbb{Z}}$ is a linear process of the form

$$X_i = \sum_{j \in \mathbb{Z}} a_j \zeta_{i-j}, \ i \in \mathbb{Z} \ , \tag{9.13}$$

where a_i's are real numbers satisfying $a_i = O(\exp(-C|i|))$ as $|i| \to \infty$ for some $C \in (0, \infty)$ and $\{\zeta_i\}_{i \in \mathbb{Z}}$ is a collection of iid random variables with $E\zeta_1 = 0$, $E\zeta_1^2 = 1$, $E\zeta_1^3 = 0$ and $E\zeta_1^8 < \infty$.

(C.2) (i) The spectral density $f(w) \equiv (2\pi)^{-1} \Big| \sum_{j \in \mathbb{Z}} a_j \exp(\iota w j) \Big|^2$, $|w| \leq \pi$ satisfies

$$\inf_{w \in [0, \pi]} f(w) > 0 \ . \tag{9.14}$$

(ii) The estimator $\hat{f}_n$ of f used in the FDB method is uniformly strongly consistent, i.e.,

$$\sup_{w \in [0, \pi]} \Big| \hat{f}_n(w) - f(w) \Big| \to 0 \quad \text{as} \quad n \to \infty, \text{ a.s.} \tag{9.15}$$

(C.3) The function $\xi = (\xi_1, \ldots, \xi_p) : [0, \pi] \to \mathbb{R}^p$ is of bounded variation (component wise) and

$$\|\xi^\dagger(k)\| = O\big(\exp(-C|k|)\big) \quad \text{as} \quad |k| \to \infty \tag{9.16}$$

for some $C \in (0, \infty)$ where $\xi^\dagger(k) = 2\int_0^\pi \xi(w) \cos kw dw$ is the Fourier coefficient of ξ (extended as a symmetric function over $[-\pi, \pi]$).

(C.4) (i) $(\zeta_1, \zeta_1^2)'$ satisfies Cramer's condition, i.e.,

$$\lim_{\|t\| \to \infty} \Big| E \exp\big((\zeta_1, \zeta_1^2)t\big)\Big| < 1\,. \tag{9.17}$$

(ii) Let W_n denote the eight-dimensional finite Fourier transforms

$$n^{-1/2}\left(d_n\left(\frac{2\pi j_1}{n}\right), \ldots, d_n\left(\frac{2\pi j_8}{n}\right)\right)'$$

for $\{j_1, \ldots, j_8\} \subset \{1, \ldots, n_0 - 1\}$ or the $(d+1)$ dimensional spectral mean estimator $\int (\xi', 1)' I_n$. Then, $\Sigma = \lim_{n\to\infty} \mathrm{Cov}(W_n)$ exists and is nonsingular in each case. Further, $\Sigma_1 \equiv \int (\xi', 1)'(\xi', 1) f^2$ is nonsingular.

Next, we briefly comment on the conditions. The exponential decay of the coefficients $\{a_j\}_{j\in\mathbb{Z}}$ in (9.13) is required for establishing valid Edgeworth expansions for the normalized ratio estimator T_n, based on the work of Götze and Hipp (1983) and Janas (1994). It can be replaced by a suitable polynomial decay condition if one is to have only consistency of the FDB for T_n. The condition on the first two moments the innovations ζ_i's is standard. However, the requirement that $E\zeta_1^3 = 0$ is very stringent. Dahlhaus and Janas (1996) point out (in their Remark 5, page 1942) that the rate of FDB approximation is only $O(n^{-1/2})$, i.e., the FDB is only first-order accurate when this condition fails. Condition (C.2)(i) ensures a nondegenerate limit distribution of the periodogram $I_n(w)$ at each $w \in [0, \pi]$. The uniform strong consistency of $\hat{f}_n$ is known when $\hat{f}_n$ is a kernel spectral density estimator of f. See, for example, Theorem A1, Franke and Härdle (1992). Exponential decay of the Fourier coefficients $\xi^\dagger(k)$ in (9.16) of Condition (C.3) is again required for establishing a valid Edgeworth expansion for the ratio statistic T_n (cf. Janas (1994)). This condition does not hold for the ξ-function of Example 9.2, corresponding to the spectral distribution function estimator. The technical conditions of Condition (C.4) are needed to establish valid Edgeworth expansions for T_n. As mentioned in Chapter 6, (9.17) holds if ζ_1 has an absolutely continuous component with respect to the Lebesgue measure on $\mathbb{R}$.

Theorem 9.1 *Suppose that Conditions (C.1)–(C.4) hold. Then,*

$$\sup_{A \in \mathcal{C}} \Big| P(\Lambda_n T_n \in A) - P_*(\hat{\Lambda}_n T_n^* \in A)\Big| = o(n^{-1/2}) \quad \textit{as} \quad n \to \infty, \ \textit{a.s.}\,,$$

where $\mathcal{C}$ is the collection of all convex measurable sets in $\mathbb{R}^p$, and Λ_n and $\hat{\Lambda}_n$ are symmetric $p \times p$ matrices satisfying

$$\Lambda_n^2 = [\mathrm{Var}(T_n)]^{-1} \quad \textit{and} \quad \hat{\Lambda}_n^2 = [Var_*(T_n^*)]^{-1} \,,$$

respectively.

Proof: This is a special case of Theorem 1 of Dahlhaus and Janas (1996), when no data-tapers are used. See Dahlhaus and Janas (1996) for a proof. □

Thus, under the conditions of Theorem 9.1, the FDB provides a better approximation to the distribution of the normalized ratio statistic than the normal approximation, which has an error of the order $O(n^{-1/2})$. Dahlhaus and Janas (1996) prove the second-order correctness of the FDB for normalized ratio statistics in the more general case where the periodogram is defined using a data-taper. Furthermore, they also establish the superiority of the FDB for the normalized Whittle estimator over normal approximation.

The results on the FDB for ratio statistics are valid under the assumption that $EX_1 = 0$. If the mean of the stationary process $\{X_i\}_{i\in\mathbb{Z}}$ is indeed unknown and estimated explicitly, say, by using $X_i - \bar{X}_n$ in place of X_i for the calculation of the periodogram I_n (cf. (9.1)), the FDB has an error of the order $O(n^{-1/2})$ for ratio statistics. As a result, in this case, the FDB no longer possesses the superiority over the normal approximation (cf. Remark 4, Dahlhaus and Janas (1996)). Furthermore, as pointed out earlier, the superiority of the FDB is also lost if the third moment of the innovation variables does *not* vanish, i.e., if $E\zeta_1^3 \neq 0$. Hence, it appears that the superiority of the FDB approximation for ratio statistics is rather sensitive to violations of the model assumptions.

9.3 Bootstrapping Spectral Density Estimators

In one of the early works on the FDB, Franke and Härdle (1992) studied the FDB for spectral density estimation. In this section, we describe the FDB method for kernel estimators of the spectral density $f(\cdot)$ of a stationary process $\{X_i\}_{i\in\mathbb{Z}}$, and consider its consistency properties. We continue to suppose that $\{X_i\}_{i\in\mathbb{Z}}$ is a linear process, given by (9.13) for some constants $a_i \in \mathbb{R}$, $i \in \mathbb{Z}$ with $\sum_{i\in\mathbb{Z}} |ia_i| < \infty$, and for some iid random variables $\{\zeta_i\}_{i\in\mathbb{Z}}$ with $E\zeta_1 = 0$, $E\zeta_1^2 = 1$. In particular, we suppose that $EX_1 = 0$ and the spectral density $f(\cdot)$ of $\{X_i\}_{i\in\mathbb{Z}}$ is given by

$$f(w) = (2\pi)^{-1} \Big| \sum_{j\in\mathbb{Z}} a_j \exp(\iota jw) \Big|^2, \; |w| \leq \pi \,. \tag{9.18}$$

Define the raw periodogram of $X_1, \ldots, X_n$ by

$$I_{1n}(w) = n^{-1}\Big|\sum_{j=1}^{n} X_j \exp(\iota w j)\Big|^2, \ |w| \leq \pi \ . \tag{9.19}$$

Thus, $I_{1n}(\cdot)$ is related to the periodogram $I_n(\cdot)$ of $X_1, \ldots, X_n$, defined in (9.1), by

$$I_{1n}(w) = (2\pi) I_n(w), \ |w| \leq \pi \ . \tag{9.20}$$

Let $K : \mathbb{R} \to (0, \infty)$ be a symmetric function satisfying $\int_{-\infty}^{\infty} K(x)dx = 2\pi$. Then, a kernel estimator of f with bandwidth $h \equiv h_n > 0$ is given by

$$\hat{f}_n(w; h) = (nh)^{-1} \sum_{j=-n_0}^{n_0} K\Big(\frac{w - \lambda_{jn}}{h}\Big) I_{1n}(\lambda_{jn}), \ w \in [-\pi, \pi] \ , \tag{9.21}$$

where $\lambda_{jn} \equiv 2\pi j/n$, $-n_0 \leq j \leq n_0$ and $n_0 = \lfloor n/2 \rfloor$. Performance of $\hat{f}_n(\cdot; \cdot)$ as an estimator of $f(\cdot)$ crucially depends on the bandwidth h. Under some suitable assumption on the process $\{X_i\}_{i \in \mathbb{Z}}$ and on the kernel $K(\cdot)$, the relative mean square error (RMSE) of $\hat{f}_n(\cdot; \cdot)$ at a point $w \in [\pi, \pi]$ admits the following expansion (cf. Franke and Härdle (1992), p. 122):

$$\begin{aligned} \text{RMSE}(w; h) &\equiv E\Big[\hat{f}_n(w; h) - f(w)\Big]^2 \Big/ f(w)^2 \\ &= \Big(h^2 f''(w)/[2f(w)]\Big)^2 + (2\pi)^{-1} \int_{-\infty}^{\infty} K^2(x)dx \cdot (nh)^{-1} \\ &\quad + o\Big([nh]^{-1} + h^{-4}\Big) \end{aligned} \tag{9.22}$$

as $n \to \infty$, $h \to 0$ such that $nh \to \infty$. Thus, the optimal h that minimizes the RMSE in (9.22) is asymptotically equivalent to $C_0 \cdot n^{-1/5}$ for some suitable constant $C_0 \in (0, \infty)$. In the sequel, we suppose that bandwidth h for the spectral density estimator $\hat{f}_n(\cdot; h)$ lies in an interval of the form $[\delta n^{-1/5}, \delta^{-1} n^{-1/5}]$ for some arbitrarily small $\delta > 0$. In the next section, we describe the FDB for $\hat{f}_n(\cdot; h)$ under this restriction, although the bootstrap algorithm itself may be stated almost without any changes for other values of h.

9.3.1 *Frequency Domain Bootstrap for Spectral Density Estimation*

In the problems encountered so far, the level-1 parameters were finite dimensional and their estimators converge at the rate $O_p(n^{-1/2})$. In contrast, here the level-1 parameter of interest is a function, namely, $f(\cdot)$, and the estimator $\hat{f}_n(\cdot; \cdot)$ has a slower rate of convergence than the standard $O_p(n^{-1/2})$ rate. For a bandwidth $h_n \sim C n^{-1/5}$, the estimator $\hat{f}_n(\cdot; h_n)$

has a bias that is of the same order as its standard deviation. As a result, for a valid approximation, the bootstrap algorithm must implicitly correct for the effect of the bias. A similar situation arises in the context of density estimation (cf. Romano (1988), Faraway and Jhun (1990), Hall (1992), Hall, Lahiri and Truong (1995)) and regression function estimation (cf. Härdle and Bowman (1988), Hall, Lahiri and Polzehl (1995)) with both independent and dependent data. Since $I_{1n}(\lambda_{jn})/f(\lambda_{jn})$'s are approximately independent, this leads to the "approximate" multiplicative regression model

$$I_{1n}(\lambda_{jn}) = f(\lambda_{jn}) \cdot \epsilon_{jn} \ , \tag{9.23}$$

with "approximately" independent error variables ϵ_{jn}'s and with the "regression function" $f(\cdot)$. The FDB for the spectral density estimation makes use of two bandwidths h_{1n} and h_{2n} of different orders as in bootstrapping a nonparametric regression model with independent errors (cf. Hall (1992)). The main steps of the bootstrap procedure are as follows:

Step 1: Define the residuals $\hat{\epsilon}_{jn} = I_{1n}(\lambda_{jn})/\hat{f}_n(\lambda_{jn}; h_{1n})$, $j = 1, \ldots, n_0$ with an initial bandwidth $h_{1n} > 0$, where $\hat{f}_n(\cdot;\cdot)$ is as defined in (9.21).

Step 2: Rescale the residuals $\hat{\epsilon}_{jn}$'s to get

$$\tilde{\epsilon}_{jn} = \hat{\epsilon}_{jn}/\bar{\epsilon}_{\cdot n}, \ 1 \le j \le n_0 \ ,$$

where $\bar{\epsilon}_{\cdot n} = n_0^{-1} \sum_{j=1}^{n_0} \hat{\epsilon}_{jn}$.

Step 3: Draw a sample $\epsilon_{1n}^*, \ldots, \epsilon_{n_0 n}^*$ of size n_0, randomly, with replacement from $\{\tilde{\epsilon}_{jn} : j = 1, \ldots, n_0\}$.

Step 4: Define the bootstrap periodogram values

$$I_{1n}^*(\lambda_{jn}) = \hat{f}_n(\lambda_{jn}; h_{2n}) \cdot \epsilon_{jn}^*, \ 1 \le j \le n_0 \tag{9.24}$$

using a second bandwidth $h_{2n} > 0$ (typically, different from h_{1n}). Then, the FDB version of the estimator $\hat{f}_n(w; h)$ is given by (cf. (9.21))

$$f_n^*(w; h) = (nh)^{-1} \sum_{j=-n_0}^{n_0} K\Big(\frac{w - \lambda_{jn}}{h}\Big) I_{1n}^*(\lambda_{jn}), \ w \in [-\pi, \pi] \ . \tag{9.25}$$

As in the FDB for ratio statistics, the rescaling of the "residuals" in Step 2 ensures that ϵ_{in}^*'s have mean 1, and this avoids an additional bias at the resampling stage. Indeed, the FDB may fail without the rescaling. As the regularity conditions below show, the two bandwidths h_{1n} and h_{2n} used in the FDB above are required to satisfy different decay conditions. The initial bandwidth h_{1n} may be of the same order as the given bandwidth $h = h_n$, which is assumed to have the RMSE-optimal order, $n^{-1/5}$. However, the second bandwidth h_{2n}, employed in Step 4 to define the bootstrap raw periodogram $I_{1n}^*(\cdot)$, is required to go to zero at a rate slower than $n^{-1/5}$. Thus, the estimator $\hat{f}_n(\cdot; h_{2n})$ is smoother than the given estimator $\hat{f}_n(\cdot; h)$. The over-smoothing at the bootstrap level is needed to ensure consistent estimation of the bias of the estimator $\hat{f}_n(\cdot; h)$, and is a standard device in bootstrapping nonparametric regression models (cf. Hall (1992)). In the next section, we show that the above version of the FDB provides a valid approximation to the distribution of centered and scaled spectral density estimator $\hat{f}_n(\cdot; h)$ for any sequence $\{h\} \equiv \{h_n\}_{n\geq 1}$ that decreases to zero at the optimal rate $n^{-1/5}$.

9.3.2 Consistency of the FDB Distribution Function Estimator

We shall make use of the following conditions for proving the results of this section.

Conditions:

(C.5) *$\{X_i\}_{i\in\mathbb{Z}}$ is a linear process, given by (9.13) where the constants $\{a_i\}_{i\in\mathbb{Z}}$ satisfy $\sum_{j=-\infty}^{\infty} |j|\,|a_j| < \infty$ and where the iid random variables $\{\zeta_i\}_{i\in\mathbb{Z}}$ satisfy $E\zeta_1 = 0$, $E\zeta_1^2 = 1$, and $E|\zeta_1|^5 < \infty$. Furthermore, sup$\{|E\exp(\iota t\zeta_1)| : |t| \geq \delta\} < 1$ for all $\delta > 0$.*

(C.6) *The spectral density $f(\cdot)$ of $\{X_i\}_{i\in\mathbb{Z}}$ is nonzero and twice continuously differentiable on $[-\pi, \pi]$.*

(C.7) *$K(\cdot)$ is a symmetric, nonnegative kernel on $(-\infty, \infty)$ satisfying $\int_{-\infty}^{\infty} K(x)dx = 2\pi$, $\int_{-\infty}^{\infty} x^2 K(x)dx = 2\pi$. Furthermore, K has a compact support and $K(\cdot)$ is Lipschitz.*

(C.8) (i) *$\{h_n\}_{n\geq 1}$ is a sequence of positive real numbers such that there exists $\delta \in (0,1)$ such that $h_n \in [\delta n^{-1/5}, \delta^{-1} n^{-1/5}]$ for all $n \geq \delta^{-1}$.*

(ii) *$h_{1n} \to 0$ and $(nh_{1n}^4)^{-1} = O(1)$ as $n \to \infty$.*

(iii) *$h_{2n} \to 0$ and $h_n/h_{2n} \to 0$ as $n \to \infty$.*

Conditions (C.5) and (C.6) are, respectively, similar to Conditions (C.1) and (C.2)(i), used in the case of ratio statistics. However, here the constants a_i's in (C.5) are allowed to go to zero at a polynomial rate. Condition (C.7) on the kernel K requires that $K(\cdot)/2\pi$ be a symmetric probability density function (with respect to the Lebesgue measure) with mean zero and variance equal to one. Finally, Condition (C.8) requires the initial bandwidth h_{1n} to go to zero at a rate not faster than $n^{-1/4}$, i.e., h_{1n} can be taken as $C \cdot n^{-\theta}$ for some $0 < \theta \leq 1/4$. In particular, h_{1n} can be of the same order (viz. $n^{-1/5}$) as the given bandwidth h_n. The second bandwidth h_{2n} must go to zero at a rate slower than h_n. Thus, a set of permissible values of h_{2n} is given by $h_{2n} = C \cdot n^{-\beta}$ for $C \in (0, \infty)$ and $0 < \beta < 1/5$.

Next, define the centered and scaled version of the spectral density $\hat{f}_n(w; h_n)$ as

$$R_n(w; h_n) \equiv \sqrt{nh_n}\Big(\hat{f}_n(w; h_n) - f(w)\Big) \Big/ f(w) \,. \tag{9.26}$$

The bootstrap version of $R_n(\cdot; \cdot)$ is given by

$$R_n^*(w; h_n) = \sqrt{nh_n}\Big(f_n^*(w; h_n) - \hat{f}_n(w; h_{2n})\Big) \Big/ \hat{f}_n(w; h_{2n}) \,. \tag{9.27}$$

Note that we use $\hat{f}_n(w; h_{2n})$ to center and scale the FDB version $f_n^*(w; h_n)$ in (9.27). That this is the appropriate quantity for normalizing $f_n^*(w; h_n)$ follows from (9.24) in Step 4 of the FDB. Since the bootstrap periodogram values were generated with $\hat{f}_n(\lambda_{jn}; h_{2n})$, by comparing the relations between the pairs of equations (9.21) and (9.23) in the unbootstrapped case with their bootstrap analogs (9.24) and (9.25), we see that $\hat{f}_n(\cdot; h_{2n})$ plays the role of the true density $f(\cdot)$ for the bootstrap spectral density estimator $f_n^*(w; h_n)$.

The following result shows that the FDB provides a valid approximation to the distribution of the normalized spectral density estimator $\hat{f}_n(w; h_n)$ for any given $w \in [-\pi, \pi]$ and any h_n satisfying (C.8)(i).

Theorem 9.2 *Suppose that Conditions (C.5)–(C.8) hold. Then, for any* $w \in [-\pi, \pi]$,

$$\sup_{x \in \mathbb{R}} \Big| P(R_n(w; h_n) \leq x) - P_*(R_n^*(w; h_n) \leq x) \Big| \longrightarrow 0 \quad \textit{in probability as} \quad n \to \infty \,, \tag{9.28}$$

where $R_n(\cdot; \cdot)$ *and* $R_n^*(\cdot; \cdot)$ *are as given by (9.26) and (9.27), respectively.*

Proof: Theorem 9.2 is a version of Theorem 1 of Franke and Härdle (1992), where their Condition (C.4) has been dropped and where the distance between the probability distributions of R_n and R_n^* in Mallow's metric has been replaced by the sup-norm distance. Note that if $(2\pi)^{-1}K^\dagger(\cdot)$ denotes the characteristic function corresponding to the probability density $(2\pi)^{-1}K(\cdot)$, then

$$\begin{aligned} &\lim_{t \to 0} \frac{K^\dagger(t) - K^\dagger(0)}{t^2} \\ &= (2\pi) \lim_{t \to 0} \Big[(2\pi)^{-1}K^\dagger(t) - 1 - it \int x\Big\{(2\pi)^{-1}K(x)\Big\}dx\Big] \Big/ t^2 \end{aligned}$$

$$\begin{aligned} &= (2\pi) \cdot -\frac{1}{2} \cdot \int x^2 \Big\{ (2\pi)^{-1} K(x) \Big\} dx \\ &= -\pi \; . \end{aligned} \tag{9.29}$$

This shows that Condition (C.4) of Franke and Härdle (1992) follows from Condition (C.7) above, which is a restatement of their Condition (C.3). Hence, Theorem 9.2 follows from Theorem 1 of Franke and Härdle (1992), in view of (9.29) and in view of the fact that convergence in Mallow's metric implies weak convergence. □

As in the case of the ratio statistics, consistent estimators of the sampling distribution of $R_n(w; h_n)$ can be generated by replacing the variables ϵ_j^*'s in Steps 1–4 above with iid exponentially distributed random variables ϵ_j^{**}'s. See Theorem 1 of Franke and Härdle (1992) for the validity of this variant of the FDB, which also holds under Conditions (C.5)–(C.8). Thus, both variants of the FDB can be used for setting confidence intervals for the unknown spectral density $f(\cdot)$, where the quantiles of $R_n(w; h_n)$ are replaced by the corresponding bootstrap quantiles. Accuracy of these CIs and relative merits of the two versions of the FDB CIs for $f(w)$ are unknown at this time.

9.3.3 Bandwidth Selection

Franke and Härdle (1992) also consider an important application of the FDB to the problem of choosing optimal bandwidths for spectral density estimation. To describe their results, we suppose that the optimality of a spectral density estimator $\hat{f}_n(\cdot;\cdot)$ is measured by the relative mean-square error (RMSE) of Section 9.3.2 (cf. (9.22)):

$$\text{RMSE}(w; h) \equiv E\Big(\hat{f}_n(w; h) - f(w)\Big)^2 \Big/ f^2(w) \; .$$

Furthermore, following Rice's (1984) approach, we restrict attention to an interval $\mathcal{H}_n \equiv [\delta n^{-1/5}, \delta^{-1} n^{-1/5}]$ (for a suitably small $\delta \in (0,1)$) of possible bandwidths that go to zero at the optimal rate $n^{-1/5}$. Then, the theoretical RMSE optimal bandwidth $h_n^0 \equiv h_n^0(w)$ for estimating the spectral density $f(w)$ at w is defined by

$$\text{RMSE}(w; h_n^0) = \inf_{h \in \mathcal{H}_n} \text{RMSE}(w; h) \; . \tag{9.30}$$

Note that, in view of (9.22), the optimal bandwidth h_n^0 satisfies the relation

$$\begin{aligned} h_n^0 &= n^{-1/5} \Big[(2\pi)^{-1} \int_{-\infty}^{\infty} K^2(x) dx \cdot \{ f(w) / f''(w) \}^2 \Big]^{1/5} \big(1 + o(1)\big) \\ &\qquad\qquad \text{as} \quad n \to \infty \; , \\ &\equiv n^{-1/5} c_0 \big(1 + o(1)\big), \ \text{say} \; , \end{aligned} \tag{9.31}$$

provided $f''(w) \neq 0$, and $\delta \in (0,1)$ is small enough to satisfy $\delta < c_0 < \delta^{-1}$. Thus, the optimal bandwidth h_n^0 depends on the unknown spectral density $f(\cdot)$ and its second derivative. For a data-based choice of the optimal bandwidth, we first define an estimated version of the RMSE criterion function using the FDB, and minimize the resulting function to obtain the FDB estimator of the level-2 parameter h_n^0. Let $f_n^*(w;h)$ be the FDB version of $\hat{f}_n(w;h)$, given by (9.25). Then, the estimated criterion function is given by

$$\widehat{\text{RMSE}}(w;h) = E_*\Big[\{f_n^*(w;h) - \hat{f}_n(w;h_{2n})\}/\hat{f}_n(w;h_{2n})\Big]^2 , \tag{9.32}$$

$h \in \mathcal{H}_n$, where E_* denotes the conditional expectation given $\{X_i\}_{i\in\mathbb{Z}}$. The bootstrap estimator of h_n^0 is given by a bandwidth $\hat{h}_n^0$ that minimizes $\widehat{\text{RMSE}}(w;h)$, i.e.,

$$\hat{h}_n^0 \equiv \text{argmin}\{\widehat{\text{RMSE}}(w;h) : h \in \mathcal{H}_n\} . \tag{9.33}$$

An important feature of the FDB-based estimated criterion function $\widehat{\text{RMSE}}(\cdot;\cdot)$ is that no Monte-Carlo computation is necessary for the evaluation of $\widehat{\text{RMSE}}$. An explicit formula for $\widehat{\text{RMSE}}(\cdot;\cdot)$ can be written down using the linearity of the bootstrapped estimator $f_n^*(\cdot;\cdot)$ and the independence of the resampled variables ϵ_j^*'s, similar to the MBB estimator of the variance of the sample mean, given by (3.9). Indeed, straightforward algebra yields

$$\begin{aligned}\widehat{\text{RMSE}}(w;h) \;\equiv\; & \hat{f}_n^{-2}(w;h_{2n}) \\ & \times \Big[(nh)^{-2}\text{Var}_*(\epsilon_1^*) \sum_{j=-n_0}^{n_0} K^2\Big(\frac{w-\lambda_{jn}}{h}\Big)\hat{f}_n^2(\lambda_{jn};h_{2n}) \\ & + \Big\{(nh)^{-1}\sum_{j=-n_0}^{n_0} K\Big(\frac{w-\lambda_{jn}}{h}\Big)\hat{f}_n(\lambda_{jn};h_{2n}) \\ & - \hat{f}_n(w;h_{2n})\Big\}^2\Big] .\end{aligned} \tag{9.34}$$

Thus, one may find the FDB estimator of the optimal bandwidth h_n^0 by equivalently minimizing the explicit expression (9.34). The following result shows that $\hat{h}_n^0$ is consistent for h_n^0. Furthermore, the estimated criterion function at $\hat{h}_n^0$ attains the optimal theoretical RMSE level over the set $\mathcal{H}_n$, asymptotically, in probability.

Theorem 9.3 *Assume that the conditions of Theorem 9.2 hold and that $f''(w) \neq 0$ and $\delta < c_0 < \delta^{-1}$. Then, for h_n^0 and $\hat{h}_n^0$, respectively defined by (9.30) and (9.33),*

(i) $n^{1/5}(\hat{h}_n^0 - h_n^0) \longrightarrow_p 0$ *as* $n \to \infty$,

(ii) $\frac{\widehat{RMSE}(w;\hat{h}_n^0)}{RMSE(w;h_n^0)} \longrightarrow_p 1 \quad as \quad n \to \infty$.

Proof: See Theorem 3, Franke and Härdle (1992). □

9.4 A Modified FDB

In this section, we describe a modified version of the FDB based on the work of Kreiss and Paparoditis (2003). The modified FDB removes some of the limitations of the FDB and provides valid approximations to the distributions of a larger class spectral mean estimators than the class of ratio statistics (cf. Section 9.2). Furthermore, the modified FDB continues to provide a valid approximation in the spectral density estimation problems considered above. Let $\{X_i\}_{i\in\mathbb{Z}}$ be a causal linear process, given by

$$X_i = \sigma \sum_{j=0}^{\infty} a_j \zeta_{i-j},\ i \in \mathbb{Z}\ , \tag{9.35}$$

where $a_0 = 1$ and $\{a_i\}_{i\geq 1}$ is a sequence of real numbers satisfying $\sum_{i=1}^{\infty} i^2|a_i| < \infty$ and $\{\zeta_i\}_{i\in\mathbb{Z}}$ is a sequence of iid zero mean, unit variance random variables with $E\zeta_1^4 < \infty$. Also, let $I_n(\cdot)$ denote the periodogram of $X_1, \ldots, X_n$, defined by (9.1), i.e.,

$$I_n(w) = (2\pi n)^{-1}\Big|\sum_{t=1}^{n} X_t \exp(-\iota wt)\Big|^2,\ w \in [-\pi, \pi]\ , \tag{9.36}$$

and let $f(\cdot)$ denote the spectral density of $\{X_i\}_{i\in\mathbb{Z}}$. It is known (cf. Priestley (1981), Chapter 6) that at the discrete frequencies $\lambda_{jn} \equiv 2\pi j/n$, $1 \leq j \leq n_0$,

$$EI_n(\lambda_{jn}) = f(\lambda_{jn}) + O(n^{-1}) \tag{9.37}$$

and

$$\mathrm{Cov}\big(I_n(\lambda_{jn}), I_n(\lambda_{kn})\big) = \begin{cases} f^2(\lambda_{jn}) + O(n^{-1}) & \text{if} \quad j = k \\ n^{-1} f(\lambda_{jn}) f(\lambda_{kn})\kappa_4 + o(n^{-1}) & \text{if} \quad j \neq k \end{cases} \tag{9.38}$$

for all $1 \leq j,\ k \leq n_0$, where $n_0 = \lfloor n/2 \rfloor$ and where $\kappa_4 = (E\zeta_1^4 - 3)$ denotes the fourth cumulant of the innovation ζ_1. Thus, if $\kappa_4 \neq 0$, then the periodogram at distinct ordinates λ_{jn} and λ_{kn} have a nonzero correlation and are *dependent*. Although the dependence of the periodogram values $I_n(\lambda_{jn})$ and $I_n(\lambda_{kn})$ vanishes asymptotically, the aggregated effect of this dependence on the limit distribution of a spectral mean estimator may not be negligible. Indeed, as noted in Section 9.2 (cf. (9.6)), for the spectral mean $A(\xi; f) \equiv \int_0^{\pi} \xi f$ and its canonical estimator $A(\xi; I_n) \equiv \int_0^{\pi} \xi I_n$

corresponding to a function $\xi : [0, \pi] \to \mathbb{R}$ of bounded variation, we have

$$\sqrt{n}\big(A(\xi; I_n) - A(\xi; f)\big) \longrightarrow^d N\Big(0, 2\pi \int \xi^2 f^2 + \kappa_4\big(\int \xi f\big)^2\Big) . \tag{9.39}$$

The second term (i.e., $\kappa_4(\int \xi f)^2$) in the asymptotic variance of $\sqrt{n}(A(\xi; I_n) - A(\xi; f))$ results from the combined effect of the nonzero correlations among the $I_n(\lambda_{jn})$'s. The standard version of the FDB fails in such cases due to the fact that the bootstrap periodogram values I_{jn}^*'s generated by the FDB algorithm (cf. Step 4, Section 9.2.2) are *independent* and, hence, do not have the same correlation structure as the periodogram variables $\{I_n(\lambda_{jn}) : 1 \leq j \leq n_0\}$. The modified version of the FDB, proposed by Kreiss and Paparoditis (2003) gets around this problem by fitting an autoregressive process to the variables $\{X_1, \ldots, X_n\}$ first and then scaling the periodogram values of the fitted autoregressive process to mimic the covariance structure of the $I_n(\lambda_{jn})$'s. As a result, the modified FDB captures the dependence structure of $I_n(\lambda_{jn})$'s adequately and provides a valid approximation to the distribution of $\sqrt{n}(A(\xi; I_n) - A(\xi; f))$ even when the term $\kappa_4(\int \xi f)^2$ in (9.39) is nonzero.

9.4.1 Motivation

We now describe the intuitive reasoning behind the formulation of the modified FDB. Let $\{Y_i\}_{i\in\mathbb{Z}}$ be a stationary autoregressive process of order p, fitted to $\{X_i\}_{i\in\mathbb{Z}}$ by minimizing the distance $E(X_i - \sum_{j=1}^p \beta_j X_{i-j})^2$ over $\beta_1, \ldots, \beta_p$. Write $\gamma(k) = \text{Cov}(X_1, X_{1+k})$, $k \in \mathbb{Z}$. Then, the $\{Y_i\}_{i\in\mathbb{Z}}$-process is given by

$$Y_i = \sum_{j=1}^{p} \breve{\beta}_j Y_{i-j} + \breve{\sigma}_p \breve{\zeta}_i, \; i \in \mathbb{Z} , \tag{9.40}$$

where $\breve{\beta} \equiv (\breve{\beta}_1, \ldots, \breve{\beta}_p)' = \Gamma_p^{-1}\gamma_p$, $\breve{\sigma}_p^2 = \gamma(0) - \breve{\beta}'\Gamma_p\breve{\beta}$ and $\{\breve{\zeta}_i\}_{i\in\mathbb{Z}}$ is a sequence of iid random variables with $E\breve{\zeta}_1 = 0$ and $E\breve{\zeta}_1^2 = 1$. Here, Γ_p is the $p \times p$ matrix with (i,j)th element $\gamma(i-j)$, $1 \leq i, j \leq p$ and $\gamma_p = (\gamma(1), \ldots, \gamma(p))'$. As $\gamma(k) \to 0$ as $|k| \to \infty$, by Proposition 5.1.1 of Brockwell and Davis (1991), for every $p \in \mathbb{N}$, Γ_p^{-1} exists. For the rest of this section, suppose that $E\breve{\zeta}_1^4 < \infty$. Let

$$f_{AR}(w) = \breve{\sigma}_p^2 \Big/ \Big|1 - \sum_{j=1}^{p} \breve{\beta}_j \exp(-\iota j w)\Big|^2, \; w \in [-\pi, \pi] \tag{9.41}$$

denote the spectral density of the fitted autoregressive process $\{Y_i\}_{i\in\mathbb{Z}}$. Next, define the variables $W_n(\lambda_{jn})$, $1 \leq j \leq n_0$ by

$$W_n(\lambda_{jn}) = q(\lambda_{jn}) I_n^{AR}(\lambda_{jn}) , \tag{9.42}$$

where $I_n^{AR}(w) \equiv (2\pi n)^{-1}|\sum_{t=1}^{n} Y_t \exp(-\iota wt)|^2$, $w \in [-\pi, \pi]$ is the periodogram of $Y_1, \ldots, Y_n$ and where the multiplicative factor $q(\cdot)$ is defined as

$$q(w) = f(w)\Big/ f_{AR}(w), \ w \in [-\pi, \pi] \ . \tag{9.43}$$

Note that the periodogram I_n^{AR} of the fitted autoregressive process satisfies relations (9.37) and (9.38) with f replaced by f_{AR} and κ_4 replaced by $\breve{\kappa}_4 = (E\breve{\zeta}_1^4 - 3)$, the fourth cumulant of $\breve{\zeta}_1$. As a result, by (9.42) and (9.43), it follows that the variables $W_n(\lambda_{jn})$'s satisfy

$$EW_n(\lambda_{jn}) = f(\lambda_{jn}) + O(n^{-1})$$

and

$$\mathrm{Cov}\big(W_n(\lambda_{jn}), W_n(\lambda_{kn})\big) = \begin{cases} f^2(\lambda_{jn}) + O(n^{-1}) & \text{if} \quad j = k \\ n^{-1} f(\lambda_{jn}) f(\lambda_{kn}) \breve{\kappa}_4 & \text{if} \quad j \neq k \end{cases}$$

for all $1 \leq j, k \leq n_0$. Thus, the covariance structure of $W_n(\lambda_{jn})$, $1 \leq j \leq n_0$ closely mimics that of the periodogram variables $I_n(\lambda_{jn})$, $1 \leq j \leq n_0$, provided $\breve{\kappa}_4$ is close to κ_4. The modified version of the FDB, proposed by Kreiss and Paparoditis (2003), fits an autoregressive process empirically and replaces the multiplicative factor $q(\cdot)$ by a data-based version. In the next section, we describe the details of this modified FDB method, known as the *autoregressive-aided FDB* (or the ARFDB) method.

9.4.2 The Autoregressive-Aided FDB

Suppose that a finite stretch $X_1, \ldots, X_n$ of the series $\{X_i\}_{i \in \mathbb{Z}}$ is observed and that we want to approximate the distribution of the centered and scaled spectral mean estimator

$$T_n \equiv \sqrt{n}(A(\xi; I_n) - A(\xi; f)) \ ,$$

where $A(\xi; I_n)$ and $A(\xi; f)$ are as in relation (9.39), i.e., $A(\xi; I_n) = \int_0^{\pi} \xi I_n$ and $A(\xi; f) = \int_0^{\pi} \xi f$ for a given function $\xi : [0, \pi] \to \mathbb{R}$ of bounded variation. Extensions to the vector-valued case is straightforward and is left out in the discussion below.

The basic steps in the ARFDB are as follows:

Step (I): Given $X_1, \ldots, X_n$, fit an autoregressive process $\{Y_i\}_{i \in \mathbb{Z}} \equiv \{Y_{in}\}_{i \in \mathbb{Z}}$ of order p ($\equiv p_n$). Let $(\hat{\beta}_{1n}, \ldots, \hat{\beta}_{pn})$ and $\hat{\sigma}_n^2$ denote the estimated parameter values, obtained by using the Yule-Walker equations (cf. Chapter 8, Brockwell and Davis (1991)). Let

$$\hat{\zeta}_{tn} = X_t - \sum_{j=1}^{p} \hat{\beta}_{jn} X_{t-j}, \ t = p + 1, \ldots, n$$

denote the residuals and let

$$\tilde{\zeta}_{tn} = (\hat{\zeta}_{tn} - \bar{\zeta}_n)/s_n, \; t = p+1, \ldots, n \tag{9.44}$$

be the standardized residuals, where $\bar{\zeta}_n = (n-p)^{-1} \sum_{t=p+1}^{n} \hat{\zeta}_{tn}$ and $s_n^2 = (n-p)^{-2} \sum_{t=p+1}^{n} (\hat{\zeta}_{tn} - \bar{\zeta}_n)^2$. Write $\tilde{F}_n$ for the empirical distribution function of $\{\tilde{\zeta}_{tn} : t = p+1, \ldots, n\}$, i.e. ,

$$\tilde{F}_n(x) = (n-p)^{-1} \sum_{t=p+1}^{n} \mathbb{1}(\tilde{\zeta}_{tn} \leq x), \; x \in \mathbb{R} \,. \tag{9.45}$$

Step (II): Generate the bootstrap variables $X_1^*, \ldots, X_n^*$ from the autoregression model

$$X_i^* = \sum_{j=1}^{p} \hat{\beta}_{jn} X_{i-j}^* + \hat{\sigma}_n \cdot \zeta_i^*, \; i \in \mathbb{Z} \,, \tag{9.46}$$

where $\{\zeta_i^*\}_{i \in \mathbb{Z}}$ is a sequence of iid random variables with common distribution $\tilde{F}_n$.

Step (III): Compute the periodogram of $X_1^*, \ldots, X_n^*$ as

$$I_n^{AR*}(w) = (2\pi n)^{-1} \Big| \sum_{t=1}^{n} X_t^* \exp(-\iota w t) \Big|^2, \; w \in [-\pi, \pi] \,. \tag{9.47}$$

Step (IV): Let $\hat{f}_{n,AR}(w) = \frac{\hat{\sigma}_n^2}{2\pi} \Big| 1 - \sum_{j=1}^{p} \hat{\beta}_{jn} e^{-\iota w j} \Big|^{-2}$, $w \in [-\pi, \pi]$ denote the spectral density of $\{X_i^*\}_{i \in \mathbb{Z}}$. Also, let $K : [-\pi, \pi] \to [0, \infty]$ be a probability density function. Define the nonparametric estimator of the function $q(\cdot)$ by

$$\hat{q}_n(w) = (nh_n)^{-1} \sum_{i=-n_0}^{n_0} K\Big(\frac{w - \lambda_{jn}}{h_n}\Big) \Big\{ I_n(\lambda_{jn}) / \hat{f}_{n,AR}(\lambda_{jn}) \Big\} \,, \tag{9.48}$$

$w \in [-\pi, \pi]$, where $h_n > 0$ is a bandwidth.

Step (V): Finally, define the bootstrap version of the periodogram $I_n(w)$ by rescaling the periodogram $I_n^{AR*}(\cdot)$ of the X_i^*'s by $\hat{q}_n(\cdot)$, as

$$I_n^*(w) = I_n^{AR*}(w) \cdot \hat{q}_n(w), \; w \in [-\pi, \pi] \,. \tag{9.49}$$

The ARFDB version of the centered and scaled spectral mean estimator $T_n = \sqrt{n}(A(\xi; I_n) - A(\xi; f))$ is given by

$$T_n^* = \sqrt{n}\Big(B(\xi; I_n^*) - B(\xi; \hat{f}_n) \Big) \,, \tag{9.50}$$

where $B(\xi; I_n^*) = \int_0^\pi \xi(w) I_n^*(w) dw$ and $B(\xi; \hat{f}_n) = \int_0^\pi \xi(w) \hat{f}_n(w) dw$, with $\hat{f}_n(w) \equiv \hat{q}_n(w) \hat{f}_{n,AR}(w)$, $w \in [-\pi, \pi]$. As an alternative, the integrals in the definitions of $B(\xi; I_n^*)$ and $B(\xi; \hat{f}_n)$ may be replaced by a sum over the frequencies $\{\lambda_{jn} : 1 \leq j \leq n_0\}$ as in the case of the FDB of Section 9.2. The conditional distribution of T_n^* given $X_1, \ldots, X_n$ now gives the ARFDB estimator of the distribution of T_n.

Remark 9.1 One may use alternative methods for estimating the parameters $\beta_1, \ldots, \beta_p$ and σ^2 in Step (I) of the ARFDB. However, an advantage of using the Yule-Walker equations to estimate the parameters $\beta_1, \ldots, \beta_p$ and σ^2 of the fitted autoregression model in Step (I) is that all the roots of the polynomial $1 - \sum_{j=1}^p \hat{\beta}_{jn} z^j$ lie outside the unit circle $\{z \in \mathbb{C} : |z| \leq 1\}$ and hence, the spectral density function $\hat{f}_{n,AR}(\cdot)$ of Step (IV) is well defined.

Remark 9.2 In practice, one generates the variables $X_1^*, \ldots, X_n^*$ from the "estimated" autoregression model (9.46) by using the recursion relation (9.46) with some initial values $X_{1-p}^*, \ldots, X_0^*$ and running the chain for a long-time until stationarity is reached (cf. Chapter 8). Kreiss and Paparoditis (2003) also point out that the order p of the fitted model may be chosen using some suitable data-based criteria, such as, the Akaike's Information Criterion.

For establishing the validity of the ARFDB, we shall make use of the following conditions, as required by Kreiss and Paparoditis (2003).

Conditions:

(C.9) The linear process $\{X_i\}_{i\in\mathbb{Z}}$ of (9.35) is invertible and has an infinite order autoregressive process representation

$$X_i = \sum_{j=1}^{\infty} \beta_j X_{i-j} + \sigma\zeta_i, \ i \in \mathbb{Z}$$

where $\sum_{j=1}^\infty j^{1/2}|\beta_j| < \infty$ and $1 - \sum_{j=1}^\infty \beta_j z^j \neq 0$ for all complex z with $|z| \leq 1$.

(C.10) $\{\zeta_i\}_{i\in\mathbb{Z}}$ is a sequence of iid random variables with $E\zeta_1 = 0$, $E\zeta_1^2 = 1$, and $E\zeta_1^4 < \infty$. Further, $\sigma \in (0, \infty)$.

(C.11) The spectral density f of $\{X_i\}_{i\in\mathbb{Z}}$ is Lipschitz continuous and satisfies

$$\inf_{w\in[0,\pi]} f(w) > 0 \ .$$

(C.12) (i) The characteristic function $K^\dagger(\cdot)$ of the kernel $K(\cdot)$, given by $K^\dagger(u) \equiv \int_{-\infty}^\infty \exp(\iota ux) K(x) dx$, is a nonnegative even function with $K^\dagger(u) = 0$ for $|u| > 1$.

(ii) The bandwidth sequence $\{h_n\}_{n\geq 1}$ satisfies

$$h_n + (nh_n)^{-1} \to 0 \quad \text{as} \quad n \to \infty .$$

(C.13) The function $\xi : [0, \pi] \to \mathbb{R}$ is a function of bounded variation.

(C.14) There exist two sequences of real numbers $\{p_{1n}\}_{n\geq 1}$ and $\{p_{2n}\}_{n\geq 1}$ such that $p_{1n}^{-1} = o(1)$ and $p_{2n} = O([n/\log n]^{1/5})$ as $n \to \infty$ and the order p of the fitted autoregression model satisfies $p = p_n \in [p_{1n}, p_{2n}]$ for all $n \geq 1$.

Under Conditions (C.9)–(C.14), the ARFDB provides a valid approximation to the distribution of T_n, as shown by the next result.

Theorem 9.4 *Suppose that Conditions (C.9)–(C.14) hold. Then, with* $\kappa_4 = (E\zeta_1^4 - 3)$,

$$T_n^* \longrightarrow^d N\Big(0, \Big[2\pi \int_0^\pi \xi^2 f^2 + \kappa_4 \big(\int \xi f\big)^2\Big]\Big) , \quad \textit{in probability}$$

and, hence, by (9.39),

$$\sup_{x \in \mathbb{R}} \Big| P_*(T_n^* \leq x) - P(T_n \leq x) \Big| \longrightarrow_p 0 \quad \textit{as} \quad n \to \infty ,$$

where $T_n = \sqrt{n}(A(\xi; I_n) - A(\xi; f))$ *and* T^* *is as defined in (9.50).*

Proof: See Theorem 3.1, Kreiss and Paparoditis (2003). □

Theorem 9.4 shows that under suitable regularity conditions, the modified version of the FDB provides a valid approximation to a wider class of spectral mean estimators than the standard version of the FDB, which is applicable only to the class of ratio statistics. However, the validity of the ARFDB crucially depends on the additional requirement of invertibility (cf. Condition (C.1)), which narrows the class of linear processes $\{X_i\}_{i\in\mathbb{Z}}$ to some extent. Kreiss and Paparoditis (2003) point out that this restriction may be dispensed with, if one modifies the FDB by fitting a finite-order moving average model to the data instead of fitting an autoregressive process and then by using a suitable version of the correction factor $\hat{q}_n(\cdot)$ in Step (IV) for the moving average case. Because of these additional tuning-up-steps involved in the autoregressive-aided or the moving-average-aided versions, the modified FDB is expected to have a better finite sample performance than the usual FDB, even when such modifications are not needed for its asymptotic validity, i.e., when the methods are applied to ratio-statistics. A similar remark applies on the finite sample performance of the ARFDB in the spectral density estimation problems considered in Section 9.3. We refer the interested reader to Kreiss and Paparoditis (2003) for a discussion of these issues, for guidance on the choice of the smoothing parameters p and h, and for numerical results on finite sample performance of the ARFDB.

10 Long-Range Dependence

10.1 Introduction

The models considered so far in this book dealt with the case where the data can be modeled as realizations of a *weakly* dependent process. In this chapter, we consider a class of random processes that exhibit long-range dependence. The condition of long-range dependence in the data may be described in more than one way (cf. Beran (1994), Hall (1997)). For this book, an operational definition of long-range dependence for a second-order stationary process is that the sum of the (lag) autocovariances of process diverges. In particular, this implies that the variance of the sample mean based on a sample of size n from a long-range dependent process decays at a rate slower than $O(n^{-1})$ as $n \to \infty$. As a result, the scaling factor for the centered sample mean under long-range dependence is of smaller order than the usual scaling factor $n^{1/2}$ used in the independent or weakly dependent cases. Furthermore, the limit distribution of the normalized sample mean can be nonnormal. In Section 10.2, we describe the basic framework and review some relevant properties of the sample mean under long-range dependence. In Section 10.3, we investigate properties of the MBB approximation. Here the MBB provides a valid approximation if and only if the limit law of the normalized sample mean is normal. In Section 10.4, we consider properties of the subsampling method under long-range dependence. We show that unlike the MBB, the subsampling method provides valid approximations to the distributions of normalized and studentized versions of the sample mean for both normal and nonnormal limit cases. In Section

10.5, we report the results from a small simulation study on finite sample performance of the subsampling method.

10.2 A Class of Long-Range Dependent Processes

Let $\{Z_i\}_{i \in \mathbb{Z}}$ be a stationary Gaussian process with $EZ_1 = 0$, $EZ_1^2 = 1$ and autocovariance function

$$r(i) = EZ_1 Z_{1+i},\ i \in \mathbb{Z} .$$

We shall suppose that the autocovariance function $r(\cdot)$ can be represented as

$$r(k) = k^{-\alpha} L(k),\ k \geq 1 \tag{10.1}$$

for some $0 < \alpha < 1$ and for some function $L : (0, \infty) \to \mathbb{R}$ that is slowly varying at ∞, i.e.,

$$\lim_{t \to \infty} \frac{L(at)}{L(t)} = 1 \quad \text{for all} \quad a \in (0, \infty) . \tag{10.2}$$

Note that under (10.1), $\sum_{k=1}^{\infty} r(k)$ diverges and, hence, the process $\{Z_i\}_{i \in \mathbb{Z}}$ exhibits long-range dependence. Here we consider stationary processes that are generated by instantaneous transformations of the Gaussian process $\{Z_i\}_{i \in \mathbb{Z}}$, including many nonlinear transformations of $\{Z_i\}_{i \in \mathbb{Z}}$. Let $G_1 : \mathbb{R} \to \mathbb{R}$ be a Borel measurable function satisfying $EG_1(Z_1)^2 < \infty$. We suppose that the observations are modeled as realizations of the random variables $\{X_i\}_{i \in \mathbb{Z}}$ that are generated by the relation

$$X_i = G_1(Z_i),\ i \in \mathbb{Z} . \tag{10.3}$$

In spite of its simple form, this formulation is quite general. It allows the one-dimensional marginal distribution of X_1 to be any given distribution on $\mathbb{R}$ with a finite second moment. To appreciate why, let F be a distribution function on $\mathbb{R}$ with $\int x^2 dF(x) < \infty$. Set $G_1 = F^{-1} \circ \Phi$ in (10.3), where Φ denotes the distribution function of $N(0, 1)$ and F^{-1} is the quantile transform of F, given by

$$F^{-1}(u) = \inf\{x \in \mathbb{R} : F(x) \geq u\},\ u \in (0, 1) .$$

Then,

$$EG_1(Z_1)^2 = E\Big[F^{-1}(\Phi(Z_1))\Big]^2 = \int_0^1 \Big[F^{-1}(u)\Big]^2 du = \int x^2 dF(x) < \infty .$$

Furthermore, it readily follows that $X_1 = F^{-1}(\Phi(Z_1))$ has distribution F. Thus, relation (10.3) yields a stationary process $\{X_i\}_{i \in \mathbb{Z}}$ with one-dimensional marginal distribution F. The dependence structure of $\{X_i\}_{i \in \mathbb{Z}}$

is determined by the function G_1 and by the autocovariance function $r(\cdot)$ of the underlying Gaussian process. In a series of important papers, Taqqu (1975, 1979) and Dobrushin and Major (1979) investigated limit distributions of the normalized sample mean of the X_i's under model (10.3). Let $\mu = EX_1$ be the level-1 parameter of interest and let $\bar{X}_n = n^{-1}\sum_{i=1}^{n} X_i$ denote the sample mean based on $X_1, \ldots, X_n$. Even under long-range dependence, $\bar{X}_n$ is a consistent estimator of the level-1 parameter μ. However, the rate of convergence of $\bar{X}_n$ to μ may no longer be $O_p(n^{-1/2})$ and the asymptotic distribution of $(\bar{X}_n - \mu)$, when it exists, may be nonnormal. The limit behavior of $(\bar{X}_n - \mu)$ heavily depends on the *Hermite Rank* of the function

$$G(y) \equiv G_1(y) - \mu, \; y \in \mathbb{R} . \tag{10.4}$$

Recall that for $k \in \mathbb{Z}_+$,

$$H_k(x) = (-1)^k \exp(x^2/2) \frac{d^k}{dx^k}\left[\exp(-x^2/2)\right], \; x \in \mathbb{R}$$

denotes the kth order Hermite polynomial. Then, the Hermite rank q of $G(\cdot)$ is defined as

$$q = \inf\left\{k \in \mathbb{N} : E\Big(H_k(Z_1)G(Z_1)\Big) \neq 0\right\} . \tag{10.5}$$

Let $A = 2\Gamma(\alpha)\cos(\alpha\pi/2)$ and $c_q = E(H_q(Z_1)G(Z_1))$. Also, for $n \in \mathbb{N}$, let $d_n = [n^{2-q\alpha}L^q(n)]^{1/2}$. The following result gives the asymptotic distribution of the sample mean $\bar{X}_n$.

Theorem 10.1 *(Taqqu (1975, 1979), Dobrushin and Major (1979)). Assume that G has Hermite rank q, and that $r(\cdot)$ admits the representation at (10.1) with $0 < \alpha < q^{-1}$. Then, $n(\bar{X}_n - \mu)/d_n \longrightarrow^d W_q$ in distribution, where W_q is defined in terms of a multiple Wiener-Itô integral with respect to the random spectral measure W of the Gaussian white-noise process as*

$$W_q = \frac{c_q}{A^{q/2}} \int \frac{\exp\{\iota(x_1 + \cdots + x_q)\} - 1}{\iota(x_1 + \cdots + x_q)} \prod_{k=1}^{q} |x_k|^{(\alpha-1)/2} dW(x_1)\ldots dW(x_q) . \tag{10.6}$$

When $q = 1$, W_q has a normal distribution with mean zero and variance $2c_q^2/\{(1-\alpha)(2-\alpha)\}$, but for $q \geq 2$, the distribution of W_q is nonnormal (Taqqu (1975)). For details of the representation of W_q in (10.6), and the concept of a multiple Wiener-Itô integral with respect to the random spectral measure of a stationary process, see Dobrushin and Major (1979) and Dobrushin (1979), respectively. The complicated form of the limit distribution in (10.6) makes it difficult to use the traditional approach where large sample inference about the level-1 parameter μ is based on the limit distribution. In the next section, we consider the MBB method of Künsch (1989) and Liu and Singh (1992) and investigate its consistency properties for approximating the distribution of the normalized sample mean.

10.3 Properties of the MBB Method

10.3.1 Main Results

Let $X_1^*, \ldots, X_n^*$ denote the MBB sample based on $b \equiv n/\ell$ resampled blocks of size ℓ (cf. Section 2.5), where, for simplicity of exposition, we suppose that ℓ divides n. Define

$$T_n^* = n(\bar{X}_n^* - \hat{\mu}_n)/d_n, \; n \geq 1 ,$$

where $\bar{X}_n^* = n^{-1} \sum_{i=1}^n X_i^*$ denotes the bootstrap sample mean and $\hat{\mu}_n = E_* \bar{X}_n^*$. Then, T_n^* gives the bootstrap version of the normalized sample mean $T_n = n(\bar{X}_n - \mu)/d_n$. Although under the conditions of Theorem 10.1, T_n converges in distribution to a *nondegenerate* distribution for all values of $q \geq 1$, it turns out that the conditional distribution of T_n^* has a *degenerate* limit distribution. The following result characterizes the asymptotic behavior of $P_*(T_n^* \leq x)$ for $x \in \mathbb{R}$.

Theorem 10.2 *Assume that the conditions of Theorem 10.1 hold for some $q \geq 1$ and that $n^\epsilon \ell^{-1} + \ell n^{\epsilon - 1} = O(1)$ as $n \to \infty$, for some $\epsilon \in (0,1)$. Then,*

$$\sup_{x \in \mathbb{R}} \left| P_*(T_n^* \leq x) - \Phi\big(d_n (b d_\ell^2)^{-1/2} x / \sigma_q\big) \right| = o_p(1) \quad as \quad n \to \infty , \tag{10.7}$$

where $b = n/\ell$ and where

$$\sigma_q^2 = 2(2 - q\alpha)^{-1}(1 - q\alpha)^{-1}(q!) c_q^2 . \tag{10.8}$$

Proof: See Section 10.3.2 below. □

Note that by the definition of d_n and by the "slowly-varying" property of the function $L(\cdot)$ (cf. relation (9.9), Chapter 8, Feller (1971b)),

$$b d_\ell^2 = o(d_n^2) . \tag{10.9}$$

Hence, from (10.7), it follows that

$$P_*(T_n^* \leq x) \to 0, \; 2^{-1} \text{ or } 1 \text{ in probability}$$

according as $x < 0$, $x = 0$, or $x > 0$. Thus, the conditional distribution of T_n^* converges weakly to δ_0, the probability measure degenerate at zero, in probability. This shows that the MBB procedure fails to provide a valid approximation to the distribution of the normalized sample mean under long-range dependence.

In contrast to certain other applications, where the naive bootstrap approximations suffer from inadequate or wrong centerings (e.g., bootstrapping M-estimators, see Section 4.3), here the failure of the MBB is primarily due to wrong scaling. The natural choice of the scaling factor d_n^{-1} used in the definition of the bootstrap variable T_n^* tends to zero rather fast and thus forces T_n^* to converge to a degenerate limit. Intuitively, this may be explained by noting that by averaging *independent* bootstrap blocks to define the bootstrap sample mean, we destroy the strong dependence of the underlying observations $X_1, \ldots, X_n$ in the bootstrap samples. As a result, the variance of the bootstrap sample sum $n\bar{X}_n^*$ has a substantially *slower* growth rate (viz., bd_ℓ^2) compared to the growth rate d_n^2 for $\mathrm{Var}(n\bar{X}_n)$. When the unbootstrapped mean $\bar{X}_n$ is asymptotically normal, one can suitably redefine the scaling constant in the bootstrap case to recover the limit law. However, for nonnormal limit distributions of $\bar{X}_n$, the MBB fails rather drastically; the bootstrap sample mean is asymptotically normal irrespective of the nonnormal limit law of normalized $\bar{X}_n$. For a rigorous statement of the result, define the modified MBB version of T_n as

$$\tilde{T}_n^* = (bd_\ell)^{-1/2} n(\bar{X}_n^* - \hat{\mu}_n) \ .$$

Then, we have the following result on $\tilde{T}_n^*$.

Theorem 10.3 *Assume that the conditions of Theorem 10.2 hold. Let σ_q^2 be as in (10.8). Then,*

(i) $\sup_{x \in \mathbb{R}} \left| P_*(\tilde{T}_n^* \leq x) - \Phi(x/\sigma_q) \right| = o_p(1)$;

(ii) $\sup_{x \in \mathbb{R}} \left| P_*(\tilde{T}_n^* \leq x) - P(T_n \leq x) \right| = o_p(1)$ *as* $n \to \infty$

if and only if $q = 1$.

Proof: See Section 10.3.2 below. □

Thus, Theorem 10.3 shows that with the modified scaling constants, the MBB provides a valid approximation to the distribution of the normalized sample mean T_n only in the case where T_n is asymptotically normal. The independent resampling of blocks under the MBB scheme fails to reproduce the dependence structure of the X_i's for transformations G with Hermite rank $q \geq 2$. As a consequence, the modified MBB version $\tilde{T}_n^*$ of T_n fails to emulate the large sample behavior of T_n in the nonnormal limit case. A similar behavior is expected if, in place of the MBB, other variants of the block bootstrap method based on independent resampling (e.g., the NBB or the CBB, are employed. Theorems 10.2 and 10.3 are due to Lahiri (1993b). Lahiri (1993b) also shows that using a resample size other than the sample size n also does not fix the inconsistency problem in the nonnormal limit case, as long as the number of resampled blocks tend to infinity. As a result, the "m out of n" bootstrap is not effective in this problem if the number of resampled blocks is allowed to go to infinity with n. However, if *repeated* independent resampling in the MBB method is dropped and only a *single* block is resampled, i.e., if in place of the MBB, the subsampling method is used, then consistent approximations to the distribution of T_n can be generated (cf. Section 10.4). For some numerical results on the MaBB method of Carlstein et al. (1998), see Hesterberg (1997). We now give a proof of Theorems 10.2 and 10.3 in Section 10.3.2 below.

10.3.2 Proofs

Define $\tilde{G}(y) = G(y) - c_q H_q(y)$, $y \in \mathbb{R}$. For $i, j \in \mathbb{Z}$, let $\delta_{ij} = 1$ or 0 according as $i = j$ or $i \neq j$. Also, recall that $x \vee y = \max\{x, y\}$, $x, y \in \mathbb{R}$, $N = n - \ell + 1$, and $U_i^* = \ell^{-1} \sum_{j=(i-1)\ell+1}^{i\ell} X_j^*$, $1 \leq i \leq b$.

Lemma 10.1 *Suppose that $r(\cdot)$, $L(\cdot)$, α, and q satisfy the requirements of Theorem 10.1. Assume that $\ell = O(n^{1-\epsilon})$ for some $0 < \epsilon < 1$, and that $\ell^{-1} = o(1)$. Then,*

$$\ell(\hat{\mu}_n - \mu) = o_p(d_\ell) .$$

Proof: Without loss of generality, assume that $\mu = 0$. Then, by (2.14),

$$\ell(\hat{\mu}_n - \mu)/d_\ell = (N d_\ell)^{-1} \left[n\ell \bar{X}_n + \sum_{j=1}^{\ell-1} (j - \ell)(X_j + X_{n-j+1}) \right] .$$

Note that $EH_k(Z_i)H_m(Z_j) = [r(i-j)]^k \delta_{k,m}$ for all $i, j \in \mathbb{Z}, k, m \in \mathbb{N}$. Hence, by Corollary 3.1 of Taqqu (1977), it follows that for any real numbers $a_1, a_2, \ldots, a_\ell \in [0, \ell]$,

$$\begin{aligned}
&(N d_\ell)^{-2} E\Big(\sum_{j=1}^{\ell} a_j X_j \Big)^2 \\
&\leq \ (N d_\ell)^{-2} \left[c_q^2 \left| \sum_{i=1}^{\ell} \sum_{j=1}^{\ell} a_i a_j [r(i-j)]^q \right| + \ell^2 \sum_{i=1}^{\ell} \sum_{j=1}^{\ell} \left| E\tilde{G}(Z_i)\tilde{G}(Z_j) \right| \right] \\
&\leq \ C(c_q)(n d_\ell)^{-2} \ell^2 \left[\sum_{i=1}^{\ell} \sum_{j=1}^{\ell} |r(i-j)|^q \right] (1 + o(1)) \\
&= \ O(\ell^{2+\epsilon} n^{-2}) = o(1) .
\end{aligned}$$

Therefore, by stationarity,

$$(N d_\ell)^{-1} \sum_{j=1}^{\ell-1} (j - \ell)(X_j + X_{n-j+1}) = o_p(1) .$$

Similarly, $\ell = O(n^{1-\epsilon})$ implies that

$$(N d_\ell)^{-2} E(n\ell \bar{X}_n)^2 = O(n^{-2} \ell^2 d_\ell^{-2} d_n^2) = o(1) .$$

Hence, Lemma 10.1 follows. □

Lemma 10.2 *Let $\hat{\sigma}_n^2 = \ell^2 E_*(U_1^* - \hat{\mu}_n)^2/d_\ell^2$ and let σ_q^2 be as defined in (10.8). Assume that the conditions of Theorem 10.2 hold. Then,*

$$\hat{\sigma}_n^2 = \sigma_q^2 + o_p(1) .$$

Proof: Define

$$\begin{aligned}
\hat{\sigma}_{1n}^2 &= (Nd_\ell^2)^{-1} \sum_{t=1}^{N} \Big(\sum_{j=t}^{t+\ell-1} c_q H_q(Z_j) \Big)^2 , \quad \text{and} \\
\hat{\sigma}_{2n}^2 &= (Nd_\ell^2)^{-1} \sum_{t=1}^{N} \Big(\sum_{j=t}^{t+\ell-1} \tilde{G}(Z_j) \Big)^2 .
\end{aligned}$$

Then, by Cauchy-Schwarz inequality,

$$|\hat{\sigma}_n^2 - \hat{\sigma}_{1n}^2| \leq \hat{\sigma}_{2n}^2 + 2|\hat{\sigma}_{2n}\hat{\sigma}_{1n}| . \tag{10.10}$$

By Corollary 3.1 of Taqqu (1977),

$$E\hat{\sigma}_{2n}^2 = d_\ell^{-2} E\Big(\sum_{i=1}^{\ell} \tilde{G}(Z_i) \Big)^2 = o(1) . \tag{10.11}$$

Hence, to prove the lemma, by (10.10) and (10.11), it is now enough to show that

$$\hat{\sigma}_{1n}^2 = \sigma_q^2 + o_p(1) . \tag{10.12}$$

Note that by Lemma 3.2 of Taqqu (1977) and the stationarity of the Z_i's,

$$\begin{aligned}
&\operatorname{Var}(\hat{\sigma}_{1n}^2) \\
&\leq C c_q^4 (Nd_\ell^2)^{-2} \sum_{j=0}^{N-1} (N-j) \Big| \operatorname{Cov}\Big(\Big[\sum_{i=1}^{\ell} H_q(Z_i) \Big]^2 , \\
&\qquad \Big[\sum_{i=1}^{\ell} H_q(Z_{i+j-1}) \Big]^2 \Big) \Big| \\
&\leq C(c_q)(Nd_\ell^4)^{-1} \sum_{j=0}^{N-1} \sum_{i_1=1}^{\ell} \sum_{i_2=1}^{\ell} \sum_{i_3=j}^{j+\ell-1} \sum_{i_4=j}^{j+\ell-1} \\
&\qquad \Big| (q!)^4 2^{-2q} (2q!)^{-1} \sum_1 \prod_{k=1}^{2q} r(m_k - j_k) - (q!)^2 r(i_1 - i_2)^q r(i_3 - i_4)^q \Big| ,
\end{aligned} \tag{10.13}$$

where $\sum_1$ extends over all $m_1, j_1; \ldots; m_{2q}, j_{2q} \in \{i_1, \ldots, i_4\}$ such that

(a) $m_k \neq j_k$ for all $k = 1, \ldots, 2q,$ and

(b) there are exactly q indices among $\{m_k, j_k : 1 \leq k \leq 2q\}$ that are equal to i_t for each $t = 1, 2, 3, 4$. (10.14)

Next write $\sum_1 = \sum_{11} + \sum_{12}$, where $\sum_{11}$ extends over all indices $\{m_k, j_k : 1 \le k \le 2q\}$ under $\sum_1$ for which $|m_k - j_k| = |i_1 - i_2|$ for exactly q pairs and $|m_k - j_k| = |i_3 - i_4|$ for the remaining q pairs, and where $\sum_{12}$ extends over the rest of the indices under $\sum_1$. Clearly, for any $\{m_k, j_k : 1 \le k \le 2q\}$ appearing under $\sum_{11}$, $\prod_{k=1}^{2q} r(m_k - j_k) = r(i_1 - i_2)^q r(i_3 - i_4)^q$. We claim that the number of such indices is precisely $(2q!)2^{2q}(q!)^{-2}$. Hence, assuming the claim, one gets

$$\left|(q!)^4 2^{-2q}(2q!)^{-1} \sum_1 \prod_{k=1}^{2q} r(m_k - j_k) - (q!)^2 r(i_1 - i_2)^q r(i_3 - i_4)^q\right|$$
$$\le C(q) \sum_{12} \prod_{k=1}^{2q} |r(m_k - j_k)| \,. \tag{10.15}$$

To prove the claim, note that, for any $\{m_1, j_1; \ldots; m_{2q}, j_{2q}\}$ under $\sum_1$, if $|m_k - j_k| = |i_1 - i_2|$ for some $k_1, \ldots, k_q \in \{1, \ldots, 2q\}$, then, by (10.14),

(a) $|m_k - j_k| = |i_3 - i_4|$ for all $k \in \{1, \ldots, 2q\} \backslash \{k_1, \ldots, k_q\}$, and

(b) exactly q of $\{m_{k_1}, j_{k_1}; \ldots; m_{k_q}, j_{k_q}\}$ are i_k, $k = 1, 2$ and exactly q of the remaining $2q$ integers are i_k, $k = 3, 4$.

Using this, one can check that the set of all indices $\{m_1, j_1; \ldots; m_{2q}, j_{2q}\}$ under $\sum_{11}$ can be obtained by first selecting a subset $\{k_1, \ldots, k_q\}$ of size q from $\{1, \ldots, 2q\}$, and then setting $(m_k, j_k) = (i_1, i_2)$ or (i_2, i_1) for $k \in \{k_1, \ldots, k_q\}$ and $(m_k, j_k) = (i_3, i_4)$ or (i_4, i_3) for $k \in \{1, \ldots, 2q\} \backslash \{k_1, \ldots, k_q\}$. Hence, it follows that the number of terms under $\sum_{11}$ is $\binom{2q}{q} \cdot 2^q \cdot 2^q$, proving the claim.

Next define $N_0 = \lceil \ell n^\delta \rceil$, where δ is any real number satisfying $0 < \delta < \epsilon(5 - 2q\alpha)^{-1}$ and where ϵ is as in the statement of Theorem 10.2. Let $\tilde{r}(j) = |j|^{-\alpha}(1 + |L(|j|)|)$, $j \in \mathbb{Z}$, and $M_n = \max\{1 + |L(j)| : 1 \le j \le n\}$. Then, it is easy to check that for n large,

$$\max\{|r(k)| : N_0 \le k \le n\} \le N_0^{-\alpha} M_n \le C(\alpha, \epsilon)\tilde{r}(j) \cdot n^{-\delta\alpha} M_n \tag{10.16}$$

uniformly in $1 \le j \le \ell$.

Note that given any i_1, i_2, i_3, i_4, for every multi-index $\{m_k, j_k : 1 \le k \le 2q\}$ under $\sum_{12}$, $|m_k - j_k| \ge \min\{|i_1 - i_3|, |i_1 - i_4|, |i_2 - i_3|, |i_2 - i_4|\} \ge (j - \ell)$ for at least one $k \in \{1, \ldots, 2q\}$. Hence, by (10.13), (10.15), and (10.16), it follows that

$$\begin{aligned}
&\operatorname{Var}(\hat{\sigma}_{1n}^2) \\
&\le \; C(c_q, q)(N d_\ell^4)^{-1} \sum_{j=0}^{N-1} \sum_{i_1=1}^{\ell} \sum_{i_2=1}^{\ell} \sum_{i_3=j}^{j+\ell-1} \sum_{i_4=j}^{j+\ell-1} \sum_{12} \prod_{k=1}^{2q} |r(m_k - j_k)| \\
&\le \; C(c_q, q)(N d_\ell^4)^{-1} I_{1n} +
\end{aligned}$$

$$
\begin{aligned}
& C(c_q, \alpha, q, \epsilon)(n^{\delta\alpha} N d_\ell^4)^{-1} M_n \\
& \qquad \times \sum_{j=N_0}^{N-1} \sum_{i_1=1}^{\ell} \sum_{i_2=1}^{\ell} \sum_{i_3=1}^{\ell} \sum_{i_4=1}^{\ell} \tilde{r}(i_1 - i_2)^q \tilde{r}(i_3 - i_4)^q \\
\leq \; & C(c_q, q)(N d_\ell^4)^{-1} I_{1n} + o(n^{-\delta\alpha/2}) ,
\end{aligned}
$$

where $I_{1n} = \sum_{j=0}^{N_0} \sum_{i_1=1}^{\ell} \sum_{i_2=1}^{\ell} \sum_{i_3=j}^{j+\ell-1} \sum_{i_4=j}^{j+\ell-1} \sum_{12} \prod_{k=1}^{2q} |r(m_k - j_k)|$. It is now easy to see that for any i_1, i_2, i_3, i_4, by (10.14)(b),

$$
\begin{aligned}
\sum_{12} \prod_{k=1}^{2q} |r(m_k - j_k)| \; & \leq \; C(q) \sum^{*} \tilde{r}(i_1 - i_2)^{q_1} \tilde{r}(i_1 - i_3)^{q_2} \tilde{r}(i_1 - i_4)^{q_3} \times \\
& \qquad \tilde{r}(i_2 - i_3)^{q_4} \tilde{r}(i_2 - i_4)^{q_5} \tilde{r}(i_3 - i_4)^{q_6} ,
\end{aligned}
$$

where $\sum^*$ extends over all nonnegative integers $q_1, \ldots, q_6$ satisfying $q_1 + \cdots + q_6 = 2q$ and

$$
\begin{aligned}
q_1 + q_2 + q_3 &= q \\
q_1 + q_4 + q_5 &= q \\
q_2 + q_4 + q_6 &= q \\
q_3 + q_5 + q_6 &= q \, .
\end{aligned}
$$

The set of four equations are determined by the frequencies of the indices i_1, i_2, i_3, and i_4 on the right side of the inequality above. Next, write $a = (1 - q\alpha)/q$ and $d = (2 - q\alpha)/q$. Now, using Hölder's inequality and the conditions on $q_1, \ldots, q_6$, one can show that for any $0 \leq j \leq N_0$,

$$
\begin{aligned}
& \sum_{i_3=j}^{j+\ell-1} \sum_{i_4=j}^{j+\ell-1} \sum_{i_2=1}^{\ell} \sum_{i_1=1}^{\ell} \tilde{r}(i_1 - i_2)^{q_1} \tilde{r}(i_1 - i_3)^{q_2} \tilde{r}(i_1 - i_4)^{q_3} \times \\
& \qquad\qquad \tilde{r}(i_2 - i_3)^{q_4} \tilde{r}(i_2 - i_4)^{q_5} \tilde{r}(i_3 - i_4)^{q_6} \\
\leq \; & \sum_{i_3=j}^{j+\ell-1} \sum_{i_4=j}^{j+\ell-1} \sum_{i_2=1}^{\ell} \tilde{r}(i_2 - i_3)^{q_4} \tilde{r}(i_2 - i_4)^{q_5} \tilde{r}(i_3 - i_4)^{q_6} \times \\
& \qquad\qquad \prod_{k=2}^{4} \left(\sum_{i_1=1}^{\ell} \tilde{r}(i_1 - i_k)^q \right)^{q_{k-1}/q} \\
\leq \; & C(\alpha, q) M_n^q \sum_{i_3=j}^{j+\ell-1} \sum_{i_4=j}^{j+\ell-1} \tilde{r}(i_3 - i_4)^{q_6} ((\ell - i_3) \vee i_3)^{a q_2} ((\ell - i_4) \vee i_4)^{a q_3} \\
& \quad \times \left(\sum_{i_2=1}^{\ell} \tilde{r}(i_2 - i_3)^q \right)^{q_4/q} \left(\sum_{i_2=1}^{\ell} \tilde{r}(i_2 - i_4)^q \right)^{q_5/q} \\
& \qquad\qquad \times \left(\sum_{i_2=1}^{\ell} [(\ell - i_2) \vee i_2]^{aq} \right)^{q_1/q}
\end{aligned}
$$

$$
\begin{aligned}
&\le C(\alpha, q) M_n^{2q} (j+\ell)^{d(q_1+q_3+q_5)} \sum_{i_3=j}^{j+\ell-1} \Big\{ ((\ell - i_3) \vee i_3)^{a(q_2+q_4)} \\
&\qquad\qquad \times (j+\ell-i_3)^{aq_6} \Big\} \\
&\le C(\alpha, q) M_n^{2q} (j+\ell)^{2dq} ,
\end{aligned}
$$

since $q_1 + \ldots + q_6 = 2q$ and $q_2 + q_4 + q_6 = q$ implies $q_1 + q_3 + q_5 = q$. Hence,

$$
(Nd_\ell^4)^{-1} I_{1n} \le C(\alpha, q)(Nd_\ell^4)^{-1} N_0 M_n^{2q} N_0^{2dq} = o(1) ,
$$

implying that

$$
\mathrm{Var}(\hat{\sigma}_{1n}^2) = o(1) .
$$

Since $E\hat{\sigma}_{1n}^2 \to \sigma_q^2$, (10.12) follows. This completes the proof of the lemma □

Lemma 10.3 *Let $W_1, W_2, \ldots, W_n$ be n iid random variables with $EW_1 = 0$, and $EW_1^2 = 1$. Then, for any $\eta > 0$, and every $n \in \mathbb{N}$ satisfying $n^{-1} + \delta_n(1) < 1$,*

$$
\begin{aligned}
&\sup_{x \in \mathbb{R}} \big| P(W_1 + \cdots + W_n \le \sqrt{n}x) - \Phi(x) \big| \\
&\le (1 + \beta_n)\delta_n(1) + 22[\eta + \delta_n(\eta)]\beta_n^3 ,
\end{aligned}
$$

where $\delta_n(a) \equiv EW_1^2 \mathbb{1}(|W_1| > a\sqrt{n})$, $a > 0$, and $\beta_n \equiv |1 - n^{-1} - \delta_n(1)|^{-1/2}$.

Proof: Define $\tilde{W}_i = W_i \mathbb{1}(|W_i| \le \sqrt{n})$, and $V_i = \tilde{W}_i - E\tilde{W}_i$, $1 \le i \le n$. Then, it is easy to check that

$$
|E\tilde{W}_1| \le \delta_n(1)/\sqrt{n}, \;\; EV_1^2 \ge 1 - n^{-1} - \delta_n(1) ,
$$

and for any $\eta > 0$,

$$
E|\tilde{W}_1|^3 \le (\eta + \delta_n(\eta))\sqrt{n} .
$$

By the Berry-Esseen Theorem (cf. Theorem A.6, Appendix A),

$$
\begin{aligned}
&\sup_{x \in \mathbb{R}} \Big| P(W_1 + \cdots + W_n \le \sqrt{n}x) - \Phi(x) \Big| \\
&\quad \le \sup_{x \in \mathbb{R}} \Big| P(V_1 + \cdots + V_n \le \sqrt{n}x(EV_1^2)^{1/2}) - \Phi(x) \Big| \\
&\qquad + \sup_{x \in \mathbb{R}} \Big| \Phi(x) - \Phi(x - \sqrt{n}E\tilde{W}_1(EV_1^2)^{-1/2}) \Big| \\
&\qquad + nP(|W_1| > \sqrt{n}) \\
&\quad \le (2.75)E|V_1|^3 (EV_1^2)^{-3/2} (\sqrt{n})^{-1} + \Big(1 + (2\pi EV_1^2)^{-1/2}\Big)\delta_n(1) .
\end{aligned}
$$

This proves Lemma 10.3. □

Proof of Theorem 10.2: Let $\breve{\sigma}_n^2 \equiv \ell^2 E_*(U_1^* - \hat{\mu}_n)^2/d_\ell^2$. Then, by Lemma 10.3 with $\eta = b^{-1/4}$,

$$\sup_{x\in\mathbb{R}} \left| P_*\left(\sum_{i=1}^{b} (\ell U_i^* - \ell\hat{\mu}_n) \leq \sqrt{b} d_\ell \breve{\sigma}_n x \right) - \Phi(x) \right|$$
$$\leq C\Big[\hat{\delta}_n(1 + |1 - b^{-1} - \hat{\delta}_n|^{-1/2}) + (b^{-1/4} + \hat{\delta}_n)|1 - b^{-1} - \hat{\delta}_n|^{-3/2}\Big] , \tag{10.17}$$

where $\hat{\delta}_n \equiv (d_\ell \breve{\sigma}_n)^{-2} E_*(\ell U_1^* - \ell\hat{\mu}_n)^2 \mathbb{1}(|\ell U_1^* - \ell\hat{\mu}_n| > 2b^{1/4} d_\ell \breve{\sigma}_n)$. We shall show that $\hat{\delta}_n \to 0$ in probability. Without loss of generality, let $\mu = 0$. Then, by Lemmas 10.1 and 10.2, it follows that

$$d_\ell^{-1}\ell\hat{\mu}_n = o_p(1) \quad \text{and} \quad \breve{\sigma}_n^2 = \sigma_q^2 + o_p(1) .$$

Hence, with $S_1^* \equiv \ell U_1^*$, we get

$$\hat{\delta}_n = O_P(d_\ell^{-2} E_* S_1^{*2} \mathbb{1}(|S_1^*| > b^{1/4} d_\ell \sigma_q) + o_p(1) .$$

Now, using arguments similar to those used in the proof of Lemma 10.2, one can show that $E(\sum_{i=1}^{\ell} H_q(Z_i))^4 = O(d_\ell^4)$. Hence, with $\tilde{S}_\ell = \sum_{i=1}^{\ell} c_q H(Z_i)$,

$$\begin{aligned}
& E\big[d_\ell^{-2} E_* S_1^{*2} \mathbb{1}(|S_1^*| > b^{1/4} d_\ell \sigma_q)\big] \\
&= d_\ell^{-2} E[\ell\bar{X}_\ell]^2 \mathbb{1}(|\ell\bar{X}_\ell| > b^{1/4} d_\ell/\sigma_q) \\
&\leq 4d_\ell^{-2}\left[E\tilde{S}_\ell^2 \mathbb{1}(|\tilde{S}_\ell| > b^{1/4} d_\ell/2\sigma_q) + E\left(\sum_{i=1}^{\ell} \tilde{G}(Z_i)\right)^2\right] \\
&= o(1) .
\end{aligned}$$

Consequently, $\hat{\delta}_n = o_p(1)$, and the theorem follows from (10.17) and Lemmas 10.1 and 10.2. □

Proof of Theorem 10.3: (i) follows from (10.7). As for (ii), the "if" part follows from (i) and Theorem 10.1. To prove the converse, suppose $q \neq 1$. Then, by Theorem 10.1, $n(\bar{X}_n - \mu)/d_n$ converges in distribution to a nonnormal limit while by (i), $\tilde{T}_n^*$ is asymptotically normal, implying the "only if" part. □

10.4 Properties of the Subsampling Method

In Section 10.3, we have shown that the MBB fails to approximate the distribution of the normalized sample mean, when the latter has a nonnormal limit distribution. In this section we show that even in such situations, valid approximations can be generated if, instead of the MBB, we employ the subsampling method. In Section 10.4.1 we present the results on subsampling approximations to the distribution of the normalized sample mean. In Section 10.4.2, we describe a method of studentizing the sample mean under long-range dependence, following the work of Hall, Jing and Lahiri (1998). The main results of Section 10.4.2 assert the validity of the subsampling method for the studentized sample mean. An outline of the proofs of the results of Sections 10.4.1 and 10.4.2 are given in Section 10.4.3. As mentioned earlier, numerical examples on the finite sample performance of the subsampling method are given in Section 10.5. All through Section 10.4, we work under the basic framework of Section 10.2.

10.4.1 Results on the Normalized Sample Mean

Let $T_n = n(\bar{X}_n - \mu)/d_n$ denote the normalized sample mean based on n observations from the stationary process $\{X_i\}_{i\in\mathbb{Z}}$, where $d_n^2 = n^{2-q\alpha}L^q(n)$ is as in Section 10.2. Let $\mathcal{B}_i = (X_i, \ldots, X_{i+\ell-1})$, $1 \le i \le N$ denote the collection of overlapping blocks of length ℓ for some given integer $\ell = \ell_n \in (1, n)$ and let $S_{\ell i} = \sum_{j=i}^{i+\ell-1} X_j$ denote the sum of the elements in the block $\mathcal{B}_i$, $1 \le i \le N$, where $N = n - \ell + 1$. Then, a "subsample copy" of T_n over $\mathcal{B}_i$ is given by

$$T_{\ell i} = (S_{\ell i} - \ell \bar{X}_n)/d_\ell, \ 1 \le i \le N \ . \tag{10.18}$$

The subsampling estimator of $Q_n(x) \equiv P(T_n \le x)$, $x \in \mathbb{R}$, based on subsamples of length ℓ, is given by

$$\hat{Q}_n(x) \equiv \hat{Q}_n(x;\ell) = N^{-1}\sum_{i=1}^{N} \mathbb{1}(T_{\ell i} \le x), \ x \in \mathbb{R} \ . \tag{10.19}$$

Next we state the conditions needed for establishing consistency of $\hat{Q}_n$. For clarity of exposition, here we state the key regularity condition on the spectral density $g(\cdot)$ of the Gaussian process $\{Z_i\}_{i\in\mathbb{Z}}$, rather than on its autocovariance function $r(\cdot)$. By using Abelian-Tanberian Theorems (cf. Bingham, Goldie and Teugels (1987)), one can show that if (10.1) holds, then

$$\frac{g(x)}{x^{\alpha-1}L(1/x)} \to C(\alpha) \quad \text{as} \quad x \to 0 \tag{10.20}$$

for some constant $C(\alpha) \in (0, \infty)$. Conversely, if (10.20) holds for a slowly varying function $L(\cdot)$, then $r(\cdot)$ admits the representation (10.1) with (without loss of generality) the same function $L(\cdot)$. Thus, the requirement (10.20) on the spectral density function $g(\cdot)$ and the condition (10.1) on the autocovariance function $r(\cdot)$ are equivalent. Because $g(\cdot)$ is a symmetric function, the Fourier series of $\log g(\cdot)$ is a pure cosine series. Replacing each cosine function by the corresponding sine function, we obtain the *harmonic conjugate* of $\log g(\cdot)$. Let $\widetilde{\log} g(\cdot)$ denote the harmonic conjugate of $\log g(\cdot)$. The key regularity condition on $g(\cdot)$ is that $\widetilde{\log} g(\cdot)$ is continuous on $[-\pi, \pi]$.

While $\log g(\cdot)$, being unbounded in every neighborhood of the origin, is not continuous on $[-\pi, \pi]$, an appropriately chosen branch of $\widetilde{\log} g$ can be continuous on $[-\pi, \pi]$. The following result on $\hat{Q}_n$ is due to Hall, Jing and Lahiri (1998).

Theorem 10.4 *Suppose that $\{X_i\}_{i\in\mathbb{Z}}$ is generated by relation (10.3) and that the function G has Hermite rank $q \in \mathbb{N}$. Also, suppose that*

(i) $g(x) = |x|^{\alpha-1}L_1(|x|)$, $0 < |x| \le \pi$ where $0 < \alpha < q^{-1}$ and where $L_1(\cdot)$ is slowly varying at 0 and of bounded variation on every closed subinterval of $[0, \pi]$;

(ii) a branch of $\widetilde{\log} g(\cdot)$ is continuous on $[-\pi, \pi]$; and

(iii) $n^{\epsilon}\ell^{-1} + n^{-(1-\epsilon)}\ell = O(1)$ for some $\epsilon \in (0, 1)$.

Then,

$$\sup_{x\in\mathbb{R}} \left|\hat{Q}_n(x) - Q_n(x)\right| \longrightarrow_p 0 \quad as \quad n \to \infty\,. \tag{10.21}$$

Proof: See Section 10.4.3 below. □

Thus, it follows from Theorem 10.4 that under appropriate regularity conditions, the subsampling estimator of the distribution of the normalized sample mean T_n is consistent for both normal and nonnormal limits of T_n. In particular, by avoiding repeated independent resampling of the blocks, the subsampling method overcomes the inconsistency problem associated with the MBB method in the nonnormal limit case. In the next section, we show that the subsampling method continues to provide valid approximations, when the scaling constants d_n's are replaced by certain data-based scaling factors.

10.4.2 Results on the Studentized Sample Mean

Note that the scaling constant $d_n = (n^{2-q\alpha}L^q(n))^{1/2}$ that yields a proper limit distribution for the centered sample mean $(\bar{X}_n - \mu)$ under long-range dependence depends on the unknown population quantities q, α, and $L(\cdot)$. As a consequence, one needs to estimate these quantities consistently to be able to construct large sample confidence intervals for the level-1 parameter μ. Here, we describe an empirical device for producing suitable scaling factors for $(\bar{X}_n - \mu)$ for all possible values of the Hermite rank q. Let $m_1 \equiv m_{1n}$ and $m_2 \equiv m_{2n} \in [1, n]$ be integers such that for some $\epsilon \in (0, 1)$,

$$\frac{m_1^2}{m_2} = n(1+o(1)) \quad \text{and} \quad n^{\epsilon}m_k^{-1} + m_k n^{\epsilon-1} = O(1) \quad \text{as} \quad n \to \infty\,, \tag{10.22}$$

for $k = 1, 2$. For example, we may take $m_1 = \lfloor n^{(1+\theta)/2} \rfloor$, $m_2 = \lfloor n^{\theta} \rfloor$ for some $\theta \in (0, 1)$. Next define $\check{e}_m^2 = (n - m + 1)^{-1} \sum_{i=1}^{n-m+1} (S_{im} - m\bar{X}_n)^2$

and let

$$\hat{e}_n^2 = \check{e}_{m_1}^4 / \check{e}_{m_2}^2 , \tag{10.23}$$

where $S_{im} \equiv (X_i + \cdots + X_{i+m-1})/m$, $m \geq 1$, $i \geq 1$. Under the condition of Theorem 10.5 below, $\hat{e}_n^2/[\text{Var}(n\bar{X}_n)] \to 1$ in probability as $n \to \infty$. We use $n^{-1}\hat{e}_n$ for scaling the sample mean $\bar{X}_n$ and define the "studentized" sample mean

$$T_{1n} = n(\bar{X}_n - \mu)/\hat{e}_n . \tag{10.24}$$

Let $Q_{1n}(x) = P(T_{1n} \leq x)$, $x \in \mathbb{R}$ denote the distribution function of T_{1n}. To define the subsampling estimator of $Q_{1n}(\cdot)$ based on blocks of length ℓ, let $\hat{e}_{i\ell}$ denote the subsample version of $\hat{e}_n$, obtained by replacing $\{X_1, \ldots, X_n\}$ and n in the definition of $\hat{e}_n$ by $\{X_i, \ldots, X_{i+\ell-1}\}$ and ℓ, respectively. In particular, this requires replacing $m_k \equiv m_{kn}$ by $m_{k\ell}$, $k = 1, 2$. Let $T_{1\ell,i} = (S_{i\ell} - \ell\bar{X}_n)/\hat{e}_{i\ell}$, $1 \leq i \leq N$ denote the subsample copies of T_{1n}. Then, the subsampling estimator of Q_{1n} is given by

$$\hat{Q}_{1n}(x) \equiv \hat{Q}_{1n}(x;\ell) = N^{-1} \sum_{i=1}^{N} \mathbb{1}(T_{1\ell,i} \leq x), \;\; x \in \mathbb{R} . \tag{10.25}$$

Theorem 10.5 *Assume that the conditions of Theorem 10.1 hold for some $q \in \mathbb{N}$, that m_1 and m_2 satisfy (10.22), that $n^\epsilon \ell^{-1} + n^{-(1-\epsilon)}\ell = O(1)$ for some $\epsilon \in (0,1)$ and that*

$$L^2(xy)/\{L(x^2)L(y^2)\} \to 1 \quad \text{as} \quad x, y \to \infty . \tag{10.26}$$

Then,

(a) $\hat{e}_n^2/\text{Var}(n\bar{X}_n) \longrightarrow_p 1$ *as* $n \to \infty$;

(b) $T_{1n} \longrightarrow^d W_q/\sigma_q$ *as* $n \to \infty$, *where* σ_q^2 *is as defined in (10.8);*

(c) $\sup_{x\in\mathbb{R}} \left|\hat{Q}_{1n}(x) - Q_{1n}(x)\right| \longrightarrow_p 0$ *as* $n \to \infty$.

Proof: See Section 10.4.3 below. □

Theorem 10.5 shows that the empirical device yields a consistent estimator of the variance of the sample sum $n\bar{X}_n = \sum_{i=1}^n X_i$ for all values of the Hermite rank $q \geq 1$. Furthermore, the subsampling method provides a valid approximation to the distribution of the studentized sample mean T_{1n} for both normal and nonnormal limits of T_{1n}. Consequently, we may use Theorem 10.5 to set approximate confidence intervals for the level-1 parameter μ that attain the nominal coverage levels asymptotically, for all $q \geq 1$. An advantage of this approach over the traditional large sample theory is that the subsampling confidence intervals may be constructed without making explicit adjustments for the Hermite rank q and without estimating the covariance parameter α. For $\gamma \in (0,1)$, let $\hat{q}_\gamma$ denote the

$\lfloor N\gamma \rfloor$-th order statistic of the subsample copies $T_{1\ell,i}$, $1 \le i \le N$. Then, an approximate $(1-\gamma)100\%$ two-sided equal-tailed confidence interval for μ is given by

$$\hat{I}_n(\gamma) = \left(\bar{X}_n - \hat{q}_{1-\frac{\gamma}{2}} \cdot \frac{\hat{e}_n}{n},\ \bar{X}_n - \hat{q}_{\frac{\gamma}{2}} \cdot \frac{\hat{e}_n}{n}\right) . \tag{10.27}$$

Then, under the conditions of Theorem 10.5,

$$P\big(\mu \in \hat{I}_n(\gamma)\big) \to 1-\gamma \quad \text{as} \quad n \to \infty .$$

Although the subsampling confidence interval $\hat{I}_n(\gamma)$ attains the desired coverage level $(1-\gamma)$ in the limit, its finite sample accuracy depends on various factors, notably on the block size ℓ and on the integers m_1, m_2. If we take m_1 and m_2 to be of the form $m_1 = \lfloor n^{(1+\theta)/2} \rfloor$ and $m_2 = \lfloor n^{\theta} \rfloor$ for some $\theta \in (0,1)$, a value of θ close to 1 is suggested by Hall, Jing and Lahiri (1998) to ensure better bias properties of the estimator $\check{e}^2_{m_k}$, $k = 1, 2$. We consider numerical properties of the subsampling method in Section 10.5, following the proofs of Theorems 10.4 and 10.5 given below.

10.4.3 Proofs

Proof of Theorem 10.4: By Theorem 5.2.24 of Zygmund (1968), $r(k) \sim k^{-\alpha} L_1(1/k)$ as $k \to \infty$, and hence, by Theorem 10.1,

$$n(\bar{X}_n - \mu)/d_n \longrightarrow^d W_q , \tag{10.28}$$

where, for $n \in \mathbb{N}$, the normalizing constant d_n is now defined by $d_n^2 = n^{2-q\alpha} L_1(1/n)^q$. Consequently, by the slow variation of $L_1(\cdot)$,

$$(\ell d_n)/(n d_\ell) = o(1) . \tag{10.29}$$

In view of (10.28) and (10.29), this implies that $\ell(\bar{X}_n - \mu)/d_\ell = o_p(1)$. Hence, it is enough to show that

$$\sup_{x \in \mathbb{R}} |\tilde{Q}_n(x) - Q_n(x)| = o_p(1) , \tag{10.30}$$

where $\tilde{Q}_n(x) = N^{-1} \sum_{1 \le i \le N} \mathbb{1}\{(S_{i\ell} - \ell\mu)/d_\ell \le x\}$. Since the distribution of W_q is continuous, (10.30) holds provided $\tilde{Q}_n(x) - Q_n(x) = o_p(1)$ for each $x \in \mathbb{R}$.

Note that

$$\begin{aligned} E\{\tilde{Q}_n(x) - Q_n(x)\}^2 &\le (2\ell+1)N^{-1} \\ &\quad + \frac{2}{N} \sum_{i=\ell+1}^{N-1} \left| P\{S_{1\ell}/d_\ell \le x, S_{(i+1)\ell}/d_\ell \le x\} - \{Q(x)\}^2 \right| , \end{aligned} \tag{10.31}$$

where, for simplicity of notation, we have set $\mu = 0$ in the last line. Now by Theorem 5.5.7 of Ibragimov and Rozanov (1978), the second term on the right side of (10.31) tends to zero. Hence, Theorem 10.4 is proved. □

Lemma 10.4 *Suppose that the function $G(\cdot)$ has Hermite rank $q \in \mathbb{N}$, that $0 < \alpha < q^{-1}$ and that $n^{\epsilon}\ell^{-1} + n^{-(1-\epsilon)}\ell = O(1)$ for some $\epsilon \in (0,1)$. Then, for any $\delta \in (0,1)$, $a, b \in \mathbb{N}$,*

$$\max_{\delta n \le i \le n} \left|\mathrm{Cov}(\tilde{S}_{1\ell}^{a}, \tilde{S}_{i\ell}^{b})\right| = o(d_\ell^{a+b}) \quad as \quad n \to \infty$$

where $\tilde{S}_{im} = \sum_{j=i}^{i+m-1} c_q H_q(Z_j)$.

Proof of Lemma 10.4: The proof of Lemma 10.4 is somewhat long and hence, is omitted. We refer the interested reader to the proof of relation (4.8), page 1201–1202 of Hall, Jing and Lahiri (1998) for details. □

Proof of Theorem 10.5: Let $e_n = d_n \sigma_q$, $M = n - m + 1$, and $\tilde{e}_m^2 = M^{-1}\sum_{i=1}^{M}(S_{im} - m\mu)^2$. In view of (10.29), if m denotes either m_1 or m_2, then

$$\begin{aligned}
& E\left|\tilde{e}_m^2 - \check{e}_m^2\right| \\
\le\ & 4M^{-1}\sum_{i=1}^{M} m\Big[E(\bar{X}_n - \mu)^2\Big\{E(S_{im} - m\mu)^2 + m^2 E(\bar{X}_n - \mu)^2\Big\}\Big]^{1/2} \\
=\ & 4m(n^{-1}e_n)\big[e_m^2 + m^2 n^{-2} e_n^2\big]^{1/2} \\
=\ & o(d_n^2)\ . \qquad (10.32)
\end{aligned}$$

Next write $\bar{e}_m^2 = M^{-1}\sum_{i=1}^{M}\tilde{S}_{im}^2$. By Corollary 3.1 of Taqqu (1975) and Cauchy-Schwarz inequality, for $m = m_1, m_2$

$$\begin{aligned}
& E\left|\tilde{e}_m^2 - \bar{e}_m^2\right| \\
\le\ & E\Bigg[\Big\{M^{-1}\sum_{i=1}^{M}(S_{im} - m\mu + \tilde{S}_{im})^2\Big\}^{1/2} \\
& \times \Big\{M^{-1}\sum_{i=1}^{M}(S_{im} - m\mu - \tilde{S}_{im})^2\Big\}^{1/2}\Bigg] \\
\le\ & \Big\{2E(S_{1m} - m\mu)^2 + 2E\tilde{S}_{1m}^2\Big\}^{1/2}\Big\{E(S_{1m} - m\mu - \tilde{S}_{1m})^2\Big\}^{1/2} \\
=\ & o(d_m^2)\ .
\end{aligned}$$

By Lemma 10.4, for $m = m_1$ or m_2,

$$E\big(\bar{e}_m^2 - E\bar{e}_m^2\big)^2$$

$$
\begin{aligned}
&= O\left(n^{-2} \sum_{j_1=1}^{M} \sum_{j_2=1}^{M} \left|\operatorname{Cov}\left\{(\tilde{S}_{j_1 m})^2, (\tilde{S}_{j_2 m})^2\right\}\right|\right) \\
&= o(d_m^4) \, .
\end{aligned}
$$

Because $Ee_{m_k}^2 = e_{m_k}^2\{1 + o(1)\}$ for $k = 1, 2$, it follows that $\hat{e}_n / e_n \to 1$ in probability. This proves part (a) of the theorem. Part (b) now follows by applying part (a) and Theorem 10.4. This completes the proof of Theorem 10.5. □

10.5 Numerical Results

In this section, we consider finite sample performance of the subsampling method for the renormalized sample mean

$$
T_{1n} = n(\bar{X}_n - \mu)/\hat{e}_n
$$

of (10.24). For this, as in Hall, Jing and Lahiri (1998), we generated stationary increments of a self-similar process with self-similarity parameter (or Hurst constant) $H = \frac{1}{2}(2 - \alpha)$, and took a suitable linear transformation of these data to produce a realization of a long-range dependent process with Hermite rank q. The details of the relevant steps are as follows:

Step 1: Generate a random sample $\mathcal{Z}_{n0} = (Z_{10}, \ldots, Z_{n0})'$ of size n from the standard normal distribution.

Step 2: Let $R = ((r_{ij}))$ denote the correlation matrix defined by

$$
r_{ij} = \frac{1}{2}\left\{(k+1)^{2H} + (k-1)^{2H} - 2k^{2H}\right\} ,
$$

for $k = |j - i|$ and $\frac{1}{2} < H < 1$. Express R as $R = U'U$ by Cholesky factorization.

Step 3: Define $\mathcal{Z}_n \equiv (Z_1, \ldots, Z_n)' = U'\mathcal{Z}_{n0}$. Then $\mathcal{Z}_n$ may be considered as a finite segment of a stationary Gaussian process with zero mean, unit variance, and autocovariance function

$$
\begin{aligned}
r(k) &= \frac{1}{2}\left\{(k+1)^{2H} + (k-1)^{2H} - 2k^{2H}\right\} \\
&\sim Ck^{-\alpha} \quad \text{as} \quad k \to \infty \, , \qquad (10.33)
\end{aligned}
$$

where $\alpha = 2 - 2H \in (0, 1)$. Note that the $r(k)$'s are the autocovariances of the stationary increments of a self-similar process with self-similarity parameter H (cf. Beran (1994), p. 50).

Step 4: Define $X_i = H_q(Z_i)$, for $i = 1, \ldots, n$, where H_q is the qth Hermite polynomial. Then $\mathcal{X}_n = \{X_1, \ldots, X_n\}$ is a long-range dependent series with Hermite rank q.

We now report finite sample coverage accuracy of the subsampling method for confidence intervals for μ, produced using the renormalized statistic T_{1n} from Section 10.4.2. We consider the long-range dependent time series $\{X_1, \ldots, X_n\}$ of Hermite rank $q \in \{1, 2, 3\}$, as generated above, for $n = 100, 400, 900$. Throughout, we take the nominal confidence level to be 0.90. The empirical approximations to coverage probabilities reported here were derived by averaging over $K = 1000$ independent simulations.

For the block size ℓ, we used $\ell = cn^{1/2}$ for $c = .5, 1, 2$. Although the best choice of ℓ is unknown, this choice is based on the intuition (cf. Hall, Jing and Lahiri (1998), page 1155) that the optimal size should be greater than that for the weakly dependent case, where $\ell \sim cn^{\beta}$ for $\beta \leq \frac{1}{3}$ is generally appropriate (cf. Hall, Horowitz and Jing (1995) and Hall and Jing (1996)). In choosing m_1 and m_2 for the renormalization procedure mentioned in Section 10.4.2, we took

$$m_1 = \lfloor n^{(1+\theta)/2} \rfloor \quad \text{and} \quad m_2 = \lfloor n^{\theta} \rfloor$$

with $\theta = 0.8$.

Results from the simulation study are summarized in Table 10.1. In the table, the headings "Lower" and "Upper" represent coverage probabilities of the lower and the upper 90% one-sided confidence intervals, respectively, while q denotes the Hermite rank. The formula for the $100(1-\alpha)\%$, $0 < \alpha < 1$, lower and upper confidence limits are, respectively, given by

$$L_\alpha = n^{-1}(S_{1n} - \hat{e}_n \hat{t}_{1-\alpha}) \tag{10.34}$$

and

$$U_\alpha = n^{-1}(S_{1n} - \hat{e}_n \hat{t}_\alpha) \ , \tag{10.35}$$

where $\hat{t}_\beta$ is the β-quantile of the subsampling estimator $\hat{Q}_{1n}$, $0 < \beta < 1$, and $S_{1n} = \sum_{i=1}^{n} X_i$. It appears that for $\alpha = 0.5, 0.9$, the choice $\ell = 2\sqrt{n}$ leads to more accurate results while for $\alpha = 0.1$, $\ell = 0.5n^{1/2}$ works better. For each value of $\alpha = 0.1, 0.5, 0.9$ and for each value of $q = 1, 2, 3$, coverage accuracy with these "optimal" choices of the subsampling parameter ℓ increases with the sample size. Interestingly, the Hermit rank $q \in \{1, 2, 3\}$ seems to have little effect on accuracy of the subsampling method. We also repeated the whole simulation study with $\theta = 0.9$ in the definitions of $m_1 = \lfloor n^{(1+\theta)/2} \rfloor$ and $m_2 = \lfloor n^{\theta} \rfloor$, as considered by Hall, Jing and Lahiri (1998). The choice $\theta = 0.8$ had slightly better performance than $\theta = 0.9$ for the combinations of the factors q, α, n and ℓ considered here.

TABLE 10.1. Coverage probabilities of 90% lower and upper confidence limits, given by (10.34) and (10.35) respectively, based on $K = 1000$ simulation runs. Here n denotes the sample size, ℓ is the length of subsamples, q denotes the Hermite rank of $\{X_i\}_{i\in\mathbb{Z}}$ and α is as in (10.33).

(a): $\ell = \frac{1}{2} \cdot n^{1/2}$

		$\alpha = 0.1$		$\alpha = 0.5$		$\alpha = 0.9$	
	q	Lower	Upper	Lower	Upper	Lower	Upper
	1	87.0	86.1	95.0	93.1	97.0	94.2
$n = 100$	2	91.9	94.2	99.2	95.9	99.6	93.9
	3	96.3	93.3	98.0	97.0	97.5	95.7
	1	89.0	88.8	94.9	95.4	95.7	95.8
$n = 400$	2	90.0	96.6	98.0	95.5	99.1	93.5
	3	95.1	93.5	96.9	96.7	97.0	95.8
	1	91.7	92.0	96.8	95.8	97.6	96.6
$n = 900$	2	91.5	96.5	99.2	97.5	98.9	95.1
	3	97.6	97.2	97.5	97.5	98.1	97.1

(b): $\ell = n^{1/2}$

		$\alpha = 0.1$		$\alpha = 0.5$		$\alpha = 0.9$	
	q	Lower	Upper	Lower	Upper	Lower	Upper
	1	82.0	81.2	91.1	89.7	93.1	90.9
$n = 100$	2	81.6	91.9	95.7	93.9	96.6	91.4
	3	90.6	87.4	93.7	92.8	92.8	91.4
	1	87.6	86.6	93.4	93.2	94.2	94.0
$n = 400$	2	84.1	95.4	95.7	94.8	97.1	92.3
	3	92.1	90.0	94.6	94.9	95.3	94.7
	1	87.5	88.2	93.9	93.3	94.2	93.3
$n = 900$	2	84.9	94.4	96.9	94.9	97.0	92.7
	3	92.7	93.5	95.2	95.3	94.5	94.3

(c): $\ell = 2n^{1/2}$

		$\alpha = 0.1$		$\alpha = 0.5$		$\alpha = 0.9$	
	q	Lower	Upper	Lower	Upper	Lower	Upper
	1	78.7	76.0	89.4	85.1	90.8	86.9
$n = 100$	2	76.2	88.4	90.0	91.0	92.2	88.7
	3	86.2	81.5	90.6	88.3	89.5	87.8
	1	82.9	81.4	90.5	89.2	92.0	91.7
$n = 400$	2	77.8	92.9	91.4	92.2	93.2	89.4
	3	87.5	85.6	92.3	90.2	91.8	90.3
	1	82.6	82.2	89.9	90.3	91.3	92.0
$n = 900$	2	79.3	91.7	93.5	92.2	93.8	91.1
	3	88.7	89.0	93.0	91.7	91.8	90.6

11

Bootstrapping Heavy-Tailed Data and Extremes

11.1 Introduction

In this chapter, we consider two topics, viz., bootstrapping heavy-tailed time series data and bootstrapping the extremes (i.e., the maxima and the minima) of stationary processes. We call a random variable heavy-tailed if its variance is infinite. For iid random variables with such heavy tails, it is well known (cf. Feller (1971b), Chapter 17) that under some regularity conditions on the tails of the underlying distribution, the normalized sample mean converges to a stable distribution. Similar results are also known for the sample mean under weak dependence. In Section 11.2, we introduce some relevant definitions and review some known results in this area. In Sections 11.3 and 11.4, we present some results on the performance of the MBB for heavy-tailed data under dependence. Like the iid case, here the MBB works if the resample size is of a smaller order than the original sample size. Consistency properties of the MBB are presented in Section 11.3, while its invalidity for a resample size equal to the sample size is considered in Section 11.4.

In Sections 11.5–11.7, we consider the extremes of stationary processes. This is another classic example where the "fewer than n" resampling works better. In Section 11.5, we review some relevant definitions and results on extremes of dependent data. Results on bootstrapping the extremes are presented in Sections 11.6 and 11.7 respectively for the cases where the normalizing constants are known and where they are estimated.

11.2 Heavy-Tailed Distributions

Let $\{X_n\}_{n\geq 1}$ be a sequence of stationary random variables defined on a probability space $(\Omega, \mathcal{F}, P)$. Let F be the marginal distribution function of X_1. Also, let $S_n = X_1 + \cdots + X_n$, $n \geq 1$ denote the partial sums of the process $\{X_n\}_{n\geq 1}$. Under weak dependence, it is known (cf. Ibragimov and Linnik (1971)) that the possible limit distributions of the normalized sum process $\{a_n^{-1}(S_n - b_n)\}_{n\geq 1}$ for constants $a_n > 0$, $b_n \in \mathbb{R}$ are infinitely divisible. In particular, for stationary heavy-tailed random variables heaving an infinite second moment, the normalized sum may converge to a nonnormal limit. For independent random variables, necessary and sufficient conditions for convergence of the partial sum to a given infinitely divisible distribution are known (cf. Feller (1971b), Chapter 17). However, for dependent random variables, in addition to these tail conditions on the marginal distribution function F, some additional conditions on the dependence structure of the process $\{X_n\}_{n\geq 1}$ are imposed to guarantee convergence to an infinitely divisible law. Here, we collect some relevant definitions and results on weak convergence to infinitely divisible distributions and consider certain dependence structures that are suitable for studying theoretical properties of the MBB approximation in the heavy-tail case. For more details on the properties of the sum of dependent heavy-tailed random variables, we refer the interested readers to the papers by Davis (1983), Samur (1984), Jakubowski and Kobus (1989), Denker and Jakubowski (1989), and the references therein.

Definition 11.1 *A random variable W (or its distribution function) is said to be infinitely divisible if its characteristic function is given by*

$$\xi(t) = \exp\Big(\int [e^{\iota t x} - 1 - \iota t \tau_c(x)] x^{-2} M(dx)\Big), \quad t \in \mathbb{R}\ , \tag{11.1}$$

for some $0 \leq c \leq \infty$ and some measure M on $(\mathbb{R}, \mathcal{B}(\mathbb{R}))$, where $\iota = \sqrt{-1}$, $\tau_c(x)$ is the function $\tau_c(x) \equiv x \mathbb{1}(|x| \leq c)$, $x \in \mathbb{R}$, and where M is a canonical measure *on $\mathbb{R}$, i.e., M is a measure on $(\mathbb{R}, \mathcal{B}(\mathbb{R}))$, $M(I) < \infty$ for any bounded interval $I \subset \mathbb{R}$ and $M^+(x) \equiv \int_{(x,\infty)} y^{-2} M(dy) < \infty$ and $M^-(x) \equiv \int_{(-\infty,-x]} y^{-2} M(dy) < \infty$ for all $x > 0$.*

At $x = 0$, the expression $[e^{\iota t x} - 1 - \iota t \tau_c(x)] x^{-2}$ in (11.1) is replaced by its limit (as $x \to 0$), i.e., by $-t^2/2$. Some common examples of infinitely divisible distributions include the normal distribution with mean zero and variance $\sigma^2 \in (0, \infty)$ (with $M(A) = \sigma^2 \mathbb{1}_A(0)$, $A \in \mathcal{B}(\mathbb{R})$), the Poisson distribution with mean $\lambda \in (0, \infty)$, (with $c = 0$ and $M(A) = \lambda \mathbb{1}_A(1)$, $A \in \mathcal{B}(\mathbb{R})$), and the nonnormal stable laws of order $\alpha \in (0, 2)$, where for a given $\alpha \in (0, 2)$, the canonical measure $M = M_\alpha$ associated with the stable law of order α is given by

$$M_\alpha(A) = C_0\Big[p\int_{(0,\infty)\cap A} x^{1-\alpha}dx + q\int_{(-\infty,0)\cap A} |x|^{1-\alpha}dx\Big], \quad A \in \mathcal{B}(\mathbb{R}) , \tag{11.2}$$

for some constants $C_0 \in (0, \infty)$, $p \geq 0$, $q \geq 0$ with $p + q = 1$.

Next, let $\{\tilde{X}_n\}_{n\geq 1}$ be a sequence of iid random variables with common distribution function F, where F is the common marginal distribution function of the given stationary sequence $\{X_n\}_{n\geq 1}$. Then, the sequence $\{\tilde{X}_n\}_{n\geq 1}$ will be referred to as the *associated iid sequence* to the given sequence $\{X_n\}_{n\geq 1}$. We shall establish validity of the bootstrap approximation for sums of X_n's in a general set up where the sequence of partial sums, suitably centered and scaled, may have different limits along different subsequences, as made precise in the following definition.

Definition 11.2 *Let $\{X_n\}_{n\geq 1}$ be a sequence of stationary random variables with one-dimensional marginal distribution F and let W be an infinitely divisible random variable with distribution F_0. Then, we say that F belongs to the domain of partial attraction of F_0 if there exists a subsequence $\{n_i\}_{i\geq 1}$ and constants $a_{n_i} > 0$, $b_{n_i} \in \mathbb{R}$, $i \geq 1$ such that*

$$a_{n_i}^{-1}\Big(\tilde{X}_1 + \cdots + \tilde{X}_{n_i} - b_{n_i}\Big) \longrightarrow^d W , \tag{11.3}$$

where $\{\tilde{X}_n\}_{n\geq 1}$ is the associated iid sequence to the given sequence $\{X_n\}_{n\geq 1}$.

For the associated iid sequence $\{\tilde{X}_n\}_{n\geq 1}$, convergence of the normalized sum $a_{n_i}^{-1}(\tilde{X}_1 + \cdots + \tilde{X}_{n_i} - b_{n_i})$ to the infinitely divisible distribution of (11.1) holds solely under some regularity conditions on the marginal distribution function F. Let $\mathcal{C}(M)$ denote the set of all continuity points of (the distribution function of a) measure M, i.e., $\mathcal{C}(M) = \{x \in \mathbb{R} : M(\{x\}) = 0\}$. Let W be a random variable with the characteristic function $\xi(\cdot)$ of (11.1) and let c and M of (11.1) be such that $c \in \mathcal{C}(M)$. Then, a set of necessary and sufficient conditions for (11.3) is that (cf. Feller (1971b), Chapter 17), as $i \to \infty$,

$$n_i\big(1 - F(a_{n_i}x)\big) \longrightarrow M^+(x), \quad \text{for all} \quad x \in (0,\infty)\cap\mathcal{C}(M) , \tag{11.4}$$

$$n_i F(a_{n_i}x) \longrightarrow M^-(x), \quad \text{for all} \quad x \in (-\infty,0)\cap\mathcal{C}(M) , \tag{11.5}$$

$$n_i \mathrm{Var}\Big(\tau_c(a_{n_i}^{-1}X_1)\Big) \longrightarrow M([-c,c]) , \tag{11.6}$$

$$a_{n_i}^{-1}b_{n_i} - n_i E\tau_c(X_1/a_{n_i}) \longrightarrow 0 . \tag{11.7}$$

When (11.7) holds, we may replace b_{n_i} with $n_i E\tau_c(X_1/a_{n_i})$ and get the convergence of $\sum_{j=1}^{n_i}[\tilde{X}_j - E\tau_c(X_1/a_{n_i})]$ to W. In general, it is not possible to further replace $E\tau_c(X_1/a_{n_i})$ with EX_1/a_{n_i}. However, if

$$\lim_{\lambda\to\infty}\limsup_{i\to\infty} n_i a_{n_i}^{-1} E|X_1|\mathbb{1}(|X_1| > \lambda a_{n_i}) = 0 , \tag{11.8}$$

and (11.4)–(11.7) hold, then it can be shown that

$$a_{n_i}^{-1}(\tilde{X}_1 + \cdots + \tilde{X}_{n_i} - n_i\mu) \longrightarrow^d W_0 , \tag{11.9}$$

where $\mu = EX_1$ and W_0 is a random variable having the characteristic function (11.1) with $c = +\infty$. Note that under (11.8), it is now meaningful to consider statistical inference regarding the population mean μ on the basis of the variables $\tilde{X}_1, \ldots, \tilde{X}_n$.

Next we turn our attention to conditions that ensure a weak convergence result similar to (11.9) for the given dependent sequence $\{X_n\}_{n\geq 1}$. In addition to the above assumptions on the tails of F, an additional set of weak-dependence conditions is typically assumed to prove such a weak convergence result. As the required set of weak-dependence conditions depend on the form of the canonical measure M of the limiting infinitely divisible distribution, for simplicity, we shall restrict attention to the "purely nonnormal" case where $M(\{0\}) = 0$. (See the references cited above for conditions when $M(\{0\}) \neq 0$.) With this restriction on M, we shall assume the following regularity conditions on the dependence structure of the process $\{X_n\}_{n\geq 1}$:

$$\Psi^* \equiv \limsup_{x\to\infty} \sup_{n\geq 1} \frac{P(X_1 > x, X_{n+1} > x)}{[P(x_1 > x)]^2} < \infty \tag{11.10}$$

and

$$\sum_{n=1}^{\infty} \rho(2^n) < \infty , \tag{11.11}$$

where $\rho(\cdot)$ denotes the ρ-mixing coefficient of the process $\{X_n\}_{n\geq 1}$. Recall that for $n \geq 1$, we define

$$\rho(n) = \Big\{|Efg|/(Ef^2Eg^2)^{1/2} : f \in \mathcal{L}_2'(\mathcal{F}_1^{k+1}),\ g \in \mathcal{L}_2'(\mathcal{F}_{k+n+1}^{\infty}),\ k \geq 1\Big\} , \tag{11.12}$$

where $\mathcal{F}_i^j$ = the σ-field generated by $\{X_k : k \in \mathbb{N}, i \leq k < j\}$, $1 \leq i \leq j \leq \infty$ and $\mathcal{L}_2'(\mathcal{F}_i^j) = \{f : \Omega \to \mathbb{R} \mid \int f^2 dP < \infty, \int f dP = 0$ and f is $\mathcal{F}_i^j$-measurable$\}$.

Condition (11.11) is quite common for proving Central Limit Theorems for ρ-mixing random variables (cf. Peligrad (1982)). The quantity Ψ^* in (11.10) is closely related to the well known Ψ-mixing coefficient, defined by

$$\begin{aligned}\Psi(n) \;=\; \sup\Big\{&|P(A\cap B) - P(A)P(B)|/P(A)P(B) : \\ &A \in \mathcal{F}_1^{k+1},\ B \in \mathcal{F}_{k+n+1}^{\infty},\ k \geq 1\Big\},\ n \geq 1 .\end{aligned}$$

Together, conditions (11.10) and (11.11) specify the dependence structure of the sequence $\{X_n\}_{n\geq 1}$ that yields the following analog of (11.9) for the given sequence $\{X_n\}_{n\geq 1}$.

Theorem 11.1 *Assume that (11.4)–(11.6), (11.8), (11.10), and (11.11) hold for some subsequence* $\{n_i\}_{i\geq 1}$. *Then*

$$T_{n_i} \equiv a_{n_i}^{-1}(S_{n_i} - n_i\mu) \longrightarrow^d W_0 \ ,$$

where $S_n = X_1 + \cdots + X_n$, $n \in \mathbb{N}$, $\mu = EX_1$, *and* W_0 *has characteristic function* $\xi(t)$ *given by (11.1) with* $c = +\infty$, *and* $M(\{0\}) = 0$.

Proof: See Lemma 3.5, Lahiri (1995). □

In the next section, we consider properties of the bootstrap approximation to the normalized sum T_n.

11.3 Consistency of the MBB

Let $\mathcal{X}_n = \{X_1, \ldots, X_n\}$ denote the sample at hand and let $\{X_1^*, \ldots, X_m^*\}$ denote the MBB resample of size $m \equiv m_n$ based on blocks of size ℓ. For simplicity, we suppose that $m = k\ell$ for some integer k, so that the bootstrap approximation is generated using k "complete" blocks. Furthermore, we suppose that the block length variable ℓ satisfies the requirement that $\ell = o(n)$, but may or may *not* tend to infinity. Thus, for $\ell \equiv 1$, this also covers the IID resampling scheme of Efron (1979) where single data-values, rather than blocks of them, are resampled at a time. Let $S_{m,n}^* = X_1^* + \cdots + X_m^*$ and $\hat{\mu}_n = E_*[S_{m,n}^*/m]$. Then, the bootstrap version of $T_n \equiv a_n^{-1}(S_n - n\mu)$ is given by

$$T_{m,n}^* \equiv a_m^{-1}(S_{m,n}^* - m\hat{\mu}_n) \ .$$

The main result of this section says that bootstrap approximations to the distribution of T_n are valid along every subsequence n_i for which $T_{n_i} \longrightarrow^d W_0$, provided $k \to \infty$ and the resample size $m = o(n)$ as $n \to \infty$.

Theorem 11.2 *Suppose that* $1 \leq \ell \ll n$ *and that* $k \to \infty$ *such that* $m \equiv k\ell = o(n)$ *as* $n \to \infty$. *Also, suppose that the conditions of Theorem 11.1 hold for some subsequence* $\{n_i\}_{i\geq 1}$. *If the subsequence* $\{m_{n_i} : i \geq 1\}$ *is contained in* $\{n_i : i \geq 1\}$ *and* $k_{n_i}^{-1/2}(m_{n_i} n_i^{-1})[a_{n_i}/a_{m_{n_i}}] \to 0$ *as* $n \to \infty$, *then*

$$\varrho(\hat{\Gamma}_{m_{n_i}, n_i}, \Gamma_{n_i}) \longrightarrow_p 0 \quad as \quad i \to \infty \ ,$$

where $\hat{\Gamma}_{m,n}(x) = P_*(T_{m,n}^* \leq x)$ *and* $\Gamma_n(x) = P(T_n \leq x)$, $x \in \mathbb{R}$, $n \geq 1$, *and* ϱ *is a metric that metricizes the topology of weak convergence of probability measures on* $(\mathbb{R}, \mathcal{B}(\mathbb{R}))$.

Proof: See Theorem 2.1 of Lahiri (1995). □

Thus, Theorem 11.2 asserts the validity of the MBB approximation along every subsequence for which the limit distribution of the normalized sum is a nonnormal infinitely divisible distribution (which may be different for different subsequences). Theorem 11.2 shows that the MBB adapts itself to the form of the true distribution of the normalized sum so well that it captures all subsequential limits of S_n and provides a valid approximation along every convergent subsequence, provided the resample size m grows slowly compared to the sample size n. See Lahiri (1995) for more details. For *independent* random variables, a similar result has been proved by Arcones and Giné (1989) for the IID bootstrap method of Efron (1979).

Next we comment on the block size ℓ that leads to a valid bootstrap approximation. For simplicity of exposition, here and in the rest of this section, we suppose that the subsequential limits of normalized S_n are the same. Then, there exists an $\alpha \in (1, 2)$ and scaling constants $a_n > 0$ such that

$$T_n = a_n^{-1}(S_n - n\mu) \longrightarrow^d W_\alpha \,, \tag{11.13}$$

where W_α has a (nonnormal) stable law of order α, having characteristic function $\xi(t)$ of (11.1) with canonical measure M_α of (11.2). In this case, the variance of X_1 is *infinite* and Theorem 11.2 shows that the MBB approximation works for the normalized sum T_n with a nonnormal limit, provided the resample size m is of *smaller* order than the sample size. It is interesting to note that the block length parameter ℓ here need *not* go to infinity in order to provide a valid approximation even though the random variables $\{X_n\}_{n\geq 1}$ are dependent. Thus, Efron's (1979) IID resampling scheme, which uses no blocking and resamples a single observation at a time (i.e., $\ell \equiv 1$ for all n), also provides a valid approximation to the distribution of the normalized sum of heavy-tailed random variables under dependence. Validity of the IID bootstrap here is in sharp contrast with the finite variance dependent case, where the example of Singh (1981) (cf. Section 2.3) shows that the IID bootstrap fails to capture the effect of dependence on the distribution of the sum converging to a normal limit. An intuitive justification for this fact may be given by noting that under the conditions of Theorem 11.1, the limit distribution of T_n in the heavy-tail case depends only on the marginal distribution F of the sequence $\{X_n\}_{n\geq 1}$. As a consequence, the resampling of single data-values capture adequate information about the distribution of the X_n's to produce a valid approximation to the distribution of the normalized sum.

If the constants $\{a_n\}_{n\geq 1}$ in (11.13) are known, then the MBB can be used to construct asymptotically valid confidence intervals for the parameter μ. Let $\hat{q}_m(\gamma)$ denote the γ-quantile of $\hat{\Gamma}_{m,n}$, $0 < \gamma < 1$. Then, an equal tailed $(1-\gamma)$ bootstrap confidence interval for the parameter μ is given by

$$\hat{I}_{m,n}(1-\gamma) \equiv \Big[n^{-1}\big(S_n - a_n\hat{q}_m(1-[\gamma/2])\big),\; n^{-1}\big(S_n - a_n\hat{q}_m(\gamma/2)\big)\Big] \,. \tag{11.14}$$

Under (11.13) and the regularity conditions of Theorem 11.2,

$$P\big(\mu \in \hat{I}_{m,n}(1-\gamma)\big) \to 1-\gamma \quad \text{as} \quad n \to \infty \,. \tag{11.15}$$

When the scaling constants $\{a_n\}_{n\geq 1}$ in (11.13) are unknown, we may instead consider a "studentized" statistic of the form $T_{1n} \equiv \hat{a}_n(S_n - n\mu)$, where $\hat{a}_n$ is an estimator of a_n satisfying

$$\hat{a}_n / a_n \longrightarrow_p 1 . \tag{11.16}$$

For example, when X_1 has a stable distribution of order $\alpha \in (1, 2)$, the scaling constants $\{a_n\}_{n\geq 1}$ are given by $a_n = n^{1/\alpha}$, $n \geq 1$. In this case, we may take $\hat{a}_n = n^{1/\hat{\alpha}_n}$, where $\{\hat{\alpha}_n\}_{n\geq 1}$ is a sequence of estimators of the tail index α that satisfies $\hat{\alpha}_n - \alpha = o_p\big((\log n)^{-1}\big)$ as $n \to \infty$. See Hsing (1991), Resnick and Stărciă (1998), and the references therein. For an iid sequence $\{X_n\}_{n\geq 1}$, when F lies in the domain of attraction of a stable law of order $\alpha \in (1, 2)$, Athreya, Lahiri and Wei (1998) developed a *self-normalization* technique for the sum S_n. A similar approach may be applied in the dependent case. See also Datta and McCormick (1998) for related work on data based normalization of the sum when $\{X_n\}_{n\geq 1}$ is a linear process.

Returning to our discussion of the studentized estimator T_{1n} with a given scaling sequence $\{\hat{a}_n\}_{n\geq 1}$, the MBB version of T_{1n} based on a resample of size m is given by

$$T^*_{1,m,n} = a^*_{m,n}(S^*_{m,n} - m\hat{\mu}_n) ,$$

where $a^*_{m,n}$ is obtained by replacing $X_1, \ldots, X_m$ in the definition of $\hat{a}_m$ by the MBB samples $X^*_1, \ldots, X^*_m$. And a "hybrid" MBB version of T_{1n} may be defined as

$$\tilde{T}^*_{1,m,n} = \hat{a}_m(S^*_{m,n} - m\hat{\mu}_n) ,$$

where the same data-based scaling sequence $\{\hat{a}_n\}_{n\geq 1}$ that appears in the definition of T_{1n} is also used to define the bootstrap version of T_{1n}. Write $\hat{G}_{m,n}$ and $\tilde{G}_{m,n}$ for the conditional distributions of $T^*_{1,m,n}$ and $\tilde{T}^*_{1,m,n}$ respectively. Also, let $G_n(x) = P(T_{1n} \leq x)$, $x \in \mathbb{R}$. Then, using Lemma 4.1, it is easy to show that if the conditions of Theorem 11.2, (11.13), and (11.16) hold, then

$$\varrho(\hat{G}_{m,n}, G_n) \longrightarrow_p 0 \tag{11.17}$$

provided, for every $\epsilon > 0$,

$$P\Big(|a^*_{m,n}/a_n - 1| > \epsilon \mid \mathcal{X}_\infty\Big) \to 0 \quad \text{in probability as} \quad n \to \infty . \tag{11.18}$$

On the other hand, consistency of the "hybrid" estimator $\tilde{G}_{m,n}$ holds without the additional condition (11.18). Indeed, under the conditions of Theorem 11.2, (11.13), and (11.16), it is easy to show that

$$\varrho(\tilde{G}_{m,n}, G_n) \longrightarrow_p 0 . \tag{11.19}$$

Both (11.17) and (11.19) can be used to construct bootstrap confidence intervals for μ when the scaling constants $\{a_n\}_{n\geq 1}$ are not completely known.

Let $\hat{t}_m(\gamma)$ and $\tilde{t}_m(\gamma)$ denote the γ-quantile $(0 < \gamma < 1)$ of $\hat{G}_{m,n}$ and $\tilde{G}_{m,n}$, respectively. Then, a $(1-\gamma)$ equal-tailed two-sided bootstrap confidence interval for μ is given by

$$\hat{J}_{m,n}(1-\gamma) = \Big[n^{-1}\big(S_n - \hat{a}_n \hat{t}_m(1-\gamma/2)\big),\ n^{-1}\big(S_n - \hat{a}_n \hat{t}_m(\gamma/2)\big)\Big] . \quad (11.20)$$

Similarly, a $(1-\gamma)$ equal-tailed two-sided bootstrap confidence interval $\tilde{J}_m(1-\gamma)$ (say), for μ, based on the "hybrid" version of T_{1n}, is obtained by replacing $\hat{t}_m(\cdot)$'s in (11.20) by $\tilde{t}_m(\cdot)$'s. Both $\hat{J}_m(1-\gamma)$ and $\tilde{J}_m(1-\gamma)$ attain the nominal converge probability $(1-\gamma)$ in the limit. However, magnitudes of the errors in the coverage probabilities of all three bootstrap confidence intervals $\hat{I}_m(1-\gamma)$, $\hat{J}_m(1-\gamma)$, and $\tilde{J}_m(1-\gamma)$ are unknown at this stage. Note that the rates of approximations in (11.15), (11.17), and (11.19) depend on the resample size m as well as on the block length ℓ. The optimal choices of the resampling parameters ℓ and m are not known. In some similar problems involving *independent* data, the choice of m has been addressed *empirically* using the Jackknife-After-Bootstrap (JAB) method of Efron (1992); see Datta and McCormick (1995) and Athreya and Fukuchi (1997). A similar approach may be applied here as well. However, no theoretical result on the properties of the JAB in these problems seems to be available even under independence. See also Sakov and Bickel (2000) for the effects of m on the accuracy of the "m out of n" bootstrap for the median.

11.4 Invalidity of the MBB

In the last section, we proved consistency of the MBB approximation under the restriction that the resample size m be of smaller order than the sample size n. It is natural to ask the question: What happens when this condition is violated? For independent random variables, Athreya (1987) showed that the IID bootstrap method of Efron (1979) fails drastically in the heavy-tail case if one chooses $m = n$. In this section, we show that in the dependent case, a similar result holds for the MBB and the IID bootstrap method of Efron (1979). Thus, we cannot expect the bootstrap approximation to work for heavy-tailed data if the condition '$m = o(n)$ as $n \to \infty$' is violated. Further ramifications of this phenomenon have been studied, among others, by Arcones and Giné (1989, 1991), Giné and Zinn (1989, 1990) in the independent case and by Lahiri (1995) in the dependent case.

Let $\{X_n\}_{n\geq 1}$ be a sequence of stationary random variables with common marginal distribution F and let $\{\tilde{X}_n\}_{n\geq 1}$ be the associated iid sequence. For simplicity, we describe the asymptotic behavior of the MBB for heavy-tailed dependent random variables when the resample size m equals the sample size n and the normalized sum has a nonnormal stable limit law as in (11.13). Suppose that F lies in the domain of attraction of a stable law

F_α (say) of order $\alpha \in (1,2)$, i.e., there exist constants $a_n > 0$, $b_n \in \mathbb{R}$ such that

$$a_n^{-1}(\tilde{X}_1 + \cdots + \tilde{X}_n - b_n) \longrightarrow^d W_\alpha ,$$

where W_α has distribution F_α. It is well known (cf. Feller (1971b), Chapter 17) that in this case, the tails of F must satisfy the growth conditions:

$$F(x) \sim px^{-\alpha}L(x) \quad \text{as} \quad x \to \infty \tag{11.21}$$

$$1 - F(x) \sim qx^{-\alpha}L(x) \quad \text{as} \quad x \to \infty \tag{11.22}$$

for some $p \geq 0$, $q \geq 0$ with $p + q = 1$ and for some slowly varying function $L(\cdot)$. Recall that a function $L(\cdot)$ is called slowly-varying (at infinity) if

$$\lim_{x \to \infty} L(ax)/L(x) = 1 \quad \text{for all} \quad a > 0 . \tag{11.23}$$

Because $\alpha \in (1,2)$, $E|X_1| < \infty$. Let $\{a_n\}_{n \geq 1}$ be a sequence of constants satisfying

$$nL(a_n)/a_n^\alpha \to 1 \quad \text{as} \quad n \to \infty . \tag{11.24}$$

Then, under the dependence conditions of Theorem 11.1,

$$T_n \equiv a_n^{-1}(S_n - n\mu) \longrightarrow^d W_\alpha ,$$

where W_α has characteristic function (11.1) with $c = +\infty$ and with the canonical measure $M_\alpha(\cdot)$ of (11.2) with $C_0 = \alpha$, i.e., W_α has the characteristic function

$$\xi_\alpha(t) = \exp\Big(\int (e^{\iota tx} - 1 - \iota tx) d\lambda_\alpha(x) \Big), \quad t \in \mathbb{R} \tag{11.25}$$

with

$$\lambda_\alpha(A) = \alpha \Big[\int_{A \cap (0,\infty)} px^{-1-\alpha} dx + \int_{A \cap (-\infty,0)} q|x|^{-1-\alpha} dx \Big] \tag{11.26}$$

for any Borel subset A of $\mathbb{R}$.

Next, define the bootstrap version of $T_n = a_n^{-1}(S_n - n\mu)$ based on a MBB resample of size $m = n$ and block length ℓ as before, by $T_{n,n}^* = a_n^{-1}(S_{n,n}^* - n\hat{\mu}_n)$. Also, let $\hat{\Gamma}_{n,n}(x) = P_*(T_{n,n}^* \leq x)$, $x \in \mathbb{R}$. We shall show that, unlike the $m = o(n)$ case treated in Theorem 11.2, the bootstrap estimator $\hat{\Gamma}_{n,n}$ converges in distribution to a *random* limit distribution $\hat{\Gamma}$, say, and therefore, fails to provide an approximation to the *nonrandom*, exact distribution Γ_n of T_n. The random limit distribution $\hat{\Gamma}$ is defined in terms of a Poisson random measure $\tilde{N}(\cdot)$ on $(\mathbb{R}, \mathcal{B}(\mathbb{R}))$ having mean measure λ_α of (11.26). Recall that $\tilde{N}(\cdot)$ is called a Poisson random measure on $(\mathbb{R}, \mathcal{B}(\mathbb{R}))$ with mean measure λ_α (cf. Kallenberg (1976) if

(i) $\{\tilde{N}(A) : A \in \mathcal{B}(\mathbb{R})\}$ is a collection of random variables defined on some probability space $(\tilde{\Omega}, \tilde{\mathcal{F}}, \tilde{P})$, such that for each $\tilde{w} \in \tilde{\Omega}$, $\tilde{N}(\cdot)(\tilde{w})$ is a measure on $(\mathbb{R}, \mathcal{B}(\mathbb{R}))$, and

(ii) for every *disjoint* collection of sets $A_1, \ldots, A_k \in \mathcal{B}(\mathbb{R})$, $2 \le k < \infty$, the random variables $\tilde{N}(A_1), \ldots, \tilde{N}(A_k)$ are independent Poisson random variables with respective means $\lambda_\alpha(A_1), \ldots, \lambda_\alpha(A_k)$, i.e., for any $x_1, x_2, \ldots, x_k \in \{0, 1, 2, \ldots\}$,

$$\tilde{P}\Big(\tilde{N}(A_1) = x_1, \ldots, \tilde{N}(A_k) = x_k\Big) = \prod_{j=1}^{k} \exp\big(-\lambda_\alpha(A_j)\big)\frac{[\lambda_\alpha(A_j)]^{x_j}}{x_j!} .$$

For simplicity of exposition, here we describe the random probability measure $\hat{\Gamma}$ in terms of the corresponding (random) characteristic function $\hat{\xi}(t) \equiv \int \exp(\iota t x)\hat{\Gamma}(dx)$, $t \in \mathbb{R}$. The characteristic function $\hat{\xi}$ of the random limit $\hat{\Gamma}$ is given by

$$\hat{\xi}(t) = \exp\Big(\int (e^{\iota t x} - 1 - \iota t x)\tilde{N}(dx)\Big), \quad t \in \mathbb{R} . \tag{11.27}$$

Note that as a consequence of the "inversion formula" (cf. Chow and Teicher (1997)), $\hat{\xi}(\cdot)$ uniquely determines the probability measure $\hat{\Gamma}$. With this, we have the following result.

Theorem 11.3 *Suppose that (11.10), (11.11), (11.21), (11.22), and (11.24) hold. Also, suppose that ℓ is such that n/ℓ is an integer, $\ell^{-1} + n^{-1/2}\ell = o(1)$ as $n \to \infty$, and $n\alpha(\ell)/\ell = O(1)$ as $n \to \infty$, where $\alpha(\cdot)$ denotes the strong mixing coefficient of $\{X_n\}_{n\ge 1}$. Then, for any $x_1, \ldots, x_k \in \mathbb{R}$, $1 \le k < \infty$,*

$$\Big(\hat{\Gamma}_{n,n}(x_1), \ldots, \hat{\Gamma}_{n,n}(x_k)\Big) \longrightarrow^d \Big(\hat{\Gamma}(x_1), \ldots, \hat{\Gamma}(x_k)\Big)$$

as $n \to \infty$, where $\hat{\Gamma}$ is defined via its characteristic function $\hat{\xi}$ given by (11.27).

Proof: See Theorem 2.2, Lahiri (1995). □

Theorem 11.3 shows that, with the resample size $m = n$, the MBB estimator $\hat{\Gamma}_{n,n}(x)$ converges in distribution to a random variable $\hat{\Gamma}(x)$ for every $x \in \mathbb{R}$ and, hence, is an inconsistent estimator of the nonrandom level-2 parameter $\Gamma_n(x)$. Indeed, for any real number x, if n is large, the bootstrap probability $\hat{\Gamma}_{n,n}(x)$ behaves like the random variable $\hat{\Gamma}(x)$, having a *nondegenerate* distribution on the interval $[0, 1]$, rather than coming close to the desired target $\Gamma_n(x)$ or to the nonrandom limiting value $\Gamma_\alpha(x) \equiv P(W_\alpha \le x) = \lim_{n\to\infty} \Gamma_n(x)$. From a practical point of view, this implies that the bootstrap approximations generated with $m = n$ would have a nontrivial variability even for arbitrarily large sample sizes, and hence, would not be a reliable estimate of the target probability even for large n. We point out that the conclusions of Theorem 11.3 remain true in a slightly more general setting, where n/ℓ is not necessarily an integer and the MBB is applied with the standard choice of the resample size, viz., $m = n_1 \equiv \ell \lfloor n/\ell \rfloor$.

In the next two sections, we describe some results on bootstrapping the extremes of a stationary process.

11.5 Extremes of Stationary Random Variables

Let $\{X_n\}_{n\geq 1}$ be a sequence of stationary random variables with one-dimensional marginal distribution function F, and let $\{\tilde{X}_n\}_{n\geq 1}$ be the associated iid sequence, i.e., $\{\tilde{X}_i\}_{i\geq 1}$ is a sequence of iid random variables with common distribution function F. For each $n \geq 1$, let $X_{1:n} \leq \cdots \leq X_{n:n}$ denote the order-statistics corresponding to $X_1, \ldots, X_n$. Define $\tilde{X}_{1:n} \leq \cdots \leq \tilde{X}_{n:n}$ similarly. In this section, we review some standard results on the maximum order-statistic $X_{n:n}$ under dependence. By considering the sequence $Y_n = -X_n$, $n \geq 1$, and using the relation $X_{1:n} = -\max_{1\leq i\leq n} Y_i$, one can carry out a parallel development for the minimum $X_{1:n}$.

In this and in the next sections, we shall assume that the process $\{X_n\}_{n\geq 1}$ satisfies a strong-mixing type condition (known as Condition $D(u_n)$), introduced by Leadbetter (1974).

Definition 11.3 *Let $\{u_n\}_{n\geq 1}$ be a sequence of positive real numbers. Then $\{X_n\}_{n\geq 1}$ is said to satisfy Condition $D(u_n)$ if there is a sequence $r_n = o(n)$ such that*

$$\begin{aligned}
\alpha_{n,r_n}(u_n) \equiv & \sup\Big\{\big|P(X_j \leq u_n \text{ for } j \in A \cup B) - \\
& \qquad P(X_j \leq u_n \text{ for } j \in A) \cdot P(X_j \leq u_n \text{ for } j \in B)\big| \\
& : A \subset \{1, \ldots, k\},\ B \in \{k + r_n, \ldots, n\},\ 1 \leq k \leq n - r_n\Big\} \\
& \to 0 \quad as \quad n \to \infty .
\end{aligned}$$

It is clear that if the sequence $\{X_n\}_{n\geq 1}$ is strongly mixing, then it satisfies Condition $D(u_n)$ for any sequence of real numbers $\{u_n\}_{n\geq 1}$. A result of Chernick (1981b) shows that if $\{X_n\}_{n\geq 1}$ is a stationary Markov chain, then it satisfies $D(u_n)$ for any sequence $\{u_n\}_{n\geq 1}$ with $\lim_{n\to\infty} F(u_n) = 1$. The following result, due to Leadbetter (1974), specifies possible types of limit distributions of the normalized sample maximum under Condition $D(u_n)$. Here and elsewhere, we say that the distribution function of a random variable V is of the *type* of a given distribution function G on $\mathbb{R}$ if there exist constants $a > 0$, $b \in \mathbb{R}$ such that $P(V \leq x) = G\big(a^{-1}(x - b)\big)$, $x \in \mathbb{R}$.

Theorem 11.4 *Suppose that there exist constants $a_n > 0$ and $b_n \in \mathbb{R}$ such that*

$$a_n^{-1}(X_{n:n} - b_n) \longrightarrow^d V \tag{11.28}$$

for some nondegenerate random variable V. Also suppose that Condition $D(u_n)$ is satisfied for $u_n = a_n x + b_n$, $n \geq 1$, for each $x \in \mathbb{R}$. Then, the distribution function of V is of the type of one of the following distribution functions:

(I) $\check{\Lambda}(x) = \exp(-e^{-x}), \quad x \in \mathbb{R}$;

(II) $\check{\Phi}_\alpha(x) = \begin{cases} 0 & \text{if } x \leq 0 \\ \exp(-x^{-\alpha}) & \text{if } x > 0 \end{cases}$
for some $\alpha > 0$;

(III) $\check{\Psi}_\alpha(x) = \begin{cases} \exp(-|x|^\alpha) & \text{if } x \leq 0 \\ 1 & \text{if } x > 0 \end{cases}$
for some $\alpha > 0$.

The classes of distributions (I), (II), and (III) above are known as the extreme-value distributions. The distribution function of the limiting random variable V in Theorem 11.4 is necessarily one of these extreme-value distributions up to a suitable translation and scaling. In the iid case, i.e., for the sequence $\{\tilde{X}_n\}_{n\geq 1}$, Gnedenko (1943) gives necessary and sufficient conditions on the tails of F for weak convergence of the normalized maximum $a_n^{-1}(\tilde{X}_{n:n} - b_n)$ to a given extreme-value distribution. In line with the iid case, we say that F belongs to the *extremal-domain of attraction* of an extreme value distribution G, and write $F \in \mathcal{D}(G)$, if there exist constants $a_n > 0$ and $b_n \in \mathbb{R}$ such that

$$a_n^{-1}(\tilde{X}_{n:n} - b_n) \longrightarrow^d V \ , \tag{11.29}$$

the distribution function of V is G. In the iid case, a set of possible choices of the constants $\{a_n\}_{n\geq 1}$ and $\{b_n\}_{n\geq 1}$ for the three extremal classes are given by (cf. Gnedenko (1943), de Haan (1970)):

$$\begin{array}{lll} \text{(i)} & \text{For} & F \in \mathcal{D}(\check{\Lambda}),\ a_n = F^{-1}\big(1 - [en]^{-1}\big) - c_n,\ b_n = c_n, \\ \text{(ii)} & \text{For} & F \in \mathcal{D}(\check{\Phi}_\alpha),\ a_n = c_n,\ b_n = 0, \\ \text{(iii)} & \text{For} & F \in \mathcal{D}(\check{\Psi}_\alpha),\ a_n = M_F - c_n,\ b_n = M_F, \end{array} \tag{11.30}$$

where $F^{-1}(u) = \inf\{x \in \mathbb{R} : F(x) \geq u\}$, $u \in (0,1)$, $e = \sum_{k=0}^{\infty} 1/k!$, $c_n = F^{-1}(1 - n^{-1})$, and $M_F = \sup\{x : F(x) < 1\}$ is the upper endpoint of F. Under suitable conditions on the dependence structure of the sequence $\{X_n\}_{n\geq 1}$, we may employ the normalizing constants $\{a_n\}_{n\geq 1}$ and $\{b_n\}_{n\geq 1}$, specified by (11.30), in the dependent case as well.

An important result of Chernick (1981a) (see also Loynes (1965)) says that if for each $\tau > 0$, there is a sequence $u_n \equiv u_n(\tau)$, $n \geq 1$, such that

$$\lim_{n\to\infty} n\Big(1 - F\big(u_n(\tau)\big)\Big) = \tau \quad \text{for all} \quad \tau \in (0, \infty) \ , \tag{11.31}$$

$$\text{Condition} \quad D\big(u_n(\tau)\big) \quad \text{holds for all} \quad \tau \in (0, \infty) , \tag{11.32}$$

and $\lim_{n\to\infty} P\big(X_{n:n} \le u_n(\tau_0)\big)$ exists for some $\tau_0 > 0$, then there exists a constant $\theta \in [0,1]$ such that

$$\lim_{n\to\infty} P\big(X_{n:n} \le u_n(\tau)\big) = e^{-\theta\tau} \quad \text{for all} \quad \tau \in (0, \infty) . \tag{11.33}$$

This result leads to the following definition.

Definition 11.4 *A stationary process $\{X_n\}_{n\ge1}$ is said to have extremal index θ if conditions (11.31)–(11.33) hold.*

When the extremal index $\theta > 0$, both $X_{n:n}$ and its iid counterpart $\tilde{X}_{n:n}$ have extremal limit distributions of the *same type*. However, for $\theta = 0$, $X_{n:n}$ and $\tilde{X}_{n:n}$ may have different asymptotic behaviors. Here, we shall restrict our attention only to the case $\theta > 0$, covered by the following result.

Theorem 11.5 *Suppose that the sequence $\{X_n\}_{n\ge1}$ has extremal index $\theta > 0$ and that $F \in \mathcal{D}(G)$ for some extreme value distribution G. Let $\{a_n\}_{n\ge1}$ and $\{b_n\}_{n\ge1}$ be the sequences of constants specified by (11.30) for the class containing G. Then*

$$a_n^{-1}(X_{n:n} - b_n) \longrightarrow^d V ,$$

where the distribution function of V is of the type G, i.e., $P(V \le x) = G\big((x-b)/a\big)$, $x \in \mathbb{R}$ for some $a > 0$, $b \in \mathbb{R}$.

Proof: Follows from Corollary 3.7.3 of Leadbetter, Lindgren and Rootzen (1983) and the discussion above. □

The extremal index θ is a parameter whose value is determined by the joint distribution of the sequence $\{X_n\}_{n\ge1}$. Theorem 11.5 shows that for $\theta > 0$, both $X_{n:n}$ and its iid counterpart $\tilde{X}_{n:n}$ may be normalized by the same sequences of constants $\{a_n\}_{n\ge1}$ and $\{b_n\}_{n\ge1}$, and the limit distributions of the normalized maxima are of the same type but not necessarily identical. When $0 < \theta < 1$, the two limit laws are related by a *nontrivial* linear transformation in the sense that if $a_n^{-1}(\tilde{X}_{n:n} - b_n) \longrightarrow^d \tilde{V}$, then $a_n^{-1}(X_{n:n} - b_n) \longrightarrow^d [a\tilde{V} + b]$ for some $(a,b) \ne (1,0)$. Furthermore, the values of (a,b) depend on θ. Thus, for $0 < \theta < 1$, the limit distribution in the dependent case is *different* from that in the iid case, and the effect of the dependence of $\{X_n\}_{n\ge1}$ shows up in the limit through the extremal index θ. In contrast, when $\theta = 1$, both $X_{n:n}$ and $\tilde{X}_{n:n}$ have the *same* limit distribution. In this case, the effect of the dependence of $\{X_n\}_{n\ge1}$ vanishes asymptotically. This observation has an important implication regarding validity of the bootstrap methods for dependent random variables. In the next section, we shall show that with a proper choice of the resampling size, the MBB provides a valid approximation for all $\theta \in (0,1]$, while the IID-bootstrap method of Efron (1979) is effective only in the case $\theta = 1$.

Because of the special role played by the case $\theta = 1$, we now briefly describe a general regularity condition on the sequence $\{X_n\}_{n\geq 1}$ that leads to the extremal index $\theta = 1$.

Definition 11.5 *Let $\{u_n\}_{n\geq 1}$ be a sequence of real numbers. Then, $\{X_n\}_{n\geq 1}$ is said to satisfy Condition $D'(u_n)$ if*

$$\lim_{k\to\infty} \limsup_{n\to\infty} n \Bigg[\sum_{2\leq j\leq n/k} P(X_1 > u_n, X_j > u_n) \Bigg] = 0 \ .$$

To get some idea about the class of processes for which Condition $D'(u_n)$ holds, suppose that $\{X_n\}_{n\geq 1}$ are iid and that $nP(X_1 > u_n) = O(1)$. Then it is easy to check that Condition $D'(u_n)$ holds. However, condition $D'(u_n)$ need not hold for a sequence $\{u_n\}_{n\geq 1}$ with $nP(X_1 > u_n) = O(1)$, even when $\{X_n\}_{n\geq 1}$ are m-dependent with $m \geq 1$. The following result shows that $X_{n:n}$ and $\tilde{X}_{n:n}$ have the same limit law when Condition $D'(u_n)$ holds.

Theorem 11.6 *Suppose that $a_n^{-1}(\tilde{X}_{n:n} - b_n) \longrightarrow^d V$ for some constants $a_n > 0$, $b_n \in \mathbb{R}$, $n \geq 1$ where V is a nondegenerate random variable. Also, suppose that Conditions $D(u_n)$ and $D'(u_n)$ hold for all $u_n \equiv a_n x + b_n$, $n \geq 1$, $x \in \mathbb{R}$. Then,*

$$a_n^{-1}(X_{n:n} - b_n) \longrightarrow^d V \ .$$

Proof: See Theorem 3.5.2, Leadbetter, Lindgren and Rootzen (1983). □

In the next section, we describe properties of the MBB and the IID bootstrap of Efron (1979) for stationary random variables under Conditions like $D(u_n)$ and $D'(u_n)$.

11.6 Results on Bootstrapping Extremes

First we consider consistency properties of the MBB approximation to the distribution of a normalized maximum. Suppose that $\{X_n\}_{n\geq 1}$ is stationary and has an extremal index $\theta \in (0,1]$. Let $\mathcal{X}_n = \{X_1, \ldots, X_n\}$ denote the observations and let $\mathcal{X}_{m,n}^* = \{X_1^*, \ldots, X_m^*\}$ denote the MBB resample of size m based on k resampled blocks of length ℓ. Thus, here $m = k\ell$. Let $X_{1:m}^* \leq \cdots \leq X_{m:m}^*$ denote the corresponding bootstrap order-statistics. To define the MBB version of the normalized maximum $V_n \equiv a_n^{-1}(X_{n:n} - b_n)$, here we suppose that the constants $\{a_n\}_{n\geq 1}$ and $\{b_n\}_{n\geq 1}$ are known. Then, the bootstrap version of V_n is given by

$$V_{\ell,m,n}^* = a_m^{-1}(X_{m:m}^* - b_m) \ . \qquad (11.34)$$

For proving consistency of the MBB approximation, in addition to Condition $D(u_n)$, we shall make use of the following weaker version of the strong mixing condition.

Definition 11.6 *For $n \geq 1$, let*

$$\begin{aligned} \tilde{\alpha}(n) \; = \; & \sup\Big\{ \big| P(X_j \in I_u, \; j \in A \cup B) \\ & - P(X_j \in I_u, \; j \in A)\, P(X_j \in I_u, \; j \in B) \big| : A \subset \{1, \ldots, k\}, \\ & B \in \{k+n, \ldots\}, \; k \geq 1, \; I_u \in \big\{(-\infty, u], (u, \infty)\big\}, \; u \in \mathbb{R} \Big\} . \end{aligned} \tag{11.35}$$

It is clear that for all $n \geq 1$, $\tilde{\alpha}(n) \leq \alpha(n)$, the strong-mixing coefficient of the sequence $\{X_n\}_{n \geq 1}$. Here we shall require that $\tilde{\alpha}(n)$ decreases at a polynomial rate as $n \to \infty$. The following result proves the validity of the MBB approximation.

Theorem 11.7 *Suppose that $\{X_n\}_{n \geq 1}$ is a stationary process with extremal index $\theta \in (0, 1]$ (as defined in Definition 11.4) and that $\tilde{\alpha}(r) \leq r^{-\eta}$, $r \geq 1$, for some $\eta > 0$. Further suppose that (11.28) holds and the MBB block size variable ℓ and the number of resampled blocks k satisfy $\ell = \lfloor n^{\epsilon} \rfloor$, $k = \lfloor n^{\delta} \rfloor$ for some $0 < \epsilon < 1$, $0 < \delta < \min\{\epsilon, \frac{1-\epsilon}{2}\}$. Then,*

$$\sup_{x \in \mathbb{R}} \big| P_*(V^*_{\ell,m,n} \leq x) - P(V_n \leq x) \big| \longrightarrow_p 0 \quad as \quad n \to \infty \, , \tag{11.36}$$

where $V^*_{\ell,m,n}$ is defined by (11.34).

Proof: Follows from Theorem 3.8 and Corollary 3.2 of Fukuchi (1994). □

Thus, under the conditions of Theorem 11.7, the MBB approximation to the distribution of V_n is consistent for all values of the extremal index θ in the interval $(0, 1]$. In addition to the regularity of the upper tail of F (cf. (11.31)), this requires the dependence structure of the process $\{X_n\}_{n \geq 1}$ to satisfy Condition $D\big(u_n(\tau)\big)$ in (11.32) and the weak mixing condition on $\tilde{\alpha}(\cdot)$. Furthermore, the conditions on the block length variable ℓ and the resample size m require that $\ell \to \infty$ and $m = o(n)$ as $n \to \infty$. Both of these conditions are necessary for the validity of the MBB method. When the extremal index θ lies in the interval $(0, 1)$, it is produced by the dependence among all X_n's, and as a consequence, the block length ℓ must grow to infinity with the sample size in order to capture this effect of the dependence on the limit distribution of V_n. For $\theta = 1$, the limit distribution of the normalized maximum is essentially determined by the one-dimensional marginal distribution function F of X_1. As a result, one may have consistency even when ℓ does not go to infinity (see the discussion on the IID resampling scheme of Efron (1979) below). On the other hand, the condition "$m = o(n)$" on the resample size m is needed to ensure that the conditional distribution of $V^*_{\ell,m,n}$ (given the X_n's) converges to the correct *nonrandom* limit. For the extremes, bootstrap approximations tend to behave in a way similar to the case of bootstrapping the normalized sample sum of heavy-tailed data. Indeed, when the resample size $m = n$, a result of Fukuchi (1994) shows that the bootstrap approximation generated by the IID resampling scheme of Efron (1979) has a random limit (see Theorem 11.9 below).

Next we briefly state the results on the IID resampling scheme of Efron (1979) as alluded to in the above paragraph. Let $V^*_{m,n} \equiv V^*_{1,m,n}$ denote the IID-bootstrap version of V_n as defined by (11.34) with $\ell \equiv 1$ and $k = m$. The first result gives conditions for the consistency of $\mathcal{L}(V^*_{m,n} \mid \mathcal{X}_n)$, where $\mathcal{X}_n = \{X_1, X_2, \ldots, X_n\}$.

Theorem 11.8 *Let $\{X_n\}_{n\geq 1}$ be a stationary process such that (11.28) holds. Suppose that Conditions $D(u_n)$, $D'(u_n)$ hold with $u_n = u_n(x) = a_n x + b_n$, $n \geq 1$, for all $x \in \mathbb{R}$ and that $\tilde{\alpha}(n) = O(n^{-\delta})$ as $n \to \infty$ for some $\delta \geq 2$. If, in addition, $m = o(n)$ as $n \to \infty$, then*

$$\sup_x \left| P_*(V^*_{m,n} \leq x) - P(V_n \leq x) \right| \longrightarrow_p 0 \quad \textit{as} \quad n \to \infty .$$

Proof: Follows from Theorem 3.4 and Corollary 3.1 of Fukuchi (1994), by noting that his mixing coefficient $\alpha_j^-(u)$ is bounded above by the coefficient $\tilde{\alpha}(j)$ of (11.35) for all $j \geq 1$, $u \in \mathbb{R}$. □

As pointed out in Section 11.4 (cf. Theorem 11.6), under Condition $D'(u_n)$ of Theorem 11.8, the extremal index $\theta = 1$, and V_n has the same limit distribution as the normalized maximum $\tilde{V}_n$ of the associated iid sequence. As a result, under the conditions of Theorem 11.8, the dependence of the X_n's does not have any effect on the limit law and the IID bootstrap method provides a valid approximation to the distribution of the normalized sample maximum even for such dependent random variables, provided that the resample size $m = o(n)$. However, if the resample size $m = n$, the consistency is no longer guaranteed, as shown by the next result.

Theorem 11.9 *Let $\{X_n\}_{n\geq 1}$ be a stationary sequence such that (11.28) holds for some $a_n > 0$, $b_n \in \mathbb{R}$, and for some nondegenerate random variable V. Suppose that Condition $D'(u_n)$ holds for $u_n = a_n x + b_n$, $n \geq 1$, for all $x \in \mathbb{R}$ and that*

$$\lim_{n\to\infty} \alpha_n^+(p_n; \boldsymbol{x}_r) = 0 \qquad (11.37)$$

for some sequence $\{p_n\}_{n\geq 1}$ of positive integers satisfying $p_n = o(n)$ for every $\boldsymbol{x}_r = (x_1, \ldots, x_r)' \in \mathbb{R}^r$, $r \geq 1$, where

$$\alpha_n^+(p_n; \boldsymbol{x}_r) \equiv \sup\Big\{ \big| P(X_j \in I_j,\ j \in A \cup B) - P(X_j \in I_j,\ j \in A) \cdot P(X_j \in I_j,\ j \in B) \big| :$$

$$I_j \in \{(-\infty, a_n x_i + b_n] : 1 \le i \le r\} \quad \textit{for} \quad j \in A \cup B ,$$
$$A \in \{1, \ldots, k\},\ B \in \{k + p_n, \ldots, n\},\ 1 \le k \le n - p_n\Big\} . \tag{11.38}$$

Then, for $m = n$,

$$P_*(V^*_{n:n} \le x) \longrightarrow^d \exp(-\Gamma_x)$$

for every $x \in \mathbb{R}$, *where* Γ_x *is a Poisson random variable with the mean* $-\log\big[P(V \le x)\big]$.

Thus, under the conditions of Theorem 11.9, the bootstrap distribution function at any given $x \in \mathbb{R}$, being a random variable with values in the interval $[0, 1]$, converges in distribution to a nondegenerate random variable $\exp(-\Gamma_x)$. As a consequence, when the resample size m equals n, the resulting bootstrap estimator of the target probability $P(V_n \le x)$ fluctuates around the true value even for arbitrarily large sample sizes. Like the heavy-tail case, a similar behavior is expected of the MBB even when the block length $\ell \to \infty$, if the resample size m grows at the rate n, i.e., if $m \sim n$. However, a formal proof of this fact is not available in the literature.

We conclude the discussion of the asymptotic properties of the IID bootstrap of Efron (1979) by considering the case where $\{X_n\}_{n \ge 1}$ has an extremal index $\theta \in (0, 1)$. In this case, Fukuchi (1994) (cf. p. 47) shows that under regularity conditions similar to those of Theorem 11.8, for $m = o(n^{1/2})$,

$$P_*(V^*_{m,n} \le x) \longrightarrow_p \exp\big(-\gamma(x)\big)$$

while

$$P(V_n \le x) \longrightarrow \exp\big(-\theta\gamma(x)\big)$$

for each $x \in \mathbb{R}$, where $\gamma(x) \equiv \lim_{n\to\infty} n\big[1 - P(X_1 \le a_n x + b_n)\big]$. Thus, even with a resample size m that grows at a slower rate than the sample size n, the IID resampling scheme of Efron (1979) fails. As explained earlier, the reason behind this is that the value of $\theta \in (0, 1)$ is determined by the joint distribution of the X_i's, but when a single observation is resampled at a time, this information is totally lost. As a consequence, the limit distribution of the IID bootstrap version $V^*_{m,n}$ of V_n coincides with the limit distribution of the normalized sample maximum $\tilde{V}_n$ of the associated iid sequence $\{\tilde{X}_n\}_{n \ge 1}$.

11.7 Bootstrapping Extremes With Estimated Constants

For many applications, the assumption that the normalizing constants $\{a_n\}_{n \ge 1}$, $\{b_n\}_{n \ge 1}$ are known is very restrictive. In such situations, we may be interested in bootstrapping the sample maximum where some random normalizing factors may be used in place of $\{a_n\}_{n\geq 1}$ and $\{b_n\}_{n\geq 1}$ to yield a nondegenerate limit distribution. Accordingly, let $\{\hat{a}_n\}_{n\geq 1}$ and $\{\hat{b}_n\}_{n\geq 1}$ be random variables with $\hat{a}_n \in (0,\infty)$ and $\hat{b}_n \in \mathbb{R}$ for all $n \geq 1$ such that

$$\frac{\hat{a}_n}{a_n} \longrightarrow_p a_0 \tag{11.39}$$

and

$$a_n^{-1}(\hat{b}_n - b_n) \longrightarrow_p b_0 \tag{11.40}$$

for some constants $a_0 \in (0,\infty)$ and $b_0 \in \mathbb{R}$. Here, we do allow the possibility that $\hat{b}_n$ or $\hat{a}_n$ be a function of a population parameter and be nonrandom, so that the bootstrap approximation may be used to construct inference procedures like tests and confidence intervals for the parameter involved. For example, we may be interested in setting a confidence interval for the upper endpoint $M_F \equiv \sup\{x : F(x) < 1\}$ of the distribution function F of X_1 when $F \in \mathcal{D}(\check{\Psi}_\alpha)$ (cf. Theorem 11.4). In this case we would set $\hat{b}_n = M_F$ and replace the corresponding scaling constant $a_n = M_F - F^{-1}(1-1/n)$ of (11.30) by a random scaling constant $\hat{a}_n$ that is a suitable function of M_F and the empirical quantile function F_n^{-1}. Then, we may apply the MBB to the pivotal quantity $\hat{V}_n \equiv \hat{a}_n^{-1}(X_{n:n}-M_F)$ and construct bootstrap confidence intervals for the parameter M_F. In general, consider the normalized sample maximum with "estimated" constants

$$\hat{V}_n = \hat{a}_n^{-1}(X_{n:n} - \hat{b}_n)\ , \quad n \geq 1\ . \tag{11.41}$$

Let a_m^*, b_m^* be some suitable functions of the MBB sample $\{X_1^*, \ldots, X_m^*\}$, based on blocks of length ℓ, and of the data $X_1, \ldots, X_n$, such that for every $\epsilon > 0$,

$$P\Big(\big|a_m^{-1}a_m^* - a_0\big| > \epsilon \mid \mathcal{X}_\infty\Big) + P\Big(\big|a_m^{-1}(b_m^* - b_m) - b_0\big| > \epsilon \mid \mathcal{X}_\infty\Big)$$
$$\to 0 \quad \text{in probability as} \quad n \to \infty\ , \tag{11.42}$$

where $a_0 \in (0,\infty)$ and $b_0 \in \mathbb{R}$ are as in (11.39) and (11.40), respectively, and where $\mathcal{X}_\infty = \sigma\langle\{X_1, X_2, \ldots\}\rangle$. Then, a bootstrap version of $\hat{V}_n$ is given by

$$\hat{V}^*_{\ell,m,n} = a_m^{*-1}(X^*_{m:m} - b_m^*)\ . \tag{11.43}$$

As in the definition of $\hat{V}_n$ (cf. (11.41)), here we do allow the possibility that a_m^* or b_m^* in (11.42) and (11.43) be just a function of $X_1, \ldots, X_n$ and do not involve the X_i^*'s. A prime example of this is the "hybrid" MBB version of $\hat{V}_n$, given by

$$\check{V}^*_{\ell,m,n} = \hat{a}_m^{-1}(X^*_{m:m} - \hat{b}_m)\ , \tag{11.44}$$

which corresponds to (11.43) with

$$a_m^* = \hat{a}_m \quad \text{and} \quad b_m^* = \hat{b}_m\ .$$

The following result shows that both $\hat{V}^*_{\ell,m,n}$ and $\check{V}^*_{\ell,m,n}$ provide a valid approximation to the distribution of $\hat{V}_n$.

Theorem 11.10 *Suppose that the conditions of Theorem 11.7 hold and that relations (11.39) and (11.40) hold. Then,*

$$\sup_x \left| P_*(\check{V}^*_{\ell,m,n} \le x) - P(\hat{V}_n \le x) \right| \longrightarrow_p 0 \quad \textit{as} \quad n \to \infty \,. \tag{11.45}$$

If, in addition, relation (11.42) holds, then

$$\sup_x \left| P_*(\hat{V}^*_{\ell,m,n} \le x) - P(\hat{V}_n \le x) \right| \longrightarrow_p 0 \quad \textit{as} \quad n \to \infty \,. \tag{11.46}$$

Proof: We consider (11.46) first. Note that by (11.39) and (11.40) and Slutsky's theorem,

$$\hat{V}_n \longrightarrow^d a_0^{-1}(V - b_0) \quad \text{as} \quad n \to \infty \tag{11.47}$$

where V is as in Theorem 11.4 (cf. (11.28)). With $V^*_{\ell,m,n}$ given by (11.34), we may write

$$\hat{V}^*_{\ell,m,n} = [a_m / a^*_m] V^*_{\ell,m,n} + a^{*-1}_m (b_m - b^*_m) \,. \tag{11.48}$$

From (11.42), it easily follows that for each $\epsilon > 0$,

$$\begin{aligned} &P\Big(\Big| \frac{a_m}{a^*_m} - a_0^{-1} \Big| > \epsilon \mid \mathcal{X}_\infty \Big) + P\Big(\Big| \frac{b_m - b^*_m}{a^*_m} + a_0^{-1} b_0 \Big| > \epsilon \mid \mathcal{X}_\infty \Big) \\ &\to 0 \quad \text{in probability as} \quad n \to \infty \,. \end{aligned} \tag{11.49}$$

Hence, by Lemma 4.1, (11.46) follows from (11.47)–(11.49) and Theorem 11.7.

Next consider (11.45). Because $\hat{a}_m$ and $\hat{b}_m$ are $\mathcal{X}_\infty$-measurable, with $a^*_m = \hat{a}_m$ and $b^*_m = \hat{b}_m$ in (11.42), for any $\epsilon > 0$, we get

$$\begin{aligned} &\text{the left side of (11.42)} \\ &= \mathbb{1}\big(|a_m^{-1} \hat{a}_m - a_0| > \epsilon \big) + \mathbb{1}\big(|a_m^{-1}(\hat{b}_m - b_m) - b_0| > \epsilon \big) \,, \end{aligned}$$

which goes to zero in L^1 and, hence, in probability as $n \to \infty$, by (11.39) and (11.40). Hence, (11.45) follows from (11.46). □

A similar result may be proved for the IID bootstrap of Efron (1979) under the regularity conditions of Theorem 11.8. Theorem 11.10 and its analog for the IID bootstrap in the "unknown normalizing constant" case may be used for statistical inference for dependent random variables. For results along this line for independent data, see Athreya and Fukuchi (1997) who apply the IID bootstrap of Efron (1979) to construct CIs for the end-points of the distribution function F of X_1, when the random variables X_n's are iid. For results on bootstrapping the joint distribution of the sum and the maximum of a stationary sequence, see Mathew and McCormick (1998).

12 Resampling Methods for Spatial Data

12.1 Introduction

In this chapter, we describe bootstrap methods for spatial processes observed at finitely many locations in a sampling region in $\mathbb{R}^d$. Depending on the spatial sampling mechanism that generates the locations of these data-sites, one gets quite different behaviors of estimators and test statistics. As a result, formulation of resampling methods and their properties depend on the underlying spatial sampling mechanism. In Section 12.2, we describe some common frameworks that are often used for studying asymptotic properties of estimators based on spatial data. In Section 12.3, we consider the case where the sampling sites (also referred to as data-sites in this book) lie on the integer grid and describe a block bootstrap method that may be thought of as a direct extension of the MBB method to spatial data. Here, some care is needed to handle sampling regions that are not rectangular. We establish consistency of the bootstrap method and give some numerical examples to illustrate the use of the method. Section 12.4 gives a special application of the block resampling methods. Here, we make use of the resampling methods to formulate an asymptotically efficient least squares method of estimating spatial covariance parameters, and discuss its advantages over the existing estimation methods. In Section 12.5, we consider irregularly spaced spatial data, generated by a stochastic sampling design. Here, we present a block bootstrap method and show that it provides a valid approximation under nonuniform concentration of sampling sites even in presence of infill sampling. It may be noted that infill sampling leads to conditions of long-range dependence in the data, and thus, the block bootstrap method presented here provides a valid approximation under this *form* of long-range dependence. Resampling methods for spatial prediction are presented in Section 12.6.

12.2 Spatial Asymptotic Frameworks

In this section, we describe some spatial asymptotic frameworks that are commonly used for studying large sample properties of inference procedures. For time series data, observations are typically taken at a regular interval of time and the limiting procedure describes the long-run behavior of a system as the time approaches "infinity". Because of the unidirectional flow of time, the concept of "infinity" is unambiguously defined. For random processes observed over space (and possibly, also over time), this uniqueness of limiting procedures is lost. In this case, there are several ways of approaching the "ultimate state" or the "infinity," giving rise to different asymptotic frameworks for studying large sample properties of inference procedures, including the bootstrap. It turns out that these different asymptotic structures arise from two *basic* paradigms, known as the *increasing domain asymptotics* and the *infill asymptotics* (cf. Chapter 5, Cressie (1993)). When all sampling sites are separated by a fixed positive distance and the sampling region becomes unbounded as the sample size increases, the resulting structure leads to increasing domain asymptotics. This is the most common framework used for asymptotics for spatial data and often leads to conclusions similar to those obtained in the time series case. Processes observed over increasing and nested rectangles on the integer grid $\mathbb{Z}^d$ in the d-dimensional space provide examples of such an asymptotic structure. On the other hand, if an increasing number of samples are collected at spatial sampling sites from within a fixed *bounded* region of $\mathbb{R}^d$, the resulting structure leads to infill asymptotics. In this case, the minimum distance among the sampling sites tends to zero as the sample size increases, and typically results in very strong forms of dependence in the data. Such a structure is suitable for Mining and other Geostatistical applications where a given resource is sampled increasingly over a given region. It is well known that under infill asymptotics many standard inference procedures have drastically different large sample behaviors compared to those under increasing domain asymptotics. See, for example, Morris and Ebey (1984), Stein (1987, 1989), Cressie (1993), Lahiri (1996c), and the references therein. In some cases, a combination of the two basic asymptotic frameworks is also employed. In Sections 12.5 and 12.6, we shall consider one such structure (which we refer to as a *mixed increasing domain asymptotic structure*), where the sampling region grows to infinity and at the same time, the distance between neighboring sampling sites goes to zero. Except for some prediction problems treated in Section 12.6.2, the sampling region $R \equiv R_n$ in all other sections becomes unbounded as n increases to infinity.

We conclude this section with a description of the structure of the sampling regions R_n, $n \geq 1$. Let $\tilde{R} \subset (-\frac{1}{2}, \frac{1}{2}]^d$ be an open connected set containing the origin and let R_0 be a prototype set for the sampling regions such that $\tilde{R} \subset R_0 \subset \text{cl.}(\tilde{R})$, where $\text{cl.}(\tilde{R})$ denotes the closure of the set $\tilde{R}$. Also, let $\{\lambda_n\}_{n \geq 1} \subset [1, \infty)$ be a sequence of real numbers such that $\lambda_n \uparrow \infty$ as $n \to \infty$. We shall suppose that the sampling region R_n is obtained by "inflating" the prototype set R_0 by the scaling constant λ_n as

$$R_n = \lambda_n R_0 \ . \tag{12.1}$$

Because the origin is assumed to lie in R_0, relation (12.1) shows that the shape of the sampling region remains unchanged for different values of n. Furthermore, this formulation allows the sampling region R_n to have a wide range of (possibly irregular) shapes. Some examples of such regions are spheres, ellipsoids, polyhedrons, and star-shaped regions. Here we call a set $A \subset \mathbb{R}^d$ containing the origin *star-shaped* if for any $x \in A$, the line joining x and the origin lies in A. As a result, star-shaped regions can be nonconvex. To avoid pathological cases, we shall suppose that the prototype set R_0 satisfies the following boundary condition:

Condition B *For every sequence of positive real numbers* $\{a_n\}_{n \geq 1}$ *with* $a_n \to 0$ *as* $n \to \infty$*, the number of cubes of the form* $a_n(i + [0, 1)^d)$*,* $i \in \mathbb{Z}^d$ *that intersects both* R_0 *and* R_0^c *is of the order* $O([a_n]^{-(d-1)})$ *as* $n \to \infty$*.*

This condition is satisfied by most regions of practical interest. For example, Condition B is satisfied in the plane (i.e., $d = 2$) if the boundary ∂R_0 of R_0 is delineated by a simple rectifiable curve of finite length. When the sampling sites lie on the integer grid $\mathbb{Z}^d$, an important implication of Condition B is that the effect of the data points lying near the boundary of R_n is negligible compared to the totality of data points.

12.3 Block Bootstrap for Spatial Data on a Regular Grid

In this section, we consider bootstrapping a spatial process indexed by the integer grid $\mathbb{Z}^d$. Let R_n denote the sampling region, given by (12.1) for some prototype set R_0 satisfying the boundary Condition B. Suppose that $\{Z(s) \cdot s \in \mathbb{Z}^d\}$ is a stationary spatial process that is observed at finitely many locations $\mathcal{S}_n \equiv \{s_1, \ldots, s_{N_n}\}$, given by the part of the integer grid $\mathbb{Z}^d$ that lies inside R_n, i.e.,

$$\{s_1, \ldots, s_{N_n}\} = R_n \cap \mathbb{Z}^d \ , \tag{12.2}$$

$n \geq 1$. Note that the number N_n of elements of the set $R_n \cap \mathbb{Z}^d$ is determined by the scaling constant λ_n and the shape of the prototype set R_0. As a result, in this case, the collection $\{N_n : n \geq 1\}$ of all possible sample sizes may not equal $\mathbb{N}$, the set of all positive integers. For spatial data observed on a regular grid, this is the primary reason for using N_n to denote the sample size at stage n, instead of using the standard symbol n, which runs over $\mathbb{N}$. For notational simplicity, we shall set $N_n = N$ for the rest of this section. This N should not be confused with the N used in Chapters 2–11 to denote the number of overlapping blocks of length ℓ in a sample of size n from a *time series*.

It is easy to see that the sample size N and the volume of the sampling region R_n satisfies the relation

$$N \sim \text{vol.}(R_0) \cdot \lambda_n^d \,, \tag{12.3}$$

where, recall that, for any Borel set $A \subset \mathbb{R}^d$, vol.(A) denotes the volume (i.e., the Lebesgue measure) of A and where for any two sequences $\{r_n\}_{n\geq 1}$ and $\{t_n\}_{n\geq 1}$ of positive real numbers, we write $r_n \sim t_n$ if $r_n/t_n \to 1$ as $n \to \infty$. Let

$$T_n = t_n(\mathcal{Z}_n; \theta)$$

be a random variable of interest, where $\mathcal{Z}_n = \{Z(s_1), \ldots, Z(s_N)\}$ denotes the collection of observations and where θ is a parameter. For example, we may have $T_n = \sqrt{N}(\bar{Z}_n - \mu)$ with $\bar{Z}_n = N^{-1}\sum_{i=1}^N Z(s_i)$ denoting the sample mean and μ denoting the population mean. Our goal is to define block bootstrap estimators of the sampling distribution of T_n.

Different variants of spatial subsampling and spatial block bootstrap methods have been proposed in the literature. See Hall (1985), Possolo (1991), Politis and Romano (1993, 1994a), Sherman and Carlstein (1994), Sherman (1996), Politis, Paparoditis and Romano (1998, 1999), Politis, Romano and Wolf (1999), and the references therein. Here we shall follow a version of the block bootstrap method, suggested by Bühlmann and Künsch (1999b) and Zhu and Lahiri (2001), that is applicable to sampling regions of general shapes, given by (12.1).

12.3.1 Description of the Block Bootstrap Method

Let $\{\beta_n\}_{n\geq 1}$ be a sequence of positive integers such that

$$\beta_n^{-1} + \beta_n/\lambda_n = o(1) \quad \text{as} \quad n \to \infty \,. \tag{12.4}$$

Thus, β_n goes to infinity but at a rate slower than the scaling factor λ_n for the sampling region R_n (cf. (12.1)). Here, β_n gives the scaling factor for the blocks or subregions for the spatial block bootstrap method. Let $\mathcal{U} = [0,1)^d$ denote the unit cube in $\mathbb{R}^d$. As a first step, we partition the sampling region R_n using cubes of volume β_n^d. Let $\mathcal{K}_n = \{k \in \mathbb{Z}^d : \beta_n(k +$

$\mathcal{U}) \cap R_n \neq \emptyset\}$ denote the index set of all cubes of the form $\beta_n(k + \mathcal{U})$ that have nonempty intersections with the sampling region R_n. We will define a bootstrap version of the process $Z(\cdot)$ over R_n by defining its version on each of the subregions

$$R_n(k) \equiv R_n \cap [\beta_n(k + \mathcal{U})], \ k \in \mathcal{K}_n \ . \tag{12.5}$$

For this, we consider one $R_n(k)$ at a time and for a given $R_n(k)$, resample from a suitable collection of subregions of R_n (called subregions of "type k") to define the bootstrap version of $Z(\cdot)$ over $R_n(k)$. Let $\mathcal{I}_n = \{i \in \mathbb{Z}^d : i + \beta_n\mathcal{U} \subset R_n\}$ denote the index set of all cubes of volume β_n^d in R_n, with "starting points" $i \in \mathbb{Z}^d$. Then, $\{i + \beta_n\mathcal{U} : i \in \mathcal{I}_n\}$ gives us a collection of cubic subregions or blocks that are *overlapping* and are contained in R_n. Furthermore, for each $i \in \mathcal{I}_n$, the subsample of observations $\{Z(s) : s \in \mathbb{Z}^d \cap [i + \beta_n\mathcal{U}]\}$ is *complete* in the sense that the $Z(\cdot)$-process is observed at *every* point of the integer grid in the subregion $i + \beta_n\mathcal{U}$.

For any set $A \subset \mathbb{R}^d$, let $\mathcal{Z}_n(A) = \{Z(s) : s \in A \cap \mathcal{S}_n\}$ denote the set of observations lying in the set A, where, recall that $\mathcal{S}_n \equiv \{s_1, \ldots, s_N\}$ is the set of all sampling sites in R_n. Thus, in this notation, $\mathcal{Z}_n(R_n)$ is the entire sample $\mathcal{Z}_n = \{Z(s_1), \ldots, Z(s_N)\}$ and $\mathcal{Z}_n(R_n(k))$ denotes the subsample lying in the subregion $R_n(k)$, $k \in \mathcal{K}_n$. For the overlapping version of the spatial block bootstrap method, for each $k \in \mathcal{K}_n$, we resample one block at random from the collection $\{i + \beta_n\mathcal{U} : i \in \mathcal{I}_n\}$, independently of the other resampled blocks, and define a version of the observed process on the subregion $R_n(k)$ using the observations from the resampled subregion. To that end, let $K \equiv K_n$ denote the size of $\mathcal{K}_n$ and let $\{I_k : k \in \mathcal{K}_n\}$ be a collection of K iid random variables having common distribution

$$P(I_1 = i) = \frac{1}{|\mathcal{I}_n|}, \ i \in \mathcal{I}_n. \tag{12.6}$$

For $k \in \mathcal{K}_n$, we define the overlapping block bootstrap version $\mathcal{Z}_n^*(R_n(k))$ of $\mathcal{Z}_n(R_n(k))$ by using a part of the resampled block $\mathcal{Z}_n(I_k + \beta_n\mathcal{U})$ that is *congruent* to the subregion $R_n(k)$. More precisely, we define $\mathcal{Z}_n^*(R_n(k))$ by

$$\mathcal{Z}_n^*(R_n(k)) = \mathcal{Z}_n\Big([I_k + \beta_n\mathcal{U}] \cap [R_n(k) - k\beta_n + I_k]\Big) \ . \tag{12.7}$$

Note that the set $[R_n(k) - k\beta_n + I_k]$ is obtained by an integer translation of the subregion $R_n(k)$ that maps the starting point $k\beta_n$ of the set $(k + \mathcal{U})\beta_n$ to the starting point I_k of the resampled block $(I_k + \beta_n\mathcal{U})$. As a result, $R_n(k)$ and $(I_k + \beta_n\mathcal{U}) \cap [R_n(k) - k\beta_n + I_k]$ have the *same* shape, and the resampled observations retain the same spatial dependence structure as the original process $\mathcal{Z}_n(R_n(k))$ over the subregion $R_n(k)$. Furthermore, because of translation by integer vectors, the number of resampled observations in $\mathcal{Z}_n^*(R_n(k))$ is the same as that in $\mathcal{Z}_n(R_n(k))$, for every $k \in \mathcal{K}_n$.

To gain further insight into the structure of the resampled blocks of observations $\mathcal{Z}_n^*(R_n(k))$'s in (12.7), let $\mathcal{K}_{1n} = \{k \in \mathcal{K}_n : (k + \mathcal{U})\beta_n \subset R_n\}$

and $\mathcal{K}_{2n} = \{k \in \mathcal{K}_n : (k + \mathcal{U})\beta_n \cap R_n^c \neq \emptyset\}$, respectively, denote the index set of all *interior* cubes contained in R_n and that of all *boundary* cubes that intersect both R_n and R_n^c. See Figure 12.1. Note that for all $k \in \mathcal{K}_{1n}$, $R_n(k) = (k + \mathcal{U})\beta_n$ and, hence, it is a cubic subregion of R_n. However, for $k \in \mathcal{K}_{2n}$, $R_n(k)$ is a *proper* subset of $(k + \mathcal{U})\beta_n$ and the shape of $R_n(k)$ depends on the shape of the boundary of R_n. In particular, for $k \in \mathcal{K}_{2n}$, $R_n(k)$ need not be a cubic region. As a result, for $k \in \mathcal{K}_{1n}$, $\mathcal{Z}_n^*(R_n(k))$ contains all the observations from the resampled cubic subregion $I_k + \beta_n\mathcal{U} \subset R_n$. In contrast, for $k \in \mathcal{K}_{2n}$, $\mathcal{Z}_n^*(R_n(k))$ contains only a subset of the observed values in $I_k + \beta_n\mathcal{U}$, lying in a subregion of $I_k + \beta_n\mathcal{U}$ that is congruent to $R_n(k)$. Note that for $k \in \mathcal{K}_{1n}$, the number ℓ of observations in the resampled block $\mathcal{Z}_n^*(R_n(k))$ is precisely β_n^d. Hence, by (12.3) and (12.4), the typical block size ℓ and the original sample size N satisfies the relation

$$\ell^{-1} + N^{-1}\ell = o(1) \quad \text{as} \quad n \to \infty ,$$

as in the time series case (cf. Chapter 2). The overlapping block bootstrap version $\mathcal{Z}_n^*(R_n)$ of $\mathcal{Z}_n(R_n)$ is now given by concatenating the resampled blocks of observations $\{\mathcal{Z}_n^*(R_n(k)) : k \in \mathcal{K}_n\}$. Note that by our construction, the resample size equals the sample size. Hence, the bootstrap version of a random variable $T_n \equiv t_n(\mathcal{Z}_n; \theta)$ is given by

$$T_n^* = t_n\Big(\mathcal{Z}_n^*(R_n); \tilde{\theta}_n\Big) , \tag{12.8}$$

where the same function $t_n(\cdot;\cdot)$, appearing in the definition of T_n, is also used to define its bootstrap version. Here, $\tilde{\theta}_n$ is an estimator of θ, defined by mimicking the relation between the joint distribution of $\mathcal{Z}_n$ and θ. For an example, consider $T_n = \sqrt{N}(\bar{Z}_n - \mu)$ with $\bar{Z}_n = n^{-1}\sum_{i=1}^{N} Z(s_i)$. Then, the overlapping block bootstrap version of T_n is given by

$$T_n^* = \sqrt{N}\Big(\bar{Z}_n^* - \hat{\mu}_n\Big) ,$$

where $\bar{Z}_n^*$ is the average of the N resampled observations $\mathcal{Z}_n^*(R_n)$, $\hat{\mu}_n = E_*\bar{Z}_n^*$, and E_* denotes the conditional expectation given $\{Z(s) : s \in \mathbb{Z}^d\}$. Similarly, if $T_n = \sqrt{N}(H(\bar{Z}_n) - H(\mu))$ for some function H, then we may define T_n^* as $T_n^* = \sqrt{N}(H(\bar{Z}_n^*) - H(\hat{\mu}_n))$. Note that the block bootstrap method described above can also be applied to vector-valued spatial processes with obvious notational changes. We shall make use of the block bootstrap for the vector case later in the section where we consider M-estimators of parameters of a vector-valued spatial process $Z(\cdot)$.

Next we briefly describe the nonoverlapping version of the block bootstrap method. Let $R_n(k)$, $k \in \mathcal{K}_n$ denote the partition of the sampling region R_n given by (12.5). For the nonoverlapping version, we restrict attention to the collection of nonoverlapping cubes $\mathcal{J}_n \equiv \{j \in \mathbb{Z}^d : [j + \mathcal{U}]\beta_n \subset$

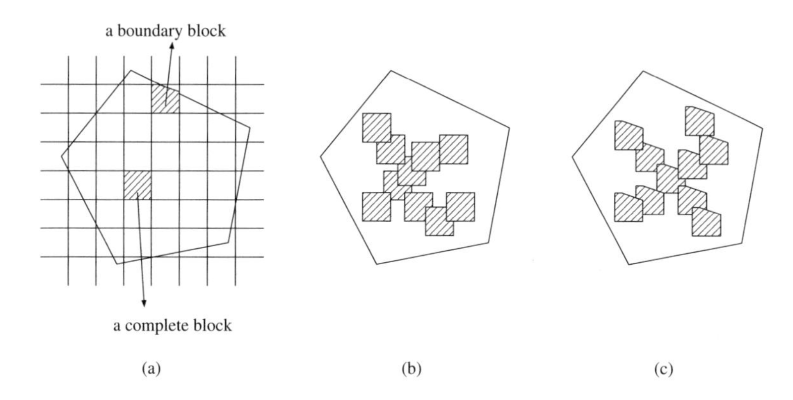

FIGURE 12.1. The blocking mechanism for the overlapping spatial block bootstrap method. (a) Partition of a pentagonal sampling region R_n by the subregions $R_n(k)$, $k \in \mathcal{K}_n$ of (12.5); (b) a set of overlapping "complete" blocks; (c) a set of overlapping copies of the "boundary" block shown in (a). Bootstrap versions of the spatial process $Z(\cdot)$ over the shaded "complete" and the shaded "boundary" blocks in (a) are, respectively, obtained by resampling from the observed "complete" blocks in (b) and the observed "boundary" blocks in (c).

$R_n\}$ and generate K iid random variables $\{J_k : k \in \mathcal{K}_n\}$ with common distribution

$$P(J_1 = j) = \frac{1}{|\mathcal{J}_n|}, \quad j \in \mathcal{J}_n, \tag{12.9}$$

where $K = K_n$ is the size of $\mathcal{K}_n$. Then, the nonoverlapping bootstrap version of the spatial process $\mathcal{Z}_n(R_n(k))$ over the subregion $R_n(k)$ is given by

$$\mathcal{Z}_n^{*(2)}\big(R_n(k)\big) = \mathcal{Z}_n\Big(\big[(J_k + \mathcal{U})\beta_n\big] \cap \big[R_n(k) - k\beta_n + J_k\beta_n\big]\Big), \; k \in \mathcal{K}_n .$$

This is equivalent to selecting a random sample of size K with replacement from the collection of all nonoverlapping cubes $\{(j + \mathcal{U})\beta_n : j \in \mathcal{J}_n\}$ and defining a version of the $Z(\cdot)$-process on each subregion $R_n(k)$ by considering all data-values that lie on a congruent part of the resampled cube. The nonoverlapping block bootstrap version of $T_n = t_n(\mathcal{Z}_n; \theta)$ is now given by

$$T_n^{*(2)} = t_n\Big(\mathcal{Z}_n^{*(2)}(R_n); \tilde{\theta}_n\Big), \tag{12.10}$$

where $\mathcal{Z}_n^{*(2)}(R_n)$ is obtained by concatenating the resampled blocks $\mathcal{Z}_n^{*(2)}(R_n(k))$, $k \in \mathcal{K}_n$, and $\tilde{\theta}_n$ is a suitable estimator of θ that is defined by mimicking the relation between θ and $\mathcal{Z}(R_n)$, as before.

For both versions of the spatial block bootstrap, we may define a "blocks of blocks" version for random variables that are (symmetric) functions of p-dimensional vectors of the form $Y(s) = \big((Z(s + h_1), \ldots, Z(s + h_p)\big)'$,

$s \in R_{n,p} \cap \mathbb{Z}^d$ for some $p \in \mathbb{N}$, where $h_1, \ldots, h_p \in \mathbb{Z}^d$ are given lag vectors and

$$R_{n,p} \equiv \{s \in \mathbb{R}^d : s + h_1, \ldots, s + h_p \in R_n\} .$$

For example, consider the centered and scaled estimator

$$T_n = |N_n(h)|^{1/2}(\hat{\theta}_n - \theta) ,$$

where $\theta = \text{Cov}(Z(0), Z(h))$ denotes the autocovariance of the spatial process at a given lag $h \in \mathbb{Z}^d \setminus \{0\}$, $\hat{\theta}_n = |N_n(h)|^{-1} \sum_{s \in N_n(h)} Z(s)Z(s+h) - (|N_n(h)|^{-1} \sum_{s \in N_n(h)} Z(s))^2$ is a version of the sample autocovariance estimator, and $N_n(h) = \{s \in \mathbb{Z}^d : s, s+h \in R_n\}$. Here, recall that, $|A|$ denotes the size of a set A. Then, T_n is a function of the bivariate spatial process $Y(s) = (Z(s), Z(s+h))'$, $s \in R_{n,2}$, where the set $R_{n,2}$ is given by

$$R_{n,2} = \{s \in \mathbb{R}^d : s, s+h \in R_n\} = R_n \cap (R_n - h) .$$

As in the time series case, the bootstrap version of such variables may be defined by using the vectorized process $\{Y(s) : s \in R_{n,2} \cap \mathbb{Z}^d\}$.

Next we return to the case of a general p-dimensional vectorized process $Y(\cdot)$. Let $T_{n,p} = t_n(\mathcal{Y}_n; \theta)$ be a random variable of interest, where $\mathcal{Y}_n = \{y(s) : s \in R_{n,p}\}$ and θ is a parameter. To define the overlapping bootstrap version of $T_{n,p}$, we introduce the partition $\{R_{n,p}(k) : k \in \mathcal{K}_{n,p}\}$ of $R_{n,p}$ by cubes of the form $(k + \mathcal{U})\beta_n$, $k \in \mathbb{Z}^d$ as before, where $\mathcal{K}_{n,p} = \{k \in \mathbb{Z}^d : (k + \mathcal{U})\beta_n \cap R_{n,p} \neq \emptyset\}$. Next, we resample $|\mathcal{K}_{n,p}|$-many indices randomly and with replacement from the collection $\mathcal{I}_{n,p} \equiv \{i \in \mathbb{Z}^d : i + \mathcal{U}\beta_n \subset R_{n,p}\}$, define a version of the Y-process on each subregion $R_{n,p}(k)$, $k \in \mathcal{K}_{n,p}$ as before, and then concatenate the resampled blocks of Y-values to define a version $\mathcal{Y}_n^*$ of $\mathcal{Y}_n$ over the region $R_{n,p}$. The "blocks of blocks" version of T_n is now given by

$$T_{n,p}^* = t_n(\mathcal{Y}_n^*; \tilde{\theta}_n) \tag{12.11}$$

where $\tilde{\theta}_n$ is a suitable estimator of θ.

12.3.2 Numerical Examples

In this section, we illustrate the implementation of the spatial block bootstrap method with a numerical example. Let $\{Z(s) : s \in \mathbb{Z}^2\}$ be a zero mean stationary Gaussian process with the isotropic Exponential variogram:

$$\begin{aligned} 2\gamma(h; \theta) &\equiv E\big(Z(h) - Z(0)\big)^2, \; h \in \mathbb{Z}^2 \\ &= \begin{cases} \theta_1 + \theta_2\Big(1 - \exp(-\theta_3\|h\|)\Big) , & h \neq 0 \\ 0 & h = 0 , \end{cases} \end{aligned} \tag{12.12}$$

$\theta = (\theta_1, \theta_2, \theta_3)' \in [0, \infty) \times (0, \infty) \times (0, \infty) \equiv \Theta$. The variogram provides a description of the covariance structure of the spatial process $Z(\cdot)$. The parameter θ_1 is called the "nugget" effect, which often results from an additive white noise component of $Z(\cdot)$. The "isotropy" condition on the random field means that the variogram at lag $h \in \mathbb{Z}^2 \setminus \{0\}$ depends only on the distance $\|h\|$ between the spatial indices of the variables $Z(0)$ and $Z(h)$, but not on the direction vector $h/\|h\|$. For more details on the variogram and its use in spatial statistics, see Cressie (1993) and the discussion in Section 12.4 below. Plots of the variogram (12.12) for the parameter values $\theta = (0, 2, 1)'$ (with no-nugget effects) and $\theta = (1, 1, 1)'$ are given in Figure 12.2. Realizations of the Gaussian random field $Z(\cdot)$ were generated over a rectangular region of size 20×30 using these parameter values. The corresponding data sets are shown in Figures 12.3 and 12.4, respectively. Note that the surface corresponding to the "no-nugget" effect case (viz., $\theta_1 = 0$) has lesser "small-scale variation" than the surface with a nonzero nugget effect case (viz., $\theta_1 = 1$).

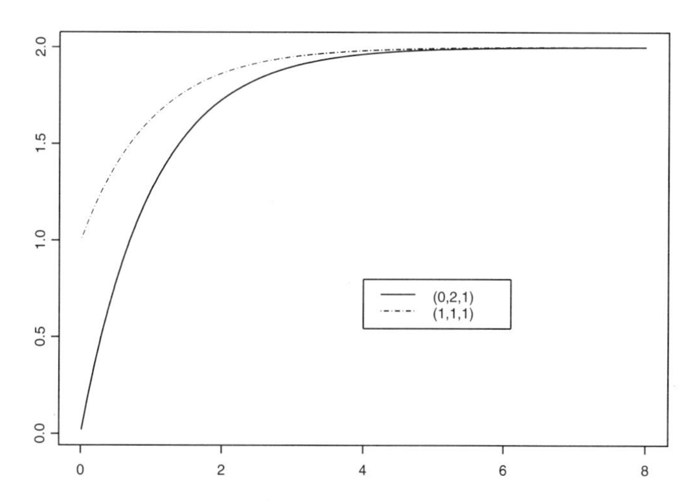

FIGURE 12.2. Plots of the isotropic variogram $2\gamma(h; \theta)$ of (12.12) against $\|h\|$ for $\theta = (0, 2, 1)'$ (shown in solid line) and for $\theta = (1, 1, 1)'$ (shown in dot-and-dash line).

To apply the spatial block bootstrap method, we identify R_n with the 20×30 rectangular region $[-10, 10) \times [-15, 15)$, and fix the scaling constant λ_n and the prototype set $R_0 \subset [-\frac{1}{2}, \frac{1}{2}]^2$ as $\lambda_n = 30$ and $R_0 = [-\frac{1}{3}, \frac{1}{3}) \times [-\frac{1}{2}, \frac{1}{2})$. Here R_0 is chosen to be a maximal set in $[-\frac{1}{2}, \frac{1}{2}]^2$ that corresponds to the given rectangular region $[-10, 10) \times [-15, 15)$ up to a scaling constant. This, in turn, determines λ_n uniquely. We applied the block bootstrap method to each of the above data sets with two choices of β_n, given by $\beta_n = 5$ and 8. In the first case, 5 divides both 20 and 30, so that the partitioning subregions $R_n(k)$'s of (12.5) are all squares (and hence, are complete). Thus, there are 24 subregions in the partition (12.5), given by $R_n(k) = [5k_1, 5k_1 + 5) \times [5k_2, 5k_2 + 5)$, $k = (k_1, k_2)' \in \mathbb{Z}^2$, $-2 \leq k_1 < 2$,

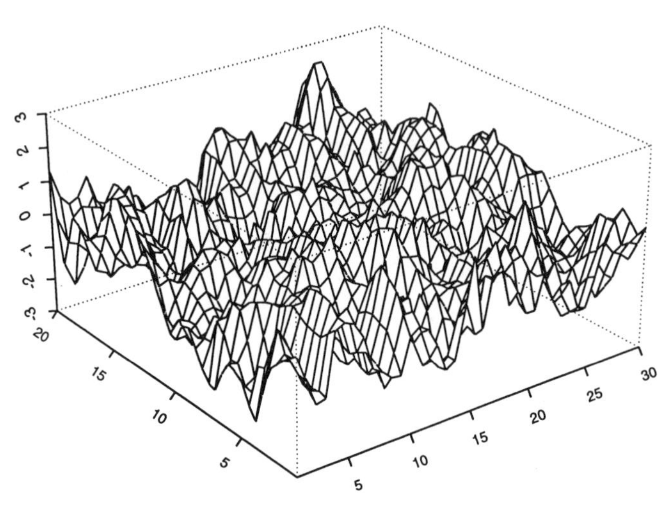

FIGURE 12.3. Realizations of a zero mean unit variance Gaussian random field with variogram (12.12) over a 20×30 region on the planar integer grid for $\theta = (0, 2, 1)'$ (with no nugget effect).

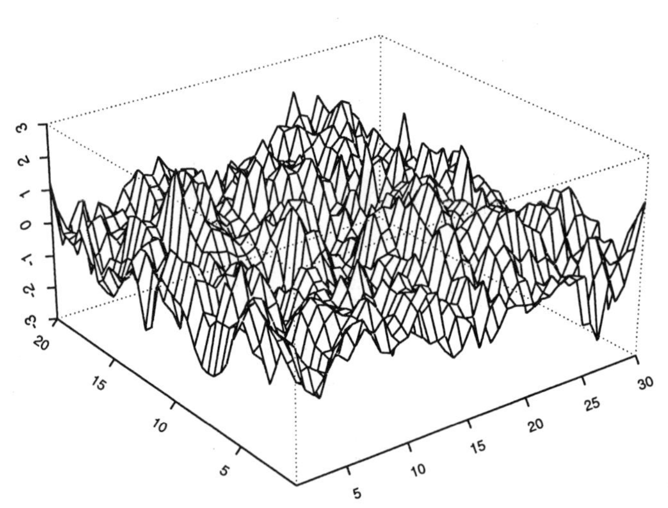

FIGURE 12.4. Realizations of a zero mean unit variance Gaussian random field with variogram (12.12) over a 20×30 region on the planar integer grid for $\theta = (1, 1, 1)'$ with nugget effect $\theta_1 = 1$.

$-3 \le k_2 < 3$. To define the overlapping block bootstrap version of the $Z(\cdot)$-process over R_n, we resample 24 times randomly, with replacement from the collection of all observed complete blocks

$$\{[i_1, i_1 + 5) \times [i_2, i_2 + 5) : (i_1, i_2)' \in \mathbb{Z}^2, \ -10 \le i_1 \le 5, \ -15 \le i_2 \le 10\} \ .$$

For $\beta_n = 8$, there are 4 interior blocks of size 8×8, while 12 rectangular boundary blocks, 4 of size 8×7, 4 of size 2×8, and 4 of size 2×7. To define the bootstrap version of the $Z(\cdot)$-process over these 16 subregions, we resample 16 blocks randomly with replacement from the collection

$$\{[i_1, i_1 + 8) \times [i_2, i_2 + 8) : (i_1, i_2)' \in \mathbb{Z}^2, \ -10 \leq i_1 \leq 2, \ -15 \leq i_2 \leq 7\}$$

and use *all* observations from 4 of these for the 4 "complete" blocks of size 8×8 and use suitable parts of the remaining 12 blocks for the 12 boundary regions. For example, for the 8×7 region $[0, 8) \times [8, 15)$, we would use only the observations lying in $[i_1^0, i_1^0 + 8) \times [i_2^0, i_2^0 + 7)$ if the selected block is given by $[i_1^0, i_1^0 + 8) \times [i_2^0, i_2^0 + 8)$. Similarly, for the 2×8 region $[-10, -8) \times [-8, 0)$, we would use the observations lying in $[i_1^0 + 6, i_1^0 + 8) \times [i_2^0, i_2^0 + 8)$ only, when the selected block is given by $[i_1^0, i_1^0 + 8) \times [i_2^0, i_2^0 + 8)$. When $\cup\{(k + \mathcal{U})\beta_n : k \in \mathcal{K}_n\} \neq R_n$, a simpler and *valid* alternative (*not* described in Section 12.3.1) is to use the complete sets of observations in all K (= 16 in the example, for $\beta_n = 8$) resampled blocks and define the bootstrap version of a random variable $T_n = t(N; \{Z(s_1), \ldots, Z(s_N)\}, \theta)$ as

$$T_n^{**} = t(M; \{Z^*(s_1), \ldots, Z^*(s_M)\}, \tilde{\theta}_n) ,$$

where $\{Z^*(s_1), \ldots, Z^*(s_M)\}$ is the collection of all observations in the K-many resampled complete blocks, and where $\tilde{\theta}_n$ is an estimator of θ based on $\{Z(s_1), \ldots, Z(s_N)\}$. However, for the rest of this section, we continue to work with the original version of the block bootstrap method described in Section 12.3.1.

First we consider the problem of variance estimation by the overlapping block bootstrap method. Suppose that the level-2 parameter of interest is given by $\sigma_n^2 = \mathrm{Var}(T_{1n})$, the variance of the centered and scaled sample mean

$$T_{1n} \equiv \lambda_n^{d/2}(\bar{Z}_n - \mu)$$

(note that here $d = 2$ and $\mu = 0$). To find the block bootstrap estimator $\hat{\sigma}_n^2(\beta_n)$ of the parameter σ_n^2, note that by the linearity of the sample mean in the observations, we can write down an exact formula for $\hat{\sigma}_n^2(\beta_n)$, as in the time series case. For later reference, we state the formula for the general case of a $\mathbb{R}^d$-valued random field $\{Z(s) : s \in \mathbb{R}^d\}$. Let $S_n(i; k)$ denote the sum of all observations in the ith block of "type k," $\mathcal{B}_n(i; k) \equiv [R_n(k) - k\beta_n + i] \cap [i + \mathcal{U}\beta_n]$, for $i \in \mathcal{I}_n \equiv \{j \in \mathbb{Z}^d : j + \mathcal{U}\beta_n \subset R_n\}$, $k \in \mathcal{K}_n$. Then, the spatial bootstrap estimator of σ_n^2 is given by

$$\begin{aligned} \hat{\sigma}_n^2(\beta_n) \quad = \quad & N^{-2}\lambda_n^d \Big[|\mathcal{K}_{1n}| \Big\{ |\mathcal{I}_n|^{-1} \sum_{i \in \mathcal{I}_n} S_n(i; 0)^2 \Big\} \\ & + \sum_{k \in \mathcal{K}_{2n}} \Big\{ |\mathcal{I}_n|^{-1} \sum_{i \in \mathcal{I}_n} S_n(i; k)^2 \Big\} - N^2 \hat{\mu}_n^2 \Big] , \quad (12.13) \end{aligned}$$

where $\hat{\mu}_n = N^{-1}|\mathcal{I}_n|^{-1}\{|\mathcal{K}_{1n}|\sum_{i\in\mathcal{I}_n} S_n(i;0) + \sum_{k\in\mathcal{K}_{2n}}\sum_{i\in\mathcal{I}_n} S_n(i;k)\}$, and where $\mathcal{K}_{1n}$ and $\mathcal{K}_{2n}$ denote the set of all interior and all boundary blocks, respectively. For the block size parameter $\beta_n = 5$, $|\mathcal{K}_{1n}| = 24$, $|\mathcal{K}_{2n}| = 0$, while for $\beta_n = 8$, $|\mathcal{K}_{1n}| = 4$, and $|\mathcal{K}_{2n}| = 12$ in our example. The corresponding block bootstrap estimates are reported in Table 12.1. The true values of the level-2 parameter $\sigma_n^2 \equiv \lambda_n^d \text{Var}(\bar{Z}_n)$ and its limit σ_∞^2 are given by 8.833 and 9.761 under the variogram model (12.12) with $\theta = (0, 2, 1)'$. The corresponding values of σ_n^2 and σ_∞^2 under $\theta = (1, 1, 1)'$ are given by 5.167 and 5.630, respectively.

TABLE 12.1. Bootstrap estimates $\hat{\sigma}_n^2(\beta_n)$ of the level-2 parameter $\sigma_n^2 = \lambda_n^2 \text{Var}(\bar{Z}_n)$ for the data sets of Figures 12.3 and 12.4 with block size parameter $\beta_n = 5, 8$.

	$\theta = (0,2,1)'$	$\theta = (1,1,1)'$
β_n	$\hat{\sigma}_n^2(\beta_n)$	$\hat{\sigma}_n^2(\beta_n)$
5	5.950	4.469
8	7.811	5.590

Next we apply the bootstrap method to estimate the distribution function of T_{1n}. Note that under both θ-values, the true distribution of T_{1n} is given by $N(0, \sigma_n^2)$, where $\sigma_n^2 = 8.833$ for $\theta = (0, 2, 1)'$ and $\sigma_n^2 = 5.167$ for $\theta = (1, 1, 1)'$. Unlike the variance estimation case, the bootstrap estimators of $P(T_{1n} \leq \cdot)$ do not admit a closed-form formula like (12.13) and have to be evaluated by the Monte-Carlo method, as in the time series case (cf. Section 4.5). Histograms corresponding to the block bootstrap distribution function estimators with block size parameter $\beta_n = 5, 8$ for the data set of Figure 12.3 are shown in the upper panel of Figure 12.5. The corresponding distribution functions are shown in the lower panel of Figure 12.5. Figure 12.6 gives the histograms and the distribution functions of the bootstrap estimates of $P(T_{1n} \leq \cdot)$ for the data set of Figure 12.4. In both cases, we used $B = 1000$ bootstrap replicates to generate the Monte-Carlo approximation to $P_*(T_{1n}^* \leq \cdot)$.

In the next three sections, we study some theoretical properties of the variance and the distribution function estimators, generated by the spatial block bootstrap method.

12.3.3 Consistency of Bootstrap Variance Estimators

In this section, we show that the spatial block bootstrap method can be used to derive consistent estimators of the variance of the sample mean, and more generally, of statistics that are smooth functions of sample means. Suppose that the random field $\{Z(i) : i \in \mathbb{Z}^d\}$ is m-dimensional. Let

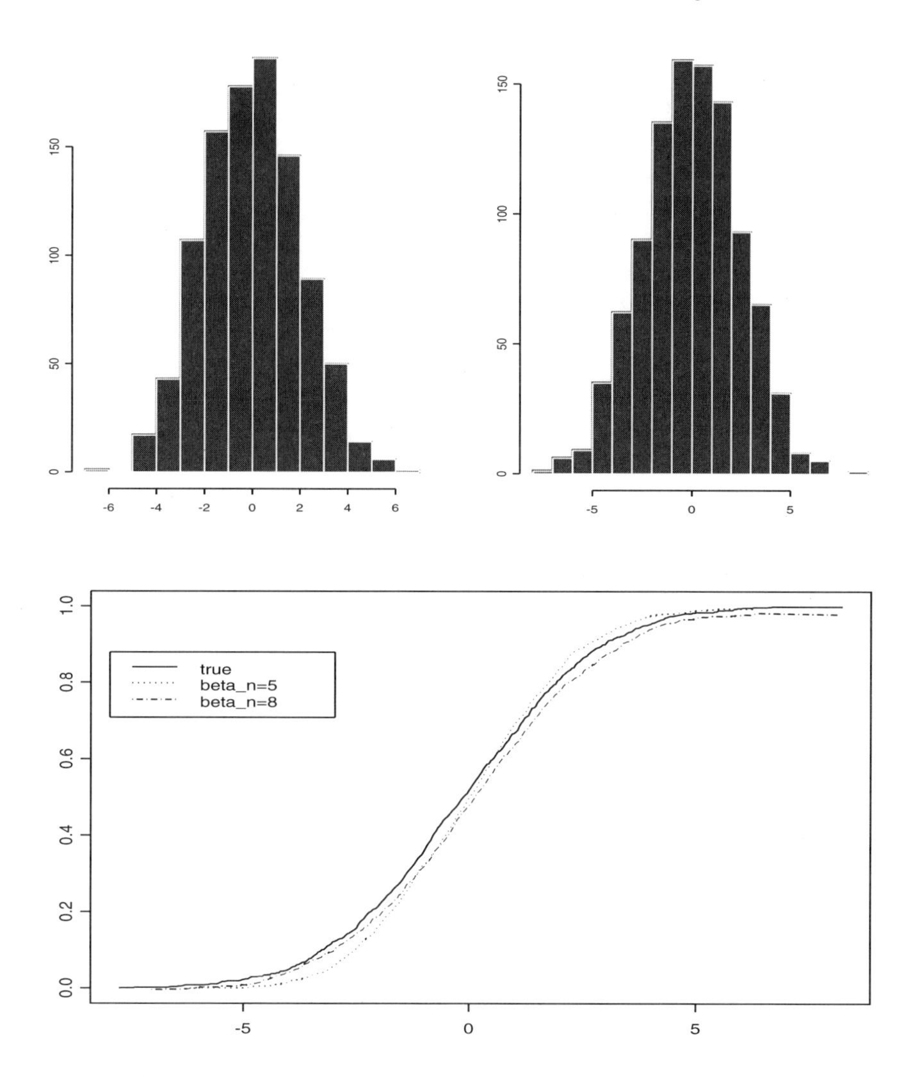

FIGURE 12.5. Histograms and cumulative distribution functions of the block bootstrap estimates of $P(\lambda_n(\bar{Z}_n - \mu) \leq \cdot)$ for the data set of Figure 12.3. The left histogram corresponds to $\beta_n = 5$ and the right one to $\beta_n = 8$. The true distribution of $\lambda_n(\bar{Z}_n - \mu)$ is given by $N(0, \sigma_n^2)$ with $\sigma_n^2 = 8.833$, and is denoted by the solid line in the lower panel.

$\bar{Z}_n = N^{-1} \sum_{i=1}^{N} Z(s_i)$ denote the sample mean and let $\hat{\theta}_n = H(\bar{Z}_n)$ be an estimator of a level-1 parameter of interest $\theta = H(\mu)$, where $\mu = EZ(0)$ and $H : \mathbb{R}^m \rightarrow \mathbb{R}$ is a smooth function. As in the time series case, by considering suitable transformations of the observations, one can express many common estimators under this spatial version of the Smooth Function Model. Let $\bar{Z}_n^*$ denote the block bootstrap sample mean based on a block-size pa-

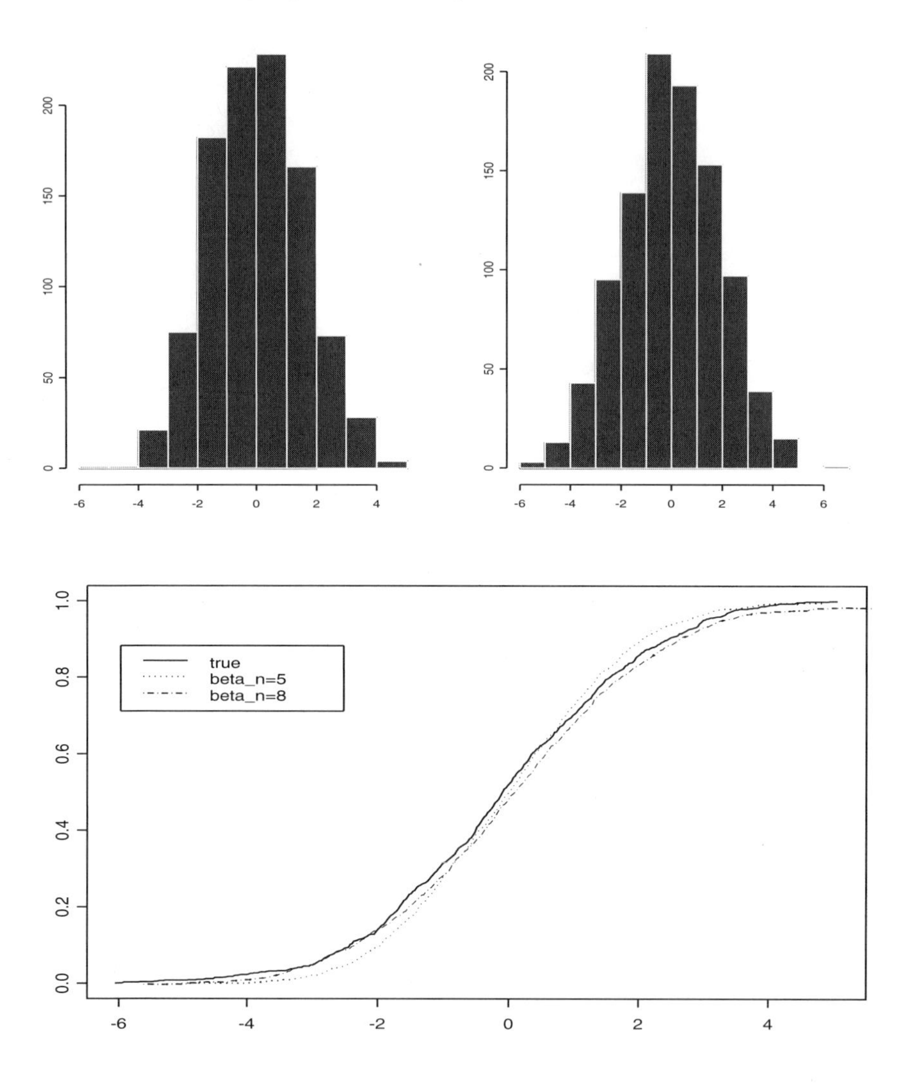

FIGURE 12.6. Histograms and cumulative distribution functions of the block bootstrap estimates of $P(T_{1n} \leq \cdot)$ for the data set of Figure 12.4. The left histogram corresponds to $\beta_n = 5$ and the right one to $\beta_n = 8$. The true distribution is given by $N(0, \sigma_n^2)$ with $\sigma_n^2 = 5.167$, and is denoted by the solid line in the lower panel.

rameter β_n. Then, the bootstrap version of $\hat{\theta}_n$ is given by $\theta_n^* = H(\bar{Z}_n^*)$, and the bootstrap estimator of the level-2 parameter $\sigma_n^2 \equiv \lambda_n^d \text{Var}(\hat{\theta}_n)$ is given by

$$\hat{\sigma}_n^2 \equiv \hat{\sigma}_n^2(\beta_n) = \lambda_n^d \text{Var}_*(\theta_n^*) \ . \tag{12.14}$$

For establishing consistency of $\hat{\sigma}_n^2$, we shall assume that the random field $\{Z(s) : s \in \mathbb{R}^d\}$ satisfies certain weak dependence condition. The weak-dependence condition will be specified through a spatial version of the strong-mixing condition. For $S_1, S_2 \subset \mathbb{R}^d$, let

$$\alpha_1(S_1, S_2) = \sup\Big\{|P(A \cap B) - P(A)P(B)| : A \in \mathcal{F}_Z(S_1),\ B \in \mathcal{F}_Z(S_2)\Big\}, \tag{12.15}$$

where $\mathcal{F}_Z(S)$ is the σ-algebra generated by the random variables $\{Z(s) : s \in S \cap \mathbb{Z}^d\}$, $S \subset \mathbb{R}^d$. Let $\text{dist.}(S_1, S_2) = \inf\{|x-y| : x \in S_1, y \in S_2\}$ denote the distance between the sets S_1 and S_2 in the ℓ^1-norm on $\mathbb{R}^d$, defined by $|x| = |x_1| + \cdots + |x_d|$ for $x = (x_1, \ldots, x_d)' \in \mathbb{R}^d$. The mixing coefficient of $Z(\cdot)$ is defined as

$$\alpha(a; b) = \sup\{\alpha_1(S_1, S_2) : S_1, S_2 \in \mathcal{R}(b),\ \text{dist.}(S_1, S_2) \geq a\} \tag{12.16}$$

for $a > 0$, $b \geq 1$, where $\mathcal{R}(b)$ is the collection of all sets in $\mathbb{R}^d$ that have a volume of b or less and that can be represented as unions of up to $\lceil b \rceil$ many cubes.

Many variants of the strong-mixing coefficient have been proposed and used in the literature, where the suprema in (12.16) are taken over various classes of sets S_1, S_2. In (12.16), we restrict attention to sets S_1 and S_2 that are finite unions of d-dimensional cubes and have a finite volume. As a result, the sets S_1 and S_2 are bounded subsets of $\mathbb{R}^d$. This restriction is important in dimensions $d \geq 2$. Some important results of Bradley (1989, 1993) show that a random field in $\mathbb{R}^d$, $d \geq 2$ with a strong mixing coefficient satisfying

$$\lim_{a \to \infty} \alpha(a; \infty) = 0 \tag{12.17}$$

is also ρ-mixing. Thus, if one allows unbounded sets S_1, S_2 in (12.16), then random fields satisfying (12.17) necessarily belong to the smaller class of ρ-mixing random fields. For more discussion of various mixing coefficients for random fields, see Doukhan (1994).

The following result proves consistency of the bootstrap variance estimators.

Theorem 12.1 *Suppose that the random field $\{Z(i) : i \in \mathbb{Z}^d\}$ is stationary with $E\|Z(0)\|^{6+\delta} < \infty$ and with the strong mixing coefficient $\alpha(a, b)$ satisfying*

$$\alpha(a; b) \leq C \cdot a^{-\tau_1} b^{-\tau_2} \quad \textit{for} \quad a \geq 1,\ b \geq 1 \tag{12.18}$$

for some $\delta > 0$, $\tau_1 > 5d(6+\delta)/\delta$, and $0 \leq \tau_2 \leq \tau_1/d$. Also, suppose that H is continuously differentiable and the partial derivatives $D^\alpha H(\cdot)$, $|\alpha| = 1$, satisfy a Hölder's condition of order $\eta \in (0, 1]$. If, in addition, $\beta_n^{-1} + \lambda_n^{-1}\beta_n = o(1)$, then

$$\hat{\sigma}_n^2(\beta_n) \longrightarrow_p \sigma_\infty^2 \quad \textit{as} \quad n \to \infty\ , \tag{12.19}$$

where $\hat{\sigma}_n^2(\beta_n)$ *is as defined in (12.14),* $\sigma_\infty^2 \equiv \lim_{n\to\infty} \lambda_n^d \text{Var}(\hat{\theta}_n) = \frac{1}{\text{vol.}(R_0)} \sum_{i\in\mathbb{Z}^d} EW(0)W(i)$ *and where* $W(i) = \sum_{|\alpha|=1} D^\alpha H(\mu)(Z(i)-\mu)^\alpha$.

For proving the result, we need the following moment bound on partial sums of a possibly nonstationary random field that is a special case of a result stated in Doukhan (1994).

Lemma 12.1 *Let* $\{Y(i),\ i \in \mathbb{Z}^d\}$ *be a random field with* $EY(i) = 0$ *and* $E|Y(i)|^{2q+\delta} < \infty$ *for all* $i \in \mathbb{Z}^d$ *for some* $q \in \mathbb{N}$ *and* $\delta \in (0,\infty)$. *Let* $\alpha_Y(a;b)$ *denote the strong-mixing coefficient of* $Y(\cdot)$, *defined by (12.15) and (12.16) with* $Z(i)$*'s replaced by* $Y(i)$*'s. Suppose that*

$$\sum_{r=0}^{\infty} (r+1)^{d(2q-1)-1} [\alpha_Y(r;2q)]^{\delta/(2q+\delta)} < \infty \ . \tag{12.20}$$

Then, for any subset A *of* $\mathbb{Z}^d$,

$$\begin{aligned} & E\Big|\sum_{i\in A} Y(i)\Big|^{2q} \qquad\qquad (12.21)\\ & \le C(q,\delta) \max\Big\{ \Big[\sum_{i\in A} (E|Y(i)|^{2+\delta})^{\frac{2}{2+\delta}}\Big]^q, \sum_{i\in A} (E|Y(i)|^{2q+\delta})^{\frac{2q}{2q+\delta}} \Big\} , \end{aligned}$$

where $C(q,\delta)$ *is a constant that depends only on* q, δ, d, *and the* α*-mixing coefficient* $\alpha_Y(\cdot;\cdot)$, *but not on the subset* A.

Proof: Follows from Theorem 1.4.1.1, Doukhan (1994). □

Proof of Theorem 12.1: Let $\hat{\mu}_n = E_* \bar{Z}_n^*$. By Taylor's expansion, we get

$$\begin{aligned} \theta_n^* &= H(\bar{Z}_n^*) \\ &= H(\hat{\mu}_n) + \sum_{|\nu|=1} D^\nu H(\hat{\mu}_n)(\bar{Z}_n^* - \hat{\mu}_n)^\nu + Q_{1n}^* \\ &= H(\hat{\mu}_n) + \sum_{|\nu|=1} D^\nu H(\mu)(\bar{Z}_n^* - \hat{\mu}_n)^\nu + Q_{2n}^* \ , \end{aligned} \tag{12.22}$$

where, by the Hölder's condition on the partial derivatives of H,

$$\begin{aligned} |Q_{1n}^*| &\le C\|\bar{Z}_n^* - \hat{\mu}_n\|^{1+\eta} \ , \\ |Q_{2n}^*| &\le |Q_{1n}^*| + C\|\hat{\mu}_n - \mu\|^{1+\eta} \end{aligned} \tag{12.23}$$

for some nonrandom $C = C(\eta, d) \in (0,\infty)$.

Next, for $i \in \mathcal{I}_n$, $k \in \mathcal{K}_n$, let $\mathcal{B}_n(i;k)$ denote the ith block of "type k" defined by $\mathcal{B}_n(i;k) = R_n(k) - k\beta_n + i$. Also, let $S_n(i;k)$ denote the sum over all $W(j)$ with $j \in \mathcal{B}_n(i;k)$, $i \in \mathcal{I}_n$, $k \in \mathcal{K}_n$, and let $S_n^*(k)$ denote the sum of the $W(j)$'s in the kth resampled block $\mathcal{B}_n(I_k;k)$, $k \in \mathcal{K}_n$. Write L_k

for the number of data sites in $R_n(k)$, $k \in \mathcal{K}_n$. Then, using (12.22) and the independence of the resampled blocks, we get

$$\begin{aligned}
\text{Var}_*(\theta_n^*) &= \text{Var}_*\Big(\sum_{|\nu|=1} D^\nu H(\mu)(\bar{Z}_n^* - \hat{\mu}_n)\Big) + \hat{Q}_{3n} \\
&= N^{-2}\text{Var}_*\Big(\sum_{k\in\mathcal{K}_n} S_n^*(k)\Big) + \hat{Q}_{3n} \\
&= N^{-2}\sum_{k\in\mathcal{K}_n} \text{Var}_*\big(S_n^*(k)\big) + \hat{Q}_{3n} \\
&= N^{-2}\Big[\sum_{k\in\mathcal{K}_n} E_* S_n^*(k)^2\Big] + \hat{Q}_{4n} , \qquad (12.24)
\end{aligned}$$

where, by Cauchy-Schwarz inequality, the remainder terms $\hat{Q}_{3n}$ and $\hat{Q}_{4n}$ satisfy the inequalities

$$\begin{aligned}
|\hat{Q}_{3n}| &\le E_*[Q_{2n}^*]^2 + 2\{E_* Q_{2n}^{*2}\}^{1/2}\Big\{N^{-2}\sum_{k\in\mathcal{K}_n} E_* S_n^*(k)^2\Big\}^{1/2} , \\
|\hat{Q}_{4n}| &\le |\hat{Q}_{3n}| + C(d)\cdot\Big[N^{-2}\sum_{k\in\mathcal{K}_n} L_k^2\Big]\cdot\|\hat{\mu}_n - \mu\|^2 . \qquad (12.25)
\end{aligned}$$

Note that by (12.3) and the fact that $|\mathcal{K}_{1n}| \sim \text{vol.}(R_0)\cdot(\lambda_n/\beta_n)^d$, we have

$$\begin{aligned}
&\lambda_n^d E\Big[N^{-2}\sum_{k\in\mathcal{K}_{1n}} E_* S_n^*(k)^2\Big] \\
&= \lambda_n^d N^{-2}\cdot|\mathcal{K}_{1n}|\cdot E\Big[|\mathcal{I}_n|^{-1}\sum_{i\in\mathcal{I}_n} S_n(i;0)^2\Big] \\
&= \lambda_n^d|\mathcal{K}_{1n}|\cdot N^{-2}\cdot E\Big[\sum_{i\in\beta_n\mathcal{U}\cap\mathbb{Z}^d} W(i)\Big]^2 \\
&\to \sigma_\infty^2 \quad \text{as} \quad n\to\infty . \qquad (12.26)
\end{aligned}$$

Also, by the boundary condition on R_0, $|\mathcal{K}_{2n}| = O([\lambda_n/\beta_n]^{d-1})$ as $n\to\infty$. Hence, by Lemma 12.1,

$$\begin{aligned}
&\lambda_n^d E\Big[N^{-2}\sum_{k\in\mathcal{K}_{2n}} E_* S_n^*(k)^2\Big] \\
&= \lambda_n^d N^{-2}\sum_{k\in\mathcal{K}_{2n}} E\Big[\frac{1}{|\mathcal{I}_n|}\sum_{i\in\mathcal{I}_n} S_n(i;k)^2\Big] \\
&= O\Big(\lambda_n^d N^{-2}\cdot|\mathcal{K}_{2n}|\cdot\beta_n^d\Big) \\
&= O(\beta_n/\lambda_n) \quad \text{as} \quad n\to\infty . \qquad (12.27)
\end{aligned}$$

Next, define $\tilde{S}_n^*(k)$ and $\tilde{S}_n(k;i)$ by replacing the $W(i)$'s in the definition of $S_n^*(k)$ and $S_n(k;i)$ by $Z(i)$'s. Note that $\hat{\mu}_n$ can be expressed as $\hat{\mu}_n = N^{-1}\sum_{r=1}^N \omega_{rn} Z(s_r)$ for some nonrandom weights $\omega_{rn} \in [0,1]$. Then, by Lemma 12.1, (5.11), and arguments similar to (12.26) and (12.27),

$$E\|\hat{\mu}_n - \mu\|^4 = O(N^{-2}) ,$$

$$\begin{aligned}
&E\{E_*\|\bar{Z}_n^* - \hat{\mu}_n\|^4\} \\
&\leq C(d)N^{-4}E\Big\{\sum_{k\in\mathcal{K}_n} E_*\|\tilde{S}_n^*(k) - L_k\hat{\mu}_n\|^4 \\
&\qquad\qquad + \Big(\sum_{k\in\mathcal{K}_n} E_*\|\tilde{S}_n^*(k) - L_k\hat{\mu}_n\|^2\Big)^2\Big\} \\
&\leq C(d)N^{-4}|\mathcal{K}_n|^2 \max\Big\{E\|\tilde{S}_n(k;0) - L_k\mu\|^4 \\
&\qquad\qquad + L_k^4 E\|\hat{\mu}_n - \mu\|^4 : k \in \mathcal{K}_n\Big\} \\
&\leq C(d)N^{-4}|\mathcal{K}_n|^2\beta_n^{2d} \\
&= O(\lambda_n^{-2d}) .
\end{aligned} \tag{12.28}$$

Hence, $E[E_* Q_{2n}^{*2}] \leq C\big[(E\{E_*\|\bar{Z}_n^* - \hat{\mu}_n\|^4\})^{(1+\eta)/2} + (E\|\hat{\mu}_n - \mu\|^4)^{(1+\eta)/2}\big] = O(\lambda_n^{-d(1+\eta)})$ as $n\to\infty$. Also, note that $\sum_{k\in\mathcal{K}_n} L_k^2 \leq \max\{L_k : k\in\mathcal{K}_n\} \times [\sum_{k\in\mathcal{K}_n} L_k] \leq \beta_n^d N$. Hence, by (12.23), (12.26), (12.27), and (12.28), it follows that

$$\begin{aligned}
E|\hat{Q}_{3n}| &\leq E\{E_*(Q_{2n}^*)^2\} \\
&\quad + 2\big[E\{E_*(Q_{2n}^*)^2\}\big]^{1/2}\Big[N^{-2}E\Big\{\sum_{k\in\mathcal{K}_n} E_* S_n^*(k)^2\Big\}\Big]^{1/2} \\
&= O(\lambda_n^{-d(1+[\eta/2])}) ;
\end{aligned}$$

which in turn implies that

$$E|\hat{Q}_{4n}| = O(\lambda_n^{-d(1+[\eta/2])}) + O(N^{-2}\beta_n^d) . \tag{12.29}$$

As a result, by (12.24), (12.26), (12.27), and (12.29), it remains to show that

$$N^{-1}\sum_{k\in\mathcal{K}_{1n}} \Big[E_* S_n^*(k)^2 - E\{E_* S_n^*(k)^2\}\Big] \longrightarrow_p 0 \quad \text{as} \quad n\to\infty . \tag{12.30}$$

As in the time series case, we do this by regrouping the squared block sums $S_n(i;k)^2$, $i\in\mathcal{I}_n$, $k\in\mathcal{K}_{1n}$. Note that $E_* S_n^*(k)^2 = E_* S_n^*(0)^2 = |\mathcal{I}_n|^{-1}\sum_{i\in\mathcal{I}_n} S_n(i;0)^2$ for all $k\in\mathcal{K}_{1n}$. Let $J_n = \{j\in\mathbb{Z}^d : [j+\mathcal{U}]2\beta_n \cap R_n \neq \emptyset\}$ and for $h\in\{0,1\}^d$, define $J_n(h) = \{j\in J_n : (j-h)/2 \in \mathbb{Z}^d\}$. Thus,

J_n is the index set of a partition of R_n by cubes of sides $2\beta_n$ and for each $h \in \{0,1\}^d$, $J_n(h)$ is the subset of J_n consisting of integral vectors of "type h". For example, with $h = 0$, $J_n(0)$ is the set of all vectors i in J_n such that all d coordinates of i are *even* integers. Similarly, with $h = (1, 0, \ldots, 0)'$, every $i \in J_n((1, 0, \ldots, 0)')$ has an *odd* integer in its first coordinate and *even* integers in the remaining $(d-1)$ coordinates. For $j \in J_n$, let $V_n(j)$ denote the sum over all $[S_n(i;0)^2 - ES_n(i;0)^2]$ such that $i \in [j + \mathcal{U}]2\beta_n$. Set $V_n(j) = 0$ if $\mathcal{I}_n \cap [j + \mathcal{U}]2\beta_n = \emptyset$. Note that the ℓ^1-distance between the regrouped blocks $\cup\{\mathcal{B}_n(i;0) : i \in [j + \mathcal{U}]2\beta_n\}$ and $\cup\{\mathcal{B}_n(i;0) : i \in [k + \mathcal{U}]2\beta_n\}$ of "type h" for any two distinct indices $j \neq k \in J_n(h)$ is at least $|j - k| \cdot \beta_n$. Hence, using Hölder's inequality and Lemma 12.1, we have

$$
\begin{aligned}
& E\Big[N^{-1} \sum_{k \in \mathcal{K}_{1n}} \{E_* S_n^*(k)^2 - E(E_* S_n^*(k)^2)\}\Big]^2 \\
&= N^{-2}|\mathcal{K}_{1n}|^2 |\mathcal{I}_n|^{-2} E\Big\{\sum_{h \in \{0,1\}^d} \sum_{j \in J_n(h)} V_n(j)\Big\}^2 \\
&\leq C(d) N^{-2}|\mathcal{K}_{1n}|^2 |\mathcal{I}_n|^{-2} \sum_{h \in \{0,1\}^d} E\Big\{\sum_{j \in J_n(h)} V_n(j)\Big\}^2 \\
&\leq C(d) N^{-2}|\mathcal{K}_{1n}|^2 |\mathcal{I}_n|^{-2} |J_n| \max\big\{(E|V_n(j)|^3)^{2/3} : j \in J_n\big\} \\
&\qquad \times \sum_{r=0}^{\infty} (r+1)^{d-1} \alpha(r\beta_n; 2\beta_n^d)^{1/3} \\
&\leq C(d) N^{-2}|\mathcal{K}_{1n}|^2 |\mathcal{I}_n|^{-2} |J_n| (2\beta_n^d)^2 \max\big\{E|S_n(i,0)|^6 : i \in \mathcal{I}_n\big\}^{2/3} \\
&\qquad \times \sum_{r=0}^{\infty} (r+1)^{d-1} (r+1)^{-\tau_1/3} (\beta_n^{\tau_2 d - \tau_1})^{1/3} \\
&= O\Big(N^{-2} \cdot [\lambda_n^d/\beta_n^d]^2 \cdot (\lambda_n^d)^{-2} \cdot [\lambda_n^d/(2\beta_n)^d](2\beta_n^d)^2 \cdot \{\beta_n^{3d}\}^{2/3}\Big) \\
&= O(\lambda_n^{-d}\beta_n^d) .
\end{aligned}
$$

This proves (12.30). Hence, the proof of Theorem 12.1 is completed. □

An inspection of the proofs of Theorem 12.1 and Theorem 3.1 (on consistency of the MBB variance estimator for time series data) shows that the consistency of the spatial block bootstrap variance estimator may be established under reduced moment conditions by using suitable truncations of the variables $S_n(i;k)$'s in the proof of (12.30) and elsewhere. However, we avoid the truncation step here in order to keep the proof simple. It follows that the spatial block bootstrap variance estimator is consistent whenever the block-size parameter β_n satisfies $\beta_n^{-1} + \lambda_n^{-1}\beta_n = o(1)$ as $n \to \infty$. Going through the proof of Theorem 12.1, we also see that the leading term in the variance part of the bootstrap variance estimator, $\hat{\sigma}_n^2(\beta_n) \equiv \lambda_n^d \mathrm{Var}_*(\theta_n^*)$, is determined by $\mathrm{Var}(\lambda_n^d N^{-2} \sum_{k \in \mathcal{K}_{1n}} E_* S_n^*(k)^2)$, where $S_n^*(k)$ is the sum of the variables $\sum_{|\nu|=1} D^\nu H(\mu)(Z(s_i)-\mu)^\nu$ over s_i in the resampled block $\mathcal{B}_n(I_k;k)$. As in the time series case, this term increases as the block size parameter β_n increases. On the other hand, the leading term in the bias part of $\hat{\sigma}_n^2(\beta_n)$ is determined by the difference

$$\begin{aligned} & N^{-2}\lambda_n^d E\Big[\sum_{k\in\mathcal{K}_n} E_* S_n^*(k)^2\Big] - \sigma_\infty^2 \\ = \; & N^{-2}\lambda_n^d E\Big[\sum_{k\in\mathcal{K}_{1n}} E_* S_n^*(k)^2 + \sum_{k\in\mathcal{K}_{2n}} E_* S_n^*(k)^2\Big] - \sigma_\infty^2 \; . \end{aligned}$$

As (12.27) shows, the contribution from the *boundary* subregions to the bootstrap variance estimator, viz., $B_{2n} \equiv N^{-2}\lambda_n^d \sum_{k\in\mathcal{K}_{2n}} E\{E_* S_n^*(k)^2\}$ vanishes asymptotically, at the rate $O(\beta_n/\lambda_n)$ as $n\to\infty$. However, the exact rate at which B_{2n} goes to zero heavily depends on the geometry of the boundary of R_0 and is difficult to determine without additional restrictions on the prototype set R_0 when $d\geq 2$. To appreciate why, note that in the one-dimensional case, the number of boundary blocks is at most two (according to our formulation here) and hence, is bounded. However, in dimensions $d\geq 2$, it grows to infinity at a rate $O([\lambda_n/\beta_n]^{d-1})$. As a result, the contribution from the "incomplete" boundary blocks play a nontrivial role in higher dimensions. In contrast, the behavior of the first term arising from the interior blocks, viz., $B_{1n} \equiv N^{-2}\lambda_n^d E\{\sum_{k\in\mathcal{K}_{1n}} E_* S_n^*(k)^2\}$, can be determined for a general prototype set R_0, solely under the boundary condition, Condition B.

The discussion of the previous paragraph suggests that we may settle for an alternative bootstrap variance estimator of σ_∞^2 that is based on the "bootstrap observations" over the *interior* blocks $\{R_n(k) : k \in \mathcal{K}_{1n}\}$ only. Let $N_1 \equiv N_{1n} = |\mathcal{K}_{1n}|\beta_n^d$ denote the total number of data-values in the resampled "complete" blocks $\mathcal{B}_n(I_k;k)$, $k\in\mathcal{K}_{1n}$ and let $\bar{Z}_n^{**}$ be the average of these N_1 resampled values. Then, we define the bootstrap version of $\hat{\theta}_n$ based on the complete blocks as $\theta_n^{**} = H(\bar{Z}_n^{**})$ and the corresponding variance estimator of σ_∞^2 as

$$\hat{\sigma}_{1n}^2(\beta_n) \equiv \lambda_n^d \mathrm{Var}_*(\theta_n^{**}) \; . \tag{12.31}$$

In the context of applying the MBB to a time series data set of size n, this definition corresponds to the case where we resample $b = \lfloor n/\ell \rfloor$ "complete" blocks of length ℓ and define the bootstrap variance estimator in terms of a resample of size $n_1 = b\ell$ only, ignoring the last few boundary values (if any) in the bootstrap reconstruction of the chain. For the modified estimator $\hat{\sigma}_{1n}^2(\beta_n)$, we can refine the error bounds in the proof of Theorem 12.1 to obtain an expansion for its MSE. Indeed, applying the results of Nordman and Lahiri (2003a, 2003b) to the leading term in the variance of $\hat{\sigma}_{1n}^2(\beta_n)$, we get

$$\mathrm{Var}\Big(\hat{\sigma}_{1n}^2(\beta_n)\Big) \;=\; \mathrm{Var}\Big(N_1^{-2}\lambda_n^d |\mathcal{K}_{1n}| E_* S_n^*(k)^2\Big)(1+o(1))$$

$$
\begin{aligned}
&= \frac{\beta_n^d}{\lambda_n^d}\Big[\Big(\frac{2}{3}\Big)^d \cdot \frac{2\sigma_\infty^4}{(\text{vol.}(R_0))^3}\Big](1+o(1)) \\
&\equiv \frac{\beta_n^d}{\lambda_n^d} \cdot \gamma_1^2(1+o(1)), \text{ say} . \qquad (12.32)
\end{aligned}
$$

Next, using arguments as in (12.26), we see that the bias part of $\hat{\sigma}_{1n}^2(\beta_n)$ is given by

$$
\begin{aligned}
E\hat{\sigma}_{1n}^2(\beta_n) - \sigma_\infty^2 &= -\frac{1}{\beta_n \text{vol.}(R_0)} \sum_{i \in \mathbb{Z}^d} |i| \sigma_W(i) + o(\beta_n^{-1}) \\
&\equiv \beta_n^{-1}\gamma_2 + o(\beta_n^{-1}), \text{ say} , \qquad (12.33)
\end{aligned}
$$

where $\sigma_W(i) = \text{Cov}(W(0), W(i))$, $i \in \mathbb{Z}^d$ and $|i| = i_1 + \cdots + i_d$ for $i = (i_1, \ldots, i_d)' \in \mathbb{Z}^d$. Combining these, we have

$$
\text{MSE}\Big(\hat{\sigma}_{1n}^2(\beta_n)\Big) = \lambda_n^{-d}\beta_n^d\gamma_1^2 + \beta_n^{-2}\gamma_2^2 + o\Big(\lambda_n^{-d}\beta_n^d + \beta_n^{-2}\Big) . \qquad (12.34)
$$

Now, minimizing the leading terms in the expansion above, we get the first-order optimal block size for estimating σ_∞^2 (or σ_n^2) as

$$
\beta_n^0 = \Big[2\gamma_2^2/d\gamma_1^2\Big]^{\frac{1}{d+2}} \cdot \lambda_n^{\frac{d}{d+2}}(1+o(1)) . \qquad (12.35)
$$

Note that for $d = 1$ and $R_0 = (-\frac{1}{2}, \frac{1}{2}]$, the constants γ_1^2 and γ_2 in (12.32) and (12.33) are respectively given by $\gamma_1^2 = \frac{2}{3} \cdot [2\sigma_\infty^4]$ and $\gamma_2 = -2\sum_{i=1}^\infty i\sigma_W(i)$ and hence, the formula for the MSE-optimal block length coincides with that given in Chapter 5. In particular, the optimal block length β_n^0 for variance estimation grows at the rate $O(N^{1/3})$ for $d = 1$. For $d = 2$, the optimal rate of the *volume* of the blocks (viz., $(\beta_n^0)^d$) is $O(N^{1/2})$, while for $d = 3$ it is $O(N^{3/5})$, where N is the sample size. As $\frac{d}{d+2}$ is an increasing function of d, (12.35) shows that one must employ blocks of larger volumes in higher dimensions to achieve the best possible performance of the bootstrap variance estimators.

In the next two sections, we consider validity of approximations generated by the spatial bootstrap method for estimating the sampling distributions of some common estimators.

12.3.4 Results on the Empirical Distribution Function

We now discuss consistency properties of the spatial block bootstrap method for the empirical distribution function of the data. As in the case of time series data, many common estimators used in the analysis of spatial data may be expressed as smooth functionals of the empirical distribution of certain multidimensional spatial processes. As a result, here we suppose that the spatial process $Z(\cdot)$ is an m-dimensional ($m \in \mathbb{N}$) stationary process with components $Z_1(\cdot), \ldots, Z_m(\cdot)$. Thus, the observations are given by $Z(s_i) = \big(Z_1(s_i), \ldots, Z_m(s_i)\big)'$, $i = 1, \ldots, N$, where the data locations $\{s_1, \ldots, s_N\}$ lie on the integer grid $\mathbb{Z}^d$ inside the sampling region R_n (cf. (12.2)). Let $F_n^{(m)}(\cdot)$ denote the empirical distribution function of $Z(s_1), \ldots, Z(s_N)$, defined by

$$F_n^{(m)}(z) = N^{-1} \sum_{i=1}^{N} \mathbb{1}(Z(s_i) \le z), \; z \in \mathbb{R}^m , \tag{12.36}$$

where, recall that for two vectors $x = (x_1, \ldots, x_m)' \in \mathbb{R}^m$ and $y = (y_1, \ldots, y_m)' \in \mathbb{R}^m$, we write $x \le y$ if $x_i \le y_i$ for all $1 \le i \le m$. Let $G^{(m)}(z) = P(Z(0) \le z)$, $z \in \mathbb{R}^m$ denote the marginal distribution function of the process $Z(\cdot)$ under stationarity. Define the empirical process

$$\xi_n^{(m)}(z) = \lambda_n^{d/2}\Big(F_n^{(m)}(z) - G^{(m)}(z)\Big), \; z \in \mathbb{R}^m . \tag{12.37}$$

Because the sample size N grows at the rate $[\text{vol.}(R_0) \cdot \lambda_n^d]$ (cf. (12.3)), an alternative scaling sequence for the difference $F_n^{(m)}(\cdot) - G^{(m)}(\cdot)$ is given by the more familiar choice $\sqrt{N}$. However, in the context of spatial asymptotics, $\lambda_n^{d/2}$ happens to be the correct scaling sequence even in presence of partial infilling (cf. Zhu and Lahiri (2001)), while the scaling $N^{1/2}$ is inappropriate in presence of infilling. As a result, we shall use $\lambda_n^{d/2}$ as the scaling sequence here.

Next, we define the bootstrap version of $\xi_n^{(m)}$. Let $Z_n^*(R_n) \equiv \{Z^*(s_1), \ldots, Z^*(s_N)\}$ denote the block bootstrap version of the process $\{Z(s) : s \in R_n \cap \mathbb{Z}^d\}$, based on a block size parameter β_n. Let $F_n^{(m)*}(z) = N^{-1}\sum_{i=1}^{N} \mathbb{1}(Z^*(s_i) \le z)$, $z \in \mathbb{R}^m$, be the empirical distribution function of $\{Z^*(s_1), \ldots, Z^*(s_N)\}$. Then, the block bootstrap version of $\xi_n^{(m)}$ is given by

$$\xi_n^{(m)*}(z) = \lambda_n^{d/2}\Big(F_n^{(m)*}(z) - E_* F_n^{(m)*}(z)\Big), \; z \in \mathbb{R}^m . \tag{12.38}$$

To establish the weak convergence of the processes $\xi_n^{(m)}$ and $\xi_n^{(m)*}$, we consider the space $\mathbb{D}_m$ of real-valued functions on $[-\infty, \infty]^m$ that are continuous from above and have finite limits from below. We equip $\mathbb{D}_m$ with the extended Skorohod J_1-topology (cf. Bickel and Wichura (1971)). Then, both $\xi_n^{(m)}$ and $\xi_n^{(m)*}$ are $\mathbb{D}_m$-valued random variables. The following result asserts that under some regularity conditions, the sequence $\{\xi_n^{(m)}\}_{n \ge 1}$ converges in distribution to a nondegenerate Gaussian process as $\mathbb{D}_m$-valued random variables and that the bootstrapped empirical process $\xi_n^{(m)*}$ also has the same limit, almost surely. Let $\longrightarrow^d$ denote convergence in distribution of $\mathbb{D}_m$-valued random variables under the given extended Skorohod J_1-topology and let $\mathbb{C}_m^0$ denote the collection of all continuous functions from $[-\infty, \infty]^m$ to $\mathbb{R}$ that vanish at $(-\infty, \ldots, -\infty)'$ and $(\infty, \ldots, \infty)'$. Also, let $\alpha(a; b)$ denote the strong mixing coefficient of the vector random field $Z(\cdot)$, as defined in (12.15) and (12.16).

Theorem 12.2 *Suppose that* $\{Z(s) : s \in \mathbb{Z}^d\}$ *is a stationary vector-valued random field with components* $Z_1(s), \ldots, Z_m(s)$, $s \in \mathbb{R}^d$, *such that*

(i) $G_i(a) \equiv P(Z_i(0) \le a)$, $a \in \mathbb{R}$ *is continuous on* $\mathbb{R}$, $i = 1, \ldots, m$, *and*

(ii) $\alpha(a, b) \le C_1 \exp(-C_2 a) \cdot b^{\tau_2}$ *for all* $a \ge 1$, $b \ge 1$ *for some constants* $C_1, C_2 \in (0, \infty)$, *and* $0 \le \tau_2 \le 2$.

Also, suppose that Condition B holds and that

$$\beta_n^{-1} + \lambda_n^{-(1-\epsilon)} \beta_n = o(1) \tag{12.39}$$

for some $\epsilon \in (0, 1)$. *Let* $\mathcal{W}^{(m)}$ *be a zero mean Gaussian process on* $[-\infty, \infty]^m$ *with* $P(\mathcal{W}^{(m)} \in \mathbb{C}_m^0) = 1$ *and with covariance function*

$$\begin{aligned} &\mathrm{Cov}\Big(\mathcal{W}^{(m)}(z_1), \mathcal{W}^{(m)}(z_2)\Big) \qquad (12.40) \\ &= [\mathrm{vol.}(R_0)]^{-1} \sum_{i \in \mathbb{Z}^d} \Big\{P(Z(0) \le z_1, Z(i) \le z_2) - G^{(m)}(z_1) G^{(m)}(z_2)\Big\}, \end{aligned}$$

$z_1, z_2 \in \mathbb{R}^m$. *Then,*

$$\xi_n^{(m)} \longrightarrow^d \mathcal{W}^{(m)} \quad as \quad n \to \infty$$

and

$$\xi_n^{(m)*} \longrightarrow^d \mathcal{W}^{(m)} \quad as \quad n \to \infty, \quad a.s.$$

Proof: This is a special case of Theorem 3.3 of Zhu and Lahiri (2001), who establish the theorem under a polynomial strong-mixing condition. Here, we used the exponential mixing condition only to simplify the statement of Theorem 12.2. See Zhu and Lahiri (2001) for details. □

An immediate consequence of Theorem 12.2 is that for any Borel subset A of $\mathbb{D}_m$ with $P(\mathcal{W}^{(m)} \in \partial A) = 0$, where ∂A denotes the boundary of A, we have

$$\Big|P(\xi_n^{(m)} \in A) - P_*(\xi_n^{(m)*} \in A)\Big| \to 0 \quad \text{as} \quad n \to \infty, \quad \text{a.s.} \tag{12.41}$$

Thus, we may approximate the probability $P(\xi_n^{(m)} \in A)$ by its bootstrap estimator $P_*(\xi_n^{(m)*} \in A)$ for almost all realizations of the process $\{Z(i) : i \in \mathbb{Z}^d\}$, without having to explicitly estimate the covariance function of the limiting process $\mathcal{W}^{(m)}$, given by (12.40). In particular, if $\Upsilon : \mathbb{D}_m \to \mathbb{S}$ is a Borel-measurable function from $\mathbb{D}_m$ to some complete and separable metric space $\mathbb{S}$ that is continuous over $\mathbb{C}_m^0$, then by Theorem 12.2 and the continuous mapping theorem (cf. Billingsley (1968), Theorem 5.1),

$$\begin{aligned} &\Big|P_*(\Upsilon(\xi_n^{(m)*}) \in B) - P(\Upsilon(\xi_n^{(m)}) \in B)\Big| \\ &\to 0 \quad \text{as} \quad n \to \infty, \quad \text{a.s.} \qquad (12.42) \end{aligned}$$

for any Borel subset B of $\mathbb{S}$ with $P(\Upsilon(\mathcal{W}^{(m)}) \in \partial B) = 0$. Because the exact distribution of $\Upsilon(\xi_n^{(m)})$ or of $\Upsilon(\mathcal{W}^{(m)})$ may have a complicated form for certain functionals $\Upsilon(\cdot)$, (12.42) provides an effective way of approximating the large sample distribution of $\Upsilon(\xi_n^{(m)})$, without further analytical considerations. As an example, suppose that $m = 1$ and that we want to set a simultaneous confidence band for the unknown marginal distribution function $G(z) \equiv G^{(1)}(z) = P(Z(0) \leq z)$, $z \in \mathbb{R}$ of the process $Z(\cdot)$. Then, we take $\Upsilon(g) = \|g\|_\infty$, $g \in \mathbb{D}_1$, where for any function $g : [-\infty, \infty] \to \mathbb{R}$, we write $\|g\|_\infty = \sup\{|g(x)| : x \in [-\infty, \infty]\}$. It is easy to check that this $\Upsilon(\cdot)$ is continuous on $\mathbb{C}_1^0$ and, hence, (12.42) holds. For $0 < \alpha < 1$, let $\hat{q}_\alpha$ denote the α-th quantile of the bootstrap distribution function estimator $P_*(\|\xi_n^{*(1)}\|_\infty \leq \cdot)$, i.e.,

$$\hat{q}_\alpha = \inf\left\{a \in \mathbb{R} : P_*(\|\xi_n^{*(1)}\|_\infty \leq a) \geq \alpha\right\}.$$

Then, a $100(1-\alpha)\%$ large sample confidence region for $G(\cdot)$ is given by

$$\hat{I}_n(\alpha) = \Big\{F : F \text{ is a distribution function on } \mathbb{R} \text{ and } \|F_n^{(1)} - F\|_\infty \leq \lambda_n^{-d/2}\hat{q}_{1-\alpha}\Big\} \tag{12.43}$$

which, by (12.42), attains the desired confidence level $(1-\alpha)$ asymptotically. Note that in this case, the traditional large sample confidence region for $G(\cdot)$ uses the $(1-\alpha)$-th quantile of the distribution of $\|\mathcal{W}^{(1)}\|_\infty$, for which no closed-form expression seems to be available. In the special case where $d = 1$, G is the uniform distribution on $[0,1]$ and the $Z(i)$'s are independent, $\mathcal{W}^{(1)}$ reduces to $\widetilde{\mathcal{W}}$, the Brownian Bridge on $[0,1]$. Although an explicit form of the distribution of $\|\widetilde{\mathcal{W}}\|_\infty$ is known (cf. Chapter 11, Billingsley (1968)), it has a very complicated structure that makes computation of the quantiles an arduous task. In comparison, the block bootstrap confidence region $\hat{I}_n(\alpha)$ may be found for any $d \geq 1$ even in the presence of spatial dependence, without the analytical consideration required for deriving an explicit expression for the $(1-\alpha)$-th quantile of the distribution of $\|\mathcal{W}\|_\infty$ and without having to estimate the unknown population parameters that appear in this expression.

In the next section, we consider properties of the spatial bootstrap method for estimators that are smooth functionals of the m-dimensional empirical distribution $F_n^{(m)}$.

12.3.5 Differentiable Functionals

As in Section 4.4, validity of the spatial bootstrap method for the m-dimensional empirical process readily allows us to establish its validity for approximating the distributions of estimators that may be represented as smooth functionals of the empirical distribution. Here we will consider a form of differentiability condition, known as Hadamard differentiability, which is a weaker condition than the Fréchet differentiability condition of Section 4.4. Specialized to the $d = 1$, i.e., the time series case, these results also imply the validity of the MBB for Hadamard differentiable functionals. Here we follow van der Vaart and Wellner (1996) to define Hadamard differentiability.

Definition 12.1 *Let $\mathbb{D}^0$ be a subset of and $\mathbb{D}^+$ be a subspace of $\mathbb{D}_m$. Then, a mapping $\Upsilon : \mathbb{D}^0 \to \mathbb{R}^p$, $p \in \mathbb{N}$ is called Hadamard differentiable tangentially to $\mathbb{D}^+$ at $g_0 \in \mathbb{D}^0$ if there exists a continuous linear mapping $\Upsilon^{(1)}(g_0; \cdot) : \mathbb{D}^+ \to \mathbb{R}^p$ such that*

$$\frac{\Upsilon(g_0 + a_n f_n) - \Upsilon(g_0)}{a_n} \longrightarrow \Upsilon^{(1)}(g_0; f) \tag{12.44}$$

for all converging sequences $a_n \to 0$ and $f_n \to f$ with $f \in \mathbb{D}^+$ and $g_0 + a_n f_n \in \mathbb{D}^0$ for all $n \geq 1$. When $\mathbb{D}^+ = \mathbb{D}_m$, Υ is simply called Hadamard differentiable at g_0. The linear function $\Upsilon^{(1)}(g_0; \cdot)$ is called the Hadamard derivative of Υ at g_0.

When $\mathbb{D}^+ = \mathbb{D}_m$, i.e., when the derivative $\Upsilon^{(1)}(g_0; \cdot)$ is defined on all of $\mathbb{D}_m$, (12.44) is equivalent to requiring that for any $a_n \to 0$ and any *compact* set $\mathbb{K}^0$ of $\mathbb{D}_m$,

$$\begin{aligned} \sup_{f \in \mathbb{K}^0, g_0 + a_n f \in \mathbb{D}^0} & \left\| \Upsilon(g_0 + a_n f) - \Upsilon(g_0) - a_n \Upsilon^{(1)}(g_0; f) \right\| \\ & = o(a_n) \quad \text{as} \quad n \to \infty , \end{aligned} \tag{12.45}$$

where $\|\cdot\|$ denotes the usual Euclidean norm on $\mathbb{R}^p$. In comparison, Fréchet differentiability of Υ at g_0 requires (12.45) to be valid for all *bounded sets* $\mathbb{K}^0 \subset \mathbb{D}_m$. As a result, Fréchet differentiability of a functional is a stronger condition than Hadamard differentiability.

Hadamard differentiability of M-estimators and other important statistical functionals have been investigated by many authors; see Reeds (1976), Fernholz (1983), Ren and Sen (1991, 1995), van der Vaart and Wellner (1996), and the references therein. The following result proves the validity of the spatial bootstrap for Hadamard differentiable functionals. Here, we shall always assume that the domain $\mathbb{D}^0$ of definition of the functional Υ is large enough such that $G^{(m)}$, $F_n^{(m)}(\cdot)$, $F_n^{(m)*}(\cdot)$, $E_* F_n^{(m)*}(\cdot) \in \mathbb{D}^0$ (with probability one). This ensures that the estimators $\Upsilon(F_n^{(m)})$, $\Upsilon(E_* F_n^{(m)*})$ of the parameter $\Upsilon(G^{(m)})$ and the bootstrap version $\Upsilon(F_n^{(m)*})$ of $\Upsilon(F_n^{(m)})$ are well defined.

Theorem 12.3 *Suppose that the conditions of Theorem 12.2 hold. Let $\Upsilon : \mathbb{D}^0 \to \mathbb{R}^p$ be Hadamard differentiable at $G^{(m)}$ tangentially to $\mathbb{C}_m^0$ with derivative $\Upsilon^{(1)}(G^{(m)}; \cdot)$ for some $\mathbb{D}^0 \subset \mathbb{D}_m$.*

(a) Then,

$$\lambda_n^{d/2}\Big(\Upsilon(F_n^{(m)}) - \Upsilon(G^{(m)})\Big) \longrightarrow^d \Upsilon^{(1)}(G^{(m)}; \mathcal{W}^{(m)}) \quad \textit{as} \quad n \to \infty \,. \tag{12.46}$$

(b) Suppose that Υ *and* $\Upsilon^{(1)}(G^{(m)}; \cdot)$ *satisfy the following stronger version of (12.44): For any* $a_n \to 0$, $f_n \to f \in \mathbb{D}^+$ *and* $g_n \to G^{(m)}$ *with* g_n, $g_n + a_n f_n \in \mathbb{D}^0$ *for all* $n \geq 1$,

$$\lim_{n\to\infty} \frac{\Upsilon(g_n + a_n f_n) - \Upsilon(g_n)}{a_n} = \Upsilon^{(1)}(G^{(m)}; f)\,. \tag{12.47}$$

Then, with probability 1,

$$\lambda_n^{d/2}\Big(\Upsilon(F_n^{(m)*}) - \Upsilon(E_* F_n^{(m)*})\Big) \longrightarrow^d \Upsilon^{(1)}(G^{(m)}; \mathcal{W}^{(m)}) \ \textit{as}\ n \to \infty \,. \tag{12.48}$$

Proof: Part (a) follows from Theorem 3.9.4 of van der Vaart and Wellner (1996). Next consider part (b). Using Lemma 12.1, the Borel-Cantelli Lemma, and the arguments in the proof of the Glivenko-Cantelli Theorem, it can be shown that

$$\sup_{x\in[-\infty,\infty]^d} \Big|\hat{G}_n^{(m)}(x) - G^{(m)}(x)\Big| = o(1) \quad \text{a.s.} \quad (P) \tag{12.49}$$

for $\{\hat{G}_n^{(m)}\} = \{F_n^{(m)}\}$, $\{\tilde{F}_n^{(m)}\}$, where $\tilde{F}_n^{(m)} \equiv E_* F_n^{(m)*}$. Let

$$\begin{aligned} H_n(f) \;&\equiv\; \lambda_n^{d/2}\Big(\Upsilon(\tilde{F}_n^{(m)} + \lambda_n^{-d/2} f) - \Upsilon(\tilde{F}_n^{(m)})\Big) \\ &\qquad \times \mathbb{1}(\tilde{F}_n^{(m)} + \lambda_n^{-d/2} f \in \mathbb{D}^0)\,, \end{aligned}$$

$f \in \mathbb{D}_m$. Then, by (12.47) and (12.49), there exists a set A with $P(A) = 1$ such that on A,

$$H_n(f_n) \to \Upsilon^{(1)}(G^{(m)}; f).$$

for any $f_n \to f \in \mathbb{C}_m^0$ with $\lambda_n^{-d/2} f_n + \tilde{F}_n^{(m)} \in \mathbb{D}^0$ for all $n \geq 1$. Hence, by Theorem 12.1 and the extended continuous mapping Theorem (cf. Theorem 5.5, Billingsley (1968); Theorem 1.11.1, van der Vaart and Wellner (1996)), applied pointwise on a set of probability 1, we have

$$\begin{aligned} &\lambda_n^{d/2}\Big(\Upsilon(F_n^{(m)*}) - \Upsilon(\tilde{F}_n^{(m)})\Big) \\ &= \;\; H_n(\xi_n^{(m)*}) \longrightarrow^d \Upsilon^{(1)}(G^{(m)}; \mathcal{W}^{(m)}) \quad \text{as} \quad n \to \infty, \quad \text{a.s.} \quad (P)\,. \end{aligned}$$

This proves part (b). □

12.4 Estimation of Spatial Covariance Parameters

12.4.1 The Variogram

In this section, we describe a method for fitting variogram models to spatial data using spatial resampling methods. Suppose that $\{Z(i) : i \in \mathbb{Z}^d\}$, $d \in \mathbb{N}$ is an intrinsically stationary random field, i.e., $\{Z(i) : i \in \mathbb{Z}^d\}$ is a collection of random variables defined on a common probability space such that

$$E(Z(i) - Z(i+h)) = 0 \tag{12.50}$$

and

$$\mathrm{Var}(Z(i) - Z(i+h)) = \mathrm{Var}(Z(0) - Z(h)) \tag{12.51}$$

for all $i, h \in \mathbb{Z}^d$. The function $2\gamma(h) \equiv \mathrm{Var}(Z(0) - Z(h))$ is called the *variogram* of the process $Z(\cdot)$. Note that if the process $Z(\cdot)$ is second-order stationary with auto covariance function $\sigma(h) = \mathrm{Cov}(Z(0), Z(h))$, $h \in \mathbb{Z}^d$, then (12.50) holds and, for any $i, h \in \mathbb{Z}^d$,

$$\begin{aligned}\mathrm{Var}(Z(i) - Z(i+h)) &= \mathrm{Var}(Z(i)) + \mathrm{Var}(Z(i+h)) \\ &\qquad - 2\mathrm{Cov}(Z(i), Z(i+h)) \\ &= 2\sigma(0) - 2\sigma(h) ,\end{aligned}$$

which implies (12.51) with

$$\gamma(h) = \sigma(0) - \sigma(h), \;\; h \in \mathbb{Z}^d . \tag{12.52}$$

Thus, second-order stationarity implies intrinsic stationarity. Note that if the process $Z(\cdot)$ is regular in the sense that $\sigma(h) \to 0$ as $\|h\| \to \infty$, then, from (12.52),

$$\sigma(0) = \lim_{\|h\| \to \infty} \gamma(h) . \tag{12.53}$$

Hence, by (12.52) and (12.53), the function $\sigma(\cdot)$ can be recovered from the knowledge of the variogram $2\gamma(\cdot)$. Thus, under some mild conditions, the variogram $2\gamma(\cdot)$ provides an equivalent description of the covariance structure of the process $Z(\cdot)$ as does the autocovariance function $\sigma(\cdot)$. In spatial statistics, it is customary to describe the spatial-dependence structure of a spatial process by its variogram rather than the autocovariance function (also called the covariogram) $\sigma(\cdot)$. Like the nonnegative definiteness property of the auto-covariance function $\sigma(\cdot)$, the variogram must satisfy the following *conditional negative definiteness* property (cf. Chapter 2, Cressie (1993)): For any spatial locations $s_1, \ldots, s_m \in \mathbb{Z}^d$, $m \in \mathbb{N}$, and any real numbers $a_1, \ldots, a_m$ with $\sum_{i=1}^m a_i = 0$,

$$\sum_{i=1}^m \sum_{j=1}^m a_i a_j \gamma(s_i - s_j) \le 0 . \tag{12.54}$$

Thus, for any estimator of a spatial variogram to be valid, it must satisfy this conditional nonnegative definiteness property.

In the next section, we describe a general estimation method that produces conditionally nonnegative definite variogram estimators.

12.4.2 Least Squares Variogram Estimation

A popular approach for estimating the variogram is the *method of least squares variogram model fitting.* Initially proposed in Geostatistical literature (cf. David (1977), Journel and Huijbregts (1978)) and then further modified and studied by Cressie (1985), Zhang, van Eijkeren and Heemink (1995), Genton (1997), Barry, Crowder and Diggle (1997), Lee and Lahiri (2002), and others, this method fits a parametric variogram model by minimizing a certain quadratic distance function between a generic nonparametric variogram estimator and the parametric model using various least squares methods. Specifically, suppose that the true variogram of the spatial process $Z(\cdot)$ lies in a parametric family $\{2\gamma(\cdot;\theta) : \theta \in \Theta\}$ of valid variograms, where Θ is a subset of $\mathbb{R}^p$. Our objective here is to estimate the variogram parameter vector θ on the basis of the sample $\{Z(s) : 1 \leq i \leq N\}$ where the sampling sites $s_1, \ldots, s_N$ lie on the part of the integer grid $\mathbb{Z}^d$ inside the sampling region R_n, as specified by (12.2). Let $2\check{\gamma}_n(h)$ be a nonparametric estimator of the variogram $2\gamma(h)$ at lag h. Also, let $h_1, \ldots, h_K \in \mathbb{R}^d$, $2 \leq K < \infty$ be a given set of lag vectors and let $V(\theta)$ be a $K \times K$ positive-definite weight matrix, that possibly depends on the covariance parameter θ. Then, a least squares estimator (LSE) of θ corresponding to the weight matrix $V(\theta)$ is defined as

$$\hat{\theta}_{n,V} = \text{argmin}\{Q_n(\theta; V) : \theta \in \Theta\} , \qquad (12.55)$$

where $Q_n(\theta; V) = g_n(\theta)' V(\theta) g_n(\theta)$ and $g_n(\theta)$ is the $K \times 1$ vector with ith element $(2\check{\gamma}_n(h_i) - 2\gamma(h_i;\theta))$, $i = 1, \ldots, K$. For $V(\theta) = \mathbb{I}_K$, the identity matrix of order K, $\hat{\theta}_{n,V}$ is the ordinary least squares (OLS) estimator of θ. Choosing $V(\theta) = \Sigma(\theta)^{-1}$, the inverse of the asymptotic covariance matrix of $g_n(\theta)$, we get the generalized least squares (GLS) estimator of θ. In the same vein, choosing $V(\theta)$ to be a diagonal matrix with suitable diagonal entries, we can get the various weighted least squares (WLS) estimators proposed by Cressie (1985), Zhang, Eijkeren and Heemink (1995), and Genton (1997). In addition to guaranteeing the conditional nonnegative definiteness property (12.54) of the resulting variogram estimator, this method has a visual appeal similar to that of fitting a regression function to a scatter plot. This makes the least squares methods of variogram model fitting popular among practitioners.

Statistical properties of the LSEs heavily depend on the choice of the weighting matrix $V(\theta)$, employed in the definition of the LSE $\hat{\theta}_{n,V}$. Let $\Gamma(\theta)$ denote the $K \times q$ matrix with (i,j)-th element $-\frac{\partial}{\partial\theta_j}[2\gamma(h_i;\theta)]$ and let $A(\theta) = \Gamma(\theta)[\Gamma(\theta)'V(\theta)\Gamma(\theta)]^{-1}$. Theorem 3.2 of Lahiri, Lee and Cressie (2002) shows that under some regularity conditions, if

$$a_n g_n(\theta) \longrightarrow^d N(0, \Sigma(\theta)) \quad \text{for all} \quad \theta \in \Theta , \tag{12.56}$$

for some sequence $\{a_n\}_{n\geq 1}$ of positive real numbers and for some positive definite matrix $\Sigma(\theta)$, then for all $\theta \in \Theta$,

(i)
$$a_n(\hat{\theta}_{n,V} - \theta) \longrightarrow^d N(0, D_V(\theta)) \quad \text{as} \quad n \to \infty \tag{12.57}$$
where $D_V(\theta) = A(\theta)'V(\theta)\Sigma(\theta)V(\theta)A(\theta)$;

(ii)
$$D_V(\theta) - (\Gamma(\theta)'\Sigma(\theta)^{-1}\Gamma(\theta))^{-1} \tag{12.58}$$
is nonnegative definite for any $V(\theta)$;

(iii) for the GLS method with $V(\theta) = \Sigma(\theta)^{-1}$,
$$D_{\Sigma^{-1}}(\theta) = (\Gamma(\theta)'\Sigma(\theta)^{-1}\Gamma(\theta))^{-1} . \tag{12.59}$$

Hence, it follows from (12.57)–(12.59) that the LSE $\hat{\theta}_{n,V}$ of θ is asymptotically multivariate normal and hence, one may compare different LSEs in terms of their limiting covariance matrices. This leads to the following definition of asymptotically efficient LSEs of θ.

Definition 12.2 *A sequence $\{\hat{\theta}_{n,V_0}\}$ of LSEs of θ corresponding to a weighting matrix $V_0(\theta)$ is said to be asymptotically efficient if for any other weighting matrix $V(\theta)$, the difference $D_V(\theta) - D_{V_0}(\theta)$ is nonnegative definite for all $\theta \in \Theta$.*

This definition of asymptotic efficiency is equivalent to the requirement that for every $x \in \mathbb{R}^p$, the estimator $x'\hat{\theta}_{n,V_0}$ of the linear parametric function $x'\theta$ has the minimum asymptotic variance among the class of all LSEs, for all $\theta \in \Theta$. From (12.58) and (12.59), it follows that the optimal covariance matrix of the limiting normal distribution is given by $D_{\Sigma^{-1}}(\theta)$ and that the GLS estimator of θ is asymptotically efficient among all LSEs.

Although it is an optimal estimator from the statistical point of view, computation of the GLS estimator can be difficult in practice. To appreciate why, note that the GLS estimator $\hat{\theta}_{n,GLS} \equiv \hat{\theta}_{n,\Sigma^{-1}}$ is defined as (cf. (12.55)) $\hat{\theta}_{n,GLS} = \text{argmin}\{Q_n(\theta; \Sigma^{-1}) : \theta \in \Theta\}$, which involves minimization of the *nonlinear* criterion function $Q_n(\theta; \Sigma^{-1})$ over the parameter space $\Theta \subset \mathbb{R}^q$. Because of the computational complexity associated with a *general* optimization method for minimizing such nonlinear functions, whether iterative or grid based (cf. Dennis and Schnabel (1983)), the GLS method is computationally demanding. A second undesirable feature of the GLS method is that it requires the knowledge of the asymptotic covariance matrix of the generic variogram estimator, which must be found analytically and, therefore, may be intractable for certain non-Gaussian processes. In practice, these factors often prompt one to use other statistically inefficient LSEs, such as the OLS and WLS estimators.

Following the work of Lee and Lahiri (2002), we now describe a least squares method based on spatial resampling methods that is also asymptotically efficient within the class of all least squares methods. Furthermore, it is computationally much simpler than the standard GLS method and does not require any additional analytical consideration. The main idea behind the new method is to replace the asymptotic covariance matrix of the generic variogram estimator in the GLS criterion function by a consistent, nonparametric estimator of the covariance matrix based on spatial resampling, which can be evaluated without knowing the exact form of the covariance matrix.

12.4.3 The RGLS Method

We now describe the resampling-based least squares method (or the RGLS method, in short). Let $\hat{\Sigma}_n$ be an estimator of the asymptotic covariance matrix $\Sigma(\theta)$ (cf. (12.56)) of the normalized generic variogram estimator based on a suitable resampling method. Then, we replace the matrix $[\Sigma(\theta)]^{-1}$ in the GLS criterion function by $\hat{\Sigma}_n^{-1}$ and define the *resampling method based GLS* (RGLS) estimator of θ as

$$\hat{\theta}_{n,RGLS} = \operatorname{argmin}\Big\{ g_n(\theta)'\hat{\Sigma}_n^{-1} g_n(\theta) : \theta \in \Theta \Big\} . \tag{12.60}$$

Since $\hat{\Sigma}_n$ itself does not involve the parameter θ, computation of the RGLS estimator requires inversion of the estimated covariance matrix $\hat{\Sigma}_n$ *only once*. In contrast, for the GLS estimator, the inverse of the matrix $\Sigma(\theta)$ needs to be computed a large number of times for finding the "minimizer" of the GLS criterion function. As a result, the RGLS estimator is computationally much simpler than the GLS estimator. And as we shall show below, the RGLS estimator is also asymptotically efficient, making it "as good as" the GLS estimator from a statistical point of view.

Lee and Lahiri (2002) suggest using a spatial subsampling method to derive the estimator $\hat{\Sigma}_n$ of the asymptotic covariance matrix Σ and call it the "subsampling based GLS method" or the "SGLS" method. A second possibility is to employ the spatial block bootstrap method of the previous section to form the nonparametric estimator $\hat{\Sigma}_n$, leading to what one may refer to as the BGLS method. However, an advantage of the subsampling method over spatial bootstrap methods is that the computation of the estimator $\hat{\Sigma}_n$ does *not* require any resampling and may be found using an explicit formula given below (cf. (12.62),(12.64)).

We now briefly describe the spatial subsampling method associated with the SGLS methods of Lee and Lahiri (2002). Let $R_n = \lambda_n R_0$ be the sampling region (cf. Section 12.2) and let $\{Z(s_1), \ldots, Z(s_N)\} = \{Z(i) : i \in \mathbb{Z}^d \cap R_n\}$ be the observations. As in the spatial bootstrap method, let $\beta_n \in \mathbb{N}$ be an integer satisfying (12.4), i.e.,

$$\beta_n^{-1} + \lambda_n^{-1}\beta_n = o(1) \quad \text{as} \quad n \to \infty .$$

The subregions for the spatial subsampling method are obtained by considering suitable translates of the set $\beta_n R_0$. More specifically, we consider d-dimensional cubes of the form $i + \mathcal{U}_0\beta_n$, $i \in \mathbb{Z}^d$ that are contained in the sampling region R_n, where $\mathcal{U}_0 = (-\frac{1}{2}, \frac{1}{2}]^d$ is the unit cube in $\mathbb{R}^d$, centered at the origin. Let $\mathcal{I}_n^0 = \{i \in \mathbb{Z}^d : i + \mathcal{U}_0\beta_n \subset R_n\}$ denote the index set of such cubes. Then, we define the subregions $\{R_n^{(i)} : i \in \mathcal{I}_n^0\}$ by inscribing for each $i \in \mathcal{I}_n^0$, a translate of $\beta_n R_0$ inside the cube $i + \mathcal{U}_0\beta_n$ such that the origin is mapped onto i, i.e., we define

$$R_n^{(i)} = i + \beta_n R_0, \; i \in \mathcal{I}_n^0 . \tag{12.61}$$

Then, $\{R_n^{(i)} : i \in \mathcal{I}_n^0\}$ is a collection of overlapping subregions of R_n that are of the *same shape* as the original sampling region R_n, but are of smaller volume. Moreover, the number (say, ℓ) of observations in each subregion is the same and it grows at the rate $[\text{vol.}(R_0)\beta_n^d]$, as in the block bootstrap case.

The observations from the subregions can be used to define the subsampling estimator of the covariance matrix (and, more generally, the probability distribution) of a given K-dimensional ($K \in \mathbb{N}$) random vector of the form $T_n = t_n(\mathcal{Z}_n; \eta)$, where $\mathcal{Z}_n = \{Z(s_1), \ldots, Z(s_N)\}$ and η is a population parameter. For this, on each subregion $R^{(i)}$, we define a version $T^{(i)}$ of T_n by replacing the observed values $\mathcal{Z}_n$ with the subsample $\mathcal{Z}^{(i)} \equiv \{Z(s) : s \in R^{(i)} \cap \mathbb{Z}^d\}$ from the subregion $R^{(i)}$, and by replacing the parameter η by an estimator $\hat{\eta}_n$ based on $\mathcal{Z}_n$. Thus, we define

$$T^{(i)} \equiv t_\ell(\mathcal{Z}^{(i)}; \hat{\eta}_n), \; i \in \mathcal{I}_n^0 .$$

Note that $T^{(i)}$ is defined using the function $t_\ell(\cdot;\cdot)$, not $t_n(\cdot;\cdot)$, since the subsample $\mathcal{Z}^{(i)}$ has only ℓ observations. For example, if $t_n(\mathcal{Z}_n; \eta) = \sqrt{n}(\bar{Z}_n - \mu)$ with $\bar{Z}_n = n^{-1}\sum_{i=1}^n Z(s_i)$ and $\mu = EZ(0)$, then $T^{(i)} = \sqrt{\ell}(\bar{Z}^{(i)} - \hat{\mu}_n)$, where $\bar{Z}^{(i)}$ is the average of the ℓ observations in the subsample $\mathcal{Z}^{(i)}$ and $\hat{\mu}_n = E_*\bar{Z}_n^*$ is an estimator of μ based on $\mathcal{Z}_n$.

The subsampling estimator of the sampling distribution of T_n is now defined as the empirical distribution function of the subsample copies $\{T^{(i)} : i \in \mathcal{I}_n^0\}$. By the "plug-in" principle, the subsampling estimator of the (asymptotic) covariance matrix of T_n is given by

$$\hat{\Sigma}_n \equiv |\mathcal{I}_n^0|^{-1} \sum_{i \in \mathcal{I}_n^0} T^{(i)} T^{(i)\prime} . \tag{12.62}$$

Next we apply the subsampling method to obtain an estimator of the covariance matrix of the (scaled) variogram estimator at lags $h_1, \ldots, h_K$. Thus, the random vector T_n here is now given by

$$T_n \equiv \sqrt{n}\Big(2\check{\gamma}_n(h_1) - 2\gamma(h_1;\theta), \ldots, 2\check{\gamma}_n(h_K) - 2\gamma(h_K;\theta)\Big)' . \quad (12.63)$$

Let $2\gamma^{(i)}(h)$ denote the lag-h variogram estimator obtained by replacing $\mathcal{Z}_n$ and n in the definition of $2\check{\gamma}_n(h)$ by the subsample $\mathcal{Z}^{(i)}$ and the subsample size ℓ, respectively. Also, let $2\bar{\gamma}_n(h) \equiv |\mathcal{I}_n^0|^{-1} \sum_{i \in \mathcal{I}_n^0} 2\gamma^{(i)}(h)$. Then, the subsample version of T_n is given by

$$T^{(i)} \equiv \sqrt{\ell}\Big(2\gamma^{(i)}(h_1) - 2\bar{\gamma}_n(h_1); \ldots ; 2\gamma^{(i)}(h_K) - 2\bar{\gamma}_n(h_K)\Big)', \; i \in \mathcal{I}_n^0 . \quad (12.64)$$

The SGLS estimator of θ is now given by (12.60) with $\hat{\Sigma}_n$ defined by relations (12.62) and (12.64).

12.4.4 Properties of the RGLS Estimators

Next we prove consistency and asymptotic efficiency of the RGLS estimator $\hat{\theta}_{n,RGLS}$, of (12.60), based on *a general* resampling method.

Theorem 12.4 *Suppose that the following conditions hold:*
(C.1) $\sqrt{n} g_n(\theta_0) \longrightarrow^d N(0, \Sigma(\theta_0))$ *under* θ_0, *and* $\Sigma(\theta_0)$ *is positive definite.*
(C.2) (i) For any $\epsilon > 0$, *there exists a* $\delta > 0$ *such that*
$\inf\Big\{\sum_{i=1}^K (2\gamma(h_i;\theta_1) - 2\gamma(h_i;\theta_2))^2 : \|\theta_1 - \theta_2\| \geq \epsilon,\ \theta_1, \theta_2 \in \Theta\Big\} > \delta$.
(ii) $\sup\{\gamma(h_i;\theta) : \theta \in \Theta\} < \infty$ *for* $i = 1, \ldots, K$.
(iii) $\gamma(h_i;\theta)$ *has continuous partial derivatives respect to* θ *for* $i = 1, \ldots, K$.
(C.3) $\hat{\Sigma}_n \longrightarrow_p \Sigma(\theta_0)$ *as* $n \to \infty$.
Then,

(a) $\hat{\theta}_{n,RGLS} \longrightarrow_p \theta_0$ *as* $n \to \infty$,

(b) $\sqrt{n}(\hat{\theta}_{n,RGLS} - \theta_0) \longrightarrow^d N(0, D_{\Sigma^{-1}}(\theta_0))$ *as* $n \to \infty$.

Proof: Let $g(\theta) = \big(2\gamma(h_1;\theta_0) - 2\gamma(h_1;\theta), \ldots, 2\gamma(h_K;\theta_0) - 2\gamma(h_K;\theta)\big)'$, $Q(\theta) = g(\theta)'[\Sigma(\theta_0)]^{-1} g(\theta)$, $\hat{Q}_n(\theta) = g_n(\theta)'[\hat{\Sigma}_n]^{-1} g_n(\theta)$, and $\check{Q}_n(\theta) = g_n(\theta)'[\Sigma(\theta_0)]^{-1} g_n(\theta)$. Then,

$$\begin{aligned} \|\hat{Q}_n(\theta) - Q(\theta)\| &\leq \|\hat{Q}_n(\theta) - \check{Q}_n(\theta)\| + \|Q(\theta) - \check{Q}_n(\theta)\| \\ &\leq \|g_n(\theta)\|^2 \|[\hat{\Sigma}_n]^{-1} - \Sigma(\theta_0)^{-1}\| \\ &+ \|g_n(\theta_0)\| \; \|\Sigma(\theta_0)^{-1}\| \Big\{\|g_n(\theta)\| + \|g(\theta)\|\Big\} . \end{aligned} \quad (12.65)$$

Note that by Condition (C.1), $g_n(\theta_0) = o_p(1)$. Hence, by (12.65), Conditions (C.2) and (C.3),

$$\hat{\Delta}_n \equiv \sup\{|\hat{Q}_n(\theta) - Q(\theta)| : \theta \in \Theta\} \to 0 \quad \text{in probability as} \quad n \to \infty . \quad (12.66)$$

Now, if possible, suppose that $\hat{\theta}_n \nrightarrow \theta_0$ in probability as $n \to \infty$. Then, (by Proposition A.1, Appendix A), there exist an $\epsilon > 0$ and a subsequence $\{m_n\}_{n\geq 1}$ such that $\|\hat{\theta}_{m_n} - \theta_0\| \geq \epsilon$ for all $n \geq 1$. Now, by (12.66), there is a further subsequence $\{m_{1n}\}_{n\geq 1}$ of $\{m_n\}_{n\geq 1}$ such that $\hat{\Delta}_{m_{1n}} = o(1)$ almost surely. Also note that under the hypotheses of Theorem 12.4, $Q(\theta)$ is strictly positive on $\Theta\backslash\{\theta_0\}$, and $Q(\theta_0) = 0$. Thus, $Q(\theta)$ has a unique minimum at θ_0. However, with probability 1, $\hat{Q}_{m_{1n}}(\hat{\theta}_{m_{1n}}) - \hat{Q}_{m_{1n}}(\theta_0) \geq Q(\hat{\theta}_{m_{1n}}) - Q(\theta_0) - 2\hat{\Delta}_{m_{1n}} \geq \inf\{Q(\theta) : \|\theta - \theta_0\| > \epsilon\} - 2\hat{\Delta}_{m_{1n}} > 0$ for all $n \geq n_0$, for some $n_0 \geq 1$. This contradicts the definition of $\hat{\theta}_{m_{1n}}$ as the minimizer of $\hat{Q}_{m_{1n}}(\theta)$ for all $n \geq n_0$, proving part (a) of Theorem 12.4.

To prove the second part, let $\hat{w}_{qr}$ denote the (q, r) component of $[\hat{\Sigma}_n]^{-1}$, $1 \leq q, r \leq K$. Also, let $g_{nq}(\theta)$ denote the qth component of $g_n(\theta)$ and let $\gamma_q(\cdot;\theta) = \partial\gamma(\cdot;\theta)/\partial\theta_q$, $1 \leq q \leq p$. Since $\hat{\theta}_n$ minimizes the function $g_n(\theta)'[\hat{\Sigma}_n]^{-1}g_n(\theta)$, it satisfies the equations

$$\begin{aligned} 0 &= \frac{\partial}{\partial\theta_m}\Big(g_n(\theta)'[\hat{\Sigma}_n]^{-1}g_n(\theta)\Big)\Big|_{\theta=\hat{\theta}_n} \\ &= \sum_{q=1}^{K}\sum_{r=1}^{K}\hat{w}_{qr}g_{nr}(\hat{\theta}_n)\big(-2\gamma_m(h_q;\hat{\theta}_n)\big) \\ &\quad + \sum_{q=1}^{K}\sum_{r=1}^{K}\hat{w}_{qr}g_{nq}(\hat{\theta}_n)\big(-2\gamma_m(h_r;\hat{\theta}_n)\big) , \end{aligned}$$

$1 \leq m \leq p$. Next, let $\{e_1 \equiv (1, 0, \ldots, 0)', \ldots, e_p \equiv (0, \ldots, 0, 1)'\}$ denote the standard basis of $\mathbb{R}^p$. Hence, by a one-term Taylor series expansion of $g_{nq}(\hat{\theta}_n)$ and $g_{nr}(\hat{\theta}_n)$ around θ_0, we obtain,

$$\begin{aligned} &\sum_{q=1}^{K}\sum_{r=1}^{K}\hat{w}_{qr}\bigg\{\sum_{a=1}^{p}\bigg[\int_0^1 -2\gamma_a\big(h_r; u\theta_0 + (1-u)\hat{\theta}_n\big)du\bigg] \\ &\qquad\qquad \times(\hat{\theta}_n - \theta_0)'e_a\bigg\}\big(-2\gamma_m(h_q;\hat{\theta}_n)\big) \\ &\quad + \sum_{q=1}^{K}\sum_{r=1}^{K}\hat{w}_{qr}\bigg\{\sum_{a=1}^{p}\bigg[\int_0^1 -2\gamma_a\big(h_q; u\theta_0 + (1-u)\hat{\theta}_n\big)du\bigg] \\ &\qquad\qquad \times(\hat{\theta}_n - \theta_0)'e_a\bigg\}\big(-2\gamma_m(h_r;\hat{\theta}_n)\big) \\ &= -\sum_{q=1}^{K}\sum_{r=1}^{K}\hat{w}_{qr}g_{nr}(\theta_0)\big(-2\gamma_m(h_q;\hat{\theta}_n)\big) \\ &\quad - \sum_{q=1}^{K}\sum_{r=1}^{K}\hat{w}_{qr}g_{nq}(\theta_0)\big(-2\gamma_m(h_r;\hat{\theta}_n)\big) , \end{aligned} \tag{12.67}$$

$1 \leq m \leq p$. Then, it is easy to see that the set of p equations in (12.67) can be rewritten as

$$\left(\Gamma(\hat{\theta}_n)'[\hat{\Sigma}_n]^{-1}\Gamma_n^*\right)(\hat{\theta}_n - \theta_0) = -\Gamma(\hat{\theta}_n)'[\hat{\Sigma}_n]^{-1}g_n(\theta_0) , \tag{12.68}$$

where $\Gamma_n^* = \int_0^1 \Gamma(u\theta_0 + (1-u)\hat{\theta}_n)du$. Because $\hat{\theta}_n$ is a consistent estimator of θ and the matrix-valued function $\Gamma(\theta)$ is continuous in θ, the result follows from (12.68), Condition (C.2), and Slutsky's Theorem. □

Thus, if Conditions (C.1)-(C.3) in the statement of Theorem 12.4 hold for all $\theta_0 \in \Theta$, then the RGLS estimator $\hat{\theta}_{n,RGLS}$ has the same asymptotic covariance matrix as the GLS estimator, and hence, according to Definition 12.2, $\hat{\theta}_{n,RGLS}$ is asymptotically efficient.

Next we comment about the conditions required for the validity of Theorem 12.4. Condition (C.1) of Theorem 12.4 assumes asymptotic normality of the generic variogram estimator $2\check{\gamma}_n(\cdot)$, and can be verified for a given variogram estimator under suitable moment and mixing conditions on the process $Z(\cdot)$. Condition (C.2) (i) essentially requires that the choice of the lag vectors $h_1, \ldots, h_K$ should be such that the model variogram $2\gamma(\cdot;\theta)$ can be distinguished at distinct parameter values $\theta_1, \theta_2 \in \Theta$ by its values $(2\gamma(h_1;\theta_i), \ldots, 2\gamma(h_K;\theta_i))'$, $i = 1,2$ at $h_1, \ldots, h_K$. Condition (C.2) (ii) is stringent, as it requires the model variogram to be bounded over the parameter space at $h_1, \ldots, h_K$. If the variables $Z(s)$'s are normalized to unit variance, then $2\gamma(\cdot;\cdot)$ is bounded by 2 and this condition holds. For a spatial process $Z(\cdot)$ with an unbounded variance function (over Θ), one may apply the RGLS methodology to estimate the parameters of the scaled variogram or the *correlogram*, defined by

$$2\rho(h;\theta) \equiv 2\gamma(h;\theta)/\mathrm{Var}_\theta(Z(0)),\ h \in \mathbb{R}^d , \tag{12.69}$$

$\theta \in \Theta$, using the generic estimator $2\check{\rho}_n(h) \equiv 2\check{\gamma}_n(h)/s_n^2$, where $s_n^2 = N^{-1}\sum_{i=1}^N (Z(s_i) - \bar{Z}_n)^2$ is the sample variance of $\{Z(s_1), \ldots, Z(s_N)\}$. Then, Condition (C.2) (ii) holds for $\rho(h;\cdot)$, and the conclusions of Theorem 12.4 remain valid under conditions analogous to (C.1)-(C.3) where the function $\gamma(\cdot;\cdot)$ is replaced by $\rho(\cdot;\cdot)$. This modified approach yields estimators of those covariance parameters that determine the *shape* of the variogram.

Next consider the remaining conditions. Condition (C.2) (iii) is a smoothness condition that may be directly verified for a given variogram model. Condition (C.3) requires consistency of the covariance matrix estimator generated by the resampling method under consideration. Under mild moment and mixing condition, (C.3) typically holds for the spatial block bootstrap method of Section 12.3 and the spatial subsampling method described above. As an illustration, we now give a simple set of sufficient conditions on the process $Z(\cdot)$ under which the conclusions of Theorem 12.4 hold for the case where the subsampling method is employed to generate the covariance matrix estimator $\hat{\Sigma}_n$ and where the generic variogram estimator

$2\check{\gamma}_n(\cdot)$ is given by Matheron's (1962) method of moments estimator:

$$2\hat{\gamma}_n(h) = |N_n(h)|^{-1} \sum_{(s_i,s_j)\in N_n(h)} \Big(Z(s_i) - Z(s_j)\Big)^2 . \tag{12.70}$$

Here, $N_n(h) \equiv \{(s_i, s_j) : s_i - s_j = h,\ s_i, s_j \in R_n\}$ and, recall that, for any finite set A, $|A|$ denotes its size.

Theorem 12.5 *Suppose that (12.4) and Condition (C.2) of Theorem 12.4 hold. Also, suppose that there exists a* $\eta > 0$ *such that* $\max\{E|Z(h_j) - Z(0)|^{12+2\eta} : 1 \le j \le K\} < \infty$ *and the strong mixing coefficient* $\alpha(a, b)$ *of* $Z(\cdot)$ *(cf. (12.16), Section 12.3) satisfies the condition*

$$\alpha(a, b) \le Ca^{-\tau_1}b^{\tau_2},\ a \ge 1, b \ge 1$$

for some $C \in (0,\infty)$, $\tau_1 > 5d(6+\eta)/\eta$, *and* $0 < \tau_2 \le (\tau_1 - d)/d$. *Then, parts (a) and (b) of Theorem 12.4 hold with* $\hat{\theta}_{n,RGLS} = \hat{\theta}_{n,SGLS}$ *and* $2\check{\gamma}_n(\cdot) = 2\hat{\gamma}_n(\cdot)$. *The asymptotic covariance matrix* $D_{\Sigma^{-1}}(\theta)$ *in part (b) is given by* $D_{\Sigma^{-1}}(\theta) = (\Gamma(\theta)'\Sigma(\theta)^{-1}\Gamma(\theta))'$ *where the* (q, r)*-th element of* $\Sigma(\theta)$ *is*

$$\sigma_{qr}(\theta) \equiv \sum_{i\in\mathbb{Z}^d} \mathrm{Cov}_\theta\Big([Z(i + h_r) - Z(i)]^2, [Z(h_q) - Z(0)]^2\Big),\ 1 \le q, r \le K . \tag{12.71}$$

Proof: Follows from Theorem 5.1 and Remark 5.1 of Lee and Lahiri (2002). □

12.4.5 Numerical Examples

We now present the results of a small simulation study on the performance of the SGLS method. We consider a stationary two-dimensional Gaussian process $\{Z(i) : i \in \mathbb{Z}^2\}$ with zero mean and an "exponential" variogram, given by

$$2\gamma(h; \theta) = 2\Big(1 - \exp(-\theta_1|h_1| - \theta_2|h_2|)\Big),\ h = (h_1, h_2)' \in \mathbb{R}^2 , \tag{12.72}$$

where $\theta = (\theta_1, \theta_2)' \in (0,\infty)^2 \equiv \Theta$. The model variogram $2\gamma(\cdot;\theta)$ and its contour plot for $(\theta_1, \theta_2) = (0.10, 0.08)$ are given in Figure 12.7. Under the same θ-parameter values, a realization of the process over a 15×15 rectangular region is shown in Figure 12.8. The realization of the Gaussian process was generated by the spectral method of Shinozuka (1971) and Meijia and Rodriguez-Iturbe (1974).

For the simulation study, we considered three square-shaped sampling regions given by $(-3, 3] \times (-3, 3]$, $(-5, 5] \times (-5, 5]$, and $(-15, 15] \times (-15, 15]$. The prototype set R_0 for all three square regions was taken as $(-\frac{1}{2}, \frac{1}{2}] \times (-\frac{1}{2}, \frac{1}{2}]$, with the scaling factor λ_n being equal to 6, 10, and 30 for the three

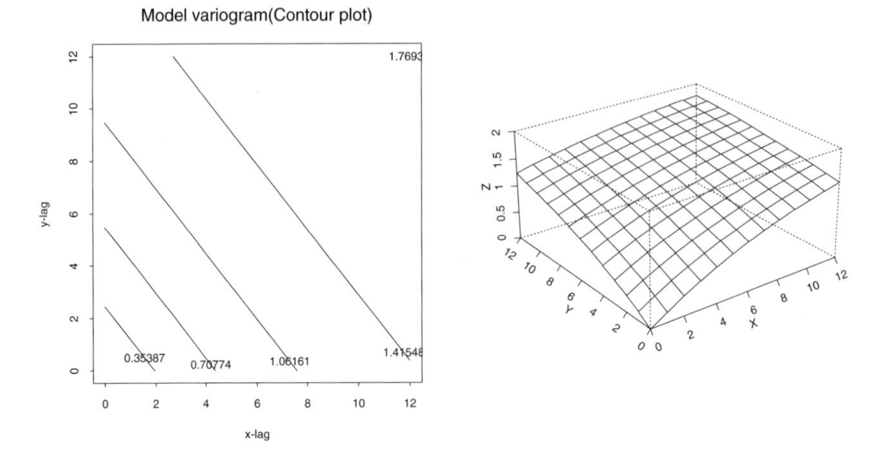

FIGURE 12.7. A plot of the "exponential" variogram $2\gamma(h;\theta)$ of (12.72) (right) and its contour plot (left) for $\theta = (0.1, 0.08)'$.

regions, respectively. The subregions were formed by considering translates of the set $\beta_n R_0 = (-\frac{\beta_n}{2}, \frac{\beta_n}{2}] \times (-\frac{\beta_n}{2}, \frac{\beta_n}{2}]$ for different choices of the subsampling block size β_n. We also considered a nonsquare-type rectangular sampling region, given by $(-5, 5] \times (-15, 15]$, with $R_0 = (-\frac{1}{6}, \frac{1}{6}] \times (-\frac{1}{2}, \frac{1}{2}]$ and $\lambda_n = 30$. Following the work of Sherman (1996) and Nordman and Lahiri (2003a) on optimal choice of the subsampling scaling factor β_n, we worked with $\beta_n = C\lambda_n^{1/2}$ for different values of the constant $C > 0$.

We took the generic variogram estimator $2\check{\gamma}_n(h)$ to be Matheron's method of moments estimator $2\hat{\gamma}_n(h)$, given by (12.70), and the lag vectors $h_1, \ldots, h_K$ as $h_1 = (1,0)'$, $h_2 = (0,1)'$, and $h_3 = (1,1)'$ with $K = 3$. For each sampling region, we considered the OLS estimator and Cressie's (1985) weighted least-squares estimator (CWLS) of $(\theta_1, \theta_2)'$. The latter is defined by (12.55) with $V(\theta) = \text{diag}(\sigma_{11}(\theta), \ldots, \sigma_{KK}(\theta))$, where $\sigma_{rr}(\theta)$ is as in (12.71), $1 \leq r \leq K$. Because of the long computation time and instability of the GLS estimators in these examples, a variation of the GLS estimator (denoted by TGLS) was used, where the matrix-valued function $\Sigma(\theta)$ of θ was substituted by the true matrix $\Sigma(\theta_0)$ for all θ. Thus, the TGLS estimators of θ are defined by minimizing the criterion function $Q(\theta, V)$ with $V(\theta) \equiv \Sigma(\theta_0)^{-1}$ for all θ. It can be shown that the TGLS estimator has the same asymptotic covariance matrix as the GLS estimator at $\theta = \theta_0$. The TGLS estimators are available only in simulation study and not in practice because the true values of the parameters are unknown in practice. Note that the TWLS, TGLS, and SGLS require only nonlinear minimizing routine of nonlinear regression type, such as a modified Gauss-Newton algorithm, which is faster and more stable than *general* nonlinear minimizing routines required for computing the GLS estimator.

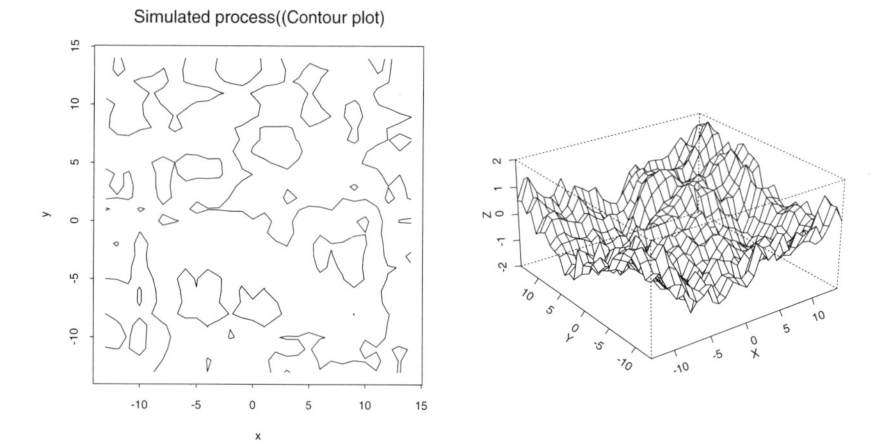

FIGURE 12.8. (Right panel) A realization of a zero mean unit variance Gaussian process with variogram $2\gamma(\cdot;\theta)$ of Figure 12.7 over the region $(-15, 15) \times (-15, 15)$. (Left panel) Contour plot of the same realization.

The results of the simulation study based on 3000 simulation runs are summarized in Table 12.2. The leading numbers in columns 4–5, respectively, denote the means for the estimators of θ_1 and θ_2, while the numbers within parentheses represent *N times the MSE*, where N denotes the sample size. The first and the third columns of Table 12.2 specify the sizes of the two sides of the rectangular sampling and subsampling regions, respectively.

From the table, it appears that the SGLS method performed better than the OLS and CWLS methods in most cases, and produced MSE values that fell between those of the CWLS and TGLS methods. Furthermore, for the nonsquare sampling region of size (10,30), the rectangular subregions of size (4,6) yielded slightly better results than the square subregions of size (4,4). See Lee and Lahiri (2002) for more simulation results under a *different* variogram model. The SGLS method has a similar performance under the variogram model treated therein.

Although the SGLS method has the same *asymptotic* optimality as the GLS method, its *finite-sample* statistical accuracy (as measured by the MSE) may not be as good as the GLS (or the idealized TGLS) method, particularly for small sample sizes. In the simulation studies carried out in Lee and Lahiri (2002), the SGLS estimators typically provided improvements over the OLS and the WLS estimators for small sample sizes and became competitive with the GLS estimators for moderately large sample sizes. A negative feature of the SGLS method is that the block size parameter β_n must be chosen by the user. A working rule of thumb is to use a β_n that is comparable to $\lambda_n^{1/2}$ in magnitude. On the other hand, as explained in the previous paragraph, the computational complexity associated with

TABLE 12.2. Mean and scaled mean squared error (within parentheses) of various least squares estimators of θ_1 and θ_2 under variogram model (12.72). Here R_n denotes the size of the rectangular sampling region, BS denotes the size of the subsampling regions.

R_n	LSEs	BS	$(\theta_1 = 0.1)$	$(\theta_2 = 0.08)$
6×6	OLS		0.10(0.14)	0.082(0.09)
	CWLS		0.10(0.14)	0.083(0.08)
	TGLS		0.10(0.12)	0.082(0.08)
	SGLS	2×2	0.10(0.12)	0.080(0.08)
		3×3	0.09(0.12)	0.075(0.07)
10×10	OLS		0.10(0.17)	0.082(0.11)
	CWLS		0.10(0.17)	0.082(0.11)
	TGLS		0.10(0.13)	0.081(0.10)
	SGLS	3×3	0.10(0.16)	0.080(0.10)
		4×4	0.10(0.15)	0.079(0.09)
10×30	OLS		0.10(0.27)	0.081(0.10)
	CWLS		0.10(0.27)	0.081(0.09)
	TGLS		0.10(0.22)	0.081(0.09)
	SGLS	4×4	0.10(0.25)	0.080(0.08)
		4×6	0.10(0.24)	0.080(0.09)
30×30	OLS		0.10(0.26)	0.080(0.14)
	CWLS		0.10(0.25)	0.080(0.13)
	TGLS		0.10(0.20)	0.080(0.13)
	SGLS	4×4	0.10(0.23)	0.080(0.13)
		5×5	0.10(0.24)	0.080(0.13)
		6×6	0.10(0.23)	0.080(0.13)

the GLS method can be much higher than that associated with the SGLS method. Table 12.3 gives a comparison of the time required for computing the SGLS and the GLS estimators, using an Alpha workstation. Here, I denotes the number of times iterations in the optimization routine for the GLS method are carried out. The reported times are obtained by averaging 100 repetitions. It follows from Table 12.3 that the SGLS method is considerably faster than the GLS method. However, the most important advantage of the SGLS and other RGLS methods is that they provide asymptotically efficient estimates of the covariance parameters even when the form of the asymptotic covariance matrix of the generic variogram estimator is unknown, in which case the GLS method is no longer applicable.

TABLE 12.3. A comparison of computation times (time in seconds).

Sample Size	Block Size	SGLS	GLS ($I = 30$)
10×10	3×3	0.019/100	1.151/100
30×30	4×4	0.088/100	93.892/100
60×60	5×5	0.258/100	1542.893/100
90×90	6×6	0.589/100	8073.810/100

12.5 Bootstrap for Irregularly Spaced Spatial Data

Let $\{Z(s) : s \in \mathbb{R}^d\}$, $d \in \mathbb{N}$, be an m-dimensional ($m \in \mathbb{N}$) random field with a continuous spatial index. In this section, we describe a bootstrap method that is applicable to irregularly spaced spatial data generated by a class of stochastic designs. We introduce the spatial sampling design in Section 12.5.1. Some relevant results on the asymptotic distribution of a class of M-estimators are presented in Section 12.5.2. The spatial block bootstrap method and its properties are described in Sections 12.5.3 and 12.5.4, respectively. Unlike the regular grid case, in this section, we use n to denote the sample size.

12.5.1 A Class of Spatial Stochastic Designs

Suppose that the process $Z(\cdot)$ is observed at finitely many locations $\mathcal{S}_n \equiv \{s_1, \ldots, s_n\}$ that lie in the sampling region R_n. We continue to use the framework of Section 12.2 for the sampling region R_n and suppose that R_n is obtained by inflating a prototype set R_0 by a scaling constant λ_n, as specified by (12.1). When the sampling sites $s_1, \ldots, s_n$ are irregularly spaced, a standard approach in the literature is to model them using a homogeneous Poisson point process. However, here, we adopt a slightly different approach and consider sampling designs driven by a collection of independent random vectors with values in the prototype set R_0. More precisely, let $f(x)$ be a probability density function (with respect to the Lebesgue measure) on R_0 and let $\{X_n\}_{n \geq 1}$ be a sequence of iid random vectors with density $f(x)$ such that $\{X_n\}_{n \geq 1}$ are independent of $\{Z(s) : s \in \mathbb{R}^d\}$. We suppose that the sampling sites $s_1, \ldots, s_n$ are obtained from a realization $x_1, \ldots, x_n$ of the random vectors $X_1, \ldots, X_n$ as

$$s_i = \lambda_n x_i, \; 1 \leq i \leq n \;, \tag{12.73}$$

where λ_n is the scaling constant associated with R_n (cf. (12.1)). We further suppose that

$$n^\delta / \lambda_n = O(1) \quad \text{as} \quad n \to \infty$$

for some $\delta > 0$. This condition is imposed for proving consistency of bootstrap approximations for almost all realizations of the random vectors $X_1, X_2, \ldots$.

In view of (12.73), a more precise notation for the sampling sites should be $s_{1n}, \ldots, s_{nn}$, but we drop the subscript n for notational simplicity. Note that as $X_1, \ldots, X_n$ takes values in R_0, the sampling sites $s_1, \ldots, s_n$ potentially take values over the entire sampling region $R_n = \lambda_n R_0$. Furthermore, by the Strong Law of Large Numbers (cf. Theorem A.3, Appendix A), the expected number of sampling sites lying over any subregion $A \subset R_n$ is given by $nP(\lambda_n X_1 \in A) = n \cdot \int_{\lambda_n^{-1}A} f(x)dx$, which may be different from $n \cdot \text{vol.}(\lambda_n^{-1}A)$ for a nonconstant design density $f(x)$. As a result, this formulation allows us to model irregularly spaced spatial data that may have different degrees of concentration over different parts of the sampling region. A second important feature of the stochastic sampling design is that it allows the sample size n and the volume of the sampling region R_n to grow at different rates. For a positive design density $f(x)$, when the sample size n grows at a rate faster than the volume of R_n, the ratio of the expected number of sampling sites in any given subregion A of R_n to the volume of A tends to infinity. Under the stochastic design framework, this corresponds to "infill" sampling of subregions of R_n (cf. Section 12.2). Thus, the stochastic sampling design presented here provides a unified framework for handling irregularly spaced spatial data with a nonuniform concentration across R_n and with a varying rate of sampling.

In the next section, we describe some results on the large sample distribution of a class of M-estimators under the stochastic sampling design.

12.5.2 Asymptotic Distribution of M-Estimators

Suppose that $\{Z(s) : s \in \mathbb{R}^d\}$ is an m-variate ($m \in \mathbb{N}$) stationary random field that is independent of the random vectors $X_1, X_2, \ldots,$ generating the sampling sites. In applications, the components of the multivariate random field $Z(\cdot)$ could be defined in terms of suitable functions of a given univariate (or a lower-dimensional) random field. Suppose that the $Z(\cdot)$ process is observed at locations $\{s_1, \ldots, s_n\} \equiv \mathcal{S}_n$, generated by the spatial stochastic design of Section 12.5.1, and that we are interested in estimating a p-dimensional ($p \in \mathbb{N}$) level-1 parameter θ based on the observations $\{Z(s_i) : 1 \le i \le n\}$. Let $\Psi : \mathbb{R}^{p+m} \to \mathbb{R}^p$ be a Borel-measurable function satisfying

$$E\Psi(Z(0); \theta) = 0 . \tag{12.74}$$

Then, an M-estimator $\hat{\theta}_n$ of θ corresponding to the score function Ψ is defined as a measurable solution to the estimating equation (in $t \in \mathbb{R}^p$)

$$\sum_{i=1}^{n} \Psi(Z(s_i); t) = 0 . \tag{12.75}$$

This class of M-estimators covers many common estimators, such as the sample moments, the maximum likelihood estimators of parameters of Gaussian random fields, and the pseudo-likelihood estimators in certain conditionally specified spatial models, like the Markov Random field models (cf. Cressie (1993); Guyon (1995)). The asymptotic distribution of $\hat{\theta}_n$ depends, among other factors, on the spatial sampling density $f(x)$ and on the relative growth rates of the sample size n and the volume of the sampling region R_n, given by vol.$(R_n) = \lambda_n^d \cdot$ vol.(R_0). Here, we suppose that

$$n/\lambda_n^d \to \Delta \quad \text{for some} \quad \Delta \in (0, \infty] \ . \tag{12.76}$$

When $\Delta \in (0, \infty)$, the sample size n and the volume of the sampling region R_n grow at the same rate. In analogy to the fixed design case (cf. (12.3)), we classify the resulting asymptotic structure as the *pure* increasing domain asymptotic structure under the stochastic design. On the other hand, when $\Delta = \infty$, the sample size n grows at a faster rate than the volume of R_n and, therefore, any given subregion of R_n of unit volume may contain an unbounded number of sampling sites as $n \to \infty$. Thus, for $\Delta = \infty$, the sampling region R_n is subjected to infill sampling, thereby resulting in a *mixed* increasing domain asymptotic structure in the stochastic design case. As we will shortly see in Theorem 12.6 below, these spatial asymptotic structures have nontrivial effects on the asymptotic distribution of the M-estimator $\hat{\theta}_n$.

To state the results, we now introduce some notation. Let $\Psi_1, \ldots, \Psi_p$ denote the components of Ψ and let D_Ψ be the $p \times p$ matrix with (q, r)th element $E[\frac{\partial}{\partial \theta_r} \Psi_q(Y(0); \theta)]$ where $1 \le r, q \le p$. For $\alpha \in \mathbb{Z}_+^p$, write $D^\alpha \Psi(Z; \theta)$ for the α-th order partial derivative in the θ-coordinates. Let

$$\Sigma_{\Psi,\Delta} = \Gamma_\Psi(0) \cdot \Delta^{-1} + \int_{\mathbb{R}^d} \Gamma_\Psi(s) ds \cdot \int_{R_0} f^2(x) dx \ , \tag{12.77}$$

$\Delta \in (0, \infty]$, where $\Gamma_\Psi(s) = E\Psi(Z(0))\Psi(Z(s))'$, $s \in \mathbb{R}^d$, and where we set $\Delta^{-1} = 0$ for $\Delta = \infty$. Let $\alpha(a; b)$ denote the strong mixing coefficient of the multivariate spatial process $\{Z(s) : s \in \mathbb{R}^d\}$, defined by (12.15) and (12.16), with $\mathcal{F}_Z(S) \equiv \sigma\langle\{Z(s) : s \in S\}\rangle$, $S \subset \mathbb{R}^d$. We shall suppose that there exist constants C, $\tau_1 \in (0, \infty)$, and $\tau_2 \in [0, \infty)$ such that

$$\alpha(a; b) \le C a^{-\tau_1} b^{\tau_2} \quad \text{for all} \quad a > 0, \ b \ge 1 \ , \tag{12.78}$$

for any $d \ge 2$ and for $d = 1$, (12.78) holds with $\tau_2 = 0$. As before, let $G^{(m)}$ denote the marginal distribution of $Z(0)$, i.e., $G^{(m)}(A) = P(Z(0) \in A)$, $A \in \mathcal{B}(\mathbb{R}^m)$. Also, recall that for a positive-definite matrix Σ of order $k \in \mathbb{N}$, $\Phi(\cdot; \Sigma)$ denotes the probability measure corresponding to the k-dimensional Gaussian distribution with mean 0 and variance matrix Σ.

Next, suppose that $\{X_n\}_{n \ge 1}$ and $\{Z(s) : s \in \mathbb{R}^d\}$ are defined on a common probability space $(\Omega, \mathcal{F}, P)$. Let $P_{\mathbf{X}}$ denote the joint probability distribution of the sequence $\{X_n\}_{n \ge 1}$ and let $P_{\cdot|\mathbf{X}}$ and $E_{\cdot|\mathbf{X}}$, respectively, denote the conditional distribution and expectation, given $\mathcal{X}_\infty \equiv \sigma\langle\{X_n : n \geq 1\}\rangle$. We now state the main result of this section that asserts consistency and asymptotic normality of the multivariate M-estimator $\hat{\theta}_n$ conditional on the sequence of iid random vectors $X_1, X_2, \ldots$.

Theorem 12.6 *Suppose that (12.75) has a unique solution, that (12.76) holds, and that the following conditions hold:*

(C.4) The density $f(x)$ of X_1 is continuous on $cl.(R_0)$.

(C.5) For some $\eta_0 \in (0, \infty)$, $\Psi(z; t)$ has continuous second-order partial derivatives with respect to t on the set $\{\|t - \theta\| \leq \eta_0\} \equiv \Theta_0$ for almost all z $(G^{(m)})$.

(C.6) The $p \times p$ matrices D_Ψ and $\int \Gamma_\Psi(s)ds$ are nonsingular.

Also suppose that with $r = 1$, the following conditions holds:

$(C.7)_r$ For some $\delta \in (0, \infty)$,

(a) $E|\Psi(Z(0); \theta)|^{2r+\delta} < \infty$,

(b) $\max_{|\alpha|=1} E|D^\alpha \Psi(Z(0); \theta)|^{2r+\delta} < \infty$,

(c) $E \sup\{|D^\alpha \Psi(Z(0) : t)|^{2r+\delta} : t \in \Theta_0,\ |\alpha| = 2\} < \infty$,

(d) $\tau_1 > (2r - 1)(2r + \delta)d/\delta$, *and*

(e) for $d \geq 2$, $0 \leq \tau_2 < (\tau_1 - d)/4d$.

Then, for both the pure- and the mixed-asymptotic structures (i.e., for '$\Delta \in (0, \infty)$' and for '$\Delta = +\infty$', respectively),

$$\sup_{A \in \mathcal{C}} \left| P_{\cdot|\mathbf{X}}\left(\lambda_n^{d/2}(\hat{\theta}_n - \theta) \in A\right) - \Phi\left(A; D_\Psi^{-1}\Sigma_{\Psi,\Delta}(D_\Psi)^{-1})'\right)\right| = o(1) \ a.s.\ (P_{\mathbf{X}}) \tag{12.79}$$

where $\mathcal{C}$ is the collection of all measurable convex sets in $\mathbb{R}^p$.

Proof: See Lahiri (2003d). □

When the solution to equation (12.75) is unique and the other conditions of Theorem 12.6 hold, (12.79) shows that $\hat{\theta}_n$ is consistent for θ for almost all realizations of the random vectors $X_1, X_2, \ldots$, i.e., $\hat{\theta}_n \to \theta$ in $P_{\cdot|\mathbf{X}}$-probability, a.s. $(P_{\mathbf{X}})$. (See Definition 12.3, (12.89), and (12.90) in Section 12.5.4 below for a precise definition of this notion of convergence and its connection with the usual notion of convergence in probability.) When the uniqueness condition on the solution to (12.75) does not hold, Lahiri (2003d) shows that a consistent sequence of solutions of (12.75) exists. For the nonunique case, conclusions of Theorem 12.6 remain valid for this sequence of solutions.

A notable feature of Theorem 12.6 is that under both types of asymptotic structures, the rate of convergence of the M-estimator is $O_p(\lambda_n^{-d/2})$. Note that this rate is comparable to the rate $O_p(n^{-1/2})$ for the "pure increasing domain" case (i.e., for $\Delta \in (0,\infty)$), but it is slower than $O_p(n^{-1/2})$ under the "mixed increasing domain" case (i.e., for $\Delta = +\infty$). The slowness of the convergence rate in the "$\Delta = +\infty$" case compared to the standard $O_p(n^{-1/2})$ rate is due to the "infill" sampling component, which leads to conditions of *long-range dependence* in the sampled data values. Also note that the asymptotic variance matrices under the two asymptotic structures are different. The infill component in the "mixed increasing domain" case leads to a reduction in the asymptotic variance matrix of $\hat{\theta}_n$ as the positive-definite matrix $\Gamma(0)$ drops out for $\Delta = \infty$. Thus, the asymptotic variance of the estimator $t_0'\hat{\theta}_n$ of the linear parametric function $t_0'\theta$, $t_0 \in \mathbb{R}^p$, $t_0 \neq 0$, becomes smaller in the mixed increasing domain case by the positive additive factor $t_0'\Gamma(0)t_0$.

Although block bootstrap methods do not always work under strong dependence (cf. Chapter 10), we now present a block bootstrap method that remains valid, not only in the "pure increasing domain" case, but also in presence of such strong dependence under the "mixed increasing domain" case.

12.5.3 A Spatial Block Bootstrap Method

Let $R_n = \lambda_n R_0$ be the sampling region. Suppose that the process $Z(\cdot)$ is observed at locations $\{s_1, \ldots, s_n\} \equiv \mathcal{S}_n$, where s_i's are generated by the stochastic design of Section 12.5.1. The block bootstrap method for the irregularly spaced sampling sites attempts to recreate a version of the process $Z(\cdot)$ over R_n using a mechanism similar to that used for the grided data. Let $\{\beta_n\}_{n\geq 1}$ be a sequence of positive real numbers satisfying (12.4), i.e., satisfying

$$\beta_n^{-1} + \lambda_n^{-1}\beta_n = o(1) \quad \text{as} \quad n \to \infty .$$

Also, let $R_n = \bigcup_{k\in\mathcal{K}_n} R_n(k)$ be the partition of the sampling region R_n by the subregions $R_n(k)$, $k \in \mathcal{K}_n$ of (12.5), where, recall that $R_n(k) \equiv (k+\mathcal{U})\beta_n \cap R_n$, $\mathcal{K}_n \equiv \{k \in \mathbb{Z}^d : (k+\mathcal{U})\beta_n \cap R_n \neq \emptyset\}$ and $\mathcal{U} = [0,1)^d$. We now define the bootstrap version of the $Z(\cdot)$-process over each of the subregions $R_n(k)$, $k \in \mathcal{K}_n$, using the collection $\{i+\mathcal{U}\beta_n : i \in \mathcal{I}_n\}$ of overlapping cubes of volume β_n^d, where as in Section 12.3, $\mathcal{I}_n \equiv \{j \in \mathbb{Z}^d : j+\mathcal{U}\beta_n \subset R_n\}$. For $k \in \mathcal{K}_n$, define $\mathcal{B}_n(i;k) = [i+\mathcal{U}\beta_n] \cap [R_n(k) - k\beta_n + i]$, $i \in \mathcal{I}_n$. Then, for any given $k \in \mathcal{K}_n$, $\mathcal{B}_n(i;k)$ has the same shape as the subregion $R_n(k)$ for all $i \in \mathcal{I}_n$. We call $\{\mathcal{B}_n(i;k) : i \in \mathcal{I}_n\}$ the collection of *observed* blocks of "type k". To define the bootstrap version of the $Z(\cdot)$-process over $R_n(k)$, we now select a block of "type k" by resampling at random (and independently of the other resampled blocks) from the collection $\{\mathcal{B}_n(i;k) : i \in \mathcal{I}_n\}$. Formally, let $\{I_k : k \in \mathcal{K}_n\}$ be (conditionally) iid random variable with common distribution (12.6), i.e.,

$$P(I_k = i) = \frac{1}{|\mathcal{I}_n|}, \; i \in \mathcal{I}_n \; ,$$

for $k \in \mathcal{K}_n$. Then the collection of resampled blocks is given by $\{\mathcal{B}(I_k; k) : k \in \mathcal{K}_n\}$. Without loss of generality, we suppose that the variables $\{I_k : k \in \mathcal{K}_n\}_{n \geq 1}$ are defined on the same probability space $(\Omega, \mathcal{F}, P)$ supporting the variables $\{Z(s) : s \in \mathbb{R}^d\}$ and $\{X_n : n \geq 1\}$. Next, write $\mathcal{Z}_n(A) = \{Z(s) : s \in \mathcal{S}_n \cap A\}$ to denote the collection of observations over a set $A \subset \mathbb{R}^d$. Then, the bootstrap version of the $Z(\cdot)$-process over the k-th subregion is given by

$$\mathcal{Z}_n^*\Big(R_n(k)\Big) \equiv \mathcal{Z}_n\Big(\mathcal{B}_n(I_k; k)\Big), \; k \in \mathcal{K}_n \; . \tag{12.80}$$

The bootstrap version $\mathcal{Z}_n^*(R_n)$ of the random field $Z(\cdot)$ over the entire sampling region R_n is now obtained by pasting together the copies $\mathcal{Z}_n^*(R_n(k))$, $k \in \mathcal{K}_n$:

$$\mathcal{Z}_n^*(R_n) = \bigcup_{k \in \mathcal{K}_n} \mathcal{Z}_n^*(R_n(k)) \; .$$

Although all the blocks $\mathcal{B}_n(i; k)$, $i \in \mathcal{I}_n$ of "type k" have the *same* shape as the subregion $R_n(k)$ in the stochastic design case, each of them may contain a *different* number of sampling sites, as the sampling sites $s_1, \ldots, s_n$ are *randomly* distributed over the sampling region R_n. Since the bootstrap version of the process over $R_n(k)$ is defined by randomly selecting one of the "type k" blocks, the number of the resampled observations over $R_n(k)$ is typically different from the number of observations in $R_n(k)$ itself. Let $L_k^* \equiv L_{k,n}^*$ denote the size of the resample $\mathcal{Z}_n^*(R_n(k))$ over the subregion $R_n(k)$, $k \in \mathcal{K}_n$. Also, let $n^* = \sum_{k \in \mathcal{K}_n} L_k^*$ denote the total number of the resampled values over the sampling region R_n, i.e., n^* is the size of $\mathcal{Z}_n^*(R_n)$. Although the resample size n^* is typically different from the original sample size n, it can be shown that

$$\frac{E_{\cdot|\mathbf{X}}(n^*)}{n} \to 1 \quad \text{as} \quad n \to \infty, \quad \text{a.s.} \quad (P_{\mathbf{X}}) \; . \tag{12.81}$$

We define the bootstrap version of a statistic $\hat{\gamma}_n = t_n(\mathcal{Z}_n(R_n))$ by

$$\gamma_n^* = t_{n^*}\left(\mathcal{Z}_n^*(R_n)\right) \; . \tag{12.82}$$

In particular, the bootstrap version of the sample mean $\bar{Z}_n = n^{-1} \sum_{i=1}^n Z(s_i)$ is given by

$$\bar{Z}_n^* = \sum_{k \in \mathcal{K}_n} S_n^*(k)/n^* \; , \tag{12.83}$$

where $S_n^*(k)$ is the sum of the L_k^*-many resampled values $\mathcal{Z}_n^*(R_n(k))$ over the subregion $R_n(k)$, $k \in \mathcal{K}_n$. For later reference, we also define the bootstrap version of the normalized M-estimator

$$T_n = \lambda_n^{d/2}(\hat{\theta}_n - \theta) \; ,$$

where θ and $\hat{\theta}_n$ are respectively given by (12.74) and (12.75). As explained in Section 4.3, there is more than one way of defining the bootstrap version of T_n. Here, we follow the approach of Shorack (1982) that centers the score function to define the bootstrap version of the M-estimator $\hat{\theta}_n$. For $k \in \mathcal{K}_n$ and $t \in \mathbb{R}^p$, let $S_n^*(k;t)$ denote the sum of all $\Psi(Z(s_i);t)$-variables corresponding to the $Z(s_i)$'s in the resample $\mathcal{Z}_n^*(R_n(k))$, i.e., $S_n^*(k;t) = \sum_{i=1}^n \Psi(Z(s_i);t)\mathbb{1}(s_i \in \mathcal{B}_n(I_k;k))$ (cf. (12.80)). Then, the bootstrap version θ_n^* of $\hat{\theta}_n$ is defined as a measurable solution (in t) of the equation

$$\sum_{k \in \mathcal{K}_n} \left[S_n^*(k;t) - E_* S_n^*(k;\hat{\theta}_n) \right] = 0 \ , \tag{12.84}$$

where, in this section, P_* and E_*, respectively, denote the conditional probability and the conditional expectation given $\mathcal{G} \equiv \sigma\langle \{X_n : n \geq 1\} \cup \{Z(s) : s \in \mathbb{R}^d\}\rangle$. The bootstrap version of T_n is now given by

$$T_n^* = \lambda_n^{d/2}(\theta_n^* - \hat{\theta}_n) \ . \tag{12.85}$$

Note that centering θ_n^* at $\hat{\theta}_n$ in the definition of T_n^* is permissible as the (conditional) expected value of the left side of (12.84) at the true "bootstrap parameter" value $t = \hat{\theta}_n$ is zero. An alternative bootstrap version of T_n is given by $T_n^{**} = \lambda_{n^*}^{d/2}(\theta_n^* - \hat{\theta}_n)$. In view of (12.81), both bootstrap versions of T_n have the same limit distribution. In this book, we shall consider T_n^* only.

In the next section, we describe the properties of the spatial block bootstrap method under the stochastic design of Section 12.5.2 in variance and distribution function estimation problems. For a closely related version of the spatial block bootstrap method and its properties under a homogeneous marked Poisson process set up, see Politis, Paparoditis and Romano (1998, 1999).

12.5.4 Properties of the Spatial Bootstrap Method

First we consider properties of the block bootstrap variance estimator for the sample mean. Suppose that $\{Z(s) : s \in \mathbb{R}^d\}$ is a univariate ($m = 1$) stationary random field. Note that in this case, the sample mean $\bar{Z}_n = n^{-1}\sum_{i=1}^n Z(s_i)$ is the M-estimator of the population mean $\mu = EZ(0)$ corresponding to the score function $\Psi(x;t) = x - t$, $x, t \in \mathbb{R}$, with $p = 1 = m$. If the conditions of Theorem 12.6 are satisfied for this choice of $\Psi(\cdot;\cdot)$, then it follows from Theorem 12.6 that for any $\Delta \in (0, \infty]$,

$$\lambda_n^{d/2}(\bar{Z}_n - \mu) \longrightarrow^d N(0; \sigma_{\infty,\Delta}^2) \quad \text{a.s.} \quad (P_{\mathbf{X}}) \ , \tag{12.86}$$

where, with $\sigma(s) = EZ(0)Z(s)$, $s \in \mathbb{R}^d$, the asymptotic variance $\sigma_{\infty,\Delta}^2$ is defined as

$$\sigma_{\infty,\Delta}^2 = \frac{\sigma(0)}{\Delta} + \int_{\mathbb{R}^d} \sigma(s)ds \int_{R_0} f^2(x)dx \ . \tag{12.87}$$

It can be shown (cf. Lahiri (2003d)) that $\sigma^2_{\infty,\Delta}$ is not only the asymptotic variance of $\lambda_n^{d/2}(\bar{Z}_n - \mu)$, but it is also the exact limit of $\lambda_n^d \cdot \mathrm{Var}_{\cdot|\mathbf{X}}(\bar{Z}_n)$, a.s. $(P_{\mathbf{X}})$. This shows that both the spatial sampling density f and the type of the asymptotic structure (viz., pure increasing domain and mixed increasing domain asymptotic structures) have nontrivial effects on the variance of $\bar{Z}_n$. Under both asymptotic structures, the variance of $\lambda_n^{d/2}\bar{Z}_n$ takes the minimum value when the design density f is uniform over R_0. On the other hand, as noted earlier, the infill component of the mixed increasing domain asymptotic structure (with "$\Delta = \infty$") leads to a reduction in the variance of the scaled sample mean $\lambda_n^{d/2}\bar{Z}_n$. Inspite of the variations in the form of the asymptotic variance due to these factors, the block bootstrap method provides a "consistent" estimator of $\sigma^2_{\infty,\Delta}$ in all cases, as shown by the following result.

Theorem 12.7 *Suppose that Conditions (C.4) and (C.7)$_r$ of Theorem 12.6 hold with $p = 1 = m$, $r = 3$ and $\Psi(x;t) = x - t$. Also, suppose that there exists $\delta \in (0,1)$ such that*

$$\beta_n^{-1} + \lambda_n^{-(1-\delta)}\beta_n = o(1) \quad as \quad n \to \infty \tag{12.88}$$

and that (12.76) holds. Then, for almost all realizations of $X_1, X_2, \ldots$ under $P_{\mathbf{X}}$,

$$\lambda_n^d . \mathrm{Var}_*(\bar{Z}_n^*) \to \sigma^2_{\infty,\Delta} \text{ in } P_{\cdot|\mathbf{X}}\text{-probability, a.s. } (P_{\mathbf{X}})$$

where $\bar{Z}_n^*$ is as defined by (12.83).

Proof: See Lahiri (2003d) □

Thus, it follows that the block bootstrap variance estimator provides a valid approximation to the variance of the sample mean under both pure- and mixed-increasing domain asymptotic structures for a large class of design densities f that can be nonuniform over R_0. Because the variance estimator can be computed without the knowledge of the sampling density f, it enjoys considerable advantage over traditional "plug-in" estimators of $\sigma^2_{\infty,\Delta}$ that require one to estimate each component of the population parameter $\sigma^2_{\infty,\Delta}$ explicitly.

Next note that the convergence of the bootstrap variance estimator in Theorem 12.7 is asserted "in $P_{\cdot|\mathbf{X}}$-probability, a.s. $(P_{\mathbf{X}})$," which is somewhat nonstandard. A precise definition of this notion of convergence is given as follows:

Definition 12.3 *Let $\{t_n(\mathbf{Z},\mathbf{X})\}_{n\geq 1}$ be a sequence of random variables on $(\Omega, \mathcal{F}, P)$ such that $t_n(\mathbf{Z},\mathbf{X})$ is measurable with respect to the σ-field $\mathcal{G} \equiv \sigma\langle\{Z(s) : s \in \mathbb{R}^d\} \cup \{X_n : n \geq 1\}\rangle$, and let $a \in \mathbb{R}$. Then, we say that $t_n(\mathbf{Z},\mathbf{X}) \to a$ in $P_{\cdot|\mathbf{X}}$-probability, almost surely $(P_{\mathbf{X}})$, if*

$$P_{\mathbf{X}}\Big(\lim_{n\to\infty} P_{\cdot|\mathbf{X}}\Big(|t_n(\mathbf{Z},\mathbf{X}) - a| > \epsilon\Big) = 0 \quad for\ all \quad \epsilon > 0\Big) = 1 \,. \tag{12.89}$$

It is easy to show that the event on the left side of (12.89) is in $\mathcal{F}$ as we may (without loss of generality) restrict attention to a countable set of $\epsilon > 0$. It is easy to see that (12.89) implies the more familiar notion of "convergence in probability," i.e., (12.89) implies that

$$\lim_{n\to\infty} P\Big(|t_n(\mathbf{Z},\mathbf{X})| > \epsilon\Big) = 0 \quad \text{for all} \quad \epsilon > 0 \ . \tag{12.90}$$

The reason for stating Theorem 12.7 using the nonstandard notion of convergence is that it allows us to interpret consistency of the bootstrap variance estimator for almost all realizations of the stochastic design vectors $X_1, X_2, \ldots$. Thus, once the values of $X_1, X_2, \ldots$ are given, i.e., once the locations of the sampling sites are given, we may concern ourselves only with the randomness arising from the random field $Z(\cdot)$ and the bootstrap variables $\{I_k : k \in \mathcal{K}_n\}_{n\geq 1}$, by treating the locations $\{s_1, \ldots, s_n\}$ as nonrandom. However, the usual notion of "convergence-in-probability" (viz., (12.90)) does not allow such an interpretation in the stochastic design case.

Next we consider properties of the bootstrap distribution function estimators. As in Section 12.5.2, here we suppose that the random field $\{Z(s) : s \in \mathbb{R}^d\}$ is stationary and m-dimensional for some $m \in \mathbb{N}$. Let $\hat{\theta}_n$ denote the M-estimator of the p-dimensional parameter θ based on $Z(s_1), \ldots, Z(s_n)$, as defined by (12.75). Let $T_n = \lambda_n^{d/2}(\hat{\theta}_n - \theta)$ be the normalized version of $\hat{\theta}_n$ and let T_n^*, defined in (12.85), be its bootstrap version. Then, we have the following result.

Theorem 12.8 *Suppose that θ_n^* is a unique solution of (12.84). Also, suppose that Conditions (C.4), (C.5), (C.6), and $(C.7)_r$ of Theorem 12.6 hold with $r = 3$ and that (12.76) and (12.88) hold. Then,*

$$\sup_{A\in\mathcal{C}} \Big| P_*(T_n^* \in A) - P_{\cdot|\mathbf{X}}(T_n \in A) \Big| \to 0 \text{ in } P_{\cdot|\mathbf{X}}\text{-probability, a.s. } (P_{\mathbf{X}}) \tag{12.91}$$

under both pure- and mixed-increasing domain asymptotic structures (i.e., for $\Delta \in (0,\infty)$ and for $\Delta = \infty$), where $\mathcal{C}$ is the collection of all measurable convex subsets of $\mathbb{R}^p$.

Consistency of the bootstrap distribution function estimator continues to hold even when the bootstrap estimating equation (12.84) has multiple solutions. In this case, there exists a sequence of solutions $\{\theta_n^*\}_{n\geq 1}$ of (12.84) that approximates $\{\hat{\theta}_n\}_{n\geq 1}$ with $O_{P_*}(\lambda_n^{-d/2}(\log n)^c)$ accuracy for some $c > 0$, and (12.91) holds for this sequence of solutions, as in Section 4.3.

Like the variance estimation problem, the spatial bootstrap provides a valid approximation to the distribution of M-estimators under the stochastic design, allowing nonuniform concentration of sampling sites and infill sampling of subregions in R_n. We point out that in contrast to the results of Chapter 10, here the block bootstrap method remains valid even in presence of a particular form of "strong" dependence in the data, engendered by the mixed increasing domain asymptotic structure.

12.6 Resampling Methods for Spatial Prediction

In this section, we consider two types of prediction problems. For the first type, we suppose that $\{Z(s) : s \in \mathbb{R}^d\}$ is a random field with a continuous spatial index and the objective is to predict $\int_{R_n} g(Z(s))ds$, the integral of a function g of the process $Z(\cdot)$ over the sampling region R_n, on the basis of a finite sample. We describe resampling methods for this type of problems in Section 12.6.1. The other type of prediction problem we consider here is in the context of best linear unbiased prediction (or Kriging) of a "new" value on the basis of a finite set of observations. This second type of problem is addressed in Section 12.6.2.

12.6.1 Prediction of Integrals

Suppose that $g : \mathbb{R} \to \mathbb{R}$ is a bounded Borel-measurable function and that we wish to predict

$$\hat{\Delta}_\infty \equiv \hat{\Delta}_{\infty,n} = \int_{R_n} g(Z(s))ds \tag{12.92}$$

on the basis of finitely many observations lying in the sampling region R_n, where $R_n = \lambda_n R_0$ is as described in Section 12.2 (cf. (12.1)). Here we use the hat $(\hat{})$ in $\hat{\Delta}_\infty$ to indicate that it is a random quantity, while we use the subscript ∞ in $\hat{\Delta}_\infty$ to indicate that it is a functional of the totality $\{Z(s) : s \in R_n\}$ of random variables in R_n and is *unobservable*. In order to predict $\hat{\Delta}_\infty$ consistently, we adopt a sampling framework that fills in any subregion of R_n with an increasing number of sampling sites. More precisely, let $\{\eta_n\}_{n\geq 1}$ be a sequence of positive real numbers such that $\eta_n \downarrow 0$ as $n \to \infty$. We suppose that the process $Z(\cdot)$ is observed at each point of the scaled down integer grid $\eta_n \cdot \mathbb{Z}^d$ that lies in the sampling region R_n. Thus, the sampling sites are given by

$$\mathcal{S}_n \equiv \{s \in \mathbb{R}^d : s \in \eta_n \mathbb{Z}^d,\ s \in \mathbb{R}_n\} . \tag{12.93}$$

Although we use the same symbol $\mathcal{S}_n$ to denote the collection of all sampling sites in Sections 12.3, 12.5, and in here, the size of the set $\mathcal{S}_n$ is different in each case, depending on the spatial design. For the rest of Section 12.6.1, we shall use N_{2n} to denote the size of $\mathcal{S}_n$. Then, under Condition B on the boundary of the prototype set R_0, the sample size N_{2n} satisfies the relation

$$N_{2n} = \text{vol.}(R_0) \cdot \eta_n^{-d} \lambda_n^d (1 + o(1)) \quad \text{as} \quad n \to \infty . \tag{12.94}$$

Since $\eta_n \downarrow 0$ as $n \to \infty$, this implies that the sample size N_{2n} grows at a faster rate than the volume of the sampling region R_n. Thus, the resulting asymptotic structure is of the "mixed increasing domain" type, with a nontrivial infill component. A predictor of $\hat{\Delta}_\infty$ based on the observations $\{Z(s) : s \in \mathcal{S}_n\}$ is given by

$$\hat{\Delta}_n = N_{2n}^{-1} \sum_{s \in \mathcal{S}_n} g\big(Z(s)\big) \; . \tag{12.95}$$

Under mild conditions on the process $Z(\cdot)$ and the function $g(\cdot)$, $\hat{\Delta}_n$ is L^2-consistent for $\hat{\Delta}_\infty$ in the sense that $E(\hat{\Delta}_n - \hat{\Delta}_\infty)^2 \to 0$ as $n \to \infty$. The rate at which the mean squared prediction error (MSPE) $E(\hat{\Delta}_n - \hat{\Delta}_\infty)^2$ goes to zero depends on both the increasing domain scaling parameter $\{\lambda_n\}_{n\geq 1}$ and the infill scaling parameter $\{\eta_n\}_{n\geq 1}$.

Lahiri (1999b) considers the spatial cumulative distribution function (SCDF)

$$\hat{F}_\infty(z_0) = \int_{R_n} \mathbb{1}(Z(s) \leq z_0) ds, \;\; z_0 \in \mathbb{R} \; , \tag{12.96}$$

corresponding to $g(\cdot) = \mathbb{1}(\cdot \leq z_0)$, $z_0 \in \mathbb{R}$ in (12.92). The corresponding predictor, given by (12.95), is

$$\hat{F}_n(z_0) = N_{2n}^{-1} \sum_{s \in \mathcal{S}_n} \mathbb{1}(Z(s) \leq z_0), \;\; z_0 \in \mathbb{R} \; , \tag{12.97}$$

the empirical cumulative distribution function (ECDF). A result of Lahiri (1999b) shows that under some regularity conditions on the process $Z(\cdot)$,

$$E\Big(\hat{F}_n(z_0) - \hat{F}_\infty(z_0)\Big)^2 = c(z_0) \cdot \eta_n^2 \lambda_n^{-d} \big(1 + o(1)\big)$$

for some constant $c(z_0) \in (0, \infty)$. Furthermore, the scaled process

$$\xi_n(z) \equiv \Big[\eta_n^{-2} \lambda_n^d\Big]^{1/2} \Big(\hat{F}_n(z) - \hat{F}_\infty(z)\Big), \;\; z \in \mathbb{R} \tag{12.98}$$

converges in distribution to a zero mean Gaussian process $\mathcal{W}$ as random elements of the Skorohod space $\mathbb{D}_1$ of right continuous functions from $[-\infty, \infty]$ to $\mathbb{R}$ with left hand limits. The covariance function of $\mathcal{W}(\cdot)$ is given by

$$\mathrm{Cov}\Big(\mathcal{W}(z_1), \mathcal{W}(z_2)\Big) = [\mathrm{vol.}(R_0)]^{-1} \sum_{|\alpha|=2} \frac{a(\alpha)}{\alpha!} \cdot \int_{\mathbb{R}^d} D^\alpha G_2(z_1, z_2; s) ds \; , \tag{12.99}$$

$z_1, z_2 \in \mathbb{R}$, where $a(\alpha) = \int_{\mathcal{U}} \int_{\mathcal{U}} \big\{(x - s)^\alpha - (x - \mathbf{1}_d)^\alpha - (\mathbf{1}_d - s)^\alpha\big\} ds dx$, $\mathcal{U} = [0, 1)^d$, $\mathbf{1}_d = (1, \ldots, 1)' \in \mathbb{R}^d$, and where $G_2(z_1, z_2; s) \equiv P(Z(0) \leq z_1, Z(s) \leq z_2)$, $z_1, z_2 \in \mathbb{R}$ denotes the bivariate joint distribution function of $(Z(0), Z(s))'$, $s \in \mathbb{R}^d$.

Lahiri (1999b, 1999d) describes subsampling and bootstrap methods for such integral-based prediction problems. First we describe the subsampling method, which allows us to describe the main ideas more transparently. Let $\{\beta_n\}_{n\geq 1}$ and $\{\gamma_n\}_{n\geq 1}$ be two sequences of positive real numbers such that γ_n is a multiple of η_n (i.e., $\eta_n^{-1}\gamma_n \in \mathbb{N}$ for all $n \geq 1$) and

$$\gamma_n + \eta_n\gamma_n^{-1} = o(1) \quad \text{as} \quad n \to \infty \tag{12.100}$$

and

$$\beta_n^{-1} + \lambda_n^{-1}\beta_n = o(1) \quad \text{as} \quad n \to \infty \,. \tag{12.101}$$

Here β_n will be used to construct the blocks or subregions of R_n, while γ_n will be used to construct a subsample version of the $Z(\cdot)$-process on the subregions at a *lower* level of resolution. As in Sections 12.3–12.5, the requirement (12.101) says that the volume of the subregions grow to infinity, but not as fast as the volume of the original sampling region R_n. Similarly, the conditions on $\{\gamma_n\}_{n\geq 1}$ given by (12.100) say that γ_n tends to zero but at a *slower* rate than the original rate η_n of infilling. Thus, the scaled grid $\gamma_n\mathbb{Z}^d$ is a subgrid of $\eta_n\mathbb{Z}^d$ for any $n \geq 1$ and, therefore, has a lower level of resolution. For a given subregion $R_{n,i}$ (say), we use the observations in $R_{n,i}$ on the finer grid $\eta_n\mathbb{Z}^d$ to define the subsample copy of the unobservable predictand $\hat{\Delta}_\infty$ and the observations in $R_{n,i}$ on the coarser grid $\gamma_n\mathbb{Z}^d$ to define the subsample copy of the predictor $\hat{\Delta}_n$. Here we only consider overlapping subregions $R_{n,i}$'s; a nonoverlapping version of the subsampling method can be defined analogously by restricting attention to the subcollection of nonoverlapping subregions only. Let $\mathcal{U}_0 = (-\frac{1}{2}, \frac{1}{2}]^d$ denote the unit cube in $\mathbb{R}^d$, with its center at the origin. Also, let $\mathcal{I}_{0n} = \{i \in \mathbb{Z}^d : \eta_n i + \mathcal{U}_0\beta_n \subset R_n\}$ be the index set of all cubes of volume β_n^d that are *centered* at $\eta_n i \in \eta_n\mathbb{Z}^d$ and are contained in R_n. Then, the subregion $R_{n,i}$ is defined by inscribing a scaled down copy of the sampling region R_n inside $i + \mathcal{U}_0\beta_n$ such that the origin is mapped onto i (cf. Section 12.4.3). Specifically, we let

$$R_{n,i} = i + \beta_n R_0, \; i \in \mathcal{I}_{0n} \,.$$

Note that $R_{n,i}$ has the same shape as the original sampling region R_n, but a smaller volume, $\beta_n^d\text{vol.}(R_0)$, than the volume $\lambda_n^d\text{vol.}(R_0)$ of R_n. Next, we define the subsample versions of $\hat{\Delta}_\infty$ and $\hat{\Delta}_n$ for each $i \in \mathcal{I}_{0n}$. To that end, note that $R_{n,i}$'s are congruent to $\beta_n R_0 \equiv R_{n,0}$ and that the numbers of sampling sites in $R_{n,i}$ over the finer grid $\eta_n\mathbb{Z}^d$ and over the coarser grid $\gamma_n\mathbb{Z}^d$ are respectively the same for all i. Let $L_n \equiv L$ and $\ell_n \equiv \ell$ denote the sizes of the sets $\beta_n R_0 \cap \eta_n\mathbb{Z}^d$ and $\beta_n R_0 \cap \gamma_n\mathbb{Z}^d$, respectively. For each $i \in \mathcal{I}_{0n}$, we think of the L observations $\{Z(s) : s \in R_{n,i} \cap \eta_n\mathbb{Z}^d\}$ on the finer grid as the analog of $\{Z(s) : s \in R_n\}$ and the ℓ observations $\{Z(s) : s \in R_{n,i} \cap \gamma_n\mathbb{Z}^d\}$ as the analog of the original sample $\{Z(s) : s \in R_n \cap \eta_n\mathbb{Z}^d\}$, at level of the subsamples. Hence, we define the subsample versions of $\hat{\Delta}_\infty$

and $\hat{\Delta}_n$ on $R_{n,i}$ as

$$\begin{aligned} \Delta^{\star}_{\infty,i} &\equiv L^{-1} \sum_{s \in \eta_n \mathbb{Z}^d \cap R_{n,i}} g(Z(s)) \\ \Delta^{\star}_{n,i} &\equiv \ell^{-1} \sum_{s \in \gamma_n \mathbb{Z}^d \cap R_{n,i}} g(Z(s)) , \end{aligned} \tag{12.102}$$

$i \in \mathcal{I}_{0n}$. Then, for a random variable of interests $T_n = t_{N_{2n}}(\hat{\Delta}_n; \hat{\Delta}_\infty)$, its subsample version on the subregion $R_{n,i}$ is defined as

$$T^{\star}_{n,i} = t_\ell(\Delta^{\star}_{n,i}; \Delta^{\star}_{\infty,i}),\ i \in \mathcal{I}_{0n} . \tag{12.103}$$

Note that we use t_ℓ in the definition of $T^{\star}_{n,i}$, as ℓ is the analogous quantity to the sample size N_{2n} at the level of subsamples. The subsample estimator of $G_n(\cdot) \equiv P(T_n \leq \cdot)$ is now given by

$$G^{\star}_n(x) = |\mathcal{I}_{0n}|^{-1} \sum_{i \in \mathcal{I}_{0n}} \mathbb{1}(T^{\star}_{n,i} \leq x),\ x \in \mathbb{R} . \tag{12.104}$$

The subsample estimator of a functional $\varphi(\cdot)$ of G_n is given by the "plug-in" rule $\varphi(G^{\star}_n)$. For example, with $T_n = a_n(\hat{\Delta}_n - \hat{\Delta}_\infty)$, where $\{a_n\}_{n \geq 1}$ is a sequence of scaling constants, the subsample estimator of the scaled MSPE of $\hat{\Delta}_n$, $ET_n^2 \equiv a_n^2 E(\hat{\Delta}_n - \hat{\Delta}_\infty)^2$, is given by

$$\widehat{\text{MSPE}}_n \equiv |\mathcal{I}_{0n}|^{-1} \sum_{i \in \mathcal{I}_{0n}} \{a_\ell(\Delta^{\star}_{n,i} - \Delta^{\star}_{\infty,i})\}^2 .$$

The subsampling method applies not only to prediction problems involving finite dimensional predictands like $\hat{\Delta}_\infty$, but it also applies to infinite dimensional predictands. Lahiri (1999b) proves the validity of the subsampling method for predicting the SCDF $\hat{F}_\infty$. For each $z_0 \in \mathbb{R}$, define the subsampling versions of $\hat{F}_\infty$ of (12.96) and of $\hat{F}_n$ of (12.97) as

$$F^{\star}_{\infty,i}(z_0) = L^{-1} \sum_{s \in \eta_n \mathbb{Z}^d \cap R_{n,i}} \mathbb{1}(Z(s) \leq z_0) \tag{12.105}$$

$$F^{\star}_{n,i}(z_0) = \ell^{-1} \sum_{s \in \gamma_n \mathbb{Z}^d \cap R_{n,i}} \mathbb{1}(Z(s) \leq z_0) , \tag{12.106}$$

$z_0 \in \mathbb{R}$, $i \in \mathcal{I}_{0n}$. Let $w : \mathbb{R} \to [0, \infty)$ be a measurable function and let

$$T_{1n} = (\eta_n^{-2} \lambda_n^d)^{1/2} \|\hat{F}_n(\cdot) - \hat{F}_\infty(\cdot)\|_{\infty,w} , \tag{12.107}$$

where for any function $h : \mathbb{R} \to \mathbb{R}$, we write $\|h\|_{\infty,w} = \sup\{|h(x)|w(x) : x \in \mathbb{R}\}$. Then, the subsampling estimator of the sampling distribution $G_{1n}(\cdot) \equiv P(T_{1n} \leq \cdot)$ is given by

$$G^{\star}_{1n}(a) = |\mathcal{I}_{0n}|^{-1} \sum_{i \in \mathcal{I}_{0n}} \mathbb{1}\Big((\gamma_n^{-2} \beta_n^d)^{1/2} \|\hat{F}^{\star}_{n,i} - F^{\star}_{\infty,i}\|_\infty \leq a\Big),\ a \in \mathbb{R} . \tag{12.108}$$

We now show that $G_{1n}^{\star}$ is a consistent estimator of G_{1n}. Recall (cf. (12.99)) that $G_2(z_1, z_2; s) \equiv P\Big(Z(0) \leq z_1,\ Z(s) \leq z_2\Big)$, $z_1, z_2 \in \mathbb{R}$ denotes the bivariate distribution function of $(Z(0), Z(s))'$. Define its two-sided variant

$$\tilde{G}_2(z_1, z_2; s) \equiv P\Big(z_1 \leq Z(0) \leq z_2,\ z_1 < Z(s) \leq z_2\Big),$$

$z_1, z_2 \in \mathbb{R}$, $s \in \mathbb{R}^d$. Also, define the ρ-mixing coefficient of the random field $Z(\cdot)$ by

$$\begin{aligned} \rho(k; m) \quad = \quad & \sup\Big\{\rho\Big(\mathcal{F}_Z(S_1), \mathcal{F}_Z(S_2)\Big) : S_1, S_2 \in \mathcal{B}(\mathbb{R}^d), \\ & \qquad \text{vol.}(S_1) \leq m,\ \text{vol.}(S_2) \leq m,\ \text{dist.}(S_1, S_2) \geq k\Big\}, \end{aligned}$$

$k, m \in (0, \infty)$, where dist.(S_1, S_2) denotes the distance between the sets A and B in the ℓ^1-norm, $\mathcal{F}_Z(T) = \sigma\langle\{Z(s) : s \in T\}\rangle$ for any $T \subset \mathbb{R}^d$ and where the ρ-mixing coefficient between two σ-fields $\mathcal{F}_1$ and $\mathcal{F}_2$ are defined as usual as in Chapter 3 (cf. (3.4)):

$$\rho(\mathcal{F}_1; \mathcal{F}_2) = \sup\left\{\frac{|\text{Cov}(X, Y)|}{\sqrt{\text{Var}(X)}\sqrt{\text{Var}(Y)}} : X \in L^2(\mathcal{F}_1), Y \in L^2(\mathcal{F}_2)\right\}.$$

Then, we have the following result:

Theorem 12.9 *Suppose that the following conditions hold:*

(C.8) There exist positive constants C, τ_1, τ_2 satisfying $\tau_1 > 3d$ and $\tau_2 < \tau_1/d$ such that

$$\rho(k; m) \leq C \cdot k^{-\tau_1} m^{\tau_2} \quad \textit{for all} \quad m, k \in [1, \infty).$$

(C.9) $\{Z(s) : s \in \mathbb{R}^d\}$ is stationary and the marginal distribution function $G(\cdot)$ of $Z(0)$ is continuous on $\mathbb{R}$.

(C.10) For each $z_1, z_2 \in \mathbb{R}$, $G_2(z_1, z_2; \cdot)$ has

(i) bounded and integrable partial derivatives of order 2 on $\mathbb{R}^d$ (with respect to the Lebesgue measure), and

(ii) for $|\alpha| = 2$, there exist nonnegative integrable functions $H_\alpha(z_1, z_2; \cdot)$ such that for all $s, t \in \mathbb{R}^d$ with $\|t\| \leq 1$,

$$\Big|D^\alpha G_2(z_1, z_2; s + t) - D^\alpha G_2(z_1, z_2; s)\Big| \leq \|t\|^\delta H_\alpha(z_1, z_2; s)$$

for some $\delta \in (0, \infty)$, (not depending on z_1, z_2).

(C.11) There exist constants $C > 0$, $\gamma \in (\frac{1}{2}, 1]$ such that

$$\sum_{|\alpha|=2} \Big| D^\alpha \tilde{G}_2(z_1, z_2; s) \Big| \le C \Big| G(z_2) - G(z_1) \Big|^\gamma$$

for all $z_1, z_2 \in \mathbb{R}$, $s \in \mathbb{R}^d$.

(C.12) $\Big[\eta_n^{(2\gamma+1)2} \lambda_n^d\Big]^{-1} + \Big[\eta_n \lambda_n / \log \lambda_n\Big]^{-1} \to 0$ *as $n \to \infty$ where γ is as in Condition (C.11).*

(a) Then, there exists a zero mean Gaussian process $\mathcal{W}$ such that

$$\eta_n^{-1} \lambda_n^{d/2} \Big(\hat{F}_n(\cdot) - \hat{F}_\infty(\cdot) \Big) \longrightarrow^d \mathcal{W} \tag{12.109}$$

as $\mathbb{D}_1$-valued random variables under the Skorohod metric, where the Gaussian process $\mathcal{W}$ has continuous sample paths with probability one, $\mathcal{W}(+\infty) = \mathcal{W}(-\infty) = 0$ a.s., and where the covariance function of $\mathcal{W}$ is given by (12.99).

(b) Suppose that, in addition to the above conditions, (12.100) and (12.101) hold and that Condition (C.12) holds with λ_n replaced by β_n. If $\|\mathcal{W}\|_{\infty,w}$ has a continuous distribution on $\mathbb{R}$, then

$$\sup_{a \in \mathbb{R}} \Big| G_{1n}^\star(a) - G_{1n}(a) \Big| \longrightarrow_p 0 \quad as \quad n \to \infty . \tag{12.110}$$

Proof: See Lahiri (1999b). □

For discussion of the Conditions (C.8)–(C.12) and their verification for instantaneous transformations of a Gaussian random field, we refer the reader to Lahiri (1999b). Theorem 12.9 can be used to construct valid simultaneous prediction intervals for the unobservable SCDF $\hat{F}_\infty$ over any given interval (a, b) by choosing the weight function $w(\cdot)$ to be $w(\cdot) \equiv \mathbb{1}_{(a,b)}(\cdot)$. In particular, to construct a prediction band over the whole real line, we set $w(x) \equiv 1$ for all $x \in \mathbb{R}$. Let $q_{\alpha,n}^\star$ denote the α-th quantile of $G_{1n}^\star$, $0 < \alpha < 1$. Note that, if $T_{1:1:I}^\star \le \cdots \le T_{1:I:I}^\star$ denote the order statistic corresponding to the subsample version $T_{1n,i}^\star$, $i \in \mathcal{I}_{0n}$ and $I \equiv |\mathcal{I}_{0n}|$, then $q_{\alpha,n}^\star$ is given by the $\lfloor I\alpha \rfloor$th order statistic $T_{1:\lfloor I\alpha \rfloor:I}^\star$. A large sample $100(1-\alpha)\%$ simultaneous prediction interval for $\hat{F}_\infty$ is given by

$$\Big\{ F \in \mathbb{P}_1 : \|F - F_n\|_{\infty,w} < (\eta_n^2 \lambda_n^{-d})^{1/2} q_{1-\alpha,n}^\star \Big\} ,$$

where $w(x) \equiv 1$ for all $x \in \mathbb{R}$ and $\mathbb{P}_1$ is the collection of all probability distribution functions on $\mathbb{R}$. Then, under the conditions of Theorem 12.9, this prediction band attains the nominal coverage probability as $n \to \infty$.

The subsampling method presented here can be extended to grided spatial data with a similar mixed asymptotic structure, where the grid is not necessarily rectangular. See Lahiri, Kaiser, Cressie and Hsu (1999) for an extension of the subsampling method to a hexagonal grid structure in $\mathbb{R}^2$. Finite sample properties of the subsampling method for the square grid is studied by Kaiser, Hsu, Cressie and Lahiri (1997).

For the sake of completeness, we next briefly describe a spatial bootstrap method for the prediction problem. Let $\{\gamma_n\}_{n \geq 1}$ and $\{\beta_n\}_{n \geq 1}$ be two sequences of positive real numbers satisfying (12.100) and (12.101), respectively. Also, suppose that β_n, $\gamma_n \eta_n^{-1} \in \mathbb{N}$ for all $n \geq 1$. Let $\mathcal{K}_n^0 = \{k \in \mathbb{Z}^d : (k + \mathcal{U}_0)\beta_n \cap R_n \neq \emptyset\}$ and let $R_n^0(k) \equiv (k + \mathcal{U}_0)\beta_n \cap R_n$, $k \in \mathcal{K}_n^0$. Then,

$$R_n = \bigcup_{k \in \mathcal{K}_n^0} R_n^0(k) \tag{12.111}$$

is a partition of the sampling region R_n based on cubes having volumes β_n^d and having *centers* at $k\beta_n \in \mathbb{Z}^d$. This differs from the partition (12.5) in that the cubes $(k + \mathcal{U}_0)\beta_n$ in (12.111) are centered at $k\beta_n \in \mathbb{Z}^d$, while the cubes $(k + \mathcal{U})\beta_n$ in (12.5) are centered at $k + (\frac{1}{2}, \cdots, \frac{1}{2})'\beta_n$. Let $\mathcal{K}_{1n}^0 \equiv \{k \in \mathcal{K}_n^0 : (k + \mathcal{U}_0)\beta_n \subset R_n\}$ denote the index set of all interior cubes. For bootstrapping in the context of the prediction problem, we define a version of the random field $Z(\cdot)$ on the finer scale $\eta_n \mathbb{Z}^d$ over each of the subregion $R_n^0(k)$, $k \in \mathcal{K}_{1n}$. Bootstrap versions over the boundary subregions $R_n^0(k)$, $k \in \mathcal{K}_n^0 \backslash \mathcal{K}_{1n}^0$ are ignored as it is difficult to assess the order of the discrepancy between the bootstrap version of the predictor and the predictand, owing to possible irregular form of the boundary. Under Condition B on the boundary of R_0, the effect of this modification is negligible, asymptotically. In the context of time series data, this is equivalent to defining the bootstrap version of a chain $X_1, \ldots, X_n$ of size n by resampling *complete* blocks of length ℓ, say, and generating a bootstrap chain $X_1^*, \ldots, X_{n_1}^*$ of length $n_1 \equiv b\ell$.

As in the case of the subsampling method, let $\mathcal{I}_{0n} = \{i \in \mathbb{Z}^d : i\eta_n + \mathcal{U}_0\beta_n \subset R_n\}$ denote the index set of all translates of the cube $\beta_n \mathcal{U}_0$ of volume β_n^d that are centered on the finer grid $\eta_n \mathbb{Z}^d$ and that are contained in the sampling region R_n. The bootstrap version of the process $Z(\cdot)$ over each $R_n^0(k)$, $k \in \mathcal{K}_{1n}^0$ is defined using a randomly selected subregion from the collection $\{i\eta_n + \mathcal{U}_0\beta_n : i \in \mathcal{I}_{0n}\}$. More precisely, let $\{I_k^0 : k \in \mathcal{K}_{1n}^0\}$ be a collection of iid random variables with common distribution

$$P(I_1^0 = i) = \frac{1}{|\mathcal{I}_{0n}|}, \; i \in \mathcal{I}_{0n} \; .$$

Then, for each $k \in \mathcal{K}_{1n}^0$, the resampled subregion is given by $R_n^*(k) \equiv I_k^0 \eta_n + \beta_n \mathcal{U}_0$. We define the bootstrap version $\Delta_{\infty,n}^*$ of $\hat{\Delta}_{\infty,n}$ by using the observations from all the subregions $R_n^*(k)$, $k \in \mathcal{K}_{1n}^0$, corresponding to data-sites located on the *finer* grid $\eta_n \mathbb{Z}^d$, and similarly, define the bootstrap version of $\hat{\Delta}_n$ using the observation from the same collection of resampled subregions, but only corresponding to those data-sites that are located on the *coarser* grid $\gamma_n \mathbb{Z}^d$. Thus,

$$\Delta^*_{\infty,n} = \sum_{k \in \mathcal{K}^0_{1n}} \sum_{s \in \eta_n \mathbb{Z}^d \cap R^*_n(k)} g\big(Z(s)\big) \Big/ \sum_{k \in \mathcal{K}^0_{1n}} \left| \eta_n \mathbb{Z}^d \cap R^*_n(k) \right| , \qquad (12.112)$$

$$\Delta^*_n = N^{-1}_{3n} \sum_{k \in \mathcal{K}^0_{1n}} \sum_{s \in \gamma_n \mathbb{Z}^d \cap R^*_n(k)} g\big(Z(s)\big) , \qquad (12.113)$$

where $N_{3n} = \sum_{k \in \mathcal{K}^0_{1n}} |\gamma_n \mathbb{Z}^d \cap R^*_n(k)|$ denotes the number of observations on the coarser grid over $\bigcup_{k \in \mathcal{K}^0_{1n}} R^*_n(k)$. With this, the bootstrap version of a random variable of interest $T_n \equiv t_{N_{2n}}(\hat{\Delta}_n, \hat{\Delta}_\infty)$ is given by

$$T^*_n = t_{N_{3n}}\left(\Delta^*_n, \Delta^*_{\infty,n}\right) . \qquad (12.114)$$

Bootstrap estimators of the sampling distribution $G_n(x) = P(T_n \le x)$, $x \in \mathbb{R}$ may be obtained as usual by considering the conditional distribution $\hat{G}_n(x) = P_*(T^*_n \le x)$, $x \in \mathbb{R}$ of T^*_n given $\{Z(s) : s \in \mathbb{R}^d\}$.

12.6.2 *Prediction of Point Values*

In this section, we consider a stationary Gaussian process $\{Z(s) : s \in \mathbb{R}^d\}$ with constant mean $\mu = EZ(0)$ and autocovariance function (or the covariogram)

$$\sigma(h) = \mathrm{Cov}\big(Z(0), Z(h)\big), \; h \in \mathbb{R}^d . \qquad (12.115)$$

Given a set of observations $\{Z(s_1), \ldots, Z(s_n)\}$ on the spatial process at a finite set of locations $s_1, \ldots, s_n \subset \mathbb{R}^d$, suppose that we are interested in predicting the value of the spatial process at a new (unobserved) location $s_0 \in \mathbb{R}^d \backslash \{s_1, \ldots, s_n\}$. (Unlike the earlier sections of this chapter, in this section we do *not* assume that $s_1, \ldots, s_n$ are necessarily specified by one of the spatial sampling designs considered so far.) A widely used method for optimal linear prediction of $Z(s_0)$ based on $\{Z(s_1), \ldots, Z(s_n)\}$ is the method of kriging, named after a South African mining engineer, D. G. Krige (cf. Chapter 3, Cressie (1993)). Given the variables $Z(s_1), \ldots, Z(s_n)$, it seeks to find an optimal predictor from the class

$$\left\{ \sum_{i=1}^n w_{in} Z(s_i) : w_{in}, \ldots, w_{nn} \in \mathbb{R}, \; \sum_{i=1}^n w_{in} = 1 \right\} \qquad (12.116)$$

of linear unbiased predictors of $Z(s_0)$, by minimizing the mean squared prediction error $E(Z(s_0) - \sum_{i=1}^n w_{in} Z(s_i))^2$. An explicit formula for the best linear unbiased predictor (BLUP) $\tilde{Z}_n(s_0)$ of $Z(s_0)$ can be worked out using standard optimization techniques from Calculus. Let Σ_n denote the

$n \times n$ matrix with (i,j)-th entry $\sigma(s_i - s_j)$, $1 \leq i, j \leq n$ and let Υ_n denote the $n \times 1$ vector with i-th entry $\sigma(s_0 - s_i)$, $1 \leq i \leq n$. Then, the BLUP $\tilde{Z}_n(s_0)$ of $Z(s_0)$ is given by

$$\tilde{Z}_n(s_0) = \big(Z(s_1), \ldots, Z(s_n)\big)'\Sigma_n^{-1}\Big(\Upsilon_n + \mathbf{1}_n(1 - \mathbf{1}_n'\Sigma_n^{-1}\mathbf{1}_n)[\mathbf{1}_n'\Sigma_n^{-1}\mathbf{1}_n]^{-1}\Big), \tag{12.117}$$

where $\mathbf{1}_n = (1, \ldots, 1)' \in \mathbb{R}^n$ is the $n \times 1$ vector of 1's. The associated prediction error is

$$\begin{aligned} \tau_n^2 &\equiv E\Big(\tilde{Z}_n(s_0) - Z(s_0)\Big)^2 \\ &= \sigma(0)^2 - \Upsilon_n'\Sigma_n^{-1}\Upsilon_n + (\mathbf{1}_n'\Sigma_n^{-1}\Upsilon_n - 1)^2(\mathbf{1}_n'\Sigma_n^{-1}\mathbf{1}_n)^{-1} . \end{aligned} \tag{12.118}$$

If the covariogram $\sigma(\cdot)$ is known, then $\tilde{Z}_n(s_0)$ can be used as a predictor of $Z(s_0)$ and a $100(1-2\alpha)\%$ $(0 < \alpha < 1/2)$ prediction interval for $Z(s_0)$ can be constructed as

$$\tilde{I}_n(\alpha) \equiv \Big[\tilde{Z}_n(s_0) - \tau_n z_\alpha, \tilde{Z}_n(s_0) + \tau_n z_\alpha\Big] , \tag{12.119}$$

where z_α is the upper α critical point of the standard normal distribution, i.e., $\Phi(z_\alpha) = 1 - \alpha$, with $\Phi(\cdot)$ denoting the distribution function of the standard normal distribution $N(0,1)$. Note that $\tilde{I}_n(\alpha)$ attains the nominal coverage probability $(1-2\alpha)$ exactly, i.e.,

$$P\Big(Z(s_0) \in \tilde{I}_n(\alpha)\Big) = 1 - 2\alpha . \tag{12.120}$$

However, in most applications, the function $\sigma(\cdot)$ is unknown and has to be estimated from the data. Here, we shall suppose that $\sigma(\cdot)$ lies in a parametric family $\{\sigma(\cdot;\theta) : \theta \in \Theta\}$ of valid covariograms. Let $\tilde{Z}_n(s_0;\theta)$ and $\tau_n^2(\theta)$ be defined by replacing $\sigma(\cdot)$ by $\sigma(\cdot;\theta)$ in (12.117) and (12.118), respectively. Also, let $\hat{\theta}_n$ denote an estimator of θ based on $Z(s_1), \ldots, Z(s_n)$, e.g., the maximum likelihood estimator of θ. When $\sigma(\cdot)$ is unknown, one often uses

$$\text{the “plug-in” predictor} \quad \tilde{Z}_n(s_0;\hat{\theta}_n)$$

as an "approximate" or "estimated" BLUP of $Z(s_0)$ and $\tau_n^2(\hat{\theta}_n)$ as an estimator of the mean squared prediction error (MSPE). Substituting these in (12.119) we get the "plug-in" prediction interval of nominal coverage level $(1-2\alpha)$

$$\hat{I}_n(\alpha) \equiv \Big[\hat{Z}_n(s_0) - \hat{\tau}_n z_\alpha, \hat{Z}_n(s_0) + \hat{\tau}_n z_\alpha\Big] , \tag{12.121}$$

where $\hat{Z}_n(s_0) \equiv \tilde{Z}_n(s_0;\hat{\theta}_n)$ and $\hat{\tau}_n^2 = \tau_n^2(\hat{\theta}_n)$. Because of the additional randomness introduced in the prediction interval $\hat{I}_n(\alpha)$ through the estimator $\hat{\theta}_n$, the actual coverage probability of the plug-in prediction interval $\hat{I}_n(\alpha)$

is typically different from the nominal level $1 - 2\alpha$. In many situations, a more accurate interval may be constructed by calibrating the plug-in interval $\hat{I}_n(\cdot)$, where a different value α_0 is used to ensure

$$P_{\theta_0}\Big(Z(s_0) \in \hat{I}_n(\alpha_0)\Big) = 1 - 2\alpha \; . \tag{12.122}$$

In general, α_0 also depends on the true parameter value θ_0 and may not be known. Sjöstedt-DeLuna and Young (2003) suggested using a parametric bootstrap to calibrate the plug-in interval. Let

$$\pi_n(\alpha; \theta_1, \theta_2) \equiv P_{\theta_1}\Big(Z(s_0) \in \tilde{I}_n(\alpha; \theta_2)\Big), \; 0 < \alpha < 1/2, \; \theta_1, \theta_2 \in \Theta, \tag{12.123}$$

where $\tilde{I}_n(\alpha; \theta_2) = [\tilde{Z}_n(s_0; \theta_2) - \tau_n(\theta_2) z_\alpha, \tilde{Z}_n(s_0; \theta_2) + \tau_n(\theta_2) z_\alpha]$. Thus, in this notation, $\tilde{I}_n(\alpha; \hat{\theta}_n) = \hat{I}_n(\alpha)$, and relation (12.122) can be rewritten as

$$G_n(\alpha_0) = 1 - 2\alpha \; , \tag{12.124}$$

where $G_n(\cdot) \equiv \pi_n(\cdot; \theta_0, \hat{\theta}_n)$, θ_0 denotes the unknown true value of the parameter θ, $\hat{\theta}_n$ is the estimator used to define the plug-in interval $\hat{I}_n(\alpha)$, and α_0 is the unknown calibration level that depends on the distribution of $\hat{\theta}_n$ and on θ_0. Because z_γ is a decreasing function of γ, it is easy to verify that $G_n(\gamma)$ is a decreasing function of γ. Hence, α_0 can be found by inverting relation (12.124).

Next we generate an estimator of $G_n(\cdot)$, using the parametric bootstrap method. Let $Z^*(s_0), Z^*(s_1), \ldots, Z^*(s_n)$ be a collection of Gaussian random variables with mean 0 and covariances

$$\text{Cov}_*\big(Z^*(s_i), Z^*(s_j)\big) = \sigma(s_i - s_j; \hat{\theta}_n), \; 0 \le i, j \le n \; , \tag{12.125}$$

where Cov_* denotes the conditional covariance given $Z(s_1), \ldots, Z(s_n)$. Let θ_n^* be defined by replacing $Z(s_1), \ldots, Z(s_n)$ in the definition of $\hat{\theta}_n$ with $Z^*(s_1), \ldots, Z^*(s_n)$. Similarly, let $\hat{Z}_n^*(s_0)$ and τ_n^* be respectively defined by replacing $Z(s_1), \ldots, Z(s_n)$ and $\hat{\theta}_n$ in the definitions of $\hat{Z}_n(s_0) \equiv \tilde{Z}_n(s_0; \hat{\theta}_n)$ and $\tau_n(\hat{\theta}_n)$ by $Z^*(s_1), \ldots, Z^*(s_n)$ and θ_n^*. Then, for $0 < \gamma < 1/2$, the bootstrap estimator $\hat{G}_n(\gamma)$ of $G_n(\gamma)$ is given by

$$\begin{aligned} \hat{G}_n(\gamma) &= \pi_n(\gamma; \hat{\theta}_n, \theta_n^*) \\ &= P_{\hat{\theta}_n}\Big(Z^*(s_0) \in \Big[\hat{Z}_n^*(s_0) - \tau_n^* z_\gamma, \hat{Z}_n^*(s_0) + \tau_n^* z_\gamma\Big]\Big) \; . \end{aligned} \tag{12.126}$$

The parametric bootstrap estimator $\hat{\alpha}_n$ of α_0 of (12.124) is now given by inverting the relation

$$\hat{G}_n(\hat{\alpha}_n) = 1 - 2\alpha \; . \tag{12.127}$$

The bootstrap calibrated prediction interval $\hat{I}_n^C(\alpha)$ of nominal coverage level $(1-2\alpha)$ is $\tilde{I}_n(\hat{\alpha}_n;\hat{\theta}_n)$, i.e.,

$$\hat{I}_n^C(\alpha) \equiv \left[\hat{Z}_n(s_0) - \hat{\tau}_n(\hat{\theta}_n)z_{\hat{\alpha}_n}, \hat{Z}_n(s_0) - \hat{\tau}_n(\hat{\theta}_n)z_{\hat{\alpha}_n}\right] . \tag{12.128}$$

In practice, one may find $\hat{\alpha}_n$ in (12.127) by Monte-Carlo simulation, by generating B iid collections of Gaussian variables $\{Z^{*i}(s_0),\ldots,Z^{*i}(s_n)\}$ of size $(n+1)$ with the covariances given by (12.125) for $i=1,\ldots,B$, then computing θ_n^{*i} based on $\{Z_n^{*i}(s_1),\ldots,Z_n^{*i}(s_n)\}$ for each i, and then solving (12.127) by replacing $\hat{G}_n(\cdot)$ with its Monte-Carlo approximation

$$G_n^*(\gamma) = B^{-1}\sum_{i=1}^{B} \mathbb{1}\left(Z^{*i}(s_0) \in \left[\hat{Z}_n^{*i}(s_0) \pm \tau_n^{*i} z_\gamma\right]\right),\ 0 < \gamma < 1/2 .$$

Sjöstedt-DeLuna and Young (2003) established superior coverage accuracy of the bootstrap calibrated prediction interval $\hat{I}_n^C(\alpha)$ under some general conditions on $\sigma(\cdot)$ and $\hat{\theta}_n$. They also extend the methodology to the case of universal kriging, where the underlying spatial process has a nonconstant mean structure. We refer the interested reader to Sjöstedt-DeLuna and Young (2003) for further discussion and numerical examples.

Appendix A

For easy reference, here we collect some standard results and definitions from Probability Theory for independent and dependent random variables that have been used in this monograph. For proofs and further discussions, see the indicated references. The first set of definitions deal with the basic convergence notions.

Definition A.1 *Let X, $\{X_n\}_{n\geq 1}$ be random variables and $a \in \mathbb{R}$ be a constant.*

(i) $\{X_n\}_{n\geq 1}$ *is said to converge in probability to a, written as $X_n \longrightarrow_p a$, if for any $\epsilon > 0$,*

$$\lim_{n\to\infty} P(|X_n - a| > \epsilon) = 0 \ .$$

(ii) Suppose that $\{X_n\}_{n\geq 1}$ and X are defined on the same probability space $(\Omega, \mathcal{F}, P)$. Then, $\{X_n\}_{n\geq 1}$ is said to converge to X in probability, denoted by $X_n \longrightarrow_p X$, if $(X_n - X) \longrightarrow_p 0$, i.e., for any $\epsilon > 0$,

$$\lim_{n\to\infty} P(|X_n - X| > \epsilon) = 0 \ .$$

(iii) Suppose that $\{X_n\}_{n\geq 1}$ and X are defined on a common probability space $(\Omega, \mathcal{F}, P)$. Then, $\{X_n\}_{n\geq 1}$ is said to converge to X almost surely (with respect to P) if there exists a set $A \in \mathcal{F}$ such that $P(A) = 0$ and

$$\lim_{n\to\infty} X_n(w) = X(w) \quad \textit{for all} \quad w \in A^c \ .$$

In this case, we write $X_n \to X$ *as* $n \to \infty$, *a.s.* (P) *or simply,* $X_n \to X$ *a.s. if the probability measure* P *is clear from the context.*

In general, if $X_n \to X$ a.s., then $X_n \longrightarrow_p X$ but the converse is false. A useful characterization of the convergence in probability is the following (cf. Section 3.3, Chow and Teicher (1997)).

Proposition A.1 *Let* $\{X_n\}_{n\geq 1}$, X *be random variables defined on a probability space* $(\Omega, \mathcal{F}, P)$. *Then* $X_n \longrightarrow_p X$ *if and only if given any subsequence* $\{n_i\}$, *there exists a further subsequence* $\{n_k\} \subset \{n_i\}$ *such that* $X_{n_k} \to X$ *as* $k \to \infty$, *a.s.*

Definition A.2 *Let* $\{X_n\}_{n\geq 1}$, X *be a collection of* $\mathbb{R}^d$*-valued random vectors. Then,* $\{X_n\}_{n\geq 1}$ *is said to converge in distribution to* X, *written as* $X_n \longrightarrow^d X$ *if*

$$\lim_{n\to\infty} P(X_n \in A) = P(X \in A) \tag{A.1}$$

for any $A \in \mathcal{B}(\mathbb{R}^d)$ *with* $P(X \in \partial A) = 0$, *where* ∂A *denotes the boundary of* A.

For $d = 1$, i.e., for random variables, a more familiar definition of "convergence in distribution" is given in terms of the distribution functions of the random variables. Suppose that X, $\{X_n\}_{n\geq 1}$ are one dimensional. Then, $X_n \longrightarrow^d X$ if and only if

$$\lim_{n\to\infty} P(X_n \leq x) = P(X \leq x)$$

for all $x \in \mathbb{R}$ with $P(X = x) = 0$, i.e., the distribution function of X_n converges to that of X at all continuity points of the distribution function of X. Convergence in distribution of random vectors can be reduced to the one-dimensional case by considering the set of all linear combinations of the given vectors, which are one dimensional. More precisely, one has the following result.

Theorem A.1 *(Cramer-Wold Device): Let* X, $\{X_n\}_{n\geq 1}$, *be* $\mathbb{R}^d$*-valued random vectors. Then,* $X_n \longrightarrow^d X$ *if and only if for all* $t \in \mathbb{R}^d$, $t'X_n \longrightarrow^d t'X$.

For a proof of this result, see Theorem 29.4 of Billingsley (1995).

The definition of convergence in distribution can be extended to more general random functions than the random vectors. Let $(\mathbb{S}, d^*)$ be a Polish space, i.e., $\mathbb{S}$ is a complete and separable metric space with metric d^* (cf. Rudin (1987)) and let $\mathcal{S}$ denote the Borel σ-field on $\mathbb{S}$, i.e., $\mathcal{S} = \sigma\langle\{G : G \text{ is an open subset of } \mathbb{S}\}\rangle$. If $(\Omega, \mathcal{F}, P)$ is a probability space and $X : \Omega \to \mathbb{S}$ is $\langle\mathcal{F}, \mathcal{S}\rangle$-measurable, then X is called an $\mathbb{S}$*-valued random variable.*

The probability distribution $\mathcal{L}(X)$ of X is the induced measure on $(\mathbb{S}, \mathcal{S})$, defined by

$$\mathcal{L}(X)(A) = P \circ X^{-1}(A) = P(X^{-1}A), \; A \in \mathcal{S} .$$

A sequence $\{X_n\}_{n\geq 1}$ of $\mathbb{S}$-valued random variables converges in distribution to an $\mathbb{S}$-valued random variable X, also written as $X_n \longrightarrow^d X$, if (A.1) holds for all $A \in \mathcal{S}$ with $P(X \in \partial A) = 0$, where ∂A denotes the boundary of A. In this case, we also say that $\mathcal{L}(X_n)$ *converges weakly* to $\mathcal{L}(X)$. Let $\mathbb{P}_{\mathbb{S}}$ denote the set of all probability measures on $\mathbb{S}$. Then, $\{\mathcal{L}(X_n)\}_{n\geq 1}$ is a sequence of elements of the set $\mathbb{P}_{\mathbb{S}}$. Weak convergence of $\{\mathcal{L}(X_n)\}_{n\geq 1}$ to $\mathcal{L}(X)$ is the same as convergence of the sequence $\{\mathcal{L}(X_n)\}_{n\geq 1}$ to $\mathcal{L}(X)$ in the following metric on $\mathbb{P}_{\mathbb{S}}$:

$$\varrho(\mu, \nu) = \inf\{\delta > 0 : \mu(A) \leq \nu(A^\delta) + \delta \quad \text{for all} \quad A \in \mathcal{S}\} , \tag{A.2}$$

$\mu, \nu \in \mathbb{P}_{\mathbb{S}}$, where for $\delta > 0$, $A^\delta = \{x \in \mathbb{S} : d^*(x, y) \leq \delta\}$ is the δ-neighborhood of the set A in S under the metric d^*. The metric $\varrho(\cdot;\cdot)$ in (A.2) is called the *Prohorov metric*, which metricizes the topology of weak convergence in $\mathbb{P}_{\mathbb{S}}$. In particular, for $\mathbb{S}$-valued random variables X, $\{X_n\}_{n\geq 1}$, $X_n \longrightarrow^d X$ if and only if

$$\varrho\big(\mathcal{L}(X_n), \mathcal{L}(X)\big) \to 0 \quad \text{as} \quad n \to \infty . \tag{A.3}$$

For more details and discussion on this topic, see Parthasarathi (1967), Billingsley (1968), Huber (1981), and the references therein.

The next set of definitions and results relate to the notion of stopping times and moments of randomly stopped sums, which play an important role in the analysis of the SB method in Chapters 3-5.

Definition A.3 *Let $(\Omega, \mathcal{F}, P)$ be a probability space and let $\{\mathcal{F}_n\}_{n\in\mathbb{N}}$ be a collection of sub-σ-fields of $\mathcal{F}$ satisfying $\mathcal{F}_n \subset \mathcal{F}_{n+1}$ for all $n \in \mathbb{N}$. Then a $\mathbb{N} \cup \{\infty\}$-valued random variable T on Ω is called a stopping time with respect to $\{\mathcal{F}_n\}_{n\in\mathbb{N}}$ if*

$$\{T = n\} \in \mathcal{F}_n \quad \textit{for all} \quad n \in \mathbb{N} . \tag{A.4}$$

T is called a proper stopping time with respect to $\{\mathcal{F}_n\}_{n\geq 1}$, if T satisfies (A.4) and $P(T < \infty) = 1$.

We repeatedly make use of the following result on randomly stopped sums of iid random variables. For a proof, see Chapter 1, Woodroofe (1982).

Theorem A.2 *(Wald's Lemmas): Let $\{X_n\}_{n\in\mathbb{N}}$ be a sequence of iid random variables and let T be a proper stopping time with respect to an increasing sequence of σ-fields $\{\mathcal{F}_n\}_{n\in\mathbb{N}}$ such that $\mathcal{F}_n$ is independent of $\{X_k; k \geq n+1\}$ for each $n \in \mathbb{N}$. Suppose that $ET < \infty$. Let $S_n = X_1 + \cdots + X_n$, $n \in \mathbb{N}$, and define the randomly stopped sum S_T by $S_T = \sum_{n\in\mathbb{N}} S_n \mathbb{1}(T = n)$.*

(a) If $E|X_1| < \infty$*, then*

$$E(S_T) = (ET)(EX_1) \ .$$

(b) If $EX_1^2 < \infty$*, then*

$$E(S_T - T \cdot EX_1)^2 = T \cdot \text{Var}(X_1) \ .$$

The next result is a Strong Law of Large Numbers (SLLN) for independent random variables.

Theorem A.3 *(SLLN): Let* $\{X_n\}_{n\geq 1}$ *be a sequence of iid random variables with* $E|X_1| < \infty$*. Then,*

$$n^{-1} \sum_{i=1}^{n} X_i \to EX_1 \quad as \quad n \to \infty, \ a.s.$$

A refinement of Theorem A.3 is given by the following result. For a proof, see Theorem 5.2.2, Chow and Teicher (1997).

Theorem A.4 *(Marcinkiewicz-Zygmund SLLN): Let* $\{X_n\}_{n\geq 1}$ *be a sequence of iid random variables and* $p \in (0, \infty)$*. If* $E|X_1|^p < \infty$*, then*

$$n^{-1/p} \sum_{i=1}^{n} (X_i - c) \to 0 \quad as \quad n \to \infty, \ a.s. \ , \tag{A.5}$$

for any $c \in \mathbb{R}$ *if* $p \in (0, 1)$ *and for* $c = EX_1$ *if* $p \in [1, \infty)$*. Conversely, if (A.5) holds for some* $c \in \mathbb{R}$*, then* $E|X_1|^p < \infty$*.*

The next result is a Central Limit Theorem (CLT) for sums of independent random vectors with values in $\mathbb{R}^d$. For a proof in the one-dimensional (i.e., $d = 1$) case, see Theorem 9.1.1 of Chow and Teicher (1997). For $d \geq 2$ it follows from the one-dimensional case and the Cramer-Wold Device (cf. Theorem A.1).

Theorem A.5 *(Lindeberg's CLT): Let* $\{X_{nj} : 1 \leq j \leq r_n\}_{n\geq 1}$ *be a triangular array where, for each* $n \geq 1$*,* $\{X_{nj} : 1 \leq j \leq r_n\}$ *is a finite collection of independent* $\mathbb{R}^d$*-valued* ($d \in \mathbb{N}$) *random vectors with* $EX_{nj} = 0$ *for all* $1 \leq j \leq r_n$ *and* $\sum_{j=1}^{r_n} EX_{nj}X_{nj}' = \mathbb{I}_d$*. Suppose that* $\{X_{nj} : 1 \leq j \leq r_n\}_{n\geq 1}$ *satisfies the Lindeberg's condition:*
for every $\epsilon > 0$*,*

$$\lim_{n\to\infty} \sum_{j=1}^{r_n} E\|X_{nj}\|^2 \mathbb{1}(\|X_{nj}\| > \epsilon) = 0 \ . \tag{A.6}$$

Then,

$$\sum_{j=1}^{r_n} X_{nj} \longrightarrow^d N(0, \mathbb{I}_d) \quad as \quad n \to \infty \ .$$

The next result is a version of the Berry-Esseen Theorem for independent random variables. For a proof, see Theorem 12.4 of Bhattacharya and Rao (1986), who also give a multivariate version of this result in their Corollary 17.2 where the supremum on the left side of (A.7) is taken over all Borel-measurable convex subsets of $\mathbb{R}^d$, and where the constant (2.75) is replaced by a different constant $C(d) \in (0, \infty)$.

Theorem A.6 *(Berry-Esseen Theorem): Let $X_1, \ldots, X_n$ be a collection of n ($n \in \mathbb{N}$) independent (but not necessarily identically distributed) random variables with $EX_j = 0$ and $E|X_j|^3 < \infty$ for $1 \leq j \leq n$. If $\sigma_n^2 \equiv n^{-1} \sum_{j=1}^n EX_j^2 > 0$, then*

$$\begin{aligned} \sup_{x \in \mathbb{R}} \left| P\Big(\frac{1}{\sqrt{n}\sigma_n} \sum_{j=1}^n X_j \leq x\Big) - \Phi(x) \right| \\ \leq (2.75) \frac{1}{n^{3/2}} \sum_{j=1}^n \left(E|X_j|^3 / \sigma_n^3 \right), \end{aligned} \tag{A.7}$$

where $\Phi(x)$ denotes the distribution function of the standard normal distribution on $\mathbb{R}$.

Next we consider the dependent random variables.

Definition A.4 *A sequence of random vectors $\{X_i\}_{i \in \mathbb{Z}}$ is called stationary if for every $i_1 < i_2 < \cdots < i_k$, $k \in \mathbb{N}$, and for every $m \in \mathbb{Z}$, the distributions of $(X_{i_1}, \ldots, X_{i_k})'$ and $(X_{i_1+m}, \ldots, X_{i_k+m})'$ are the same.*

Definition A.5 *A sequence of random vectors $\{X_i\}_{i \in \mathbb{Z}}$ is called m-dependent for some integer $m \geq 0$ if $\sigma\langle\{X_j : j \leq k\}\rangle$ and $\sigma\langle\{X_j : j \geq k + m + 1\}\rangle$ are independent for all $k \in \mathbb{Z}$.*

Definition A.6 *Let $\{X_i\}_{i \in \mathbb{Z}}$ be a sequence of random vectors. Then the strong mixing or α-mixing coefficient of $\{X_i\}_{i \in \mathbb{Z}}$ is defined as*

$$\begin{aligned} \alpha(n) &= \sup\{|P(A \cap B) - P(A)P(B)| : A \in \sigma\langle\{X_j : j \leq k\}\rangle, \\ &\qquad B \in \sigma\langle\{X_j : j \geq k + n + 1\}, \ k \in \mathbb{Z}\}, \ n \in \mathbb{N} \, . \end{aligned}$$

The sequence $\{X_i\}_{i \in \mathbb{Z}}$ is called strongly mixing (or α-mixing) if $\alpha(n) \to 0$ as $n \to \infty$.

The next two results are Central Limit Theorems (CLTs) for m-dependent random variables and strongly mixing random variables. CLTs for random vectors can be deduced from these results by using the Cramer-Wold device. For a proof of these results and references to related work, see Ibragimov and Linnik (1971), Doukhan (1994), and Lahiri (2003b).

Theorem A.7 *(CLT for m-Dependent Sequences): Let $\{X_i\}_{i \in \mathbb{Z}}$ be a sequence of stationary m-dependent random variables for some integer*

$m \geq 0$. *If* $EX_1^2 < \infty$ *and* $\sigma_m^2 \equiv \mathrm{Var}(X_1) + 2\sum_{k=1}^m \mathrm{Cov}(X_1, X_{1+k}) > 0$, *then*

$$\frac{1}{\sqrt{n}} \sum_{i=1}^n (X_i - EX_1) \longrightarrow^d N(0, \sigma_m^2) .$$

Theorem A.8 *(CLT for Strongly Mixing Sequences): Let* $\{X_i\}_{i \in \mathbb{Z}}$ *be a sequence of stationary random variables with strong mixing coefficient* $\alpha(\cdot)$.

(i) *Suppose that* $P(|X_1| \leq c) = 1$ *for some* $c \in (0, \infty)$ *and that* $\sum_{n=1}^\infty \alpha(n) < \infty$. *Then*

$$0 \leq \sigma_\infty^2 \equiv \sum_{k=-\infty}^{\infty} \mathrm{Cov}(X_1, X_{1+k}) < \infty. \tag{A.8}$$

If, in addition, $\sigma_\infty^2 > 0$, *then*

$$\frac{1}{\sqrt{n}} \sum_{i=1}^n (X_i - EX_1) \longrightarrow^d N(0, \sigma_\infty^2) . \tag{A.9}$$

(ii) *Suppose that for some* $\delta \in (0, \infty)$, $E|X_1|^{2+\delta} < \infty$ *and* $\sum_{n=1}^\infty [\alpha(n)]^{\delta/2+\delta} < \infty$. *Then (A.8) holds. If, in addition,* $\sigma_\infty^2 > 0$, *then (A.9) holds.*

Appendix B

Proof of Theorem 6.1: Theorem 6.1 is a version of Theorem 20.1 of Bhattacharya and Rao (1986) for triangular arrays. As a result, here we give an *outline* of the proof of Theorem 6.1, highlighting only the necessary modifications. Let $Y_{jn} = X_j \mathbb{1}(\|X_j\|^2 \leq n)$ and $Z_{jn} = Y_{jn} - EY_{jn}$, $1 \leq j \leq n$. Set $S_{1n} = n^{-1/2} \sum_{j=1}^{n} Z_{jn}$. Then, under (6.25), Lemmas 14.6 and 14.8 of Bhattacharya and Rao (1986) imply that

$$\begin{aligned} & \left| Ef(S_n) - \int f d\Psi_{n,s} \right| \\ \leq \quad & \left| Ef_n(S_{1n}) - \int f_n d\Psi_{1,n,s} \right| + C(s,d) u_n a_n(s) \ , \end{aligned} \tag{B.1}$$

where $f_n(x) = f(x - n^{-1/2} \sum_{j=1}^{n} EY_{jn})$, $x \in \mathbb{R}^d$, and where $\Psi_{1,n,s}$ is obtained from $\Psi_{n,s}$ by replacing the cumulants of X_j's with those of Z_{jn}'s. Since S_{1n} may not have a density, it is customary to add to S_{1n} a suitably small random vector that has a density and that is independent of S_{1n}. The additional noise introduced by this operation is then assessed using a smoothing inequality. Applying Corollary 11.2 (a smoothing inequality) and Lemma 11.6 (the inversion formula) of Bhattacharya and Rao (1986), as in their proof of Theorem 20.1, for any $0 < \epsilon < 1$, we get

$$\begin{aligned} & \left| Ef_n(S_{1n}) - \int f_n d\Psi_{1,n,s} \right| \\ \leq \quad & M_s(f) \cdot C(s,d) \sum_{|\alpha|=0}^{s+d+1} \int \left| D^\alpha H_n^\dagger(t) \right| \exp(-\epsilon \|t\|^{1/2}) dt \end{aligned}$$

$$
\begin{aligned}
&+ w(2\epsilon; f_n, |\Psi_{1,n,s}|) + \left| \int f_n d(\Psi_{1,n,s} - \Psi_{1,n,s+d+1}) \right| \\
\equiv\ & I_{1n} + I_{2n} + I_{3n}, \text{ say,} \qquad \text{(B.2)}
\end{aligned}
$$

where $H_n^\dagger(t) = E\exp(\iota t' S_{1n}) - \int \exp(\iota t' x) d\Psi_{1,n,s+d+1}(x)$, $t \in \mathbb{R}^d$, and where $|\Psi_{1,n,s}|$ denotes the total variation measure corresponding to $\Psi_{1,n,s}$. Note that here ϵ is the amount of noise introduced through smoothing and would typically depend on n. Like $\Psi_{n,s}$, the signed measure $\Psi_{1,n,j}$ has density with respect to the Lebesgue measure on $\mathbb{R}^d$ (cf. 6.16)

$$
\psi_{1,n,j}(x) = \left[1 + \sum_{r=1}^{j-2} n^{-r/2} p_r(x; \{\bar{\chi}_\nu^0\}) \right] \phi_\Xi(x) , \qquad \text{(B.3)}
$$

$x \in \mathbb{R}^d$, $j \in \mathbb{N}$, $j \geq 3$, where $\bar{\chi}_\nu^0$ is the ν-th cumulant of S_{1n} and $\Xi = \text{Cov}(S_{1n})$. Although moments of S_n of order $s+1$ and higher may not exist, all moments of S_{1n} exist, as the variables Z_{jn}'s are bounded for all $1 \leq j \leq n$. This makes $\psi_{1,n,j}$ well defined for $j \geq s+1$.

First we consider I_{1n}. As is customary, we divide the range of integration into two regions $\{t : t \in \mathbb{R}^d, \|t\| \leq a_{1n}\}$ and $\{t : t \in \mathbb{R}^d, \|t\| > a_{1n}\}$ for some suitable constant $a_{1n} > 0$ (to be specified later). For small values of $\|t\|$ (i.e., for $\|t\| \leq a_{1n}$), we use Theorem 9.11 of Bhattacharya and Rao (1986). By (6.25) and their Corollary 14.2, Ξ is nonsingular and

$$
\frac{3}{4} \leq \|\Xi\| \leq \frac{5}{4}, \quad \|\Xi^{-1}\| \leq \frac{4}{3} .
$$

Let $\Xi_1 = \Xi^{-1/2}$ and $H_{1n}^\dagger(t) = H_n^\dagger(\Xi_1 t)$, $t \in \mathbb{R}^d$. Then, it is easy to check that

$$
n^{-1} \sum_{j=1}^n E\|Z_{jn}\|^{s+d+1} \leq C(s,d) n^{(d+1)/2} \min\{\bar{\rho}_{n,s}, \tilde{\rho}_{n,s}\} ,
$$

and, uniformly in $t \in \mathbb{R}^d$,

$$
\begin{aligned}
&\max \left\{ \left| D^\alpha H_n^\dagger(t) \right| : 0 \leq |\alpha| \leq s+d+1 \right\} \\
\leq\ & C(s,d) \max \left\{ \left| (D^\alpha H_{1n}^\dagger)(\Xi_1^{-1} t) \right| : 0 \leq |\alpha| \leq s+d+1 \right\} .
\end{aligned}
$$

Hence, by Theorem 9.11 of Bhattacharya and Rao (1986), there exists a constant $C_1(s,d)$ such that with $a_{1n} = C_1(s,d)(u_n \bar{\rho}_{n,s})^{-1/s+d+1}$,

$$
\sum_{|\alpha|=0}^{s+d+1} \int_{\|t\| \leq a_{1n}} \left| D^\alpha H_n(t) \right| \exp(-\epsilon \|t\|^{1/2}) dt \leq C(s,d) u_n \tilde{\rho}_{n,s} . \qquad \text{(B.4)}
$$

Next consider the case where $\|t\| > a_{1n}$. Note that for $a > 0$,

$$
\int_a^\infty e^{-x^2/2} dx \leq a^{-1} e^{-a^2/2} \qquad \text{(B.5)}
$$

and

$$\int_a^\infty y^k e^{-y} dy = \Big\{ \sum_{j=0}^{k} a^{k-j} \frac{k!}{(k-j)!} \Big\} e^{-a} , \tag{B.6}$$

$k \in \mathbb{N}$. Now using (B.5), (B.6), the definition of the polynomials $p_r(\cdot;\cdot)$ (cf. (6.17)), and Lemma 9.5 of Bhattacharya and Rao (1986), we get

$$\begin{aligned} &\sum_{|\alpha|=0}^{s+d+1} \int_{\|t\|>a_{1n}} \Big| D^\alpha \int \exp(\iota t'x) d\Psi_{1,n,s+d+1}(x) \Big| \\ \leq\ & C(s,d) \int_{\|t\|>a_{1n}} \Big[1 + \|t\|^{3(s+d-2)+(s+d+1)} \Big] \\ & \times \Big[\sum_{r=0}^{s+d+1} n^{-r/2} (1+\bar\rho_{n,s}) n^{(r-s)_+/2} \Big] e^{-3\|t\|^2/8} dt \\ \leq\ & C(s,d)(1+\bar\rho_{n,s}) \big[a_{1n}^{-1} + a_{1n}^{4(s+2d)} \big] \exp(-3a_{1n}^2/8) . \end{aligned} \tag{B.7}$$

Next, let $a_{2n} = (16\bar\rho_{n,3})^{-1} n^{1/2}$. Then, using Lemma 14.3 of Bhattacharya and Rao (1986) for $a_{1n} \leq \|t\| \leq a_{2n}$ and using the inequality,

$$\big| E \exp(\iota t' Z_{jn}) \big| \leq \theta_{n,j}(t),\ t \in \mathbb{R}^d,\ 1 \leq j \leq n ,$$

for $\|t\| > a_{2n}$, we get

$$\begin{aligned} &\sum_{|\alpha|=0}^{s+d+1} \int_{\|t\|>a_{1n}} \big| D^\alpha E \exp(\iota t' S_{1n}) \big| \exp(-\epsilon \|t\|^{1/2}) dt \\ \leq\ & C(s,d) \int_{a_{1n}\leq\|t\|\leq a_{2n}} (1+\|t\|^{s+d+1}) \exp(-5\|t\|^2/24) dt \\ & + \sum_{r=0}^{s+d+1} \sum_{1\leq j_1,\ldots,j_r\leq n} n^{-r/2} \prod_{i=1}^{r} E|Z_{j_i n}| \\ & \times \Big[\sup \Big\{ \prod_{j\neq j_1,\ldots,j_r} \big| E\exp(\iota t' Z_{jn}/\sqrt{n}) \big| : a_{2n} \leq \|t\| \leq \epsilon^{-4} \Big\} \\ & \times \int_{a_{2n}\leq\|t\|\leq\epsilon^{-4}} \exp(-\epsilon\|t\|^{1/2}) dt \\ & + 1 \cdot \int_{\|t\|>\epsilon^{-4}} \exp(-\epsilon\|t\|^{1/2}) dt \Big] \\ \leq\ & C(s,d) a_{2n}^{s+2d} \exp(-5a_{1n}^2/24) \\ & + C(s,d) \Big[\epsilon^{-2d} \gamma_n(\epsilon) + n^{s+d+1} \epsilon^{-8d} \exp(-\epsilon^{-1}) \Big] . \end{aligned} \tag{B.8}$$

Repeating the arguments in the proof of Theorem 20.1 of Bhattacharya and Rao (1986) in a similar fashion, it can be shown that

$$I_{2n} \leq C(d,s) M_s(f) \Big[\tilde\rho_{n,s} u_n + (1+\bar\rho_{n,s}) \Big\{ a_n(s) u_n$$

$$+ n^{[3(s-2)+s]/6} \exp(-n^{1/3}/6)\Big\}\Big]$$
$$+ C(s,d)(1+\bar{\rho}_{n,s}) w(2\epsilon; f, \Phi) \tag{B.9}$$

and

$$I_{3n} \leq C(s,d) M_s(f) \tilde{\rho}_{n,s} u_n \ . \tag{B.10}$$

Theorem 6.1 now follows from the bounds (B.1), (B.2), (B.4), and (B.7)–(B.10).

Proof of Theorem 6.2: Note that by (6.26) and (6.27),

$$\begin{aligned}
\tilde{\rho}_{n,s} &= n^{-3/2} \sum_{j=1}^{n} E\|X_{n,j}\|^{s+1} \mathbb{1}\big(\|X_{n,j}\| \leq n^{1/2}\big) \\
&\leq n^{-1} \sum_{j=1}^{n} E\|X_{n,j}\|^{s} \mathbb{1}\big(\|X_{n,j}\| > n^{1/2-\delta}\big) \\
&\quad + n^{-1/2} n^{-1} \sum_{j=1}^{n} E\|X_{n,j}\|^{s+1} \mathbb{1}\big(\|X_{n,j}\| \leq n^{1/2-\delta}\big) \\
&= o(1) + O(n^{-\delta}) \ ,
\end{aligned}$$

and, for n large,

$$\begin{aligned}
a_n(s, \tfrac{2}{3}) &\leq n^{-1} \sum_{j=1}^{n} E\|X_{n,j}\|^{s} \mathbb{1}\big(\|X_{n,j}\| > n^{1/2-\delta}\big) \\
&= o(1) \ .
\end{aligned}$$

Hence, setting $\epsilon = \eta_n$ and applying (6.28) and Theorem 6.1, we get

$$\sup_{B \in \mathcal{B}} \big| P(S_n \in B) - \Psi_{n,s}(B) \big| = o(n^{-(s-2)/2}) + O\big(\sup_{B \in \mathcal{B}} w(2\eta_n; \mathbb{1}_B, \Phi) \big) \ .$$

Theorem 6.2 now follows by noting that $w(\epsilon; \mathbb{1}_B, \Phi) = \Phi\big((\partial B)^\epsilon\big)$ for all $\epsilon > 0$ (cf. Corollary 2.6, Bhattacharya and Rao (1986)).

References

Allen, M. and Datta, S. (1999), 'A note on bootstrapping M-estimators in ARMA models', *Journal of Time Series Analysis* **20**, 365–379.

Anderson, T. W. (1971), *The Statistical Analysis of Time Series*, Wiley, New York.

Andrews, D. (2002), 'Higher-order improvements of a computationally attractive k-step bootstrap for extremum estimators', *Econometrica* **70**(1), 119–162.

Arcones, M. and Giné, E. (1989), 'The bootstrap of the mean with arbitrary bootstrap sample', *Annales de l'Institut Henri Poincaré* **25**, 457–481.

Arcones, M. and Giné, E. (1991), 'Additions and corrections to "The bootstrap of the mean with arbitrary sample size"', *Annales de l'Institut Henri Poincaré* **27**, 583–595.

Arcones, M. and Yu, B. (1994), 'Central limit theorems for empirical and U-processes of stationary mixing sequences', *Journal of Theoretical Probability* **7**, 47–70.

Athreya, K. B. (1987), 'Bootstrap of the mean in the infinite variance case', *The Annals of Statistics* **15**, 724–731.

Athreya, K. B. and Fukuchi, J. I. (1997), 'Confidence intervals for endpoints of a c.d.f. via bootstrap', *Journal of Statistical Planning and Inference* **58**, 299–320.

Athreya, K. B., Lahiri, S. N. and Wei, W. (1998), 'Inference for heavy tailed distributions', *Journal of Statistical Planning and Inference* **66**, 61–75.

Bai, Z. D. and Rao, C. R. (1991), 'Edgeworth expansion of a function of sample means', *The Annals of Statistics* **19**, 1295–1315.

Barbe, P. and Bertail, P. (1995), *The Weighted Bootstrap*, Vol. 98 of *Lecture Notes in Statistics*, Springer-Verlag, New York.

Barry, J., Crowder, M. and Diggle, P. (1997), Parametric estimation of the variogram, Preprint, Department of Mathematics and Statistics, Lancaster University, United Kingdom.

Bartlett, M. S. (1946), 'On the theoretical specification of sampling properties of autocorrelated time series', *Journal of the Royal Statistical Society, Supplement* **8**, 27–41.

Basawa, I. V., Mallik, A. K., McCormick, W. P., Reeves, J. H. and Taylor, R. L. (1991), 'Bootstrapping unstable first-order autoregressive processes', *The Annals of Statistics* **19**, 1098–1101.

Basawa, I. V., Mallik, A. K., McCormick, W. P. and Taylor, R. L. (1989), 'Bootstrapping explosive autoregressive processes', *The Annals of Statistics* **17**, 1479–1486.

Beran, J. (1994), *Statistics for Long Memory Processes*, Chapman and Hall, London.

Bhattacharya, R. N. and Ghosh, J. K. (1978), 'On the validity of the formal Edgeworth expansion', *The Annals of Statistics* **7**, 434–451.

Bhattacharya, R. N. and Ghosh, J. K. (1988), 'On moment conditions for valid formal Edgeworth expansions', *Journal of Multivariate Analysis* **27**, 68–79.

Bhattacharya, R. N. and Rao, R. R. (1986), *Normal Approximations and Asymptotic Expansions*, R. E. Krieger Publishing Company, Malabar, FL.

Bickel, P., Götze, F. and van Zwet, W. (1997), 'Resampling fewer than n observations: Gains, losses, and remedies for losses', *Statistica Sinica* **7**, 1–31.

Bickel, P. J. and Wichura, M. J. (1971), 'Convergence criteria for multiparameter stochastic processes and some applications', *The Annals of Mathematical Statistics* **42**, 1656–1670.

Billingsley, P. (1968), *Convergence of Probability Measures*, Wiley, New York.

Billingsley, P. (1995), *Probability and Measure*, Wiley, New York.

Bingham, N. H., Goldie, C. M. and Teugels, J. L. (1987), *Regular Variation*, Cambridge University Press, Cambridge, U.K..

Boos, D. D. (1979), 'A differential for L-statistics', *The Annals of Statistics* **7**, 955–959.

Bose, A. (1988), 'Edgeworth correction by bootstrap in autoregressions', *The Annals of Statistics* **16**, 1709–1722.

Bose, A. (1990), 'Bootstrap in moving average models', *Annals of the Institute of Statistical Mathematics* **42**, 753–768.

Bradley, R. C. (1989), 'A caution on mixing conditions for random fields', *Statistics and Probability Letters* **8**, 489–491.

Bradley, R. C. (1993), 'Equivalent mixing conditions for random fields', *The Annals of Probability* **21**, 1921–1926.

Bretagnolle, J. (1983), 'Limit laws for the bootstrap of some special functionals (french)', *Annales de l'Institut Henri Poincare, Section B, Calcul des Probabilities et Statistique* **19**, 281–296.

Brillinger, D. R. (1981), *Time Series: Data Analysis and Theory*, Holden-Day Inc., San Francisco.

Brockwell, P. J. and Davis, R. A. (1991), *Time Series: Theory and Methods*, 2nd edn, Springer-Verlag, New York.

Bühlmann, P. (1994), 'Blockwise bootstrapped empirical process for stationary sequences', *The Annals of Statistics* **22**, 995–1012.

Bühlmann, P. (1997), 'Sieve bootstrap for time series', *Bernoulli* **3**, 123–148.

Bühlmann, P. (2002), 'Sieve bootstrap with variable-length Markov chains for stationary categorical time series', *Journal of the American Statistical Association* **97**, 443–471.

Bühlmann, P. and Künsch, H. R. (1999b), 'Comments on "Prediction of spatial cumulative distribution functions using subsampling"', *Journal of the American Statistical Association* **94**, 97–99.

Bustos, O. H. (1982), 'General M-estimates for contaminated p-th order autoregressive processes; consistency and asymptotic normality', *Zeitschrift für Wahrscheinlichkeitstheorie und Verwandte Gebiete* **59**, 491–504.

Carlstein, E. (1986), 'The use of subseries methods for estimating the variance of a general statistic from a stationary time series', *The Annals of Statistics* **14**, 1171–1179.

Carlstein, E., Do, K.-A., Hall, P., Hesterberg, T. and Künsch, H. R. (1998), 'Matched-block bootstrap for dependent data', *Bernoulli* **4**, 305–328.

Chernick, M. R. (1981a), 'A limit theorem for the maximum of autoregressive processes with uniform marginal distributions', *The Annals of Probability* **9**, 145–149.

Chernick, M. R. (1981b), 'On strong mixing and leadbetter's D condition', *Journal of Applied Probability* **18**, 764–769.

Chernick, M. R. (1999), *Bootstrap Methods: A Practitioner's Guide*, Wiley, New York.

Chibishov, D. M. (1972), 'An asymptotic expansion for the distribution of a statistic admitting an asymptotic expansion', *Theory of Probability and Its Applications* **17**, 620–630.

Choi, E. and Hall, P. (2000), 'Bootstrap confidence regions computed from autoregressions of arbitrary order', *Journal of the Royal Statistical Society, Series B* **62**, 461–477.

Chow, Y. and Teicher, H. (1997), *Probability Theory: Independence, Interchangeability, Martingales*, 2nd edn, Springer-Verlag, Berlin.

Cressie, N. (1985), 'Fitting variogram models by weighted least squares', *Journal of the International Association for Mathematical Geology* **17**, 693–702.

Cressie, N. (1993), *Statistics for Spatial Data*, 2nd edn, Wiley, New York.

Dahlhaus, R. (1983), 'Spectral analysis with tapered data', *Journal of Times Series Analysis* **4**, 163–175.

Dahlhaus, R. (1985), 'Asymptotic normality of spectral estimates', *Journal of Multivariate Analysis* **16**, 412–431.

Dahlhaus, R. and Janas, D. (1996), 'A frequency domain bootstrap for ratio statistics in time series analysis', *The Annals of Statistics* **24**, 1934–1963.

Datta, S. (1995), 'Limit theory and bootstrap for explosive and partially explosive autoregression', *Stochastic Processes and their Applications* **57**, 285–304.

Datta, S. (1996), 'On asymptotic properties of bootstrap for AR(1) processes', *Journal of Statistical Planning and Inference* **53**, 361–374.

Datta, S. and McCormick, W. P. (1995), 'Bootstrap inference for a first-order autoregression with positive innovations', *Journal of the American Statistical Association* **90**, 1289–1300.

Datta, S. and McCormick, W. P. (1998), 'Inference for the tail parameters of a linear process with heavy tail innovations', *Annals of the Institute of Statistical Mathematics* **50**, 337–359.

Datta, S. and Sriram, T. N. (1997), 'A modified bootstrap for autoregression without stationarity', *Journal of Statistical Planning and Inference* **59**, 19–30.

David, M. (1977), *Geostatistical Ore Reserve Estimation*, Elsevier, Amsterdam.

Davis, R. A. (1983), 'Stable limits for partial sums of dependent random variables', *The Annals of Probability* **11**, 262–269.

Davison, A. C. and Hinkley, D. V. (1997), *Bootstrap Methods and Their Application*, Cambridge University Press, Cambridge, UK.

de Haan, L. (1970), *On Regular Variation and Its Application to the Weak Convergence of Sample Extremes*, Vol. 32 of *Mathematical Centre Tracts*, Mathematical Centre, Amsterdam.

Denker, M. and Jakubowski, A. (1989), 'Stable limit distributions for strongly mixing sequences', *Statistics and Probability Letters* **8**, 477–483.

Dennis, J. E. and Schnabel, R. B. (1983), *Numerical Methods for Unconstrained Optimization and Nonlinear Equations*, Prentice-Hall, Englewood Cliffs, NJ.

Deo, C. (1973), 'A note on empirical processes of strong-mixing sequences', *The Annals of Probability* **1**, 870–875.

Diéudonne, J. (1960), *Foundations of Modern Analysis*, Academic Press, New York.

Dobrushin, R. L. (1979), 'Gaussian and their subordinated self-similar random fields', *The Annals of Probability* **7**, 1–28.

Dobrushin, R. L. and Major, P. (1979), 'Non-central limit theorems for non-linear functionals of Gaussian random fields', *Zeitschrift für Wahrscheinlichkeitstheorie und Verwandte Gebiete* **50**, 27–52.

Doukhan, P. (1994), *Mixing: Properties and Examples*, Vol. 85 of *Lecture Notes in Statistics*, Springer-Verlag, New York.

Efron, B. (1979), 'Bootstrap methods: Another look at the jackknife', *The Annals of Statistics* **7**, 1–26.

Efron, B. (1982), *The Jackknife, the Bootstrap and Other Resampling Plans*, SIAM, Philadelphia.

Efron, B. (1992), 'Jackknife-after-bootstrap standard errors and influence functions (with discussion)', *Journal of Royal Statistical Society, Series B* **54**, 83–111.

Efron, B. and Tibshirani, R. (1986), 'Bootstrap methods for standard errors, confidence intervals, and other measures of statistical accuracy', *Statistical Science* **1**, 54–77.

Efron, B. and Tibshirani, R. (1993), *An Introduction to the Bootstrap*, Chapman and Hall, London.

Faraway, J. J. and Jhun, M. (1990), 'Bootstrap choice of bandwidth for density estimation', *Journal of the American Statistical Association* **85**, 1119–1122.

Feller, W. (1971a), *An Introduction to Probability Theory and Its Applications*, Vol. I, Wiley, New York.

Feller, W. (1971b), *An Introduction to Probability Theory and Its Applications*, Vol. II, Wiley, New York.

Fernholz, L. T. (1983), *Von Mises Calculus for Statistical Functionals*, Vol. 19 of *Lecture Notes in Statistics*, Springer-Verlag, New York.

Filippova, A. A. and Brunswick, N. A. (1962), 'Mises' theorem on the asymptotic behavior of functionals of empirical distribution functions and its statistical applications', *Theory of Probability and Its Applications* **7**, 24–57.

Findley, D. F. (1986), On bootstrap estimates of forecast mean square errors for autoregressive processes, *in* D. M. Allen, ed., '*Computer Science and Statistics: The Interface*', Elsevier, Amsterdam.

Franke, J. and Härdle, W. (1992), 'On bootstrapping kernel spectral estimates', *The Annals of Statistics* **20**, 121–145.

Freedman, D. A. (1981), 'Bootstrapping regression models', *The Annals of Statistics* **9**, 1218–1228.

Freedman, D. A. (1984), 'On bootstrapping two-stage least-squares estimates in stationary linear models', *The Annals of Statistics* **12**, 827–842.

Freedman, D. A. and Peters, S. F. (1984), 'Bootstrapping an economic model: Some empirical results', *Journal of Business and Economic Statistics* **2**, 150–158.

Fukuchi, J. I. (1994), Bootstrapping extremes of random variables, PhD Dissertation, Iowa State University, Department of Statistics, Ames, IA.

Fuller, W. A. (1996), *Introduction to Statistical Time Series*, 2nd edn, Wiley, New York.

Genton, M. G. (1997), Variogram fitting by generalized least squares using an explicit formula for the covariance structure, Preprint, MIT, Cambridge, MA.

Giné, E. and Zinn, J. (1989), 'Necessary conditions for the bootstrap of the mean', *The Annals of Statistics* **17**, 684–691.

Giné, E. and Zinn, J. (1990), 'Bootstrapping general empirical measures', *The Annals of Probability* **18**, 851–869.

Gnedenko, B. V. (1943), 'Sur la distribution limite du terme maximum dune seriealeatoire', *Annals of Mathematics* **44**, 423–453.

Götze, F. (1987), 'Approximations for multivariate U-statistics', *Journal of Multivariate Analysis* **22**, 212–229.

Götze, F. and Hipp, C. (1983), 'Asymptotic expansions for sums of weakly dependent random vectors', *Zeitschrift für Wahrscheinlichkeitstheorie und Verwandte Gebiete* **64**, 211–239.

Götze, F. and Hipp, C. (1994), 'Asymptotic distribution of statistics in time series', *The Annals of Statistics* **22**, 2062–2088.

Götze, F. and Künsch, H. R. (1996), 'Second-order correctness of the blockwise bootstrap for stationary observations', *The Annals of Statistics* **24**, 1914–1933.

Guyon, X. (1995), *Random Fields on a Network: Modeling, Statistics, and Applications*, Springer-Verlag, Berlin; New York.

Hall, P. (1985), 'Resampling a coverage pattern', *Stochastic Processes and Their Applications* **20**, 231–246.

Hall, P. (1987), 'Edgeworth expansion for Student's t statistic under minimal moment conditions', *The Annals of Probability* **15**, 920–931.

Hall, P. (1992), *The Bootstrap and Edgeworth Expansion*, Springer-Verlag, New York.

Hall, P. (1997), Defining and measuring long-range dependence, *in* C. D. Cutler and D. T. Kaplan, eds, '*Nonlinear Dynamics and Time Series*', Vol. 11, Fields Institute Communications, pp. 153–160.

Hall, P. and Horowitz, J. L. (1996), 'Bootstrap critical values for tests based on generalized-method-of-moments estimators', *Econometrica* **64**, 891–916.

Hall, P., Horowitz, J. L. and Jing, B.-Y. (1995), 'On blocking rules for the bootstrap with dependent data', *Biometrika* **82**, 561–574.

Hall, P. and Jing, B.-Y. (1996), 'On sample re-use methods for dependent data', *Journal of the Royal Statistical Society, Series B* **58**, 727–738.

Hall, P., Jing, B.-Y. and Lahiri, S. N. (1998), 'On the sampling window method under long range dependence', *Statistica Sinica* **8**, 1189–1204.

Hall, P., Lahiri, S. N. and Polzehl, J. (1995), 'On bandwidth choice in nonparametric regression with both short- and long-range dependent errors', *The Annals of Statistics* **23**, 2241–2263.

Hall, P., Lahiri, S. N. and Truong, Y. K. (1995), 'On bandwidth choice for density estimation with dependent data', *The Annals of Statistics* **23**, 2241–2263.

Hall, P. and Wang, Q. (2003), 'Exact convergence rate and leading term in Central Limit Theorem for Student's t-statistic', *The Annals of Probability* . (To appear).

Härdle, W. and Bowman, A. (1988), 'Bootstrapping in nonparametric regression: Local adaptive smoothing and confidence bands', *Journal of the American Statistical Association* **83**, 100–110.

Heimann, G. and Kreiss, J.-P. (1996), 'Bootstrapping general first order autoregression', *Statistics and Probability Letters* **30**, 87–98.

Helmers, R. (1991), 'On the Edgeworth expansion and the bootstrap approximation for a studentized U-statistic', *The Annals of Statistics* **19**, 470–484.

Hesterberg, T. C. (1997), Matched-block bootstrap for long memory processes, Technical Report 66, Research Department, MathSoft, Inc, Seattle, WA.

Hipp, C. (1985), 'Asymptotic expansions in the central limit theorem for compound and Markov processes', *Zeitschrift für Wahrscheinlichkeitstheorie und Verwandte Gebiete* **69**, 361–385.

Hsing, T. (1991), 'On tail estimation using dependent data', *The Annals of Statistics* **19**, 1547–1569.

Huber, P. (1981), *Robust Statistics*, Wiley, New York.

Hurvich, C. M. and Zeger, S. (1987), Frequency domain bootstrap methods for time series, Statistics and Operations Research Working Paper, New York University, New York.

Ibragimov, I. A. and Hasminskii, R. Z. (1980), 'On nonparametric estimation of regression', *Soviet Mathematics (Doklady Akademii Nauk)* **21**, 810–814.

Ibragimov, I. A. and Linnik, Y. V. (1971), *Independent and Stationary Sequences of Random Variables*, Wolters-Noordhoff, Groningen.

Ibragimov, I. A. and Rozanov, Y. A. (1978), *Gaussian Random Processes*, Springer-Verlag, New York.

Inoue, A. and Kilian, L. (2002), 'Bootstraping autoregressive processes with possible unit roots', *Econometrica* **70**(1), 377–391.

Inoue, A. and Shintani, M. (2001), Bootstrapping GMM estimators for time series, Working paper, Department of Agricultural and Resource Economics, North Carolina State University, Raleigh, NC.

Jakubowski, A. and Kobus, M. (1989), 'α-stable limit theorems for sums of dependent random vectors', *Journal of Multivariate Analysis* **29**, 219–251.

Janas, D. (1993), 'A smoothed bootstrap estimator for a studentized sample quantile', *Annals of the Institute of Statistical Mathematics* **45**, 317–329.

Janas, D. (1994), 'Edgeworth expansions for spectral mean estimates with applications to whittle estimates', *Annals of the Institute of Statistical Mathematics* **46**, 667–682.

Jensen, J. L. (1989), 'Asymptotic expansions for strongly mixing Harris recurrent Markov chains', *Scandinavian Journal of Statistics* **16**, 47–63.

Journel, A. G. and Huijbregts, C. J. (1978), *Mining Geostatistics*, Academic Press, London.

Kaiser, M. S., Hsu, N.-J., Cressie, N. and Lahiri, S. N. (1997), 'Inference for spatial processes using subsampling: A simulation study', *Environmetrics* **8**, 485–502.

Kallenberg, O. (1976), *Random Measures*, Akademie Verlag, Berlin.

Kendall, M. and Stuart, A. (1977), *The Advanced Theory of Statistics. Vol. 1: Distribution Theory (4th edn)*, Charles Griffin & Co, High Wycombe, UK.

Kreiss, J. P. (1987), 'On adaptive estimation in stationary ARMA processes', *The Annals of Statistics* **15**, 112–133.

Kreiss, J. P. and Franke, J. (1992), 'Bootstrapping stationary autoregressive moving-average models', *Journal of Time Series Analysis* **13**, 297–317.

Kreiss, J. P. and Paparoditis, E. (2003), 'Autoregressive aided periodogram bootstrap for time series', *Annals of Statistics* **31**. (In press).

Künsch, H. R. (1989), 'The jackknife and the bootstrap for general stationary observations', *The Annals of Statistics* **17**, 1217–1261.

Lahiri, S. N. (1991), 'Second order optimality of stationary bootstrap', *Statistics and Probability Letters* **11**, 335–341.

Lahiri, S. N. (1992a), Edgeworth correction by 'moving block' bootstrap for stationary and nonstationary data, *in* R. Lepage and L. Billard, eds, '*Exploring the Limits of Bootstrap*', Wiley, New York, pp. 183–214.

Lahiri, S. N. (1992b), 'Bootstrapping M-estimators of a multiple linear regression parameter', *The Annals of Statistics* **20**, 1548–1570.

Lahiri, S. N. (1992c), 'On bootstrapping M-estimators', *Sankhyā, Series A, Indian Journal of Statistics* **54**, 157–170.

Lahiri, S. N. (1993a), 'Refinements in the asymptotic expansions for sums of weakly dependent random vectors', *The Annals of Probability* **21**, 791–799.

Lahiri, S. N. (1993b), 'On the moving block bootstrap under long range dependence', *Statistics and Probability Letters* **18**, 405–413.

Lahiri, S. N. (1994), 'Two term Edgeworth expansion and bootstrap approximation for multivariate studentized M-estimators', *Sankhyā, Series A* **56**, 201–226.

Lahiri, S. N. (1995), 'On the asymptotic behaviour of the moving block bootstrap for normalized sums of heavy-tail random variables', *The Annals of Statistics* **23**, 1331–1349.

Lahiri, S. N. (1996a), 'Asymptotic expansions for sums of random vectors under polynomial mixing rates', *Sankhyā, Series A* **58**, 206–225.

Lahiri, S. N. (1996b), 'On edgeworth expansion and moving block bootstrap for studentized m-estimators in multiple linear regression models', *Journal of Multivariate Analysis* **56**, 42–59.

Lahiri, S. N. (1996c), 'On inconsistency of estimators based on spatial data under infill asymptotics', *Sankhyā, Series A* **58**, 403–417.

Lahiri, S. N. (1996d), Empirical choice of the optimal block length for block bootstrap methods, Preprint, Department of Statistics, Iowa State University, Ames, IA.

Lahiri, S. N. (1999a), 'Theoretical comparisons of block bootstrap methods', *The Annals of Statistics* **27**, 386–404.

Lahiri, S. N. (1999b), 'Asymptotic distribution of the empirical spatial cumulative distribution function predictor and prediction bands based on a subsampling method', *Probability Theory and Related Fields* **114**, 55–84.

Lahiri, S. N. (1999c), On second order properties of the stationary bootstrap method for studentized statistics, *in* S. Ghosh, ed., '*Asymptotics, Nonparametrics, and Time Series*', Marcel Dekker, New York, pp. 683–712.

Lahiri, S. N. (1999d), Resampling methods for spatial prediction, *in* 'Computing Science and Statistics. Models, Predictions, and Computing', Vol. 31 of *Proceedings of the 31st Symposium on the Interface*, pp. 462–466.

Lahiri, S. N. (2001), 'Effects of block lengths on the validity of block resampling methods', *Probability Theory and Related Fields* **121**, 73–97.

Lahiri, S. N. (2002a), 'On the jackknife after bootstrap method for dependent data and its consistency properties', *Econometric Theory* **18**, 79–98.

Lahiri, S. N. (2002b), 'Comments on "Sieve bootstrap with variable length Markov chains for stationary categorical time series"', *Journal of the American Statistical Association* **97**, 460–462.

Lahiri, S. N. (2003a), 'A necessary and sufficient condition for asymptotic independence of discrete fourier transforms under short and long range dependence', *Annals of Statistics* **31**, 613–641.

Lahiri, S. N. (2003b), 'Central limit theorems for weighted sums under some stochastic and fixed spatial sampling designs', *Sankhyā, Series A* . (In press).

Lahiri, S. N. (2003c), 'Consistency of the jackknife-after-bootstrap variance estimator for the bootstrap quantiles of a studentized statistic', *Annals of Statistics* . (To appear).

Lahiri, S. N. (2003d), Validity of a block bootstrap method for irregularly spaced spatial data under nonuniform stochastic designs, Preprint, Department of Statistics, Iowa State University, Ames, IA.

Lahiri, S. N., Furukawa, K. and Lee, Y.-D. (2003), A nonparametric plug-in rule for selecting the optimal block length for block bootstrap methods, Preprint, Department of Statistics, Iowa State University, Ames, IA.

Lahiri, S. N., Kaiser, M. S., Cressie, N. and Hsu, N.-J. (1999), 'Prediction of spatial cumulative distribution functions using subsampling (with discussion)', *Journal of the American Statistical Association* **94**, 86–110.

Lahiri, S. N., Lee, Y.-D. and Cressie, N. (2002), 'Efficiency of least squares estimators of spatial variogram parameters', *Journal of Statistical Planning and Inference* **3**, 65–85.

Leadbetter, M. R. (1974), 'On extreme values in stationary sequences', *Zeitschrift fuer Wahrscheinlichkeitstheorie und Verwandte Gebiete* **28**, 289–303.

Leadbetter, M. R., Lindgren, G. and Rootzen, H. (1983), *Extremes and Related Properties of Random Sequences and Processes*, Springer-Verlag, Berlin.

Lee, Y.-D. and Lahiri, S. N. (2002), 'Least squares variogram fitting by spatial subsampling', *Journal of the Royal Statistical Society, Series B* **64**, 837–854.

Liu, R. Y. and Singh, K. (1992), Moving blocks jackknife and bootstrap capture weak dependence, *in* R. Lepage and L. Billard, eds, '*Exploring the Limits of the Bootstrap*', Wiley, New York, pp. 225–248.

Loynes, R. M. (1965), 'Extreme values in uniformly mixing stationary stochastic processes', *The Annals of Mathematical Statistics* **36**, 993–999.

Malinovskii, V. K. (1986), 'Limit theorems for Harris-Markov chains, I', *Theory of Probability and Its Applications* **31**, 269–285.

Mammen, E. (1992), *When Does Bootstrap Work? Asymptotic Results and Simulations*, Vol. 77 of *Lecture Notes in Statistics*, Springer-Verlag, New York.

Martin, R. D. and Yohai, V. J. (1986), 'Influence functionals for time series (with discussion)', *The Annals of Statistics* **14**, 781–855.

Matheron, G. (1962), *Traite de geostatistique appliquee, Tome I*, Vol. 14 of *Memoires du Bureau de Recherches Geologiques et Minieres*, Editions Technip, Paris.

Mathew, G. and McCormick, W. P. (1998), 'A bootstrap approximation to the joint distribution of sum and maximum of a stationary sequence', *Journal of Statistical Planning and Inference* **70**, 287–299.

Meijia, J. M. and Rodriguez-Iturbe, I. (1974), 'On the synthesis of random field sampling from the spectrum: An application to the generation of hydrologic spatial processes', *Water Resources Research* **10**, 705–711.

Miller, R. G. (1974), 'The jackknife – A review', *Biometrika* **61**, 1–15.

Milnor, J. W. (1965), *Topology From the Differentiable Viewpoint*, University Press of Virginia, Charlottesville.

Morris, M. D. and Ebey, S. F. (1984), 'An interesting property of the sample mean under a first-order autoregressive model', *The American Statistician* **38**, 127–129.

Nadaraya, E. A. (1964), 'On estimating regression', *Theory of Probability and Its Applications* **9**, 141–142.

Naik-Nimbalkar, U. V. and Rajarshi, M. B. (1994), 'Validity of blockwise bootstrap for empirical processes with stationary observations', *The Annals of Statistics* **22**, 980–994.

Nordgaard, A. (1992), Resampling a stochastic process using a bootstrap approach, *in* K. H. Jöckel, G. Rothe and W. Sendler, eds, '*Bootstrapping and Related Techniques*', Lecture Notes in Economics and Mathematical Systems, Springer-Verlag, Berlin, p. 376.

Nordman, D. and Lahiri, S. N. (2003a), 'On optimal spatial subsample size for variance estimation', *Annals of Statistics* . (To appear).

Nordman, D. and Lahiri, S. N. (2003b), On optimal block size for a spatial block bootstrap method, Preprint, Department of Statistics, Iowa State University, Ames, IA.

Paparoditis, E. and Politis, D. N. (2001), 'Tapered block bootstrap', *Biometrika* **88**(4), 1105–1119.

Paparoditis, E. and Politis, D. N. (2002), 'The tapered block bootstrap for general statistics from stationary sequences', *The Econometrics Journal* **5**(1), 131–148.

Parthasarathi, K. R. (1967), *Probability Measures on Metric Spaces*, Academic Press, San Diego, CA.

Peligrad, M. (1982), 'Invariance principles for mixing sequences of random variables', *The Annals of Probability* **12**, 968–981.

Peligrad, M. (1998), 'On the blockwise bootstrap for empirical processes for stationary sequences', *The Annals of Probability* **26**, 877–901.

Peligrad, M. and Shao, Q.-M. (1995), 'Estimation of the variance of partial sums for ρ-mixing random variables', *Journal of Multivariate Analysis* **52**, 140–157.

Petrov, V. V. (1975), *Sums of Independent Random Variables*, Springer-Verlag, New York.

Politis, D. N., Paparoditis, E. and Romano, J. P. (1998), 'Large sample inference for irregularly spaced dependent observations based on subsampling', *Sankhyā, Series A, Indian Journal of Statistics* **60**, 274–292.

Politis, D. N., Paparoditis, E. and Romano, J. P. (1999), Resampling marked point processes, *in* S. Ghosh, ed., '*Multivariate Analysis, Design of Experiments, and Survey Sampling: A Tribute to J. N. Srivastava*', Mercel Dekker, New York, pp. 163–185.

Politis, D. N. and Romano, J. P. (1992a), 'A general resampling scheme for triangular arrays of α-mixing random variables with application to the problem of spectral density estimation', *The Annals of Statistics* **20**, 1985–2007.

Politis, D. N. and Romano, J. P. (1992b), A circular block resampling procedure for stationary data, *in* R. Lepage and L. Billard, eds, '*Exploring the Limits of Bootstrap*', Wiley, New York, pp. 263–270.

Politis, D. N. and Romano, J. P. (1993), 'Nonparametric resampling for homogeneous strong mixing random fields', *Journal of Multivariate Analysis* **47**, 301–328.

Politis, D. N. and Romano, J. P. (1994a), 'Large sample confidence regions based on subsamples under minimal assumptions', *The Annals of Statistics* **22**, 2031–2050.

Politis, D. N. and Romano, J. P. (1994b), 'The stationary bootstrap', *Journal of the American Statistical Association* **89**, 1303–1313.

Politis, D. N. and Romano, J. P. (1995), 'Bias-corrected nonparametric spectral estimation', *Journal of Time Series Analysis* **16**, 67–103.

Politis, D. N., Romano, J. P. and Wolf, M. (1999), *Subsampling*, Springer-Verlag, Berlin; New York.

Politis, D. N. and White, H. (2003), Automatic block-length selection for the dependent bootstrap, Preprint, Department of Mathematics, University of California at San Diego, LaJolla, CA.

Possolo, A. (1991), Subsampling a random field, *in* A. Possolo, ed., '*Spatial Statistics and Imaging*', IMS Lecture Notes Monograph Series, 20, Institute of Mathematical Statistics, Hayward, CA, pp. 286–294.

Priestley, M. B. (1981), *Spectral Analysis and Time Series*, Vol. I, Academic Press, London.

Radulović, D. (1996), 'The bootstrap of the mean for strong mixing sequences under minimal conditions', *Statistics and Probability Letters* **28**, 65–72.

Rao, R. R. (1962), 'Relations between weak and uniform convergence of measures with applications', *The Annals of Mathematical Statistics* **33**, 659–680.

Reeds, J. A. (1976), On the definition of von Mises functionals, PhD thesis, Harvard University, Cambridge, MA.

Ren, J.-J. and Sen, P. K. (1991), 'On hadamard differentiability of extended statistical functional', *Journal of Multivariate Analysis* **39**, 30–43.

Ren, J.-J. and Sen, P. K. (1995), 'Hadamard differentiability on $D[0,1]^p$', *Journal of Multivariate Analysis* **55**, 14–28.

Resnick, S. and Stărciă, C. (1998), 'Tail index estimation for dependent data', *The Annals of Applied Probability* **8**, 1156–1183.

Rice, J. (1984), 'Bandwidth choice for nonparametric regression', *The Annals of Statistics* **12**, 1215–1230.

Romano, J. P. (1988), 'Bootstrapping the mode', *Annals of the Institute of Statistical Mathematics* **40**, 565–586.

Rudin, W. (1987), *Real and Complex Analysis*, McGraw Hill, New York.

Sakov, A. and Bickel, P. J. (1999), Choosing m in the m out of n bootstrap, *in* '*ASA Proceedings of the Section on Bayesian Statistical Science*', pp. 125–128.

Sakov, A. and Bickel, P. J. (2000), 'An Edgeworth expansion for the m out of n bootstrapped median', *Statistics and Probability Letters* **49**(3), 217–223.

Samur, J. D. (1984), 'Convergence of sums of mixing triangular arrays for random vectors with stationary laws', *The Annals of Probability* **12**, 390–426.

Sen, P. K. (1974), 'Weak convergence of multidimensional empirical processes for stationary ϕ-mixing processes', *The Annals of Probability* **2**, 147–154.

Serfling, R. J. (1980), *Approximation Theorems of Mathematical Statistics*, Wiley, New York.

Shao, J. and Tu, D. (1995), *The Jackknife and Bootstrap*, Springer-Verlag, New York.

Sherman, M. (1996), 'Variance estimation for statistics computed from spatial lattice data', *Journal of the Royal Statistical Society, Series B* **58**, 509–523.

Sherman, M. and Carlstein, E. (1994), 'Nonparametric estimation of the moments of a general statistic computed from spatial data', *Journal of the American Statistical Association* **89**, 496–500.

Shinozuka, M. (1971), 'Simulation of multivariate and multidimensional random processes', *Journal of the Acoustical Society of America* **49**, 357–367.

Shorack, G. R. (1982), 'Bootstrapping robust regression', *Communications in Statistics, Part A - Theory and Methods* **11**, 961–972.

Singh, K. (1981), 'On the asymptotic accuracy of the Efron's bootstrap', *The Annals of Statistics* **9**, 1187–1195.

Sjöstedt-DeLuna, S. and Young, S. (2003), 'The bootstrap and Kriging prediction intervals', *Scandinavian Journal of Statistics* **30**, 175–192.

Skovgaard, I. M. (1981), 'Transformation of an Edgeworth expansion by a sequence of smooth functions', *Scandinavian Journal of Statistics* **8**, 207–217.

Statulevicius, V. (1969a), 'Limit theorems for sums of random variables that are connected in a Markov chain, I', *Litovsk Mat Sb* **9**, 345–362.

Statulevicius, V. (1969b), 'Limit theorems for sums of random variables that are connected in a Markov chain, II', *Litovsk Mat Sb* **9**, 635–672.

Statulevicius, V. (1970), 'Limit theorems for sums of random variables that are connected in a Markov chain, III', *Litovsk Mat Sb* **10**, 161–169.

Stein, M. L. (1987), 'Minimum norm quadratic estimation of spatial variograms', *Journal of the American Statistical Association* **82**, 765–772.

Stein, M. L. (1989), 'Asymptotic distributions of minimum norm quadratic estimators of the covariance function of a Gaussian random field', *The Annals of Statistics* **17**, 980–1000.

Swanepoel, J. and van Wyk, J. (1986), 'The bootstrap applied to power spectral density function estimation', *Biometrika* **73**, 135–141.

Taqqu, M. S. (1975), 'Weak convergence to fractional Brownian motion and to the Rosenblatt process', *Zeitschrift für Wahrscheinlichkeitstheorie und Verwandte Gebiete* **31**, 287–302.

Taqqu, M. S. (1977), 'Law of the iterated logarithm for sums of non-linear functions of Gaussian variables that exhibit a long range dependence', *Zeitschrift für Wahrscheinlichkeitstheorie und Verwandte Gebiete* **40**, 203–238.

Taqqu, M. S. (1979), 'Convergence of integrated processes of arbitrary Hermite rank', *Zeitschrift für Wahrscheinlichkeitstheorie und Verwandte Gebiete* **50**, 53–83.

Taylor, C. C. (1989), 'Bootstrap choice of the smoothing parameter in kernel density estimation', *Biometrika* **76**, 705–712.

van der Vaart, A. W. and Wellner, J. A. (1996), *Weak Convergence and Empirical Processes: With Applications to Statistics*, Springer-Verlag Inc, Berlin; New York.

von Mises, R. (1947), 'On the asymptotic distribution of differentiable statistical functions', *The Annals of Mathematical Statistics* **18**, 309–348.

Wallace, D. L. (1958), 'Asymptotic approximations to distributions', *The Annals of Mathematical Statistics* **29**, 635–654.

Watson, G. S. (1964), 'Smooth regression analysis', *Sankhyā, Series A* **26**, 359–372.

Woodroofe, M. B. (1982), *Nonlinear Renewal Theory in Sequential Analysis*, SIAM, Philadelphia.

Wu, C. F. J. (1990), 'On the asymptotic properties of the jackknife histogram', *The Annals of Statistics* **18**, 1438–1452.

Yoshihara, K. (1975), 'Weak convergence of multidimensional empirical processes for strong mixing sequences of stochastic vectors', *Zeitschrift für Wahrscheinlichkeitstheorie und Verwandte Gebiete* **33**, 133–137.

Zhang, X., van Eijkeren, J. and Heemink, A. (1995), 'On the weighted least-squares method for fitting a semivariogram model', *Computers and Geosciences* **21**, 605–608.

Zhu, J. and Lahiri, S. N. (2001), Weak convergence of blockwise bootstrapped Empirical processes for stationary random fields with statistical applications, Preprint, Department of Statistics, Iowa State University, Ames, IA.

Zhurbenko, I. G. (1972), 'On strong estimates of mixed semiinvariants of random processes', *Siberian mathematics Journal* **13**, 202–213.

Zygmund, A. (1968), *Trigonometric Series*, Vol. I, II, Cambridge University Press, Cambridge, U.K.

Author Index

Subject Index

Springer Series in Statistics

(continued from p. ii)

Knottnerus: Sample Survey Theory: Some Pythagorean Perspectives.
Kolen/Brennan: Test Equating: Methods and Practices.
Kotz/Johnson (Eds.): Breakthroughs in Statistics Volume I.
Kotz/Johnson (Eds.): Breakthroughs in Statistics Volume II.
Kotz/Johnson (Eds.): Breakthroughs in Statistics Volume III.
Küchler/Sørensen: Exponential Families of Stochastic Processes.
Lahiri: Resampling Methods for Dependent Data.
Le Cam: Asymptotic Methods in Statistical Decision Theory.
Le Cam/Yang: Asymptotics in Statistics: Some Basic Concepts, 2nd edition.
Liu: Monte Carlo Strategies in Scientific Computing.
Longford: Models for Uncertainty in Educational Testing.
Manski: Partial Identification of Probability Distributions.
Mielke/Berry: Permutation Methods: A Distance Function Approach.
Pan/Fang: Growth Curve Models and Statistical Diagnostics.
Parzen/Tanabe/Kitagawa: Selected Papers of Hirotugu Akaike.
Politis/Romano/Wolf: Subsampling.
Ramsay/Silverman: Applied Functional Data Analysis: Methods and Case Studies.
Ramsay/Silverman: Functional Data Analysis.
Rao/Toutenburg: Linear Models: Least Squares and Alternatives.
Reinsel: Elements of Multivariate Time Series Analysis, 2nd edition.
Rosenbaum: Observational Studies, 2nd edition.
Rosenblatt: Gaussian and Non-Gaussian Linear Time Series and Random Fields.
Särndal/Swensson/Wretman: Model Assisted Survey Sampling.
Schervish: Theory of Statistics.
Shao/Tu: The Jackknife and Bootstrap.
Simonoff: Smoothing Methods in Statistics.
Singpurwalla and Wilson: Statistical Methods in Software Engineering: Reliability and Risk.
Small: The Statistical Theory of Shape.
Sprott: Statistical Inference in Science.
Stein: Interpolation of Spatial Data: Some Theory for Kriging.
Taniguchi/Kakizawa: Asymptotic Theory of Statistical Inference for Time Series.
Tanner: Tools for Statistical Inference: Methods for the Exploration of Posterior Distributions and Likelihood Functions, 3rd edition.
van der Laan: Unified Methods for Censored Longitudinal Data and Causality.
van der Vaart/Wellner: Weak Convergence and Empirical Processes: With Applications to Statistics.
Verbeke/Molenberghs: Linear Mixed Models for Longitudinal Data.
Weerahandi: Exact Statistical Methods for Data Analysis.
West/Harrison: Bayesian Forecasting and Dynamic Models, 2nd edition.